Monographs in Electrical and Electronic Engineering

Series editors: P. Hammond, T. J. E. Miller, and T. Kenjo

Monographs in Electrical and Electronic Engineering

Series editors: P. Hammond, T.J.E. Miller, and S. Kenjo

25. *Electrical machines and drives: a space-vector theory approach* (1992) Peter Vas
26. *Spiral vector theory of a.c. circuits and machines* (1992) Sakae Yamamura
27. *Parameter estimation, condition monitoring, and diagnosis of electrical machines* (1993) Peter Vas
28. *An introduction to ultrasonic motors* (1993) S. Sashida and T. Kenjo
29. *Ultrasonic motors: theory and applications* (1993) S. Ueha and Y. Tomikawa
30. *Linear induction drives* (1993) J. F. Gieras
31. *Switched reluctance motors and their control* (1993) T. J. E. Miller
32. *Numerical modelling of eddy currents* (1993) Andrzej Krawczyk and John A. Tegopoulos
33. *Rectifiers, cycloconverters, and AC controllers* (1994) Thomas H. Barton
34. *Stepping motors and their microprocessor controls. Second edition* (1994) T. Kenjo and A. Sugawara
35. *Inverse problems and optimal design in electricity and magnetism* (1995) P. Neittaanmäki, M. Rudnicki, and A. Savini
36. *Electric drives and their control* (1995) Richard M. Crowder
37. *Design of brushless permanent magnet motors* (1995) J. R. Hendershott and T. J. E. Miller
38. *Reluctance synchronous machines and drives* (1995) I. Boldea
39. *Geometry of electromagnetic systems* (1996) D. Baldomir and P. Hammond
40. *Permanent-magnet d.c. linear motors* (1996) Amitava Basak
41. *Vector control and dynamics of a.c. drives* (1996) D. W. Novotny and T. A. Lipo
42. *Sensorless vector and direct torque control* (1998) Peter Vas
43. *Intelligent control: power electronic systems* (1998) Yasuhiko Dote and Richard G. Hoft
44. *Time domain wave-splittings and inverse problems* (1998) Sailing He, Staffan Ström, and Vaughan H. Weston

Time Domain Wave-Splittings and Inverse Problems

Sailing He

Docent, Senior Lecturer of Electromagnetic Theory
Royal Institute of Technology, Stockholm

Staffan Ström

Professor of Electromagnetic Theory
Royal Institute of Technology, Stockholm

Vaughan H. Weston

Professor of Mathematics
Purdue University

Oxford New York Tokyo
OXFORD UNIVERSITY PRESS

This book has been printed digitally and produced in a standard specification in order to ensure its continuing availability

OXFORD
UNIVERSITY PRESS

Great Clarendon Street, Oxford OX2 6DP

Oxford University Press is a department of the University of Oxford.
It furthers the University's objective of excellence in research, scholarship,
and education by publishing worldwide in

Oxford New York

Auckland Cape Town Dar es Salaam Hong Kong Karachi
Kuala Lumpur Madrid Melbourne Mexico City Nairobi
New Delhi Shanghai Taipei Toronto

With offices in

Argentina Austria Brazil Chile Czech Republic France Greece
Guatemala Hungary Italy Japan South Korea Poland Portugal
Singapore Switzerland Thailand Turkey Ukraine Vietnam

Oxford is a registered trade mark of Oxford University Press
in the UK and in certain other countries

Published in the United States
by Oxford University Press Inc., New York

Reprinted 2011

ISBN 978-0-19-856549-9

Printed and bound in Great Britain by CPI Antony Rowe,
Chippenham and Eastbourne

PREFACE

The purpose of the present book is to give an introduction to recent developments concerning the application of wave-splitting methods to direct and inverse scattering of wave fields. Wave-splitting refers to the decomposition of the total field into two components which propagate in opposite directions. The emphasis is on time domain methods, although some applications to frequency domain problems are also included. We believe that there exist several good reasons for putting increased emphasis on time domain methods. A time domain model is very often the most direct and natural description of the way in which an actual experiment is performed and a time domain model can be expected to provide results and insights which may be hard to arrive at by frequency domain methods. We emphasize that the wave-splitting approach provides a common framework for both direct and inverse scattering problems. We use these terms in the following sense. A direct scattering problem is that of determining the scattered field (as well as the internal field when required) that arises when a known incident field impinges on a scattering region with known physical properties (the sources of the incident field typically lie in an exterior homogeneous region where the propagation is trivial). An inverse scattering problem consists of the problem of determining the physical properties of the scattering region, using the measured scattered field produced by a known incident field as data. An important aspect of an inverse scattering problem is the identification of the set of data that are sufficient for a unique determination of one or several of the physical properties in question. All the direct scattering problems we consider are linear, with only one exception (in Chapter 5). However, the associated inverse scattering problems are highly non-linear and most often ill-posed. Consequently, existing results concerning inverse scattering problems are much less complete than those for the corresponding direct scattering problems.

Depending on the application, the properties to be determined in the inverse scattering problem can be the geometry and/or the material properties (such as spatial variation or temporal dispersion) of the scattering region. As is well known, non-invasive investigations of physical properties by means of wave phenomena are crucial in a large number of industrial and other applications, such as mineral and oil prospecting, non-destructive testing of materials, imaging and source identification in biomedical engineering, crack localization, remote sensing of earth resources and environmental quality. In addition, inverse methods may be used for design problems involving, e.g. prescribed reflection or transmission properties of radomes and shieldings, radiation patterns of antennas, etc. The wave phenomena used in probing the inhomogeneous media can be acoustic, electromagnetic or elastodynamic. Consequently both scalar and vector waves are considered. Most of the models that are treated in this book concern electromagnetic waves.

We concentrate on the solution of direct and inverse scattering problems by means of wave-splitting combined with an invariant imbedding approach or a Green's function approach. Unlike methods such as Born iterative methods and optimization methods for inverse problems, the wave-splitting method is exact and non-iterative and the reconstruction is carried out in a layer-stripping manner.

Invariant imbedding has a long history and readers interested in tracing the early developments, as well as the wider contexts of these notions, are referred to Bellman and Wing (1975), where the early contributions to both particle transport theory and scattering theory are described. One may note that the relations which are crucial in the time domain approach to inverse scattering problems (exemplified by (1.80) and (1.81) in Chapter 1), are those which are called 'surprising' and are said to be of 'interesting and unusual form' by Bellman and Wing (1975) (cf. Chapter 7 of this reference). Other early books describing the origins of the imbedding approach are Chandreshakar (1960) and Scott (1973). As is evident from these books, the time domain aspect was not emphasized in the early developments. A change regarding this was brought about mainly by the work of Weston, Corones, Krueger, et al. in the mid-seventies. This development also signified a shift of emphasis away from particle transport theory towards a scattering theory, particularly an inverse scattering theory based on wave-splitting. In April 1990 a joint American Mathematical Society and Society for Industrial and Applied Mathematics meeting was held in Albuquerque, New Mexico, USA, and the proceedings from this meeting reflect several of the new trends (J. Corones, et al. 1992).

Among the early papers which formed the basis for the later developments concerning one-dimensional inverse scattering problems in the time domain we mention particularly Corones and Krueger (1983). There it is shown how, by combining the Redheffer star product with a time domain scattering matrix description of the reflection and transmission of the split components, one obtains integro-differential equations for the scattering kernels. Thereby a general framework for solving both direct and inverse one-dimensional scattering problems was established, which has been exploited extensively in the later developments, as illustrated by Corones et al. (1983a,b), Kristensson and Krueger (1986a,b, 1987, 1989), Corones et al. (1985), and Vogel (1986). Numerous additional references to more recent developments are given in the appropriate contexts in Chapters 1–6.

The basic concepts and results concerning the one-dimensional Green's function approach in a time domain wave-splitting setting were established in Krueger and Ochs (1989). A matrix of Green's operators is introduced which maps the incident field to the split fields at any point within the medium. Furthermore, the Green's operator formalism subsumes the scattering operators since the reflection and transmission operators correspond to the boundary values of the Green's operators on the two sides of the scattering region. This connection has been exploited repeatedly in the developments that followed after the appearance of

Krueger and Ochs (1989). Therefore, our presentation focuses to a large extent on the Green's function approach.

The one-dimensional formalisms referred to above have been extended to include, e.g. dispersive, anisotropic, and also more general bi-anisotropic media. We also note that scattering of other types of incident fields from plane-stratified structures can be solved within the framework of a one-dimensional formalism. Obliquely incident plane waves and axially symmetric fields are such examples (see e.g. He and Ström (1995a) and He and Karlsson (1993), respectively). Additional applications of the one-dimensional formalism that has been pursued concern transient wave propagation in cylindrical waveguides (Kristensson 1995) and propagation in media with time-varying properties (Åberg et al. 1995).

The challenge of solving fully three-dimensional inverse scattering problems by means of the wave-splitting and imbedding technique was taken up early on. Splitting with respect to both plane and curved surfaces was considered and initially the focus was on the imbedding equations (Weston 1987, 1988a,b, 1989, 1990, 1993). Subsequently the Green's function formalism was developed for scalar and electromagnetic waves and applications to inverse problems in three dimensions were carried out in a layer-stripping manner (Weston 1992a; Weston and He 1994; He and Weston 1995; Cao et al. 1996; He and Weston 1996; Cao et al. 1998). We note that a wave-splitting with respect to some suitable set of surfaces, which is exact in an exterior region, provides in effect an exact absorbing boundary condition on those surfaces (this condition is sometimes also called a radiation boundary condition or transparent boundary condition). These exact absorbing boundary conditions are usually non-local, in contradistinction to the approximate and local absorbing boundary conditions often employed (Cao and He 1996; He and Weston 1996). Thus, the basic ingredients in a three-dimensional wave-splitting approach have been established and further extensions and applications are under way. This fact, together with the comprehensive results available for the one-dimensional case, and the fact that there does not exist a unified treatment in book form of this basic material, has created a situation in which it appeared useful to summarize the main developments which have taken place to date. Hence this book.

The plan of the presentation is as follows. In Chapter 1 the main features of the wave-splitting and imbedding approaches are described as they apply to the simplest non-trivial case, namely that of scattering of scalar waves in one space dimension from a slab of finite thickness in which the wave speed varies in an arbitrary fashion. The splitting into right- and left-travelling waves is introduced. It is noted that an explicit and simple form of the physical splitting is obtained only in the homogeneous regions exterior to the slab. This, on the other hand, is seen to be sufficient for obtaining the reflected and transmitted fields. Thus the focus is shifted from the original wave field to the split components and the technique for deriving the dynamic equations for the split fields is illustrated. The fundamental relations which provide the background to most applications of the imbedding idea to one-dimensional wave scattering are contained in the

Redheffer star product. A discussion of this product, as well as differentiations of it, is therefore included. The derivation of the imbedding equations for the scattering operators is illustrated for the simple model considered. The imbedding equations are seen to retain the general features expected from the structure of the Redheffer star product. Then it is shown how, by combining the dynamical equations of the split fields with the differentiations of the defining relations for the scattering kernels, one obtains, in a time domain setting, integro-differential equations and initial conditions for the scattering kernels. Since these derivations can be used as a template for many similar derivations in the following chapters, the presentation is rather detailed. Furthermore, it is seen that the inverse scattering problem of reconstructing the wave velocity using one-sided reflection data can be solved using the same equations. The influence of a discontinuity in the derivative of the velocity is included in this first model, but the velocity itself is assumed to be continuous. Numerical algorithms relevant for the direct and inverse problems are described.

In Chapter 2 the Green's function approach for solving one-dimensional direct and inverse problems is introduced. The approach is first illustrated by an example which generalizes the model considered in Chapter 1, in as much as the medium is now characterized by a permittivity and a conductivity which both depend on the depth. In general, the Green's functions are the regular parts of the split components of the fundamental solution. The dynamic equations for the split fields, combined with differentiation of the defining relations for the Green's functions, lead to coupled linear partial differential equations for the latter, as well as to boundary and initial conditions. The numerical implementation of both the direct and inverse problems are described in some detail. The modifications required in the case of a velocity mismatch at the back side is discussed, as well as the use of two-sided excitations. The connection between the Green's functions and the reflection and transmission kernels for the whole slab is pointed out. An analysis is given of the stability, existence, uniqueness and local correctness for the Green's function approach to the one parameter reconstruction problem. If the reflection kernel is taken as given data, the inverse problem is well-posed. However, to get the reflection kernel from actual measured data, a deconvolution, i.e. an ill-posed step, is required. A similar remark applies to the two-parameter reconstruction problem. The Green's functions mentioned so far are defined through their action on the incident field. A modified set of Green's functions called *compact Green's functions* are introduced, which are defined by their action on the transmitted field. As a result they turn out to be different from zero only for times less than one round trip (in a non-dispersive medium). This fact plays a central role for the solution of the inverse problem of simultaneous reconstruction of the permittivity and conductivity. The Green's operator formalism can in turn be considered as subsumed by a more comprehensive wave propagator formalism, where propagators for the split fields are introduced which propagate these fields between two arbitrary positions in the medium. Crucial aspects of how the deconvolution should be performed, such as windowing and regularization, are discussed for a one-parameter reconstruction which uses measured reflection data. In the actual experiment, one-dimensional electromagnetic wave propagation is mimicked by a TEM mode in a coaxial cable. The permittivity of an insert is determined from reflection of a short-pulse incident field (with a half-width of approximately 60 picoseconds).

In Chapter 3 we discuss several additional examples of one-dimensional inverse problems using the Green's function, particularly the compact Green's function, approach. An example, which has important applications and which serves well to illustrate many aspects of multi-parameter inverse problems, is that of an LCRG transmission line. Here the inductance (L), capacitance (C), resistance (R), and shunt conductance (G) per unit length are all assumed to vary with the length coordinate along the line. The compact Green's function approach is described for the case of non-dispersive parameters L, C, R, and G. It is noted that at most two of the four parameters can be determined by means of any measurement of the scattering data at the two endpoints of the (finite-length) inhomogeneous part of the line. In order to achieve reconstructions, certain additional conditions must be fulfilled. Special attention is paid to the problem of obtaining reliable reconstructions when there is no discontinuity at the back end. Applications to source determination and signal restoration are also discussed. Scattering of obliquely incident plane waves from conductive and dispersive half-spaces is considered. In contradistinction to the earlier treatment of the conductive case, a splitting is derived which is exact in a homogeneous conducting medium, and this is used to solve the scattering of obliquely incident transient waves from a homogeneous conducting half-space. However, if the half-space is stratified, one can with advantage work with a splitting which corresponds to a non-conducting half-space. For a homogeneous dispersive half-space the same approach can be carried out. In particular, for the well-known Lorentz and Debye models (for solids and fluids, respectively) the splittings can be given explicitly and, e.g. the 'precursors' can be studied in detail. Also for a stratified slab or half-space, a splitting which is exact only in a homogeneous dielectric medium can be used.

Scattering of more general incident fields from a plane-stratified half-space is considered in Chapter 4. A formalism is developed which uses moments of the fields and of the reflection kernel. For convenience the incident field is assumed to have all its sources in a finite domain in the homogeneous half-space which adjoins the plane-stratified half-space. The technique is first illustrated on the telegraph equation (for a scalar field) which contains four coefficients which are functions of the depth. The zeroth, second and fourth moments of the fields and of the split components are introduced and the dynamic equations for these moments are derived. Using the imbedding procedure, the moments of the split fields are shown to be related by means of moments of the reflection kernel. The boundary values of these moments are related to measurable quantities and are used in the reconstruction of one or several of the physical parameters. The technicalities are described and it is noted that time differentiations of the moments of the reflection kernel are needed for the reconstruction. In a second example it is shown how the zeroth and second moments of the tangential electric fields, together with the corresponding moments of the reflection kernels, can be used to reconstruct the permittivity tensor for an anisotropic medium. If the exciting field exhibits some symmetry, transform methods may be useful in reducing the scattering problem to a set of one-dimensional problems analogous to those treated in Chapter 3. An exciting field which has axial symmetry around an axis perpendicular to the stratification of the inhomogeneous half-space constitutes such a case. Acoustic wave propagation in a stratified half-space is taken as an example and it is noted that the splitting operator for the Hankel-transformed fields can be obtained in explicit form for a homogeneous medium. The Green's function formalism is developed for the Hankel-transformed fields. Several aspects concerning the numerical computation of the scattered field in the direct problem and the deconvolution needed to get to the (Hankel-transformed) reflection kernel are discussed and illustrated. One- and two-parameter reconstructions of the density and/or velocity are shown. It is also noted that for a dissipative homogeneous medium one has similarly an explicit form of the splitting operator. The explicit form of the wave-splitting described above for the Hankel-transformed fields has many other applications in the wave-splitting approach to one-dimensional wave propagation. Thus, it provides the natural splitting in the study of propagation of transient waves in cylindrical waveguides, as well as in a homogeneous plasma.

Optimization techniques play an important role in most realistic parameter reconstruction problems. In Chapter 5 we illustrate how such techniques can be combined with wave-splitting. We focus on methods which allow an explicit determination of the gradient of the objective functional. The first example concerns the reconstruction of the spatial dependence of sources on a known inhomogeneous LCRG transmission line, assuming that the time dependence is known. The data are measured voltages at the endpoints. It is shown how dual ('adjoint') functions, which satisfy appropriately chosen boundary conditions, can be used to obtain an explicit expression for the gradient of the objective functional. Several numerical aspects of the resulting iterative algorithm are discussed, for both one-sided and two-sided measurements. Similarly, the reconstruction of two unknown parameters for an inhomogeneous LCRG line is studied (attempts to reconstruct more than two parameters fail, consistent with the result derived in Chapter 3 and mentioned above). For the problem of reconstructing the temporal behaviour of a dispersive bi-isotropic slab it is shown how the introduction of a matrix formalism facilitates the derivation of an explicit expression for the gradients. Numerical results for the simultaneous reconstruction of the four susceptibility kernels are given. This formalism is also used to reconstruct the susceptibility kernel of the polar liquid 1-buthanol from measured data. In this kind of application a single time domain measurement can replace a large number of time-consuming frequency domain measurements. As an example of a higher-dimensional time domain acoustic inverse problem, a numerical reconstruction of a two-dimensional density and/or sound speed profile is given.

Wave-splitting can also be applied to many non-linear wave equations and as an example a quasi-linear acoustic wave equation is considered. A conventional splitting is used and the dynamic equations for the split fields are derived and a direct solver is obtained. The Newton–Kantorovich method is then applied to the reconstruction of the non-linearity parameter using scattering data. This involves computing the Fréchet derivative, as well as its inverse, of the operator which maps the non-linearity parameter to the scattering data. The reconstruction algorithm is implemented and numerical results are given for a case with a quadratic non-linearity.

The emphasis in the present book is on time domain approaches. However, the wave-splitting and imbedding approach is of interest also for frequency domain direct and inverse scattering problems. The fields inside inhomogeneous regions may also be treated by means of a frequency domain Green's function formalism. In addition, a transmission Green's function approach, which is the frequency domain counterpart of the time domain compact Green's function approach, is described. The frequency domain methods and results presented in Chapter 6 concern mainly one-dimensional cases and the emphasis is on complex media. Chiral media are considered first. The Riccati equation for the (matrix) reflection operator relating the forward- and backward-travelling components is derived, as well as the equations for the Green's functions. The results are then used to compute the scattering from an inhomogeneous chiral slab. A modified imbedding method, which refers to an external medium which is always a vacuum, is introduced and applied to general bi-anisotropic media. It is noted that the use of this vacuum splitting method has important advantages, such as not requiring any conditions on the derivatives of the material parameters, and that a discontinuity in the parameters does not lead to any extra complications in the numerical calculations. The relation between the matrix reflection operator for the vacuum splitting and the conventional splitting is studied. They are related through a fractional linear transformation which depends on the local refractive index. It is shown that the linear transformation can be chosen so that the derivative of the refractive index does not appear. An explicit expression for the reflection operator for a multi-layered medium (i.e. homogeneous layers) using the vacuum splitting is also given. A trace formalism is introduced to derive an explicit gradient in a simple and compact form for reconstruction/design of a stratified bi-anisotropic slab/coating. In the last section of this chapter, the time-harmonic vacuum-splitting approach is generalized to a special two-dimensional case, namely, scattering from a laterally periodic inhomogeneous structure.

Chapters 7 and 8 treat the extension of the wave-splitting approach to fully three-dimensional problems, for scalar and electromagnetic fields respectively. For a scalar field, treated in Chapter 7, up-going and down-going waves are defined with respect to a plane. In general, up-going waves have sources only below the plane, and vice versa for down-going waves. The splitting operator is seen to be an operator which maps Neumann data (the values of the normal derivative of a field) to Dirichlet data (the values of the field) on the same plane.

Integral representations of the splitting operators and their inverses in a homogeneous medium are obtained. Furthermore, the operator which maps Dirichlet data from one plane to another further inside the medium, as well as the inverse of this operator, is needed. The inverse is ill-posed and a Fejer kernel regularization is suggested. The dynamic equations for the split fields in an inhomogeneous medium are derived. Besides the splitting operator, they contain also the derivative of this operator. The splitting is applied to the (four parameter) telegraph equation. The parameter reconstruction problem is first treated by means of the Green's function approach. Details are given for a one parameter reconstruction in the case when the telegraph equation reduces to the ordinary wave equation. The incident field is assumed to emanate from a point source a finite distance away from the scattering half-space. For simplicity, it is furthermore assumed to have a step function in time. A layer-stripping procedure for the reconstruction, which assumes that no caustics are encountered by the rays throughout the region where the reconstruction is attempted, is described. The usefulness of an approximate integral formulation of the dynamic equations for the split fields is demonstrated. Secondly, the basic results concerning an approach to the same reconstruction problem, which is based on field continuation, are presented. The continuation of the field to an adjacent deeper plane is stable, whereas the corresponding continuation of the normal derivative is unstable. A simplified approach to the continuation process is adopted (at the price of stronger conditions on the data). Numerical results for the data continuation and parameter reconstruction are given. The three-dimensional wave-splittings mentioned so far refer to plane surfaces. In addition, a three-dimensional wave-splitting which refers to curved surfaces is presented. The relevant formulas representing waves with sources only inside or only outside, respectively, of the surface are derived from Green's theorem for the wave equation (Kirchhoff's formula). The splitting operators contain both single and double layer integrals and the out-going and in-coming wave conditions are derived using a curvilinear coordinate system. The usefulness of the splitting in providing an exact absorbing boundary condition is emphasized.

In Chapter 8 the plane surface and curved surface splittings from Chapter 7 are extended to electromagnetic fields. Isotropic, non-dispersive constitutive relations are considered first. The general form of the splitting operator for a transversely inhomogeneous medium is studied and a useful explicit representation of these operators for a homogeneous medium is obtained. The structure of the splitting operators in a medium where both the permittivity and the permeability vary with all three coordinates is then derived. It is also shown how a non-dispersive splitting can be used in a dispersive medium. As an application of the above-mentioned results, the simultaneous reconstruction of the permittivity and conductivity in three dimensions is considered. The exciting field is assumed to be an electric or magnetic dipole, exterior to the scattering region. The data are the measured reflected field on the surface of the scattering half-space. A layer-stripping procedure for the reconstruction is described. The splitting of

Maxwell's equations with respect to a closed curved surface is derived in the last section of this chapter. Thus an exact electromagnetic absorbing boundary condition is obtained, as well as a local approximation of this condition.

A few words on the mode of presentation are in order. Our aim is to be as tutorial and as easy to understand as is reasonable. The emphasis is on the fundamentals of the concepts and methods. Their numerical implementation is described and illustrated in many examples. The presentation in Chapters 1–6 is based on the assumption that the reader has a basic operational knowledge of partial differential equations, particularly hyperbolic ones, and associated boundary and initial value problems. Some experience with Green's functions and their singularities is also helpful. Clearly, the nature of the topics treated in Chapters 7 and 8 is such that they require a somewhat higher level of mathematical erudition. A number of exercises have been included, which, if carried out, will give the reader important 'hands on' experience of the methods.

It is our hope that the present book can serve as an introduction to the area, accessible to scientists and graduate students from several different disciplines, such as electrical engineering, mechanical engineering and applied physics.

The reader should consult the work of Fishman et al. (see e.g. Fishman and McCoy (1984a, b), Fishman et al. (1997)) and de Hoop et al. (see e.g. de Hoop and de Hoop (1994), and de Hoop (1996)) for a treatment of multi-dimensional wave-splitting in the frequency domain, which has application to long range propagation problems, such as those that occur in ocean acoustics and seismic exploration. Similarly, for applications to long range tropospheric propagation, see e.g. Levy (1997) and references given there. The material mentioned in this paragraph was not included in the book since most of it depends upon the use of pseudo-differential operators and would thus require extensive background material.

All references cited in the text are collected at the end of the book. In addition, in a bibliography we have collected references which, without being cited in text, describe methods and results concerning the wave-splitting approach which complements those that are presented in this book. Equations, figures and tables are numbered by the chapter number followed by the consecutive number within the chapter. The subdivision of the chapters into sections and subsections is made in a similar way (cf. the Contents).

Two of the authors (S.H. and S.S.) gratefully acknowledge the support that the Swedish Research Council for Engineering Sciences has given over the past decade to the Department of Electromagnetic Theory, Royal Institute of Technology, Stockholm. This support has been essential for a large part of the research that has led to the present book.

Stockholm
October 1997

CONTENTS

1

INTRODUCTION TO THE TIME DOMAIN WAVE-SPLITTING AND IMBEDDING APPROACH IN ONE SPACE DIMENSION

The ideas and techniques used in the wave-splitting and invariant imbedding approach to direct and inverse scattering problems can, naturally, be presented most easily and completely in the case of one-dimensional scattering. Besides being useful, though simplified, models for important classes of physical and technical problems, the one-dimensional cases have served as guides for many aspects of the extension of the wave-splitting and invariant imbedding approach to two- and three-dimensional scattering problems. The starting point for the one-dimensional formalism is given by the general features of wave propagation in a medium whose scattering properties vary along one fixed direction in space. The most directly relevant features are those which are concisely expressed by the Redheffer star product, as well as by differentiations of this product. The one-dimensional formalism can in essence be regarded as an implementation of the relations obtained from this product in a time domain setting. Of particular importance for our purposes is the fact that in a time domain setting, the wave-splitting and imbedding approach has provided the solution to large classes of inverse problems in which time domain measurements are used as given data. A typical one-dimensional inverse problem is to reconstruct, from a measured function of time, a physical parameter which is a function of one space variable. As will be shown below, integration along characteristics in a time domain approach, together with initial conditions, provides a solution to this type of inverse problem.

1.1 Wave-splitting in the time domain: an introductory example

Several key features of the wave-splitting and invariant imbedding approach to direct and inverse scattering problems in the time domain can with advantage be introduced in the context of a suitable simple model, such as the scattering of a plane electromagnetic wave normally incident on a plane-stratified dielectric slab, whose permittivity ϵ varies with depth and which is situated between two homogeneous half-spaces. Let z denote the Cartesian coordinate in the direction perpendicular to the slab interfaces. The slab occupies the region between $z = 0$

and $z = l$. Let further $\hat{\mathbf{e}}$ be the direction of the linearly polarized electric field in a plane perpendicular to the z direction. Thus, for the electric field we write

$$\vec{\mathbf{E}} = \vec{\mathbf{E}}(z,t) = u(z,t)\hat{\mathbf{e}}. \tag{1.1}$$

We assume for simplicity that the permeability is constant everywhere and equal to its vacuum value μ_0. From Maxwell's equations in a source-free medium (where the electric field $\vec{\mathbf{E}}$, the magnetic field $\vec{\mathbf{H}}$, the magnetic induction $\vec{\mathbf{B}}$, and the electric displacement $\vec{\mathbf{D}}$ have their usual meanings, i.e. in our example $\vec{\mathbf{B}} = \mu_0\vec{\mathbf{H}}$, $\vec{\mathbf{D}} = \epsilon(z)\vec{\mathbf{E}}$)

$$\begin{aligned}
\nabla \times \vec{\mathbf{E}} &= -\frac{\partial \vec{\mathbf{B}}}{\partial t}, \\
\nabla \times \vec{\mathbf{H}} &= \frac{\partial \vec{\mathbf{D}}}{\partial t}, \\
\nabla \cdot \vec{\mathbf{D}} &= 0, \\
\nabla \cdot \vec{\mathbf{B}} &= 0,
\end{aligned}$$

it follows that ($\partial_z \equiv \dfrac{\partial}{\partial z}$, etc.)

$$(\partial_z^2 - c^{-2}(z)\partial_t^2)u(z,t) = 0, \tag{1.2}$$

where $c(z) = [\mu_0\epsilon(z)]^{-1/2}$. For $z \leq 0$, $c(z) = c_0$ and for $z \geq l$, $c(z) = c_1$, where c_0 and c_1 are (in general different) constants. In this chapter we assume that $\epsilon(z)$ is continuous everywhere and differentiable for $0 < z < l$. The continuity of the tangential electric and magnetic fields requires $u(z,t)$ and $\partial_z u(z,t)$ to be continuous at the interfaces. The direct scattering problem for the present configuration is then: for a given incident field $\vec{\mathbf{E}}^i(z,t) \equiv u^i(z,t)\hat{\mathbf{e}}$ impinging, e.g. from the left, determine the reflected and transmitted fields for $z < 0$ and $z > l$ respectively. The reflected field propagates in the negative z direction and the transmitted field propagates in the positive z direction. Experimentally both can be obtained from a measurement of the total field for $z < 0$ and $z > l$ respectively. The half-spaces to the left and right of the slab are homogeneous and the wave propagation there is well-known and trivial in character. Therefore we may assume that the measurements are made at $z = 0^- \equiv \lim_{\varepsilon \to 0} - \varepsilon$ and $z = l^+ \equiv \lim_{\varepsilon \to 0}(l + \varepsilon), (\varepsilon > 0)$.

A basic inverse scattering problem, which is naturally associated with the slab configuration, is the following: from measurements of the field at $z = 0^-$ and/or $z = l^+$, determine the permittivity $\epsilon(z)$ for $0 < z < l$. The wave-splitting and invariant imbedding approach to this inverse problem and to the direct scattering problem provides solutions which do not rely on a determination of the wave field inside the slab.

It is intuitively plausible that in a study of one-dimensional scattering, the operators which determine how the total wave field is divided into forward and backward propagating components should play a useful and indeed essential role. In the present example the wave-splitting in a homogeneous region is obtained as a consequence of the obvious factorization of the operator $\partial_z^2 - c^{-2}\partial_t^2$ in the case of a constant velocity $c = c_0$. One has

$$(\partial_z^2 - c_0^{-2}\partial_t^2) = (\partial_z + c_0^{-1}\partial_t)(\partial_z - c_0^{-1}\partial_t), \tag{1.3}$$

and an arbitrary solution to

$$(\partial_z^2 - c_0^{-2}\partial_t^2)u(z,t) = 0 \tag{1.4}$$

can be written

$$u(z,t) = u^+(z,t) + u^-(z,t), \tag{1.5}$$

where

$$u^+(z,t) = f(z - c_0 t), \qquad u^-(z,t) = g(z + c_0 t), \tag{1.6}$$

with arbitrary functions f and g. Thus $u^+(z,t)(u^-(z,t))$ represents a wave propagating in the positive (negative) z direction. In view of the further development of the wave-splitting aspect, we add to the above the relations which extract the $u^+(z,t)$ and $u^-(z,t)$ parts from a general $u(z,t)$. Evidently, $u^+(z,t)$ and $u^-(z,t)$ satisfy

$$(\partial_z \pm c_0^{-1}\partial_t)u^\pm(z,t) = 0. \tag{1.7}$$

Thus one has, e.g.

$$\begin{aligned}(\partial_z + c_0^{-1}\partial_t)u(z,t) &= (\partial_z + c_0^{-1}\partial_t)(u^+(z,t) + u^-(z,t)) \\ &= (\partial_z + c_0^{-1}\partial_t)u^-(z,t) = 2c_0^{-1}\partial_t u^-(z,t),\end{aligned}$$

i.e.

$$\partial_t u^-(z,t) = \frac{1}{2}(\partial_t u(z,t) + c_0\partial_z u(z,t)),$$

which gives

$$\int_{-\infty}^t \partial_{t'} u^-(z,t')\,dt' = \frac{1}{2}[\int_{-\infty}^t \partial_{t'} u(z,t')\,dt' + c_0 \int_{-\infty}^t \partial_z u(z,t')\,dt'].$$

If we restrict ourselves to wave phenomena for which $u(z,-\infty) = 0$, we thus obtain

$$u^-(z,t) = \frac{1}{2}[u(x,t) + c_0 \int_{-\infty}^t \partial_z u(z,t')\,dt'].$$

For $u^+(z,t)$ we obtain similarly

$$u^+(z,t) = \frac{1}{2}[u(z,t) - c_0 \int_{-\infty}^{t} \partial_z u(z,t')\,dt'].$$

In the following it will be convenient to use the notation

$$(\partial_t^{-1} u)(z,t) \equiv \int_{-\infty}^{t} u(z,t')\,dt'. \tag{1.8}$$

For wave phenomena for which $u(z,-\infty) = 0$, we then have formally

$$\partial_t \partial_t^{-1} = \partial_t^{-1} \partial_t = I, \tag{1.9}$$

where I is the identity operator. Thus, the splitting of u into right- and left-propagating components u^+ and u^- can be expressed as

$$u^{\pm}(z,t) = \frac{1}{2}[u(z,t) \mp c_0 \int_{-\infty}^{t} u_z(z,t')\,dt'] \tag{1.10}$$

(here and often in the following we use $u_z \equiv \partial_z u$, etc.). Equation (1.10) gives the first simple example of a wave-splitting.

So far the split fields $u^{\pm}$ have been defined only for a homogeneous region and the definition directly reflects the factorization of the wave operator for this case (cf. (1.3)). However, the problem of factorizing more general wave equations describing wave propagation in inhomogeneous media is a complicated one and simple expressions for the resulting operators are usually not available. In the wave-splitting approach one deals with this situation by working with a decomposition of the total field (so that $u = u^+ + u^-$ is guaranteed), but one relinquishes the requirement that the splitting should correspond to a factorization of the original wave equation satisfied by u. In the present example, the wave equation for $0 < z < l$ is given in (1.2), which is not factorized by the operators $(\partial_t \pm c(z)\partial_z)$. However, by replacing c_0 by $c(z)$ in (1.10), one arrives at the definition

$$u^{\pm}(z,t) = \frac{1}{2}[u(z,t) \mp c(z) \int_{-\infty}^{t} u_z(z,t')\,dt']. \tag{1.11}$$

It is then to be noted that a consequence of the fact that the wave equation (1.2) is not factorized by $(\partial_t \pm c(z)\partial_z)$, is that the split components $u^{\pm}$ in (1.11) do not satisfy (1.2) and they do not have an interpretation as right and left travelling waves. However, in any local region where c is constant, they have such an interpretation. Despite the lack of precise physical interpretation in an inhomogeneous region, this kind of splitting will turn out to be useful for both direct and inverse one-dimensional scattering problems. It should be emphasized that the choice of splitting is by no means unique, even if one imposes, e.g. the condition that it should factorize the wave equation in a homogeneous medium (this is exemplified in Section 3.1).

1.2 The dynamic equations for the split fields

Thus, in the development of the wave-splitting formalism interest is focused on the split fields $u^{\pm}$ rather than on the total field u. It is therefore of interest to express the dynamics of the wave propagation in terms of $u^{\pm}$ rather than in terms of u, as in (1.2). The procedure for deriving the dynamic equations for $u^{\pm}$ when u satisfies a second-order differential equation is as follows. First, rewrite the wave equation in terms of a vector with two components u and u_z according to (where the first row is an identity)

$$\partial_z \begin{bmatrix} u \\ u_z \end{bmatrix} = \begin{bmatrix} 0 & 1 \\ c^{-2}(z)\partial_t^2 & 0 \end{bmatrix} \begin{bmatrix} u \\ u_z \end{bmatrix} \equiv W(z) \begin{bmatrix} u \\ u_z \end{bmatrix}. \tag{1.12}$$

Rewriting (1.11) in two-component form gives

$$\begin{bmatrix} u^+(z,t) \\ u^-(z,t) \end{bmatrix} = \frac{1}{2} \begin{bmatrix} 1 & -c(z)\partial_t^{-1} \\ 1 & c(z)\partial_t^{-1} \end{bmatrix} \begin{bmatrix} u(z,t) \\ u_z(z,t) \end{bmatrix} \equiv T(z) \begin{bmatrix} u(z,t) \\ u_z(z,t) \end{bmatrix}, \tag{1.13}$$

where $T(z)$ has the formal inverse

$$T^{-1}(z) = \begin{bmatrix} 1 & 1 \\ -c^{-1}(z)\partial_t & c^{-1}(z)\partial_t \end{bmatrix}. \tag{1.14}$$

Next we consider

$$\begin{aligned} \partial_z \begin{bmatrix} u^+(z,t) \\ u^-(z,t) \end{bmatrix} &= \partial_z \left(T(z) \begin{bmatrix} u(z,t) \\ u_z(z,t) \end{bmatrix} \right) \\ &= (\partial_z T(z)) \begin{bmatrix} u(z,t) \\ u_z(z,t) \end{bmatrix} + T(z)\partial_z \begin{bmatrix} u(z,t) \\ u_z(z,t) \end{bmatrix}, \end{aligned}$$

and use (1.12) and (1.13) to obtain

$$\partial_z \begin{bmatrix} u^+(z,t) \\ u^-(z,t) \end{bmatrix} = [\partial_z T(z) + T(z)W(z)]T^{-1}(z) \begin{bmatrix} u^+(z,t) \\ u^-(z,t) \end{bmatrix}. \tag{1.15}$$

Substituting the expressions for $T(z)$ and $W(z)$ into the above equation, one obtains

$$\partial_z \begin{bmatrix} u^+(z,t) \\ u^-(z,t) \end{bmatrix} = \begin{bmatrix} \alpha(z) & \beta(z) \\ \gamma(z) & \vartheta(z) \end{bmatrix} \begin{bmatrix} u^+(z,t) \\ u^-(z,t) \end{bmatrix}, \tag{1.16}$$

where

$$\left. \begin{aligned} \alpha(z) &= -c^{-1}(z)\partial_t + \frac{c_z(z)}{2c(z)}, \\ \beta(z) &= \gamma(z) = -\frac{c_z(z)}{2c(z)}, \\ \vartheta(z) &= c^{-1}(z)\partial_t + \frac{c_z(z)}{2c(z)}. \end{aligned} \right\} \tag{1.17}$$

Equation (1.16) will be referred to as the dynamic equation for the split fields.

1.3 The scattering kernels

The description of the scattering properties of the slab between $z = 0$ and $z = l$ will be made in terms of different scattering kernels. Consider specifically incidence from the left side $z < 0$, so that the incident wave $u^i(z,t)$ propagates in the positive z direction, i.e. it is of the u^+ type. For $z < 0$ there is furthermore a reflected field $u^r(z,t)$, propagating in the negative z direction, i.e. it is of the u^- type. The wave equation satisfied by u is linear and it has time-independent coefficients. From this it follows (cf. the discussion of 'Duhamel's principle' in Courant and Hilbert (1962b)) that there exists a relation between u^i and u^r of the form

$$u^r(0,t) = \int_{-\infty}^{t} R(0,l;t-t')u^i(0,t')dt', \tag{1.18}$$

where the argument $(0,l;..)$ of R, the reflection kernel, indicates that $R(0,l;t-t')$ expresses the influence of the whole slab between $z = 0$ and $z = l$.

Analogously, to the right of the slab, there is a transmitted wave u^t propagating in the positive z direction. Again there exists a linear and time translation invariant relation between u^i and u^t (which are both of the u^+ type). From a physical point of view it is natural to take the travel time through the slab into account explicitly. In view of this, it is convenient to define the transmission kernel $T(0,l;t-t')$ by means of the relation

$$\begin{aligned} u^t(l,t) &= u^+(l,t+\tau(0,l)) \\ &= \left(\frac{c_1}{c_0}\right)^{1/2}\left[u^i(0,t) + \int_{-\infty}^{t} T(0,l;t-t')u^i(0,t')dt'\right], \end{aligned} \tag{1.19}$$

where $\tau(0,l)$ is defined according to

$$\tau(z_1,z_2) = \int_{z_1}^{z_2} \frac{dz'}{c(z')}. \tag{1.20}$$

Thus, $\tau(z_1,z_2)$ is the time it takes for the wave to travel from $z = z_1$ to $z = z_2 > z_1$. Note that we have introduced a time shift in the notation for $u^t(l,t)$ so that the time t starts from zero when the wave reaches the back wall $z = l$. The first term on the right hand side of (1.19) represents the directly transmitted part and the part containing the transmission kernel describes the result of all the multiple scattering effects inside the inhomogeneous slab. The factor $(c_1/c_0)^{1/2}$ in the definition (1.19) can be determined by considering the

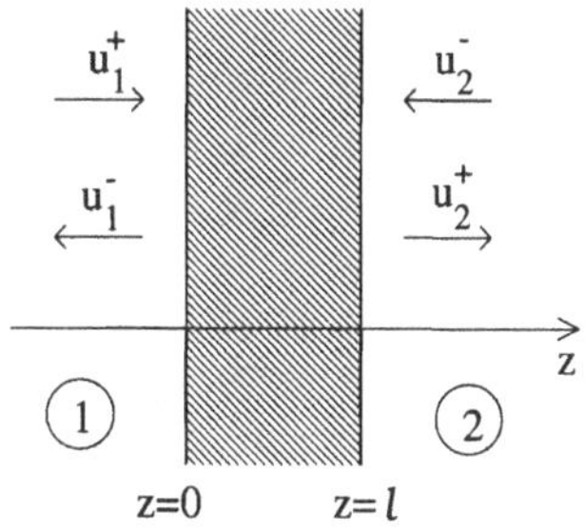

FIG. 1.1. Geometry and notation for incident and scattered fields on the two sides of a slab.

structure of the fundamental solution (i.e., the solution with a δ-function excitation) of the wave equation (cf. Section 2.1). Since the scattering process is linear, the scattering kernels are independent of the incident field u^i and they depend only on the properties of the medium between $z = 0$ and $z = l$, i.e. on $c(z)$. In order to get some background to the equations which express this dependence, we consider in the next section some basic relations which govern one-dimensional linear scattering processes.

1.4 The Redheffer star product

The Redheffer star product expresses basic relations which must be fulfilled by the elements of the scattering matrix which describes the reflection and transmission of waves in linear, one-dimensional models, irrespective of the particular model considered. This product, together with differentiations of it, gives an insight into the type of equations that are to be expected for the reflection and transmission operators for any linear model. In Sections 1.5 and 1.6 more detailed derivations, which make use of the dynamic equation for the split fields in our model, will be given and the relations to the star product will be elucidated.

Figure 1.1 describes the geometry of the configuration and the notation for the fields. Thus, waves travelling from the left (right) to the right (left) are indicated by an upper index $+(-)$. Fields to the left and right of the slab are indicated by a lower index 1 and 2 respectively. Arbitrarily prescribed fields u_1^+ and u_2^- impinge perpendicularly on the slab. The scattering process is assumed to be linear, i.e. it gives rise to reflected and transmitted fields which depend linearly on u_1^+ and u_2^-. In the notation of Fig. 1.1, both u_1^- and u_2^+ depend on both the reflection and the transmission properties of the slab. This linear relation is written

$$\begin{bmatrix} u_2^+ \\ u_1^- \end{bmatrix} = S(1,2) \begin{bmatrix} u_1^+ \\ u_2^- \end{bmatrix} \equiv \begin{bmatrix} t_{21} & r_{22} \\ r_{11} & t_{12} \end{bmatrix} \begin{bmatrix} u_1^+ \\ u_2^- \end{bmatrix}. \tag{1.21}$$

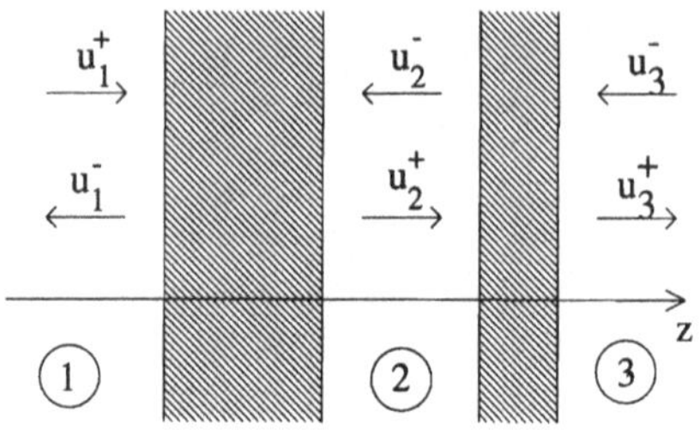

FIG. 1.2. Geometry and notation for incident and scattered fields in a two-slab configuration.

If a second slab is introduced to the right in medium 2, dividing this into two homogeneous regions, indexed 2 and 3 (see Fig. 1.2), we have the analogous relation

$$\begin{bmatrix} u_3^+ \\ u_2^- \end{bmatrix} = S(2,3) \begin{bmatrix} u_2^+ \\ u_3^- \end{bmatrix} \equiv \begin{bmatrix} \tilde{t}_{32} & \tilde{r}_{33} \\ \tilde{r}_{22} & \tilde{t}_{23} \end{bmatrix} \begin{bmatrix} u_2^+ \\ u_3^- \end{bmatrix}. \tag{1.22}$$

The two slabs in combination give rise to a linear scattering process in which the incident fields u_1^+ and u_3^- give rise to the fields u_1^- and u_3^+. This linear relationship is written

$$\begin{bmatrix} u_3^+ \\ u_1^- \end{bmatrix} = S(1,3) \begin{bmatrix} u_1^+ \\ u_3^- \end{bmatrix} \equiv \begin{bmatrix} \overset{\approx}{t}_{31} & \overset{\approx}{r}_{33} \\ \overset{\approx}{r}_{11} & \overset{\approx}{t}_{13} \end{bmatrix} \begin{bmatrix} u_1^+ \\ u_3^- \end{bmatrix}. \tag{1.23}$$

Equations (1.21)–(1.23) determine $S(1,3)$ as a function of $S(1,2)$ and $S(2,3)$. This relation is referred to as the Redheffer star product and it is denoted by $\star$. By reorganizing the relations expressed by (1.21) and (1.22) as actions on the independent incoming fields u_1^+ and u_3^-, one finds the following star product rule (cf. Exercise 1.1)

$$\begin{aligned} S(1,3) &\equiv S(1,2) \star S(2,3) \\ &= \begin{bmatrix} \tilde{t}_{32}(1 - r_{22}\tilde{r}_{22})^{-1} & \tilde{r}_{33} + \tilde{t}_{32} r_{22}(1 - \tilde{r}_{22} r_{22})^{-1} \tilde{t}_{23} \\ r_{11} + t_{12}\tilde{r}_{22}(1 - r_{22}\tilde{r}_{22})^{-1} t_{21} & t_{12}(1 - \tilde{r}_{22} r_{22})^{-1} \cdot \tilde{t}_{23} \end{bmatrix}. \end{aligned} \tag{1.24}$$

The matrix elements in $S(1,2)$, etc. depend on the scattering properties of the medium, but not on the exciting fields. It should be noted that the elements in these matrices are in general operators which do not commute. Furthermore we note that the star product is not distributive since it involves forming non-linear functions of the matrix elements. Thus in general $S_1 \star (S_2 + S_3) \neq S_1 \star S_2 + S_1 \star S_3$. The right hand side of (1.24) has several properties which have natural physical

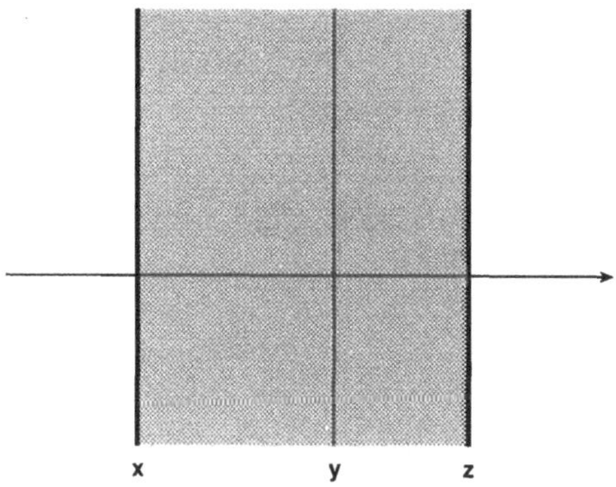

FIG. 1.3. Geometry and notation used in the discussion of the Redheffer star product.

interpretations. For instance, if a formal Neumann series expansion of the inverse operators in (1.24) is introduced, the resulting terms can be interpreted as representing multiple scatterings between the two slabs. A more detailed discussion of the star product is found in Redheffer (1962).

The above considerations apply also in the limit when the width of the region between the two slabs (region 2) goes to zero. This situation is shown in Fig. 1.3, where x, y and z denote the position of the interfaces. For simplicity we assume that the scattering properties of the medium are continuous at y. The $\star$ composition rule is written

$$S(x, z) = S(x, y) \star S(y, z). \tag{1.25}$$

The differential equations of the imbedding approach are obtained from a consideration of the effects of adding an infinitesimally thin slab to a given structure. Therefore, in Fig. 1.3 we consider the case $z = y + \Delta y$, where Δy is a small quantity. Our continuity assumption implies that

$$\lim_{\Delta y \to 0} S(y, y + \Delta y) = \begin{bmatrix} 1 & 0 \\ 0 & 1 \end{bmatrix} \equiv I. \tag{1.26}$$

Thus, for small Δy we may write

$$S(y, y + \Delta y) = I + \Delta y \cdot M(y) + o(\Delta y), \tag{1.27}$$

where the matrix M is determined by the scattering properties of the medium and where $(\Delta y)^{-1} o(\Delta y) \to 0$ as $(\Delta y) \to 0$. With the notation

$$S(x, y) \equiv \begin{bmatrix} t_{21} & r_{22} \\ r_{11} & t_{12} \end{bmatrix}, \quad M(y) = \begin{bmatrix} m_{11} & m_{12} \\ m_{21} & m_{22} \end{bmatrix}, \tag{1.28}$$

the star product gives (i.e. one first applies (1.24) and then expands in orders of Δy)

$$\begin{aligned}
&S(x, y + \Delta y) \\
&= S(x, y) \star (I_2 + \Delta y \cdot M(y) + o(\Delta y)) \\
&= S(x, y) + \Delta y \begin{bmatrix} (m_{11} + r_{22}m_{21})t_{21} & m_{12} + m_{11}r_{22} + r_{22}(m_{22} + m_{21}r_{22}) \\ t_{12}m_{21}t_{21} & t_{12}(m_{22} + m_{21}r_{22}) \end{bmatrix} \\
&\quad + o(\Delta y). \qquad (1.29)
\end{aligned}$$

Thus it follows that

$$\begin{aligned}
\frac{\partial S(x, y)}{\partial y} &\equiv \lim_{\Delta y \to 0} \frac{S(x, y + \Delta y) - S(x, y)}{\Delta y} \\
&= \begin{bmatrix} (m_{11} + r_{22}m_{21})t_{21} & m_{12} + m_{11}r_{22} + r_{22}(m_{22} + m_{21}r_{22}) \\ t_{12}m_{21}t_{21} & t_{12}(m_{22} + m_{21}r_{22}) \end{bmatrix}, \qquad (1.30)
\end{aligned}$$

which expresses the influence of the addition of a thin layer to the right hand side interface of the slab.

For a thin slab on the left hand side we write similarly

$$S(x - \Delta x, x) = I + \Delta x \cdot P(x) + o(\Delta x), \qquad (1.31)$$

where the matrix

$$P(x) = \begin{bmatrix} p_{11} & p_{12} \\ p_{21} & p_{22} \end{bmatrix}$$

is determined by the properties of the medium and where we have assumed continuity of the scattering properties at x, i.e.

$$\lim_{\Delta x \to 0} S(x - \Delta x, x) = I.$$

The star product

$$S(x - \Delta x, x) * S(x, y) = S(x - \Delta x, y) \qquad (1.32)$$

gives

$$\begin{aligned}
\frac{\partial S(x, y)}{\partial x} &\equiv \lim_{\Delta x \to 0} \frac{S(x, y) - S(x - \Delta x, y)}{\Delta x} \\
&= -\begin{bmatrix} t_{21}(p_{11} + p_{12}r_{11}) & t_{21}p_{12}t_{12} \\ p_{21} + r_{11}p_{11} + (p_{22} + r_{11}p_{12})r_{11} & (p_{22} + r_{11}p_{12})t_{12} \end{bmatrix}. \qquad (1.33)
\end{aligned}$$

If a model for the properties of the medium is given, e.g. in terms of a wave equation for a field, such as a pressure field, an electromagnetic field, an elastodynamic displacement field, etc., the form of M and P can be determined (this is illustrated in Section 1.5 below). Then, if $M(y)$ and $P(x)$ are known, (1.30) and (1.33) constitute systems of coupled differential equations for the elements of the scattering matrix $S(x, y)$. For easy reference, we list the equations for the matrix elements as separate equations:

$$\frac{\partial t_{21}}{\partial y} = (m_{11} + r_{22}m_{21})t_{21}, \tag{1.34}$$

$$\frac{\partial t_{12}}{\partial y} = t_{12}(m_{22} + m_{21}r_{22}), \tag{1.35}$$

$$\frac{\partial r_{11}}{\partial y} = t_{12}m_{21}t_{21}, \tag{1.36}$$

$$\frac{\partial r_{22}}{\partial y} = m_{12} + m_{11}r_{22} + r_{22}m_{22} + r_{22}m_{21}r_{22}, \tag{1.37}$$

$$\frac{\partial t_{21}}{\partial x} = -t_{21}(p_{11} + p_{12}r_{11}), \tag{1.38}$$

$$\frac{\partial t_{12}}{\partial x} = -(p_{22} + r_{11}p_{12})t_{12}, \tag{1.39}$$

$$\frac{\partial r_{11}}{\partial x} = -(p_{21} + r_{11}p_{11} + p_{22}r_{11} + r_{11}p_{12}r_{11}), \tag{1.40}$$

$$\frac{\partial r_{22}}{\partial x} = -t_{21}p_{12}t_{12}. \tag{1.41}$$

Note that the two equations which contain only one of the unknowns, i.e. (1.37) and (1.40), are non-linear in the unknown (r_{22} and r_{11} respectively).

Since only very general assumptions about linearity and regularity have been used in the derivation of (1.34)–(1.41), we expect that the structural features of these equations, such as the non-linearity, will be found for all wave propagation models which satisfy these general assumptions. The time domain wave-splitting and invariant imbedding approach to one-dimensional direct and inverse scattering problems can be regarded as based on an implementation, in a specific framework, of the general procedure outlined above.

1.5 The imbedding equations for the scattering operators

In analogy with the derivations which led to (1.34)–(1.41), we expect to be able to derive coupled differential equations for the scattering operators and eventually for the scattering kernels. This involves considering not only the total or physical scattering kernels $R(0, l; t)$ and $T(0, l, t)$, which refer to the whole slab between $z = 0$ and $z = l$, but also scattering kernels $R(x, y; t)$, $T(x, y; t)$

referring to a subslab between $z = x$ and $z = y$ $(0 < x < y < l)$. This subslab is assumed to be imbedded between two homogeneous half-spaces characterized by the constant velocities $c(z) = c(x)$, $z < x$, and $c(z) = c(y)$, $z > y$. The physical scattering kernels are then obtained as boundary values of these more generally defined kernels $R(x, y; t)$ and $T(x, y; t)$. We have assumed that the velocity $c(z)$ is continuous everywhere. Thus $c(0^-) = c(0^+) = c_0$ and $c(l^-) = c(l^+) = c_1$, where c_0 and c_1 are the velocities in the homogeneous half-spaces to the left and right, respectively. This method of deriving the scattering kernels is often referred to as the invariant imbedding approach. We usually refer to it simply as 'the imbedding approach'. Wave-splitting in combination with the imbedding approach provides a framework for obtaining solutions to both direct and inverse problems in the time domain, as described below.

Thus, according to the imbedding idea, the following situation is considered: the subslab between x and y, $0 \leq x \leq y \leq l$, is assumed to be imbedded between two homogeneous half-spaces characterized by $c = c(x)$ and $c = c(y)$ respectively. In these half-spaces right- and left-travelling waves $u^{\pm}(z, t)$ can be identified for $z < x$ and $z > y$ respectively. Here $u^{\pm}(z, t)$ are defined by the wave-splitting (1.11). For the subslab, $u^+(x, t)$ and $u^-(y, t)$ are then arbitrarily prescribed incoming fields and for the scattering operators which express the influence of the subslab we write

$$\begin{aligned}\begin{bmatrix} u^+(y,t) \\ u^-(x,t) \end{bmatrix} &= S(x,y) \begin{bmatrix} u^+(x,t) \\ u^-(y,t) \end{bmatrix} \\ &\equiv \begin{bmatrix} \mathcal{T}^+(x,y) & \mathcal{R}^-(x,y) \\ \mathcal{R}^+(x,y) & \mathcal{T}^-(x,y) \end{bmatrix} \begin{bmatrix} u^+(x,t) \\ u^-(y,t) \end{bmatrix}, \end{aligned} \tag{1.42}$$

where $u^-(x, t)$ and $u^+(y, t)$ denote the left- and right-travelling components at x and y respectively. $\mathcal{R}^+$ and $\mathcal{T}^+$ ($\mathcal{R}^-$ and $\mathcal{T}^-$) are the reflection and transmission operators for an incident wave from the left (right) of the subslab, respectively.

We recall that the derivation of (1.34)–(1.41), following from a differentiation of the star product, was incomplete in as much as the elements of the matrices M and P were not specified. This will be remedied by the derivation below in which differentiations of (1.42) are considered, and in which the explicit form of the dynamic equations of the split fields are taken into account.

Thus, applying ∂_x to (1.42) gives

$$\begin{bmatrix} 0 \\ \partial_x u^-(x,t) \end{bmatrix} = \partial_x S(x,y) \begin{bmatrix} u^+(x,t) \\ u^-(y,t) \end{bmatrix} + S(x,y) \begin{bmatrix} \partial_x u^+(x,t) \\ 0 \end{bmatrix}. \tag{1.43}$$

The operator relations inherent in this relation are obtained by expressing each term as an operator acting on the arbitrary incoming fields $u^+(x, t)$ and $u^-(y, t)$. The resulting equations correspond to a more explicit form of (1.34)–(1.41) for the special model considered here. The left hand side of (1.43) may be written (cf. (1.16))

$$\begin{bmatrix} 0 \\ \partial_x u^-(x,t) \end{bmatrix} = \begin{bmatrix} 0 \\ \gamma(x)u^+(x,t) + \vartheta(x)u^-(x,t) \end{bmatrix}$$

$$= \begin{bmatrix} 0 & 0 \\ \gamma(x) & \vartheta(x) \end{bmatrix} \begin{bmatrix} u^+(x,t) \\ u^-(x,t) \end{bmatrix}. \tag{1.44}$$

Invoking the second row of (1.42) one has

$$\begin{bmatrix} u^+(x,t) \\ u^-(x,t) \end{bmatrix} = \begin{bmatrix} u^+(x,t) \\ \mathcal{R}^+(x,y)u^+(x,t) + \mathcal{T}^-(x,y)u^-(y,t) \end{bmatrix}$$

$$= \begin{bmatrix} 1 & 0 \\ \mathcal{R}^+(x,y) & \mathcal{T}^-(x,y) \end{bmatrix} \begin{bmatrix} u^+(x,t) \\ u^-(y,t) \end{bmatrix}, \tag{1.45}$$

and thus the left hand side of (1.44), as an operator on the incoming fields $u^+(x,t)$ and $u^-(y,t)$, has the form

$$\begin{bmatrix} 0 \\ \partial_x u^-(x,t) \end{bmatrix} = \begin{bmatrix} 0 & 0 \\ \gamma(x) & \vartheta(x) \end{bmatrix} \begin{bmatrix} 1 & 0 \\ \mathcal{R}^+(x,y) & \mathcal{T}^-(x,y) \end{bmatrix} \begin{bmatrix} u^+(x,t) \\ u^-(y,t) \end{bmatrix}. \tag{1.46}$$

Similarly, the second term on the right hand side of (1.43) is written

$$S(x,y) \begin{bmatrix} \partial_x u^+(x,t) \\ 0 \end{bmatrix} = S(x,y) \begin{bmatrix} \alpha(x) & \beta(x) \\ 0 & 0 \end{bmatrix} \begin{bmatrix} u^+(x,t) \\ u^-(x,t) \end{bmatrix}$$

$$= S(x,y) \begin{bmatrix} \alpha(x) & \beta(x) \\ 0 & 0 \end{bmatrix} \begin{bmatrix} 1 & 0 \\ \mathcal{R}^+(x,y) & \mathcal{T}^-(x,y) \end{bmatrix} \begin{bmatrix} u^+(x,t) \\ u^-(y,t) \end{bmatrix}. \tag{1.47}$$

Thus, from (1.43), using (1.46) and (1.47), we obtain the following operator relation

$$\partial_x S(x,y)$$

$$= \left(\begin{bmatrix} 0 & 0 \\ \gamma(x) & \vartheta(x) \end{bmatrix} - S(x,y) \begin{bmatrix} \alpha(x) & \beta(x) \\ 0 & 0 \end{bmatrix} \right) \begin{bmatrix} 1 & 0 \\ \mathcal{R}^+(x,y) & \mathcal{T}^-(x,y) \end{bmatrix}$$

$$= - \begin{bmatrix} \mathcal{T}^+(x,y)\alpha(x) & \mathcal{T}^+(x,y)\beta(x) \\ \mathcal{R}^+(x,y)\alpha(x) - \gamma(x) & \mathcal{R}^+(x,y)\beta(x) - \vartheta(x) \end{bmatrix} \begin{bmatrix} 1 & 0 \\ \mathcal{R}^+(x,y) & \mathcal{T}^-(x,y) \end{bmatrix},$$

i.e. in factorized form

$$S_x(x,y)$$

$$= - \begin{bmatrix} \mathcal{T}^+(x,y) & 0 \\ \mathcal{R}^+(x,y) & 1 \end{bmatrix} \begin{bmatrix} \alpha(x) & \beta(x) \\ -\gamma(x) & -\vartheta(x) \end{bmatrix} \begin{bmatrix} 1 & 0 \\ \mathcal{R}^+(x,y) & \mathcal{T}^-(x,y) \end{bmatrix}. \tag{1.48}$$

Thus, for the elements of this relation we have

$$\begin{bmatrix} \partial_x \mathcal{T}^+(x,y) & \partial_x \mathcal{R}^-(x,y) \\ \partial_x \mathcal{R}^+(x,y) & \partial_x \mathcal{T}^-(x,y) \end{bmatrix}$$
$$= \begin{bmatrix} \mathcal{T}^+(x,y)\alpha(x) + \mathcal{T}^+(x,y)\beta(x)\mathcal{R}^+(x,y) \\ -\gamma(x) + \mathcal{R}^+(x,y)\alpha(x) - \vartheta(x)\mathcal{R}^+(x,y) + \mathcal{R}^+(x,y)\beta(x)\mathcal{R}^+(x,y) \end{bmatrix.$$
$$\begin{matrix} \mathcal{T}^+(x,y)\beta(x)\mathcal{T}^-(x,y) \\ \mathcal{R}^+(x,y)\beta(x)\mathcal{T}^-(x,y) - \vartheta(x)\mathcal{T}^-(x,y) \end{bmatrix}. \tag{1.49}$$

Similarly, by differentiating (1.42) with respect to y, one obtains

$$\begin{bmatrix} u_y^+(y,t) \\ 0 \end{bmatrix} = S_y(x,y) \begin{bmatrix} u^+(x,t) \\ u^-(y,t) \end{bmatrix} + S(x,y) \begin{bmatrix} 0 \\ u_y^-(y,t) \end{bmatrix}. \tag{1.50}$$

In complete analogy with the derivation of (1.49), this gives

$$\begin{bmatrix} \partial_y \mathcal{T}^+(x,y) & \partial_y \mathcal{R}^-(x,y) \\ \partial_y \mathcal{R}^+(x,y) & \partial_y \mathcal{T}^-(x,y) \end{bmatrix}$$
$$= \begin{bmatrix} \mathcal{R}^-(x,y)\gamma(y)\mathcal{T}^+(x,y) - \alpha(y)\mathcal{T}^+(x,y) \\ \mathcal{T}^-(x,y)\gamma(y)\mathcal{T}^+(x,y) \end{matrix}$$
$$\begin{matrix} -\beta(y) - \alpha(y)\mathcal{R}^-(x,y) + \mathcal{R}^-(x,y)\vartheta(y) + \mathcal{R}^-(x,y)\gamma(y)\mathcal{R}^-(x,y) \\ \mathcal{T}^-(x,y)\gamma(y)\mathcal{R}^-(x,y) + \mathcal{T}^-(x,y)\vartheta(y) \end{bmatrix}. \tag{1.51}$$

Equations (1.49) and (1.51) are called the imbedding equations for the scattering operators. Clearly, they correspond, in the context of the present example, to the equations (1.34)–(1.41) obtained in Section 1.4 from a consideration of the star product (1.24). In particular, from a comparison of (1.30) and (1.33) with (1.51) and (1.49), respectively, one easily identifies the elements of the matrices $M(y)$ and $P(x)$ in terms of α, β, γ, and ϑ, viz.

$$M(y) = \begin{bmatrix} -\alpha(y) & -\beta(y) \\ \gamma(y) & \vartheta(y) \end{bmatrix}, \qquad P(x) = \begin{bmatrix} -\alpha(x) & -\beta(x) \\ \gamma(x) & \vartheta(x) \end{bmatrix}. \tag{1.52}$$

The fact that M and P have the same form reflects the fact that the physical model (i.e., the wave equation) does not distinguish between left and right wave propagation.

1.6 The imbedding equations for the scattering kernels

The scattering operators $\mathcal{R}^\pm(x,y)$ and $\mathcal{T}^\pm(x,y)$ have representations in terms of convolution kernels $R^\pm(x,y;t)$ and $T^\pm(x,y;t)$ in analogy with (1.18) and (1.19). Thus we have

$$u^-(x,t) = \mathcal{R}^+(x,y)u^+(x,t) = \int_{-\infty}^{t} R^+(x,y;t-t')u^+(x,t')dt', \tag{1.53}$$

$$u^+(y,t+\tau(x,y)) = \mathcal{T}^+(x,y)u^+(x,t+\tau(x,y))$$
$$= \left[\frac{c(y)}{c(x)}\right]^{1/2}\left[u^+(x,t) + \int_{-\infty}^{t} T^+(x,y;t-t')u^+(x,t')dt'\right]. \tag{1.54}$$

For $\mathcal{R}^-, \mathcal{T}^-$, which describe the scattering properties of the slab for an incident wave $u^-(y,t)$ from the right, one has similarly

$$u^+(y,t) = \mathcal{R}^-(x,y)u^-(y,t) = \int_{-\infty}^{t} R^-(x,y;t-t')u^-(y,t')dt', \tag{1.55}$$
$$u^-(x,t+\tau(x,y)) = \mathcal{T}^-(x,y)u^-(y,t+\tau(x,y))$$
$$= \left[\frac{c(x)}{c(y)}\right]^{1/2}\cdot\left[u^-(y,t) + \int_{-\infty}^{t} T^-(x,y;t-t')u^-(y,t')dt'\right]. \tag{1.56}$$

The following remark is in order in connection with the definitions (1.54) and (1.56) of the transmission kernels $T^\pm(x,y;t)$. The two cases of right and left incidence on the slab are clearly related through reciprocity. From this a relation between $T^+(x,y;t)$ and $T^-(x,y;t)$ can be deduced (cf. e.g. the discussion in Kristensson and Krueger (1986a); note the slight difference in the definitions of the transmission kernels). With the definitions (1.54) and (1.56) one has

$$T^+(x,y;t) = T^-(x,y;t). \tag{1.57}$$

In order to make the results contained in (1.49) and (1.51) more explicit, we consider the equations for the scattering kernels which follow from them and we consider first the equations for the reflection kernel $R^+(x,y;t)$ (cf. element 21 in (1.49)), corresponding to an incident field $u^+(x,t)$ (and we put $u^-(y,t)\equiv 0$). As noted before in connection with (1.34)–(1.41), the only equations which involve only one element of $S(x,y)$ are the equations for the reflection operators and they are non-linear. In (1.49) y is fixed. Therefore, when discussing this equation, the notation is simplified by omitting y, i.e. the kernels are denoted $R^+(x;t)$, etc. From now on we shall also assume that all incident fields vanish for $t \le 0$. We recall that we have assumed that the velocity $c(z)$ is continuous everywhere and differentiable for $0 < z < l$. However, the coefficients α, β, γ and ϑ in the dynamic system (1.16) for the split fields contain $c_z(z)$, and these same coefficients appear also in (1.49) and (1.51). We should therefore expect that the continuity assumptions made so far still allow for the possibility of a 'hard' reflection from an interface where c_z has a discontinuity. Specifically, we shall assume that c_z may have a discontinuity at $z = y$ (but not for $x < z < y$). Since $\tau(x,y)$ is the travel time from x to y, the effect of such a discontinuity in the reflection kernel $R^+(x,t)$ is expected to appear at the time $t = 2\tau(x,y)$, corresponding to one round trip from x to y and then back to x. Furthermore, we note that $R^+(x,t) \equiv R^+(x,y;t)$ should be independent of y for $t < 2\tau(x,y)$. Thus, we include the possibility of a discontinuity in the reflection kernel $R^+(x,y;t)$. We assume that the discontinuity occurs for $t = g(x,y)$ and we determine $g(x,y)$

below (instead of using the expected relation $g(x,y) = 2\tau(x,y)$). Since we now consider a fixed y, we also simplify this notation, i.e. $g(x) \equiv g(x,y)$. This function is assumed to be differentiable. Since we have assumed that $u^+(x,t) = 0$, $u^-(y,t) = 0$ for $t \le 0$, we encounter mainly convolutions of the form

$$\int_0^t f_1(x,y;t-t')f_2(x,y;t')dt' \equiv (f_1 * f_2)(x,y;t) \tag{1.58}$$

(and when there is no risk of confusion we also write $(f_1 * f_2)(x,y;t) \equiv f_1(x,y;t) * f_2(x,y;t)$, with the understanding that the convolution always concerns the t variable).

The derivations and observations relevant for obtaining equations for the $R^+(x;t)$ kernel are typical for the derivations which occur in the wave-splitting and imbedding approach in the time domain and therefore they are described here in some detail. One can obtain the desired result by introducing the kernel representation for $\mathcal{R}^+$ into the result given by the element (21) of (1.49). However, we prefer to give an alternative explicit derivation which is based on combining the definition of the reflection kernel, i.e.

$$u^-(x,t) = \int_0^t R^+(x,y;t-t')u^+(x,t')dt', \tag{1.59}$$

with the dynamic equations for the split fields. In particular, the equations describing the discontinuity in $R^+(x,y;t)$ are made explicit. We proceed as follows.

Apply ∂_x to the definition (1.59) and then rewrite $\partial_x u^-(x,t)$ by means of the dynamic equation (1.16). Thus we have in the abbreviated notation

$$\partial_x u^-(x,t) = \partial_x \int_0^t R^+(x;t-t')u^+(x,t')dt'. \tag{1.60}$$

Introducing the possibility of a discontinuity at $t = g(x)$, we get for the right hand side

$$\begin{aligned}
&\partial_x\Big[\int_0^{g(x)^-} R^+(x,t-t')u^+(x,t')dt' + \int_{g(x)^+}^t R^+(x,t-t')u^+(x,t')dt'\Big] \\
&= \int_0^t R^+(x;t-t')u_x^+(x,t')dt' + \int_0^{g(x)^-} [\partial_x R^+(x;t-t')]u^+(x,t')dt' \\
&\quad + \int_{g(x)^+}^t [\partial_x R^+(x;t-t')]u^+(x,t')dt' \\
&\quad + [R^+(x,g(x)^-) - R^+(x,g(x)^+)]g_x(x)u^+(x,t-g(x)),
\end{aligned} \tag{1.61}$$

where we have used the assumption that $g(x)$ is differentiable (and that $u^+(x,t)$ is continuous so that $u^+(x;t-g(x)^-) = u^+(x;t-g(x)^+)$). We wish to rewrite all terms in the form of operators acting on the arbitrarily prescribed incoming field $u^+(x,t)$. The first term on the right hand side of (1.61) is (cf. (1.16))

$$\begin{aligned} R^+(x;t) * u_x^+(x;t) &= R^+(x;t) * [\alpha(x)u^+(x;t) + \beta(x)u^-(x;t)] \\ &= R^+(x;t) * \alpha(x)u^+(x;t) + \beta(x)R^+(x;t) * R^+(x;t) * u^+(x,t) \end{aligned} \tag{1.62}$$

(we have used the fact that according to (1.17), $\beta(x)$ can be taken out as a factor in front of the convolutions; this, of course, does not apply for $\alpha(x)$, cf. (1.17)). Introducing the explicit expression for $\alpha(x)$ from (1.17) we have

$$\begin{aligned} &R^+(x;t) * \alpha(x)u^+(x;t) \\ &= -c^{-1}(x)R^+(x;t) * \partial_t u^+(x;t) + \frac{c_x(x)}{2c(x)}R^+(x;t) * u^+(x;t). \end{aligned} \tag{1.63}$$

For the first term on the right hand side of this relation we have

$$\begin{aligned} -c^{-1}(x)R^+(x;t) * \partial_t u^+(x,t) &= -c^{-1}(x)\int_0^t R^+(x;t-t')\partial_{t'}u^+(x,t')dt' \\ = c^{-1}(x)[\int_0^{g(x)^-} & R^+(x;t')\partial_{t'}u^+(x;t-t')dt' \\ + \int_{g(x)^+}^t & R^+(x;t')\partial_{t'}u^+(x,t-t')dt'] \\ = c^{-1}(x)[R^+(x,g(x)^-)&u^+(x,t-g(x)^-) - R^+(x,0)u^+(x,t) \\ +R^+(x;t)u^+(x;0) &- R^+(x;g(x)^+)u^+(x,t-g(x)^+) \\ - \int_0^{g(x)^-} R_{t'}(x,t')&u^+(x;t-t')dt' - \int_{g(x)^+}^t R_{t'}(x,t')u^+(x;t-t')dt']. \end{aligned} \tag{1.64}$$

According to our assumptions concerning $u^+(x,t)$, we have $u^+(x,0) = 0$. For the discontinuity in $R^+(x,t)$ at $t = g(x)$ we use the notation

$$\Delta R^+(x,g(x)) \equiv R^+(x,g(x)^+) - R^+(x,g(x)^-). \tag{1.65}$$

For the last two terms within the bracket in (1.64) we write simply $R_t^+(x,t) * u^+(x,t)$, with the understanding that the integral is to be evaluated by taking the limits indicated in (1.64). A corresponding remark applies to the two integrals in (1.61) which contain $\partial_x R^+(x;t)$ and which are then written as $\partial_x R^+(x;t) * u^+(x,t)$. Thus, collecting the results given in (1.61)–(1.64), the right hand side of (1.60) may be written

$$\begin{aligned} \partial_x[R(x,t) * u^+(x,t)] &= R^+(x;t) * u_x(x,t) \\ &+R_x^+(x;t) * u^+(x,t) - \Delta R^+(x;g(x))g_x(x)u^+(x,t-g(x)) \\ &= c^{-1}(x)[-\Delta R^+(x,g(x))u^+(x,t-g(x)) - R^+(x,0)u^+(x;t) \end{aligned}$$

$$-R_t^+(x;t) * u^+(x,t)] + \frac{c_x(x)}{2c(x)} R^+(x;t) * u^+(x,t)$$
$$-\frac{c_x(x)}{2c(x)} R^+(x;t) * R^+(x;t) * u^+(x,t) + R_x^+(x;t) * u^+(x;t)$$
$$-\Delta R^+(x, g(x)) g_x(x) u^+(x, t - g(x)). \quad (1.66)$$

According to (1.60) this expression is also equal to the expression for $\partial_x u^-(x,t)$ that is obtained from the dynamic equation for the split fields (cf. (1.16), (1.17)) and again we rewrite the result in the form of operators acting on the arbitrary incoming field. We have

$$\begin{aligned} \partial_x u^-(x,t) &= \gamma(x) u^+(x,t) + \vartheta(x) u^-(x,t) \\ &= -\frac{c_x(x)}{2c(x)} u^+(x,t) + c^{-1}(x) \partial_t [R^+(x,t) * u^+(x,t)] \\ &\quad + \frac{c_x(x)}{2c(x)} R^+(x,t) * u^+(x,t). \end{aligned} \quad (1.67)$$

For the second term on the right hand side of (1.67) we have (we use the fact that $u^+(x,0) = 0$ as well as (1.64))

$$\begin{aligned} &\partial_t [R^+(x,t) * u^+(x,t)] \\ &= \partial_t \int_0^t R^+(x;t') u^+(x,t-t') dt' \\ &= R^+(x,t) u^+(x,0) + \int_0^t R^+(x,t') \partial_t u^+(x,t-t') dt' \\ &= R^+(x;t) * \partial_t u^+(x;t) \\ &= (\text{cf. } (1.64)) = \Delta R^+(x; g(x)) u^+(x, t - g(x)) \\ &\quad + R^+(x,0) u^+(x,t) + R_t^+(x;t) * u^+(x,t). \end{aligned} \quad (1.68)$$

Consequently we have

$$\begin{aligned} \partial_x u^-(x,t) = & -\frac{c_x(x)}{2c(x)} u^+(x,t) + c^{-1}(x) [\Delta R^+(x; g(x)) u^+(x, t - g(x)) \\ & + R^+(x;0) u^+(x,t) + R_t^+(x,t) * u^+(x,t)] \\ & + \frac{c_x(x)}{2c(x)} R^+(x;t) * u^+(x,t). \end{aligned} \quad (1.69)$$

We now equate the right hand sides of (1.66) and (1.69). In these expressions, the incoming field $u^+(x,t)$ appears in three different ways, namely in convolutions, as a direct factor $u^+(x,t)$ and as $u^+(x, t-g(x))$. Since $u^+(x,t)$ is arbitrary, the operators acting on each of these types of terms must be equal in the right hand sides of (1.66) and (1.69) (otherwise, the integral equation for $u^+(x,t)$ only has a trivial solution (cf. Tricomi (1957)), or a periodic (in t) solution when the convolution terms cancel each other). Thus we obtain

(i) From the convolution terms:

$$\begin{aligned} &c^{-1}(x)R_t^+(x;t) + (2c(x))^{-1}c_x(x)R^+(x;t) \\ &= -c^{-1}(x)R_t^+(x;t) + (2c(x))^{-1}c_x(x)R^+(x;t) \\ &\quad -(2c(x))^{-1}c_x(x)R^+(x;t) * R^+(x;t) + R_x^+(x;t), \end{aligned}$$

i.e.

$$R_x^+(x;t) - 2c^{-1}(x)R_t^+(x;t) - (2c(x))^{-1}c_x(x)R^+(x;t) * R^+(x;t) = 0. \tag{1.70}$$

(ii) From direct factors $u^+(x;t)$:

$$-[2c(x)]^{-1} \cdot c_x(x) + c^{-1}(x)R^+(x;0) = -c^{-1}(x)R^+(x;0),$$

i.e.

$$R^+(x;0) = \frac{1}{4}c_x(x). \tag{1.71}$$

(iii) From direct factors $u^+(x;t-g(x))$:

$$c^{-1}(x)\Delta R^+(x;g(x)) = -c^{-1}(x)\Delta R^+(x;g(x)) - \Delta R^+(x,g(x))g_x(x),$$

i.c. for $\Delta R^+ \not\to 0$

$$d_x g(x) = -2c^{-1}(x), \tag{1.72}$$

or

$$g(x) = 2\int_x^{y_0} \frac{dx'}{c(x')} = 2\tau(x, y_0), \tag{1.73}$$

where y_0 will be determined below. Here (1.70) is the non-linear equation for the kernel $R^+(x,t)$ that is anticipated by the general result (1.40) and the more explicit result (1.49) for the present example. We note that in (1.71), $R^+(x,0) \equiv R^+(x;0^+)$ and $R^+(x;0)$ is the discontinuity of R^+ at the front end $z = x$. If $c_x \neq 0$, so that $R^+(x,0) \neq 0$, we say that there is a 'hard echo' from the surface $z = x$.

As expected, the equations obtained so far do not determine the value of $\Delta R^+(x;g(x))$. ΔR^+ is expected to be related to a possible discontinuity of c_z at

$z = y$ (the relations (1.59)–(1.73) are all based on considerations of the scattering properties in the vicinity of $z = x$). We also note that the characteristics of the propagation of ΔR^+ as given by (1.72) agree with those appearing in the imbedding equation (1.70) for $R^+(x;t)$. The constant y_0 in (1.73) is determined by the fact that if we put $x = y$, the slab $x \le z \le y$ degenerates into the plane $z = x = y$ and the discontinuity then appears for $t = 0$, i.e. we have $g(y) = 0$ or $y_0 = y$.

We wish to determine $\Delta R^+(x; g(x)) \equiv \Delta R^+(x, y; g(x, y))$ and to this end we study the properties around $z = y$. However, first we note that $(R^+ * R^+)(x, y; t)$ is continuous across a discontinuity of $R^+(x, y; t)$. Consequently we obtain from (1.70) that

$$\left[R_x^+(x,y;t) - \frac{2}{c(x)} R_t^+(x,y;t)\right]_{t=g(x)^+} = \left[R_x^+(x,y;t) - \frac{2}{c(x)} R_t^+(x,y;t)\right]_{t=g(x)^-}. \tag{1.74}$$

Here $\partial_x - 2c^{-1}(x)\partial_t = \partial_x + \partial_x g(x, y)\partial_t = d_x$ and (1.74) may be written

$$d_x[R^+(x,y;g(x,y)^+) - R^+(x,y;g(x,y)^-)] = d_x \Delta R^+(x,y;g(x,y)) = 0, \tag{1.75}$$

i.e. $\Delta R^+(x, y; g(x, y))$ is independent of x and we write

$$\Delta R^+(x,y;g(x,y)) = f(y). \tag{1.76}$$

Relations which determine $f(y)$ can be expected to be obtained from equations containing $\mathcal{R}_y^+$, i.e. the (21) element in (1.51) (which was obtained from a consideration of $\partial_y S(x, y)$). Thus we consider

$$\mathcal{R}_y^+(x,y) = -\mathcal{T}^-(x,y)\gamma(y)\mathcal{T}^+(x,y), \tag{1.77}$$

where $\gamma(y) = -[2c(y)]^{-1}c_y(y)$. The kernels $T^{\pm}(x, y; t)$ associated with $\mathcal{T}^{\pm}(x, y)$ were introduced in (1.54) and (1.56). In this instance we refer directly to a particular consequence of the relations between the kernels which are obtained by introducing the definitions of the scattering kernels into (1.77). In view of the form of the left hand side of (1.77), we consider

$$\begin{aligned} &\partial_y(R^+(x,y;t) * u^+(x,t)) \\ &= \partial_y \left\{ \int_0^{g(x,y)^-} R^+(x,y;t')u^+(x,t-t')dt' \right. \\ &\quad \left. + \int_{g(x,y)^+}^{t} R^+(x,y;t')u^+(x,t-t')dt' \right\} \end{aligned}$$

$$= -\Delta R^+(x,y;g(x,y))g_y(x,y)u^+(x,t-g(x,y))$$

$$+\int_0^t R_y^+(x,y;t')u^+(x,t-t')dt'. \tag{1.78}$$

Here $g(x,y) = 2\tau(x,y)$, i.e. $g_y(x,y) = 2[c(y)]^{-1}$. We focus on the term containing ΔR^+ and this term contains the incoming field in the form $u^+(x,t-g(x,y))$. Thus we consider the corresponding terms on the right hand side of (1.77). From (1.54) and (1.56) it is clear that only the directly transmitted term has this form and this term is multiplied by a factor $-\gamma(y) = [2c(y)]^{-1}c_y(y)$. Thus, by equating the coefficients of the arbitrarily prescribed field $u^+(x,t-2\tau(x,y))$, we obtain

$$-\Delta R^+(x,y;2\tau(x,y))\frac{2}{c(y)} = \frac{c_y(y)}{2c(y)},$$

or

$$\Delta R^+(x,y;2\tau(x,y)) = -\frac{c_y(y)}{4}. \tag{1.79}$$

As has been noted already, (1.70) and (1.71) could have been derived by introducing the definition of the reflection kernel into the (21) element of (1.49). The derivation in such an approach is left as an exercise (cf. Exercise 1.2).

The relations for $R^+(x,y;t)$ obtained so far provide a framework for the solution of both the direct and inverse one-dimensional scattering problems in the time domain, which we have posed. In the direct problem, the coefficients in (1.70) are known and $R^+(x,y;t)$ can be determined from (1.70) by integration along characteristics, if appropriate boundary or initial data are given. The physical reflection kernel, i.e. the reflection kernel for the whole slab, $0 \le z \le l$, is $R^+(0,l;t)$. Sufficient initial data to determine $R^+(0,l;t)$ are provided by (1.71), which expresses how the initial data depend on the physical properties of the slab. Thus $R^+(0,l;t)$ for $0 < t < 2\tau(0,l)$ can be determined from the following equations (where z denotes an arbitrary point in the interval $(0,l)$)

$$\partial_z R^+(z,l;t) - \frac{2}{c(z)}\partial_t R^+(z,l;t) - \frac{c_z(z)}{2c(z)}(R^+ * R^+)(z,l;t) = 0, \tag{1.80}$$

$$R^+(z,l;0) = \frac{c_z(z)}{4}. \tag{1.81}$$

Figure 1.4 illustrates how the physical boundary values $R^+(0,l;t)$ for $0 < t < 2\tau(0,l)$ are obtained from the initial data $R^+(z,l;0)$, $0 \le z \le l$. For a given z, the characteristic curves have the tangent $-2c^{-1}(z)$ (cf. also the description of a numerical algorithm in Section 1.9).

After $R^+(z,l;t)$ has been computed in the region $\{(z,t)|0 \le z \le l, 0 \le t < 2\tau(z,l)\}$, $R^+(z,l;t)$ for $t > 2\tau(z,l)$ can be computed similarly by integration

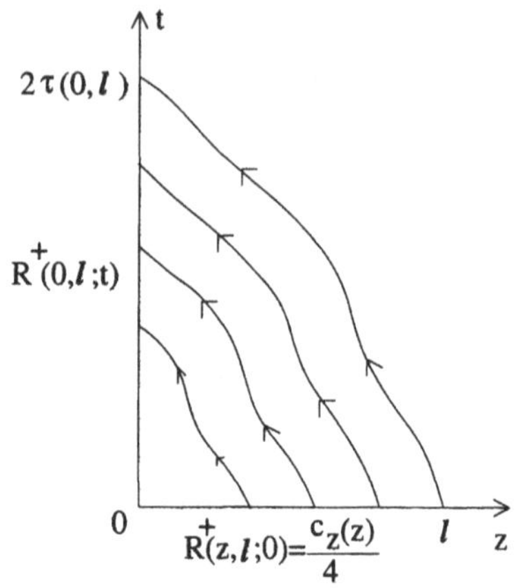

FIG. 1.4. Illustration of the integration along the characteristics of (1.80) in the direct scattering problem.

along the characteristics of (1.80), using the jump condition (1.79) (with $x = z$, $y = l$) and the following boundary condition

$$R^+(l,l;t) = 0. \tag{1.82}$$

This boundary condition follows from the fact that for $z = l$ the subslab vanishes and there is then no reflection (note that in the case of a discontinuity in c_z at $z = l^-$, the non-zero initial value $R^+(l,l;0) = \frac{1}{4}c_z(l^-)$ and the discontinuity $\Delta R^+(l,l;0) = -\frac{1}{4}c_z(l^-)$ cancel each other).

The considerations which led up to (1.80) and (1.81) are valid for an arbitrary slab thickness l and thus also in the limit $l \to \infty$, i.e. for waves incident on a stratified half-space. The reflection kernel $R^+(z,\infty;t)$ for this case satisfies (1.80) and (1.81) for arbitrary $z > 0$ and $t > 0$.

As a further illustration of the equations which can be obtained by combining the definition of the scattering kernels with the dynamic equations for the split fields, we consider also the scattering kernel $T^+(x,y;t)$. Differentiation of (1.54) with respect to x gives, in view of the time shift that has been introduced, for the left hand side

$$\begin{aligned}\partial_x u^+(y,t+\tau(x,y)) &= -c^{-1}(x)\partial_t u^+(x,t+\tau(x,y)) \\ &= -\frac{c^{1/2}(y)}{c^{3/2}(x)}[\partial_t u^+(x,t) + T^+(x,y;t) * \partial_t u^+(x,t)],\end{aligned}$$

where we have used the fact that $u^+(x,0) = 0$. The right hand side is

$$\partial_x\{[\frac{c(y)}{c(x)}]^{1/2}(u^+(x,t)+\int_0^t T(x,y;t-t')u^+(x,t')dt'\}$$

$$=-\frac{c^{1/2}(y)\partial_x c(x)}{2c^{3/2}(x)}[u^+(x,t)+T(x,y;t)*u^+(x,t)]$$

$$+[\frac{c(y)}{c(x)}]^{1/2}[\partial_x u^+(x,t)+(\partial_x T(x,y;t))*u^+(x,t)+T(x,y;t)*\partial_x u^+(x,t)].$$

Introducing the dynamic equation (1.16), with the explicit forms (1.17), as well as (1.53), one finds that the terms containing $\partial_t u^+(x;t)$ cancel and the result becomes

$$-\frac{\partial_x c(x)}{2c(x)}R^+(x,y;t)*u^+(x,t)+(\partial_x T^+(x,y;t))*u^+(x,t)$$
$$-\frac{\partial_x c(x)}{2c(x)}T^+(x,y;t)*R^+(x,y;t)*u^+(x,t)=0. \tag{1.83}$$

Here $u^+(x;t)$ is the arbitrarily prescribed incoming field. All terms are convolutions with $u^+(x,t)$, i.e. this time we obtain one functional relation from (1.83), namely

$$\partial_x T^+(x,y;t)=\frac{c_x(x)}{2c(x)}[R^+(x,y;t)+T^+(x,y;t)*R^+(x,y;t)]. \tag{1.84}$$

The boundary condition

$$T^+(y,y;t)=0 \tag{1.85}$$

follows from the fact that for $x=y$ the subslab disappears and there is only a directly transmitted part. If $R^+(x,y;t)$ is known, $T^+(x,y;t)$ can be computed from (1.84) and (1.85) by means of integration in the $-x$ direction. Note that $T^+(x,y;t)$ is continuous in the whole domain of definition (cf. also Section 2.1 for a discussion of the discontinuities in the Green's function formalism).

We have noted that with the definition (1.53) for the reflection kernel R^+, the structure of the resulting equation (1.70) resembles that given by the (21) element of (1.49). On the other hand, comparing (1.84) with the (11) element of (1.49), one notes a change in structure inasmuch as, besides the term containing both T^+ and R^+ (viz $\mathcal{T}^+$ and $\mathcal{R}^+$), there is a term R^+ in (1.84) but a term $\mathcal{T}^+$ in (1.49). As is seen in the derivation of (1.84), this is a consequence of the introduction of the time shift $\tau(x,y)$ in the definition of $T^+(x,y;t)$ in (1.54).

1.7 The 'extension of data' property

We have noted that integration along the characteristics of (1.80) from the initial values $R^+(z,l;0)$ (cf. (1.81)) allows us to determine $R^+(0,l;t)$ for $t \leq 2\tau(0,l)$. Although $R(0,l;t)$ for $t > 2\tau(0,l)$ can be obtained by means of continued integration along characteristics and using (1.82), we shall point out the important feature referred to as 'extension of data' (Kristensson and Krueger 1986a). The basic property which leads to this result is most easily obtained from the compact Green's function formalism (He 1993a) as is shown in Section 2.3. However, in view of the simple form of the results in question, we present them and their consequences here.

First we note that whereas the reflection and transmission operators are natural ones to define in the context of a direct scattering problem, there are additional operators which can be expected to be relevant for an inverse problem. These are operators which express the fields on the left hand side of the slab (i.e. the incident or the reflected field) as a function of the field on the right hand side, i.e. the transmitted field. The expression for the transmitted field in terms of the incident field is given by (1.19) and thus we wish to invert this relation. Equation (1.19) is written in a slightly abbreviated notation as

$$u^t(t) = au^i(t) + a(T^+ * u^i)(t), \tag{1.86}$$

where $a = (c_1/c_0)^{1/2}$ is a constant. To invert this relation, find a kernel W^+ such that

$$(T^+ * u^i)(t) = -W^+ * u^t(t). \tag{1.87}$$

Then the inverse of (1.86) is obtained as

$$u^i(t) = a^{-1}u^t(t) + (W^+ * u^t)(t). \tag{1.88}$$

T^+ and W^+ are determined by each other through the resolvent equation derived below. Applying the operation $W*$ to (1.86) and using (1.87) one has

$$(W^+ * u^t)(t) = a(W^+ * u^i)(t) + a(W^+ * (T^+ * u^i))(t) = -(T^+ * u^i)(t). \tag{1.89}$$

Since u^i is an arbitrarily prescribed field, it follows that

$$a^{-1}T^+ + W^+ + W^+ * T^+ = 0, \tag{1.90}$$

which is the resolvent equation. If T^+ is known, this is a second-kind Volterra equation for W^+ and conversely, if W^+ is known, it is a second-kind Volterra equation for T^+. It is well known that a Volterra equation of the second-kind has one and only one bounded solution (Tricomi 1957), and a stable numerical solution for such an equation can easily be obtained (cf. Section 3.2.4 for how a second-kind Volterra equation is solved numerically).

Similarly, we consider the relation between the transmitted and the reflected fields, expressed as (in abbreviated notation we write $V^+(0,l;t) \equiv V^+(t)$)

$$u^r(t) = (V^+ * u^t)(t). \tag{1.91}$$

Combining this with (1.18), which in the present notation reads

$$u^r(t) = (R^+ * u^i)(t), \tag{1.92}$$

and applying the operation R^+* to (1.88), wc obtain

$$(V^+ * u^t)(t) = a^{-1}(R^+ * u^t)(t) + (R^+ * W^+ * u^t)(t). \tag{1.93}$$

Here $u^t(t)$ is an arbitrary field (and all terms are of the same kind, i.e. convolutions with u^t) and it follows that

$$aV^+(t) = R^+(t) + a(R^+ * W^+)(t), \tag{1.94}$$

which is a second-kind Volterra equation for $R^+(t)$.

A crucial property of the kernels $W^+(t)$ and $V^+(t)$, first derived in Kristensson and Krueger (1986a), is that

$$W^+(0,l;t) = 0, \qquad \text{for } t > 2\tau(0,l), \tag{1.95}$$

$$V^+(0,l;t) = 0, \qquad \text{for } t > 2\tau(0,l). \tag{1.96}$$

A simplified derivation of these results has been given in He (1993a) in the context of the compact Green's functions formalism (see Section 2.3).

An important consequence of (1.95) and (1.96) is the following 'extension of data' property for the reflection and transmission kernels: if $R^+(0,l;t)$ and $T^+(0,l;t)$ are known for $t < 2\tau(0,l)$, they are known for all times. This is seen as follows. First we note that if $T^+(0,l;t)$ is known for $t < 2\tau(0,l)$ then the resolvent equation (1.90) can be solved for $W^+(0,l;t)$ for $0 < t < 2\tau(0,l)$ and in view of (1.95), $W^+(0,l;t)$ is then known for all t. Next we consider $t > 2\tau(0,l)$. Equation (1.90) then reads

$$a^{-1}T^+(0,l;t) + \int_0^{2\tau(0,l)} T^+(0,l;t-t')W^+(0,l;t')dt' = 0, \quad t > 2\tau(0,l), \tag{1.97}$$

where $W^+(0,l;t)$ is known. It is convenient to rewrite this equation somewhat so that the extension of data property becomes evident. A shift in the integration variable $t' \to t'' = t - t'$ gives an integral ($t > 2\tau(0,l)$) from $t - 2\tau(0,l)$ to t and (1.97) can be written

$$\begin{aligned} a^{-1}T^+(0,l;t) &+ \int_{2\tau(0,l)}^{t} T^+(0,l;t'')W^+(0,l;t-t'')dt'' \\ &= -\int_{t-2\tau(0,l)}^{2\tau(0,l)} T^+(0,l;t'')W^+(0,l;t-t'')dt'' \\ &= \int_{2\tau(0,l)}^{t-2\tau(0,l)} W^+(0,l;t')T^+(0,l;t-t')dt'. \end{aligned} \tag{1.98}$$

From the above expression it is clearly seen that $T^+(0,l;t)$ for $t > 2\tau(0,l)$ can be determined from a second-kind Volterra equation since the right hand side is known (the middle form shows this immediately for $2\tau(0,l) < t < 4\tau(0,l)$ and the last form shows that it vanishes for $t > 4\tau(0,l)$ since $W^+(0,l;t')$ vanishes for $t' > 2\tau(0,l)$).

The extension of data property of R^+ is obtained in a similar way. Using (1.95) and (1.96) in (1.94), we have

$$a^{-1}R^+(0,l;t) + \int_0^{2\tau(0,l)} R^+(0,l;t-t')W^+(0,l;t')dt' = 0, \quad t > 2\tau(0,l), \tag{1.99}$$

which is identical in form to (1.97). Thus, in the same way as for $T^+(0,l;t)$ it follows that if $R^+(0,l;t)$ is known for $0 \le t \le 2\tau(0,l)$, it is known for all $t > 2\tau(0,l)$.

After one round trip the wave repeats the process that it has already gone through, albeit with a different starting value. Intuitively, the subsequent behaviour should therefore be deducible from what happened during the first round trip. Equations (1.97) and (1.99) corroborate this conjecture. We note that (1.95) and (1.96) do not hold if the medium is dispersive, and thus the above-mentioned extension of data is not possible in that case. In the dispersive case, $R^+(0,l;t)$ for $t > 2\tau(0,l)$ can be obtained by integrating the imbedding equation for the reflection kernel along its characteristics, starting from the boundary data given by (1.82) (of course, after $R^+(z,l;t)$ in the region $\{(z,t)|0 \le z \le l,\ 0 \le t < 2\tau(z,l)\}$ has been computed). Direct and inverse problems for dispersive media are treated in detail in Section 3.3.

1.8 Travel time coordinates

We have noted that when the wave velocity $c(z) = (\mu_0\epsilon(z))^{-1/2}$ varies within the slab, the wave operator ${\partial^2}_z - c^2(z){\partial^2}_t$ is not factorized by the splitting (1.11). Furthermore, the characteristics are not straight lines in this case. Sometimes it is useful to be able to factorize the principal part of the relevant hyperbolic operator, if it is not possible to achieve complete factorization. In this context the travel time $\tau(z_1, z_2)$ defined in (1.20) can be a useful tool. To illustrate this, we introduce dimensionless ('normalized travel time') coordinates (ζ, s). We put

$$\zeta = \zeta(z) = \frac{\tau(0,z)}{\tau(0,l)} = \frac{1}{\tau(0,l)} \int_0^z \frac{dz'}{c(z')} \quad , \quad s = \frac{t}{\tau(0,l)}. \tag{1.100}$$

Thus $\zeta(l) = 1$ and one round trip corresponds to $s = 2$. For the inverse $z(\zeta)$ of $\zeta(z)$ we have, using the notation $c(z(\zeta)) \equiv \tilde{c}(\zeta)$,

$$z(\zeta) = \tau(0,l) \int_0^\zeta \tilde{c}(\zeta')d\zeta', \tag{1.101}$$

and $\partial_z = \dfrac{d\zeta}{dz}\partial_\zeta = [\tau(0,l)\tilde{c}(\zeta)]^{-1}\partial_\zeta$ gives $\partial_z^2 = [\tau(0,l)\tilde{c}(\zeta)]^{-2}[\partial_\zeta^2 - d_\zeta(\ln\tilde{c}(\zeta))\partial_\zeta]$.

The wave function u is now considered as a function of ζ and s and we write $u \equiv \tilde{u}(\zeta, s)$. According to (1.2) and (1.101), $\tilde{u}(\zeta, s)$ satisfies

$$[\partial_\zeta^2 - \partial_s^2 - (\ln\hat{c})_\zeta\partial_\zeta]\tilde{u}(\zeta, s) = 0, \tag{1.102}$$

where $(\ln\tilde{c}(\zeta))_\zeta \equiv d_\zeta(\ln\tilde{c}(\zeta))$. The principal part of the operator in (1.102), i.e. $\partial_\zeta^2 - \partial_s^2$, has constant coefficients and is factorized as

$$\partial_\zeta^2 - \partial_s^2 = (\partial_\zeta - \partial_s)(\partial_\zeta + \partial_s).$$

The wave-splitting and imbedding ideas applied to $\tilde{u}(\zeta, s)$ then provide an alternative to the formulation given so far concerning $u(z,t)$. The choice between using the original or the travel time coordinates is mainly a matter of convenience and has to be decided for each particular problem. In the new coordinates we put

$$\tilde{u}(\zeta, s) = u^+(\zeta, s) + \tilde{u}^-(\zeta, s), \tag{1.103}$$

where

$$\tilde{u}^\pm(\zeta, s) = \frac{1}{2}(\tilde{u}(\zeta, s) \mp \partial_s^{-1}\partial_\zeta\tilde{u}(\zeta, s)). \tag{1.104}$$

In matrix form we have

$$\begin{bmatrix} \tilde{u}^+(\zeta, s) \\ \tilde{u}^-(\zeta, s) \end{bmatrix} = \frac{1}{2}\begin{bmatrix} 1 & -\partial_s^{-1} \\ 1 & \partial_s^{-1} \end{bmatrix} \begin{bmatrix} \tilde{u}(\zeta, s) \\ \tilde{u}_\zeta(\zeta, s) \end{bmatrix} \equiv T \begin{bmatrix} \tilde{u}(\zeta, s) \\ \tilde{u}_\zeta(\zeta, s) \end{bmatrix}, \tag{1.105}$$

where the matrix T is now independent of ζ (in contradistinction to (1.13)). Corresponding to (1.12) we have

$$\partial_\zeta \begin{bmatrix} \tilde{u}(\zeta, s) \\ \tilde{u}_\zeta(\zeta, s) \end{bmatrix} = \begin{bmatrix} 0 & 1 \\ \partial_s^2 & (\ln\tilde{c}(\zeta))_\zeta \end{bmatrix} \begin{bmatrix} \tilde{u}(\zeta, s) \\ \tilde{u}_\zeta(\zeta, s) \end{bmatrix} \equiv W(\zeta) \begin{bmatrix} \tilde{u}(\zeta, s) \\ \tilde{u}_\zeta(\zeta, s) \end{bmatrix}, \tag{1.106}$$

and the dynamic equation for the split components $\tilde{u}^\pm(\zeta, s)$ becomes

$$\partial_\zeta \begin{pmatrix} \tilde{u}^+(\zeta,s) \\ \tilde{u}^-(\zeta,s) \end{pmatrix}$$
$$= TW(\zeta)T^{-1} \begin{pmatrix} \tilde{u}^+(\zeta,s) \\ \tilde{u}^-(\zeta,s) \end{pmatrix} \equiv \begin{pmatrix} \alpha(\zeta) & \beta(\zeta) \\ \gamma(\zeta) & \vartheta(\zeta) \end{pmatrix} \begin{pmatrix} \tilde{u}^+(\zeta,s) \\ \tilde{u}^-(\zeta,s) \end{pmatrix}, \quad (1.107)$$

where

$$\left.\begin{aligned} \tilde{\alpha}(\zeta) &= -\partial_s + \tfrac{1}{2}\partial_\zeta(\ln \tilde{c}(\zeta)), \\ \tilde{\beta}(\zeta) &= \tilde{\gamma}(\zeta) = -\tfrac{1}{2}\partial_\zeta(\ln \tilde{c}(\zeta)), \\ \tilde{\vartheta}(\zeta) &= \partial_s + \tfrac{1}{2}\partial_\zeta(\ln \tilde{c}(\zeta)). \end{aligned}\right\} \quad (1.108)$$

The relations between the incoming fields $u^+(0,s), u^-(1,s)$ and the scattered fields $u^+(1,s), u^-(0,s)$ are now written

$$\left.\begin{aligned} \tilde{u}^-(0,s) &= \mathcal{R}^+(0,1)\tilde{u}^+(0,s), \\ \tilde{u}^+(1,s) &= \mathcal{T}^+(0,1)\tilde{u}^+(0,s), \end{aligned}\right\} \quad (1.109)$$

$$\left.\begin{aligned} \tilde{u}^+(1,s) &= \mathcal{R}^-(0,1)\tilde{u}^-(1,s), \\ \tilde{u}^-(0,s) &= \mathcal{T}^-(0,1)\tilde{u}^-(1,s), \end{aligned}\right\} \quad (1.110)$$

for left and right hand side incidence, respectively. The imbedding equations are obtained by means of differentiations of the corresponding relations associated with a subslab (ξ,η) where $0 < \xi < \eta < 1$, i.e. (cf. (1.42))

$$\begin{bmatrix} \tilde{u}^+(\eta,s) \\ \tilde{u}^-(\xi,s) \end{bmatrix} = \begin{bmatrix} \mathcal{T}^-(\xi,\eta) & \mathcal{R}^-(\xi,\eta) \\ \mathcal{R}^+(\xi,\eta) & \mathcal{T}^+(\xi,\eta) \end{bmatrix} \begin{bmatrix} \tilde{u}^+(\xi,s) \\ \tilde{u}^-(\eta,s) \end{bmatrix}. \quad (1.111)$$

The associated kernels are denoted according to

$$\mathcal{R}^+(\xi,\eta)\tilde{u}^+(\xi,s) = \int_0^s R^+(\xi,\eta;s-s')\tilde{u}^+(\xi,s')ds',$$

etc. By repeating the steps which led to (1.79)–(1.82), we now obtain

$$\partial_\zeta R^+(\zeta,1;s) - 2\partial_s R^+(\zeta,1;s) - \frac{1}{2}\left(\partial_\zeta(\ln \tilde{c}(\zeta))\right)(R^+ * R^+)(\zeta,1;s) = 0, \quad (1.112)$$

$$R^+(\zeta,1;0) = \frac{1}{4}\partial_\zeta(\ln \tilde{c}(\zeta)), \quad (1.113)$$

$$R^+(1,1;t) = 0, \quad (1.114)$$

$$\Delta R^+(\zeta,1;2-2\zeta) = -\frac{1}{4}\left(\partial_\zeta(\ln\tilde{c}(\zeta))\right)|_{\zeta=1^-}. \qquad (1.115)$$

Thus, by replacing the original physical coordinate z by the travel time coordinate ζ one obtains an imbedding equation for which the characteristics are straight lines.

In the next two sections we summarize how the relations obtained so far are used to solve the direct and inverse scattering problems for the inhomogeneous slab.

1.9 The direct scattering problem

Consider first the determination of the reflected field in the case of incidence from the left. This is given by the (physical) reflection kernel $R^+(0,l;t)$, which is obtained as the boundary values of the reflection kernel $R^+(z,l;t)$ for the subslab between z and l. $R^+(z,l;t)$ satisfies the imbedding equation (1.80) with initial values (1.81) (similar statements of course apply to $R^+(0,1;s)$, $R^+(\zeta,1;s)$ and equations (1.112), (1.113)). When $c(z)$ is known, the initial values $R^+(z,l;0)$ and the coefficients in (1.80) are known. $R^+(z,l;0)$, i.e. the values along the horizontal axis in Fig. 1.4, provides the starting values for an integration of (1.80) along the characteristics to the boundary values $R^+(0,l;t)$ on the vertical axis for $(0<)t<2\tau(0,l)$. $R^+(0,l;t)$ for $t>2\tau(0,l)$ can be computed by either a continued integration along characteristics or the 'extension of data' property. The remaining part of the direct scattering problem for the case of incidence from the left is solved by determining the transmission kernel $T^+(0,l;t)$ as defined by (1.19) (we have added a superscript + in accordance with the notation introduced in (1.54)). When $R^+(0,l;t)$ has been determined, $T^+(0,l;t)$ can be determined from the imbedding equation (1.84) for $T^+(z,l;t)$.

Below we give a numerical algorithm for calculating $R^+(0,l;t)$ for $(0<)t<2\tau(0,l)$.

A numerical algorithm

We consider a basic example of a numerical algorithm for the solution of the discretized form of the reflection part of the direct scattering problem. It is then convenient to use the travel time formalism introduced above (cf. (1.100)–(1.113)). Equations (1.80) and (1.81) are then replaced by (1.112) and (1.113), respectively, and the characteristics of (1.112) are straight lines. In the discretization the ζ interval (0,1) is divided into N steps of length $\Delta\zeta=h$ with $Nh=1$, and according to (1.112) the associated step length in s is $\Delta s=2\Delta\zeta=2h$ (for integration along the characteristics). A general gridpoint is denoted $(\zeta_m,s_n)=(mh,2nh)$. With the notation $\partial_\zeta(\ln c(\zeta))\equiv f(\zeta)$, we thus consider a discretized form of

$$\partial_\zeta R^+(\zeta,1;s) - 2\partial_s R^+(\zeta,1;s) = \frac{1}{2} f(\zeta)(R^+ * R^+)(\zeta,1;s). \tag{1.116}$$

The characteristics are of the form $2\zeta + s =$ constant and we write $2\zeta + s = s'$, i.e. $s = s' - 2\zeta$. It is convenient to rewrite (1.116) as

$$d_\zeta R^+(\zeta,1,s'-2\zeta) = \frac{1}{2} f(\zeta)(R^+ * R^+)(\zeta,1;s'-2\zeta). \tag{1.117}$$

Integration from ζ_m to $\zeta_m + h = \zeta_{m+1}$ gives

$$\begin{aligned} &R^+(\zeta_{m+1},1,s'-2\zeta_{m+1}) - R^+(\zeta_m,1;s'-2\zeta_m) \\ &= \frac{1}{2}\int_{\zeta_m}^{\zeta_{m+1}} f(\zeta)(R*R)(\zeta,1;s'-2\zeta)d\zeta. \end{aligned} \tag{1.118}$$

We simplify the notation further by putting $R^+_{m,n} \equiv R^+(\zeta_m,1;s_n)$ for an arbitrary gridpoint (ζ_m, s_n) (here $2\zeta_m + s_n = s'$ so that, e.g. $s' - 2\zeta_{m+1} = s' - 2(\zeta_m + h) = s_n - 2h = s_{n-1}$, etc.). Thus we write

$$R^+_{m+1,n-1} - R^+_{m,n} = \frac{1}{2}\int_{\zeta_m}^{\zeta_{m+1}} f(\zeta)(R^+ * R^+)(\zeta,1;s'-2\zeta)d\zeta. \tag{1.119}$$

Using the trapezoidal approximation of the integral and the notation

$$(R^+ * R^+)(\zeta_m,1;s_n) \equiv (R^+ * R^+)_{m,n}\ ,\ f(\zeta_m) \equiv f_m,$$

we obtain

$$\begin{aligned} &R^+_{m+1,n-1} - R^+_{m,n} \\ &= \frac{h}{4}[f_{m+1}(R^+ * R^+)_{m+1,n-1} + f_m(R^+ * R^+)_{m,n}] + O(h^2) \end{aligned} \tag{1.120}$$

(where $h^{-n}O(h^n)$ is bounded as $h \to 0$). Using the trapezoidal approximation also for the convolution integral in $R^+ * R^+$, we have

$$\begin{aligned} (R*R)_{m,n} &= \int_0^{s_n} R^+(\zeta_m;1,s_n - s'')R^+(\zeta_m,1,s'')ds'' \\ &= \sum_{k=1}^{n}\int_{s_{k-1}}^{s_k} R^+(\zeta_m,1;s_n - s'')R^+(\zeta_m,1;s'')ds'' \\ &= h\sum_{k=1}^{n}\{R^+_{m,n-k}R^+_{m,k} + R^+_{m,n-(k-1)}R^+_{m,k-1}\} + O(h^2). \end{aligned} \tag{1.121}$$

In the direct scattering problem we seek to compute $R^+(0,1;s)$ and we use the known values $f_m \equiv f(\zeta_m) =$ (according to (1.113)) $= 4R^+_{m,0}$ to initialize the algorithm, which we then use in the form (cf. (1.120), (1.121)):

$$
\begin{aligned}
R^+_{m,n}& \\
&= R^+_{m+1,n-1} - \frac{h^2}{4}\{f_{m+1}\sum_{k=1}^{n-1}(R^+_{m+1,n-1-k}R^+_{m+1,k} + R^+_{m+1,n-k}R^+_{m+1,k-1}) \\
&\quad + f_m\sum_{k=1}^{n}(R^+_{m,n-k}R^+_{m,k} + R^+_{m,n-(k-1)}R^+_{m,k-1})\} + O(h^3). \qquad (1.122)
\end{aligned}
$$

We note that the right hand side also contains terms $R^+_{m,n}$. They appear in the expression for $(R^+ * R^+)_{m,n}$ for $k = n$ and $k = 1$ and these terms give a contribution

$$
-\frac{h^2}{2}f_m R^+_{m,0}R^+_{m,n}, \qquad \text{i.e.} \quad -\frac{h^2}{8}f_m^2 R^+_{m,n}. \qquad (1.123)
$$

The remaining terms in $(R^+ * R^+)$ can be expressed as

$$
f_m 2h\sum_{k=1}^{n-1} R^+_{m,n-k}R^+_{m,k}.
$$

A similar expression is obtained also for $(R^+ * R^+)_{m+1,n-1}$ and the resulting algorithm can be written

$$
\begin{aligned}
&(1 + \frac{h^2}{8}f_m^2)R^+_{m,n} = R^+_{m+1,n-1} \\
&\quad -\frac{h^2}{2}\{f_{m+1}\sum_{k=1}^{n-1} R^+_{m+1,n-1-k}R^+_{m+1,k} + f_m\sum_{k=1}^{n-1} R^+_{m,n-k}R^+_{m,k}\} + O(h^3).
\end{aligned}
\qquad (1.124)
$$

For $n = 0$ one has, to order h^2,

$$
R^+_{m,1}(1 + \frac{h^2 f_m^2}{8}) = R^+_{m+1,0} = \frac{f_{m+1}}{4}, \qquad (1.125)
$$

which illustrates how the algorithm starts from the initial values f_m and f_{m+1}. In this way all $R_{m,1}$ are obtained, and for a general $R^+_{m,n}$ it is clear from (1.124) that only terms $R_{m',n'}$ with $n' < n$ appear (cf. Fig. 1.5).

Refinements of this and associated algorithms are possible and have been studied in the literature (see e.g.. Kristensson and Krueger (1986a), Poovendran et al. (1995)). However, our purpose here is mainly to point out some basic features of the discretized system and further aspects will be dealt with in later chapters of this book.

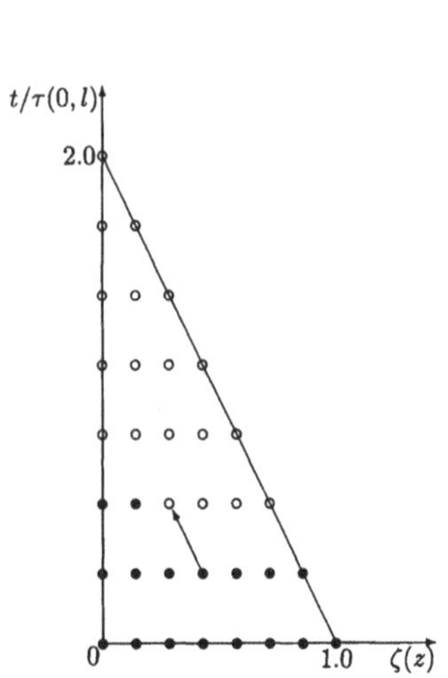

FIG. 1.5. The grid for the direct problem using the imbedding equation for the reflection kernel.

1.10 The inverse scattering problem

The interior of the slab is characterized electromagnetically by the single physical parameter $c(z) = (\epsilon(z)\mu_0)^{-\frac{1}{2}}$. Therefore a basic inverse problem associated with the slab may be stated as follows: from measurements of the reflected and/or transmitted fields outside the slab, determine $c(z)$ for $0 < z < l$. It is conceivable that it will turn out to be necessary to use incidence from both sides, but we start by investigating whether $c(z)$ can be determined from one-sided incidence, which we take to be from the left. Consider first the information contained in the measured time trace $u^r(0,t)$ of the reflected field at $z = 0^-$. The reflection kernel $R^+(0,l;t)$ is obtained from the reflected field by means of a deconvolution (cf. (1.18)). This is an ill-posed step and the question arises whether this is the only ill-posed step involved in the reconstruction of $c(z)$. This can be shown to be the case for the present problem (He et al. 1993c, 1994c; cf. also Section 2.2). Thus, if the reflection kernel, rather than the reflected field, is taken as input data, the inverse problem of reconstructing $c(z)$ is well-posed.

The key equations for the inverse problem are again (1.80) and (1.81). From (1.81) it is seen that $c_z(z)$ can be determined if we can reconstruct the initial values $R^+(z,l;0)$ of the reflection kernel for $0 \le z \le l$ ($c(z)$ is then obtained by an integration of $c_z(z)$). As noted in connection with the direct scattering problem, if $c_z(z)$ is known, (1.80) can be integrated along the characteristics. This can be done in either direction and in the direct problem we integrate from the initial values $R^+(z,l;0)$ to the physical reflection kernel $R^+(0,l;t)$. Obviously, if $c_z(z)$ is known, one can integrate in the opposite direction, starting from $R^+(0,l;t)$ and obtaining $R^+(z,l;0)$. However, these values are also determined directly by $c_z(z)$

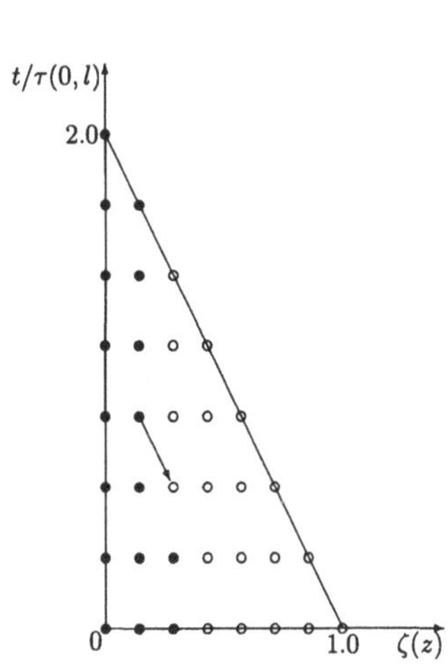

FIG. 1.6. The grid for the inverse problem using the imbedding equation for the reflection kernel.

through (1.81). In the inverse problem, these two connections between $R^+(z, l; 0)$ and $c_z(z)$, i.e. (1.80), (1.81) in conjunction, can be exploited to reconstruct $c_z(z)$ from a knowledge of $R^+(0, l; t)$ for $0 \le t \le 2\tau(0, l)$. It should be emphasized that in this reconstruction we assume that we know the form of the equation, i.e. (1.80), in which we determine the unknown coefficients.

We consider the reconstruction problem in the travel time coordinate formulation. The discussion of the discretized form of the inverse problem is also based on (1.122), which we now read in the opposite direction. In particular, we extract the additional $R^+_{m+1,n-1}$ terms from the first of the sums in that equation. They are

$$-\frac{h^2}{8} f^2_{m+1} R^+_{m+1,n-1}. \tag{1.126}$$

The remaining terms can be organized in analogy with (1.124) and we obtain to order h^2,

$$\begin{aligned} &R^+_{m+1,n-1}\left(1 - \frac{h^2}{8} f^2_{m+1}\right) \\ &= R^+_{m,n} + \frac{h^2}{2}\{f_{m+1} \sum_{k=1}^{n-2} R^+_{m+1,n-1-k} R^+_{m+1,k} + f_m \sum_{k=1}^{n} R^+_{m,n-k} R^+_{m,n}\}. \end{aligned} \tag{1.127}$$

The values to be reconstructed are $R_{m,0} = \frac{f_m}{4}$ for all m, $0 \le m \le N$, and to illustrate how the algorithm (1.127) works, we consider the case $n = 1$ for arbitrary m. We then have (1.125), which we now read as a relation which determines f_{m+1} from f_m and $R_{m,1}$. One starts from $m = 0$, where both $R_{0,0}$ and $R_{0,1}$ are known. When f_{m+1} has been determined, the algorithm is used to step upwards in n. One notes that the algorithm is such that to obtain $R_{m+1,n-1}$, one uses values $R_{m+1,n'}$ with $n' \leq n-2$, i.e. values that refer to earlier times at the same layer (as well as values $R^+_{m,n''}$ with $n'' \leq n$). The structure of the inverse algorithm is illustrated by Fig. 1.6. It is possible to modify the scheme presented here so that the computation of a value at a given layer does not involve any other values at that layer, as shown in Poovendran et al. (1995) (and this leads to some other advantages, such as being better suited for parallel computations).

Summing up the situation concerning the problem of reconstructing $\epsilon(z)$, or equivalently $c(z)$, we have that if $f(\zeta)$ is known, $\tilde{c}(\zeta)$ is obtained from $d_\zeta[ln\tilde{c}(\zeta)] = f(\zeta)$, i.e.

$$\tilde{c}(\zeta) = [\mu_0\epsilon(0)]^{-\frac{1}{2}} \exp \int_0^\zeta f(\zeta')d\zeta', \tag{1.128}$$

and

$$\tilde{\epsilon}(\zeta) = \epsilon(0) \exp\{-2 \int_0^\zeta f(\zeta')d\zeta'\}. \tag{1.129}$$

From (1.101) one has

$$z(\zeta) = \tau(0,l) \int_0^\zeta \tilde{c}(\zeta')d\zeta' = \frac{\tau(0,l)}{(\mu_0\epsilon(0))^{\frac{1}{2}}} \int_0^\zeta d(\zeta')\{\exp \int_0^{\zeta'} f(\zeta'')d\zeta''\}. \tag{1.130}$$

Thus, besides $f(\zeta)$, knowledge of $\tau(0,l)$ and $\epsilon(0)$ is needed. $\epsilon(0)$ is obtained from the continuity assumption concerning $\epsilon(z)$ at the front end ($z = 0$). $\tau(0,l)$ is half of the time for one round trip and thus we assume that $\tau(0,l)$ is available from the measurements. We note that putting $\zeta = 1$ in (1.130) gives the thickness l of the slab: $l = z(1)$. Note that if one uses the physical coordinates (z,t), one does not need to know $\tau(0,l)$ (cf. Section 2.1 below).

1.11 Extensions

There are a large number of one-dimensional inverse scattering problems which can and have been treated by the technique presented here and several such examples are presented in the following chapters. For instance, it is of interest to consider the more general case of a slab characterized by a permittivity $\epsilon(z)$, a permeability $\mu(z)$ and a conductivity $\sigma(z)$. The problem of simultaneously reconstructing the two parameters $\epsilon(z)$ and $\sigma(z)$ (assuming that $\mu(z) = \mu_0$ is a constant) has been studied thoroughly by the present technique by Kristensson and Krueger (1986a, b, 1987, 1989). Below in Chapters 2 and 3, we consider several examples of simultaneous reconstruction of parameters within the framework of the Green's functions formalism developed there.

Models which allow for a discontinuity of the medium parameters at one or both of the slab surfaces are obviously of interest. This aspect is discussed in Kristensson and Krueger (1987, 1989) and we refer to Chapters 2 and 3 for a discussion of this aspect of the Green's functions formalism.

The one-dimensional nature of the model problem treated above was an immediate consequence of the simplicity of the scattering geometry and the simple form of the incident field. If the scattering object is a plane-stratified slab, but the incident field is of a more general kind, one may still be able to reduce the inverse scattering problem to an essentially one-dimensional problem of the kind studied above (or a collection of such problems). Obliquely incident plane waves and waves generated by a point source at some distance away from the slab are cases of interest. Several examples of one-dimensional inverse scattering problems for cases of more general media properties (including dispersive media) and more general incident fields are treated in Chapters 2, 3, and 4.

Exercises for Chapter 1

Exercise 1.1. Derive the star product rule (1.24).

Exercise 1.2. Derive the imbedding equation (1.70) and the initial condition (1.71) from the (21) element of the imbedding equation (1.49) for the scattering operators.

Exercise 1.3. Consider perpendicular incidence from the left of plane electromagnetic waves on a plane-stratified slab characterized by a permeability $\epsilon(z)$ and conductivity $\sigma(z)$ (assume $\mu(z) = \mu_0$ everywhere). With $\vec{E}(z,t) \equiv u(z,t)\hat{\mathbf{e}}$ it follows from Maxwell's equations that

$$[\partial_z^2 - c^{-2}(z)\partial_t^2 - b(z)\partial_t]u(z,t) = 0, \tag{1.131}$$

where $c^{-2}(z) = \mu_0\epsilon(z)$, $b(z) = \mu_0\sigma(z)$. Assume $\sigma(z) = 0$ for $z < 0$ and $z > l$. In travel time coordinates (ζ, s) defined by (1.100) one has (and we write $\tilde{u}(\zeta,s) \equiv u(z(\zeta), t(s))$, $\tilde{c}(\zeta) \equiv c(z(\zeta))$, $\tilde{b}(\zeta) \equiv b(z(\zeta))$)

$$\tilde{u}_{\zeta\zeta} - \tilde{u}_{ss} - \partial_\zeta\left(\ln \tilde{c}(\zeta)\right)\tilde{u}_\zeta - \tau(0,l)\tilde{b}(\zeta)\tilde{c}^2(\zeta)\tilde{u}_s = 0. \tag{1.132}$$

As before, let (1.104) define the split components $\tilde{u}^{\pm}(\zeta,s)$. Show that the dynamic equation for $\tilde{u}^{\pm}(\zeta,s)$ becomes

$$\partial_\zeta \begin{bmatrix} \tilde{u}^+(\zeta,s) \\ \tilde{u}^-(\zeta,s) \end{bmatrix} = \begin{bmatrix} \tilde{\alpha}(\zeta) & \tilde{\beta}(\zeta) \\ \tilde{\gamma}(\zeta) & \tilde{\vartheta}(\zeta) \end{bmatrix} \begin{bmatrix} \tilde{u}^+(\zeta,s) \\ \tilde{u}^-(\zeta,s) \end{bmatrix}, \tag{1.133}$$

where

$$\left.\begin{aligned} \bar{\alpha}(\zeta) &= -\partial_s + \tfrac{1}{2}(f_1(\zeta) - f_2(\zeta)), \\ \bar{\beta}(\zeta) &= -\tfrac{1}{2}[f_1(\zeta) + f_2(\zeta)], \\ \bar{\gamma}(\zeta) &= -\tfrac{1}{2}[f_1(\zeta) - f_2(\zeta)], \\ \bar{\vartheta}(\zeta) &= \partial_s + \tfrac{1}{2}[f_1(\zeta) + f_2(\zeta)], \end{aligned}\right\} \tag{1.134}$$

with $f_1(\zeta) \equiv \partial_\zeta (\ln \tilde{c}(\zeta))$, $f_2(\zeta) \equiv \tau(0,1)\tilde{b}(\zeta)\tilde{c}^2(\zeta)$. Show that in this case the imbedding equation for $R^+(\zeta, 1; s)$ becomes

$$\begin{aligned} &\partial_\zeta R^+(\zeta,1;s) - 2\partial_s R^+(\zeta,1;s) - f_2(\zeta)R^+(\zeta,1;s) \\ &-\frac{1}{2}[f_1(\zeta) + f_2(\zeta)](R^+ * R^+)(\zeta,1;s) = 0, \end{aligned} \tag{1.135}$$

with the initial condition

$$R^+(\zeta,1;0) = \frac{1}{4}[f_1(\zeta) - f_2(\zeta)]. \tag{1.136}$$

Similarly, show that the imbedding equation for the reflection kernel $R^-(\zeta, 0; s)$ for incidence from the right, is given by

$$\begin{aligned} &\partial_\zeta R^-(0,\zeta;s) - 2\partial_s R^-(0,\zeta;s) + f_2(\zeta)R^-(0,\zeta;s) \\ &-\frac{1}{2}[f_1(\zeta) - f_2(\zeta)](R^- * R^-)(0,\zeta;s) = 0, \end{aligned} \tag{1.137}$$

with the initial condition

$$R^-(0,\zeta;0) = -\frac{1}{4}[f_1(\zeta) + f_2(\zeta)]. \tag{1.138}$$

Here one notes that (1.136) and (1.138) contain different linear combinations of $f_1(\zeta)$ and $f_2(\zeta)$. This means that if $R^+(\zeta,1;0)$ and $R^+(0,\zeta;0)$ can be computed from $R^+(0,1;s)$ and $R^-(0,1;s)$, $0 < s < 2$, in the inverse problem, both $f_1(\zeta)$ and $f_2(\zeta)$ can be reconstructed. We refer to Kristensson and Krueger (1986b) for a more detailed discussion of how this is done by means of an iterative algorithm, as well as for a discussion of the use of both reflection and transmission data.

Exercise 1.4. Consider a stratified dielectric half-space with varying phase velocity $c(z)$ so that the electric field component satisfies (1.2). Show that for a parabolic depth dependence, viz.

$$c(z) \equiv c_p(z) = c_0(1 - a_0 z)^2 \tag{1.139}$$

(allow a discontinuity of $\partial_z c(z)$ at the surface $z = 0$ of the half-space), the reflection kernel has the form

$$R_p(x,t) = g(x)e^{g(x)t}, \qquad x \geq 0, \tag{1.140}$$

where $x = x(z) = \int_0^z \frac{dz'}{c(z')}$ is the travel time coordinate. Show that

$$g(x) = -\frac{1}{2}c_0 a_0 (1 - a_0 z(x)). \tag{1.141}$$

The solution for the parabolic profile can be used as a tool for obtaining approximate reconstructions of general (non-parabolic) profiles. One then writes

$$c(z) = c_p(z) + \Delta c(z), \tag{1.142}$$

where $\Delta c(0) = \left(\frac{d}{dz}\Delta c\right)(0) = 0$, and

$$R(x,t) = R_p(x,t) + \Delta R(x,t). \tag{1.143}$$

Show that the imbedding equation for $\Delta R(x,t)$ becomes

$$\begin{aligned}(\partial_x - 2\partial_t)&\Delta R(x,t)\\ &= \frac{1}{2}a(x)g^2(x)e^{g(x)t} + b(x)(R_q * \Delta R)(x,t) + \frac{1}{2}b(x)(\Delta R * \Delta R)(x,t),\end{aligned} \tag{1.144}$$

with the initial condition

$$\Delta R(x,0) = \frac{1}{4}a(x). \tag{1.145}$$

Here

$$a(x) = \partial_z \Delta c(z(x)), \tag{1.146}$$
$$b(x) = \partial_z c(z(x)). \tag{1.147}$$

An approximate approach which reduces the computation time drastically for nearly parabolic velocity profiles (or when the interest is in short times only) can be obtained by setting the right hand side of (1.144) equal to zero (see Kreider (1989a) for a discussion of this aspect, including error estimates).

2

TIME DOMAIN WAVE-SPLITTING GREEN'S FUNCTION APPROACHES

The scattering kernels introduced in Chapter 1 and determined by means of the imbedding approach provide a framework for solving direct and inverse scattering problems, but the full wave propagation problem, i.e., including the determination of the field inside the scattering region, is not solved by the scattering kernels. However, during recent years the time domain wave-splitting approach has been extended to include the determination of the fields inside the inhomogeneous region. This extension, called the Green's functions approach, is one of the main themes of the presentation in the rest of this book. The term *Green's functions* refers to the regular parts of the split components of the fundamental solution. These regular parts are determined by the initial and boundary values of the propagation problem under consideration. Furthermore, the boundary values of the Green's functions are related to the physical scattering kernels studied in Chapter 1, and thus a close connection with the imbedding approach is established. The system of equations for the Green's functions is linear (and thus is one order faster to solve numerically than the non-linear imbedding equation for the reflection kernel), and is suitable for parallel processing. Thus it provides an approach to both direct and inverse scattering problems, which has some advantages as compared to the imbedding approach. In particular, in the three-dimensional case the Green's function approach allows a simpler handling of the singularities (see Chapter 7).

In this chapter the time domain Green's function approach is introduced for one-dimensional scattering. As before, the underlying model refers to the electromagnetic properties of an inhomogeneous medium. The equations for the Green's functions are derived in detail. Numerical implementations for both direct and inverse problems are given. The stability, existence, uniqueness and local correctness for this approach to inverse problems are analysed. A *compact Green's function* approach for direct and inverse problems for a slab is also introduced. In particular, this approach is shown to provide an efficient method for simultaneous reconstructions, using both the reflection and transmission data. A recent propagator approach is described, which gives many illuminating connections between various wave-splitting approaches. Finally, an experimental time domain measurement and parameter reconstruction is described, including a discussion of the properties of the deconvolution that is involved in the reconstruction process.

2.1 Time domain Green's functions

2.1.1 *Structure of the fundamental solution*

A suitable setting for the introduction of time domain Green's functions is provided by the model introduced in Exercise 1.3. Thus consider a plane wave, normally incident on a planar stratified slab in which the permittivity $\epsilon(z)$ and the conductivity $\sigma(z)$ vary with the depth z. The permeability is assumed to be a constant (μ_0) everywhere. On either side of the slab there is a homogeneous, lossless medium. The plane wave impinges normally on the surface $z = 0$ from the left at time $t = 0$. Assume that the electric field $\vec{\mathbf{E}}$ is polarized in the x-direction, and the magnetic field $\vec{\mathbf{H}}$ is polarized in the y-direction. From Maxwell's equations one has

$$\vec{\mathbf{E}}(z,t) = E(z,t)\hat{\mathbf{e}}_x, \qquad \vec{\mathbf{H}}(z,t) = H(z,t)\hat{\mathbf{e}}_y,$$

$$\frac{\partial E}{\partial z} = -\mu_0 \frac{\partial H}{\partial t}, \tag{2.1}$$

$$\frac{\partial H}{\partial z} = -\sigma(z)E - \epsilon(z)\frac{\partial E}{\partial t}. \tag{2.2}$$

It follows that $E(z,t)$ satisfies the dissipative wave equation:

$$\frac{\partial^2 E(z,t)}{\partial z^2} - \frac{1}{c^2(z)}\frac{\partial^2 E(z,t)}{\partial t^2} - \mu_0\sigma(z)\frac{\partial E(z,t)}{\partial t} = 0,$$

where the phase velocity $c(z) = [\mu_0\epsilon(z)]^{-1/2}$. In analogy with (1.11) we introduce the splitting

$$E^{\pm}(z,t) = (1/2)\left[E(z,t) \mp c(z)\int_{-\infty}^{t} \frac{\partial E(z,t')}{\partial z}\,dt'\right],$$

which, in view of (2.1), can also be written

$$E^{\pm}(z,t) = (1/2)[E(z,t) \pm Z(z)H(z,t)], \tag{2.3}$$

where the impedance $Z(z) = \sqrt{\mu_0/\epsilon(z)}$.

In a homogeneous lossless medium characterized by ϵ, μ_0 ($\sigma = 0$), the split components $E^{\pm}$ satisfy

(i) the wave equation $\dfrac{\partial^2}{\partial z^2}E^{\pm} - \dfrac{1}{c^2}\dfrac{\partial^2}{\partial t^2}E^{\pm} = 0$;

(ii) $\dfrac{\partial}{\partial z}E^{\pm} = \mp\dfrac{1}{c}\dfrac{\partial}{\partial t}E^{\pm}$;

(iii) $E^{+} + E^{-} = E$.

(iv) the Poynting vector

$$\begin{aligned}\vec{\mathbf{S}} &\equiv \vec{\mathbf{E}} \times \vec{\mathbf{H}} \\ &= [(E^+ + E^-)\hat{\mathbf{e}}_x] \times [\frac{1}{Z(z)}(E^+ - E^-)\hat{\mathbf{e}}_y] \\ &= c \cdot \left[\epsilon(E^+)^2\right] \hat{\mathbf{e}}_z - c \cdot \left[\epsilon(E^-)^2\right] \hat{\mathbf{e}}_z. \end{aligned} \tag{2.4}$$

Thus E^+ is the right-moving (along the $+z$ direction) electric field while E^- is the left-moving (along the $-z$ direction) electric field in a homogeneous lossless dielectric medium. One notes that for a general inhomogeneous lossy medium the split components $E^\pm$ no longer have a direct physical interpretation as right- and left-moving waves. Equation (2.3) is nevertheless a useful change of basis for Maxwell's equations (2.1) and (2.2) from the variables $E(z,t)$, $H(z,t)$ to $E^\pm$.

The split fields E^+ and E^- satisfy the following dynamic equation (see Chapter 1 for the derivation)

$$\frac{\partial}{\partial z}\begin{bmatrix} E^+(z,t) \\ E^-(z,t) \end{bmatrix} + \frac{1}{c(z)}\frac{\partial}{\partial t}\begin{bmatrix} E^+(z,t) \\ -E^-(z,t) \end{bmatrix} = \begin{bmatrix} \alpha(z) & \beta(z) \\ \gamma(z) & \vartheta(z) \end{bmatrix}\begin{bmatrix} E^+(z,t) \\ E^-(z,t) \end{bmatrix}, \tag{2.5}$$

where

$$\begin{cases} \alpha(z) = -\frac{1}{2}\sqrt{\mu_0/\epsilon}\,\sigma - \frac{1}{4}\frac{\epsilon_z}{\epsilon}, \\ \beta(z) = -\frac{1}{2}\sqrt{\mu_0/\epsilon}\,\sigma + \frac{1}{4}\frac{\epsilon_z}{\epsilon}, \\ \gamma(z) = \frac{1}{2}\sqrt{\mu_0/\epsilon}\,\sigma + \frac{1}{4}\frac{\epsilon_z}{\epsilon}, \\ \vartheta(z) = \frac{1}{2}\sqrt{\mu_0/\epsilon}\,\sigma - \frac{1}{4}\frac{\epsilon_z}{\epsilon}, \end{cases} \tag{2.6}$$

and where $\epsilon_z = \frac{d}{dz}\epsilon$. Note that unlike the expressions (1.17) for α and ϑ in Section 1.2, the time derivatives have been removed from the above expressions for α and ϑ (cf. the left hand side of (2.5)).

It is assumed that the permittivity $\epsilon(z)$ is continuous and piecewise differentiable within the slab, i.e., for $z \in (0, l)$. It is instructive to introduce the Green's function approach first for the simple case when the impedance $\sqrt{\epsilon/\mu_0}$ is continuous at the two boundaries $z = 0$ and $z = l$. This approach will be generalized to the impedance mismatch case in Section 2.1.6.

A fundamental solution $(\mathcal{E}, \mathcal{H})$ to Maxwell's equations is the solution due to an impulsive incident electric field impinging on the slab at $z = 0$ at time $t = 0$, and the Green's functions in the present book refer to the regular parts of the split components of the fundamental solution. Thus, the split components $\mathcal{E}^\pm \equiv (1/2)[\mathcal{E} \pm \sqrt{\mu_0/\epsilon}\mathcal{H}]$ of the fundamental solution satisfy

$$\text{(i)} \quad \frac{\partial}{\partial z}\begin{bmatrix} \mathcal{E}^+ \\ \mathcal{E}^- \end{bmatrix} + \frac{1}{c}\frac{\partial}{\partial t}\begin{bmatrix} \mathcal{E}^+ \\ -\mathcal{E}^- \end{bmatrix} = \begin{bmatrix} \alpha & \beta \\ \gamma & \vartheta \end{bmatrix}\begin{bmatrix} \mathcal{E}^+ \\ \mathcal{E}^- \end{bmatrix}, \tag{2.7}$$

$$\text{(ii)} \quad \mathcal{E}^+(0,t) = \delta(t), \tag{2.8}$$

$$\text{(iii)} \quad \mathcal{E}^-(l,t) = 0, \tag{2.9}$$

$$\text{(iv)} \quad \mathcal{E}^\pm|_{t<0} = 0, \quad z \geq 0. \tag{2.10}$$

In the above set of equations, the initial condition (2.10) is due to the assumption that the internal field is identically zero for $t \leq 0$, and the boundary condition (2.9) is due to the fact that there is no left-moving wave, but only a right-moving wave (transmitted wave), in the region $z \geq l$. This is a well-posed direct problem for the system (2.7), i.e., one can calculate $\mathcal{E}^\pm(z,t)$ for all $t \geq 0$ and $z \in [0,l]$.

Removing the singular parts from $\mathcal{E}^\pm$ and denoting the regular parts, i.e., the Green's functions, by $\mathcal{G}^\pm$, one has

$$\begin{aligned} \begin{bmatrix} \mathcal{E}^+(z,t) \\ \mathcal{E}^-(z,t) \end{bmatrix} &= \begin{bmatrix} \alpha_1(z) \\ \alpha_2(z) \end{bmatrix} \delta(t-\tau) + \begin{bmatrix} \beta_1(z) \\ \beta_2(z) \end{bmatrix} \Theta(t-\tau) + \ldots \\ &\equiv \begin{bmatrix} \alpha_1(z) \\ \alpha_2(z) \end{bmatrix} \delta(t-\tau) + \begin{bmatrix} \mathcal{G}^+(z,t) \\ \mathcal{G}^-(z,t) \end{bmatrix}, \end{aligned} \tag{2.11}$$

where α_1, α_2, etc., will be determined later, $\Theta(t)$ is the Heaviside unit step function, i.e.,

$$\Theta(t) = \begin{cases} 1, & t \geq 0, \\ 0, & t < 0, \end{cases} \tag{2.12}$$

and $\tau(z)$ is the travel time from $z = 0$ to $z > 0$, i.e., $\tau(z) = \int_0^z \frac{1}{c(z')}\, dz' = \int_0^z \sqrt{\epsilon(z')\mu_0}\, dz'$ ($\equiv \tau(0,z)$; cf. (1.20)). Substituting (2.11) into (2.7) and matching the coefficients of each singular or regular term, yields

$$-\begin{bmatrix} \alpha_1 \\ \alpha_2 \end{bmatrix} + \begin{bmatrix} 1 & 0 \\ 0 & -1 \end{bmatrix} \begin{bmatrix} \alpha_1 \\ \alpha_2 \end{bmatrix} = 0, \tag{2.13}$$

$$\frac{d}{dz}\begin{bmatrix} \alpha_1 \\ \alpha_2 \end{bmatrix} - \frac{1}{c}\begin{bmatrix} \beta_1 \\ \beta_2 \end{bmatrix} + \frac{1}{c}\begin{bmatrix} 1 & 0 \\ 0 & -1 \end{bmatrix} \begin{bmatrix} \beta_1 \\ \beta_2 \end{bmatrix} = \begin{bmatrix} \alpha & \beta \\ \gamma & \vartheta \end{bmatrix} \begin{bmatrix} \alpha_1 \\ \alpha_2 \end{bmatrix}, \tag{2.14}$$

and

$$\frac{\partial}{\partial z}\begin{bmatrix} \mathcal{G}^+ \\ \mathcal{G}^- \end{bmatrix} + \frac{1}{c}\frac{\partial}{\partial t}\begin{bmatrix} \mathcal{G}^+ \\ -\mathcal{G}^- \end{bmatrix} = \begin{bmatrix} \alpha & \beta \\ \gamma & \vartheta \end{bmatrix} \begin{bmatrix} \mathcal{G}^+ \\ \mathcal{G}^- \end{bmatrix}. \tag{2.15}$$

From (2.13) and (2.14), one obtains (noting that $\alpha_1(0) = 1$, cf. (2.8))

$$\alpha_2(z) = 0, \tag{2.16}$$

$$\alpha_1(z) = \exp[\int_0^z \alpha(z')\, dz'] = \begin{bmatrix} \frac{\epsilon(0)}{\epsilon(z)} \end{bmatrix}^{\frac{1}{4}} \exp[-\frac{1}{2}\int_0^z \sigma(z')\sqrt{\mu_0/\epsilon(z')}\, dz'], \tag{2.17}$$

$$\beta_2(z) = -\frac{1}{2}c(z)\alpha_1(z)\gamma(z). \tag{2.18}$$

Therefore, the direct problem defined by (2.7)–(2.10) is equivalent to the problem of solving the system (2.15) with the following boundary and initial conditions

$$\mathcal{G}^+(0,t) = 0, \quad t \geq 0, \tag{2.19}$$

$$\mathcal{G}^-(l,t) = 0, \quad t \geq 0, \tag{2.20}$$

$$\mathcal{G}^-(z, t = \tau(z)) = -\frac{1}{2}c(z)\left[\frac{\epsilon(0)}{\epsilon(z)}\right]^{\frac{1}{4}} \exp\left[-\frac{1}{2}\int_0^z \sigma(z')\sqrt{\mu_0/\epsilon(z')}\,dz'\right]\gamma(z). \tag{2.21}$$

Equation (2.15) together with the conditions (2.19)–(2.21) can be solved for $\mathcal{G}^\pm(z,t)$, $t \geq 0$, $z \in [0,l]$, by, e.g., the method of characteristics, when $\epsilon(z)$ and $\sigma(z)$ are known. Since the response of the medium is causal and invariant under time translation (cf. Duhamel's principle (Courant and Hilbert 1962b)), one can immediately obtain the solution for an arbitrary incident electric field $E^+(0,t) = E^i(t)$ through the Green's functions $\mathcal{G}^\pm(z,t)$:

$$E^+(z,t) = \alpha_1(z)E^i(t-\tau) + \int_0^{t-\tau} \mathcal{G}^+(z,t-t')E^i(t')\,dt', \tag{2.22}$$

$$E^-(z,t) = \int_0^{t-\tau} \mathcal{G}^-(z,t-t')E^i(t')\,dt'. \tag{2.23}$$

Equations (2.22) and (2.23) can also be viewed as a definition of the Green's functions $\mathcal{G}^\pm(z,t)$, which were introduced in Krueger and Ochs (1989).

2.1.2 *Equations for time-shifted Green's functions*

It is sometimes advantageous to modify the ansatz for the Green's functions in (2.22) and (2.23) somewhat by a time shift $\tau(z)$, and define the Green's functions $G^\pm(z,t)$ as follows

$$\begin{bmatrix} E^+ \\ E^- \end{bmatrix}(z, t+\tau(z)) = a(z)\begin{bmatrix} E^i(t) \\ 0 \end{bmatrix} + \begin{bmatrix} G^+(z,t) * E^i(t) \\ G^-(z,t) * E^i(t) \end{bmatrix}, \tag{2.24}$$

where $a(z)$ is the attenuation factor (which is equal to $\alpha_1(z)$, cf. (2.17))

$$a(z) = \exp\left[\int_0^z \alpha(z')\,dz'\right] = \left[\frac{\epsilon(0)}{\epsilon(z)}\right]^{\frac{1}{4}} \exp[-\frac{1}{2}\int_0^z \sigma(z')\sqrt{\mu_0/\epsilon(z')}\,dz'] \tag{2.25}$$

(in fact this expression for $a(z)$ can be derived together with the partial differential equations (PDEs) for the Green's functions, see the derivation below). The above definition for the Green's functions has many advantages, such as being convenient for establishing the relations between the Green's functions and the imbedding kernels (see Section 2.1.7). From the definition (2.24), one sees that the Green's functions $G^{\pm}(z,t)$ map the incident field $E^i(t)$ to the internal split fields.

The PDEs for the Green's functions $G^{\pm}$ and their initial and boundary conditions follow from the corresponding equations for $\mathcal{G}^{\pm}$. Comparing (2.22) and (2.23) with (2.24), one sees immediately that

$$G^{\pm}(z,t) = \mathcal{G}^{\pm}(z,t+\tau). \tag{2.26}$$

One can thus easily obtain the equations for $G^{\pm}$ from the corresponding equations for $\mathcal{G}^{\pm}$ given in the previous subsection, by using the above relation. However, we find it useful and instructive to derive the equations for $G^{\pm}$ directly from the definition (2.24) and the dynamic equation (2.5) for the split fields (many later results, e.g., the equations for the compact Green's functions in Section 2.3, are derived in a similar way, and the detailed derivation of this type of equations will not be given in the remaining parts of the book).

By differentiating (2.24) with respect to z, the following equation is obtained

$$\left(\frac{\partial}{\partial z}\begin{bmatrix} E^+ \\ E^- \end{bmatrix}\right)(z,t+\tau) + \frac{1}{c}\left(\frac{\partial}{\partial t}\begin{bmatrix} E^+ \\ E^- \end{bmatrix}\right)(z,t+\tau)$$
$$= \frac{da}{dz}\begin{bmatrix} E^i(t) \\ 0 \end{bmatrix} + \begin{bmatrix} \frac{\partial G^+}{\partial z} * E^i(t) \\ \frac{\partial G^-}{\partial z} * E^i(t) \end{bmatrix}.$$

The z derivative of E^+ and E^- can then be eliminated by using the dynamics of the split fields, (2.5), and one obtains (using the definition (2.24))

$$[-\frac{1}{c}\frac{\partial}{\partial t} + \alpha][aE^i(t) + G^+ * E^i(t)] + \beta[G^- * E^i(t)] + \frac{1}{c}\frac{\partial}{\partial t}[aE^i(t) + G^+ * E^i(t)]$$
$$= \frac{\partial G^+}{\partial z} * E^i(t) + \frac{da}{dz}E^i(t), \tag{2.27}$$

$$\gamma[aE^i(t) + G^+ * E^i(t)] + [\frac{1}{c}\frac{\partial}{\partial t} + \vartheta][G^- * E^i(t)] + \frac{1}{c}\frac{\partial}{\partial t}[G^- * E^i(t)]$$
$$= \frac{\partial G^-}{\partial z} * E^i(t). \tag{2.28}$$

Equation (2.27) can be rewritten in the following form

$$[\frac{\partial G^+}{\partial z} - \alpha G^+ - \beta G^-] * E^i(t) + [\frac{da}{dz} - \alpha a]E^i(t) = 0.$$

Since $E^i(t)$ is an arbitrary field, the following two equations must hold, namely

$$\frac{\partial G^+}{\partial z} = \alpha G^+ + \beta G^-, \tag{2.29}$$

and

$$\frac{da}{dz} - \alpha a = 0.$$

The latter equation leads to (2.25) (note that (2.24) implies that $a(l) = 1$). From the definition (2.24) of G^+, one has

$$G^+(0,t) = 0. \tag{2.30}$$

Similarly, from (2.28) one obtains the following equation

$$\{\frac{\partial}{\partial z}G^- - \frac{2}{c}\frac{\partial}{\partial t}G^- - \gamma G^+ - \vartheta G^-\} * E^i(t) - \{\frac{2}{c}G^-(z,0) + a\gamma\}E^i(t) = 0,$$

which gives the following PDE for G^-

$$\frac{\partial}{\partial z}G^- - \frac{2}{c}\frac{\partial}{\partial t}G^- = \gamma G^+ + \vartheta G^-, \tag{2.31}$$

and the initial condition for G^-

$$G^-(z,0) = -\frac{c}{2}a\gamma. \tag{2.32}$$

The boundary condition for G^- is

$$G^-(l,t) = 0, \tag{2.33}$$

which is due to the fact that there is no left-moving wave but only a right-moving (transmitted) wave in the region $z \geq l$. In a direct problem, (2.29)–(2.33) constitute a set of equations from which $G^+(z,t)$ and $G^-(z,t)$ can be determined for all $t \geq 0$, $z \in [0,l]$.

Integrating (2.29) along the $+z$ direction for $t = 0$ and using the conditions (2.30), (2.32), one obtains the initial condition for G^+

$$G^+(z,0) = -\frac{1}{2}a(z)\int_0^z c(z')\beta(z')\gamma(z')\,dz'. \tag{2.34}$$

Assume that σ, ϵ_z are continuous except at a finite number of positions z_m, $m = 1, 2, ..., M$. For a system of first-order PDEs, the functions are continuous along the characteristic curves but may have discontinuities across the characteristic curves (see e.g. Courant and Hilbert (1962a)). From (2.29), one sees that the characteristic curves for G^+ are $t =$ constant. Therefore, G^+ is continuous along $t =$ constant. Furthermore, G^+ is a continuous function of t at $z = 0$ (cf. (2.30)) and thus $G^+(z,t)$ is continuous in the whole domain of definition. $G^{c-}(z,t)$ is continuous everywhere except on its characteristic curves $t = 2\int_z^{z_m} \frac{dz'}{c(z')}$, $z \leq z_m$,

$m = 1, 2, ..., M$. At each discontinuity position z_m, the initial value for G^- has the following discontinuity (cf. (2.32))

$$[G^-(z,0)]_{z_m} = -\frac{1}{2}[ca\gamma]_{z=z_m},$$

where $[\cdot]_z$ denotes the discontinuity at the position z, i.e. $[\Psi(z,t)]_z = \Psi(z+0,t) - \Psi(z-0,t)$. This discontinuity in G^- will propagate along the characteristic curve Γ: $\frac{\partial t}{\partial z} = -\frac{2}{c(z)}$. From the PDE (2.31), one obtains the following equation for the discontinuity along this characteristic curve Γ (note that $[G^+]_\Gamma = 0$)

$$\frac{d}{dz}[G^-]_\Gamma = \vartheta[G^-]_\Gamma, \quad z < z_m.$$

Integrating the above equation from z_m to z, yields

$$\begin{aligned}
[G^-]_\Gamma &= [G^-(z,0)]_{z_m} \cdot \exp\left[-\int_z^{z_m} \vartheta(z')\,dz'\right] \\
&= -\frac{1}{2}[ca\gamma]_{z=z_m} \cdot \exp\left[-\int_z^{z_m} \vartheta(z')\,dz'\right], \quad z \le z_m.
\end{aligned} \tag{2.35}$$

Equation (2.35) gives the discontinuity in G^- along its characteristics. The above discontinuity condition can be used in a numerical implementation to increase the accuracy at the discontinuities (although this may not always be necessary).

The PDEs (2.29) and (2.31) for $G^\pm$ can be rewritten in the following matrix form:

$$\frac{\partial}{\partial z}\begin{bmatrix} G^+ \\ G^- \end{bmatrix} - \frac{2}{c}\frac{\partial}{\partial t}\begin{bmatrix} 0 \\ G^- \end{bmatrix} = \begin{bmatrix} \alpha & \beta \\ \gamma & \vartheta \end{bmatrix}\begin{bmatrix} G^+ \\ G^- \end{bmatrix}. \tag{2.36}$$

Note, however, if α, β, γ, or ϑ contains a time integral or convolution (e.g., when the medium is dispersive), one will not have the above form (cf. the PDEs for the Green's functions for a dispersive medium in Section 3.3.6).

The direct and inverse problems

The direct problem is to calculate the Green's functions $G^\pm(z,t)$ when the parameters $\sigma(z)$ and $\epsilon(z)$ are known. This can be achieved by propagating the initial data (cf. (2.32) and (2.34)) to the boundary data. The numerical algorithm for such a propagation can be obtained by the standard method of characteristics, i.e., integrating (2.29) along the z direction and (2.31) along the characteristic direction $dz = -c(z)dt/2$ (thus the data for the Green's functions $G^\pm(z,t)$ are built up with forward time stepping starting from time $t = 0$; see Section 2.1.3 for the numerical implementation). Once $G^\pm(z,t)$ have been computed, one knows $E^\pm(z,t)$ from the definition (2.24), and thus the internal electromagnetic fields can be obtained by

$$E(z,t) = E^+(z,t) + E^-(z,t), \tag{2.37}$$
$$H(z,t) = \sqrt{\epsilon(z)/\mu_0}[E^+(z,t) - E^-(z,t)]. \tag{2.38}$$

Furthermore, the boundary values of the Green's functions give the reflection kernel $R(t)(\equiv R^+(0,l,t))$ and the transmission kernel $T(t)(\equiv T^+(0,l,t))$, i.e., (cf. the definitions (2.24), (1.18) and (1.19) for the Green's functions and the scattering kernels)

$$G^-(0,t) = R(t), \tag{2.39}$$
$$G^+(l,t) = T(t). \tag{2.40}$$

The inverse problem is to reconstruct the permittivity $\epsilon(z)$ and/or conductivity $\sigma(z)$ using the reflection and/or transmission data. In view of (2.39) and (2.40) it follows that the inverse problem using both the reflection and transmission data via the usual Green's function approach is to solve a system of PDEs with two boundary conditions at two endpoints $z = 0$ and $z = l$. Such a two-endpoint-problem is very difficult to solve by non-iterative methods. As shown in Section 2.3 below, a compact Green's function approach will serve as an efficient method to solve the inverse problem using both reflection and transmission data. The usual Green's function approach is useful for inverse problems using only reflection data. For example, in the inverse problem of reconstructing one parameter using reflection data, one propagates the known boundary values at $z = 0$ (cf. (2.30) and (2.39)) along the characteristic curves to the initial values using the PDEs for the Green's functions. The permittivity or the conductivity can then be reconstructed by using the initial condition (2.32) (the numerical implementation for such an inverse problem is given in Section 2.1.4).

Thus, the Green's functions formalism provides an alternative to the imbedding approach to the same direct and inverse scattering problem as described in Chapter 1. In addition the Green's functions formalism provides a framework for computation of the fields inside the slab.

2.1.3 *Numerical implementation of a direct problem*

The PDEs for the Green's functions $G^\pm(z,t)$ together with the initial, boundary, and discontinuity conditions (given in the previous subsection) are suitable for numerical implementation. The system of equations for the Green's functions is linear, and thus is one order faster to solve numerically than the non-linear imbedding equation for the reflection kernel. For simplicity we assume that the conductivity is continuous and the permittivity is continuously differentiable inside the slab. To compute the Green's functions for the first round trip, one needs to solve the PDEs for $G^\pm(z,t)$ numerically with the following initial and boundary conditions

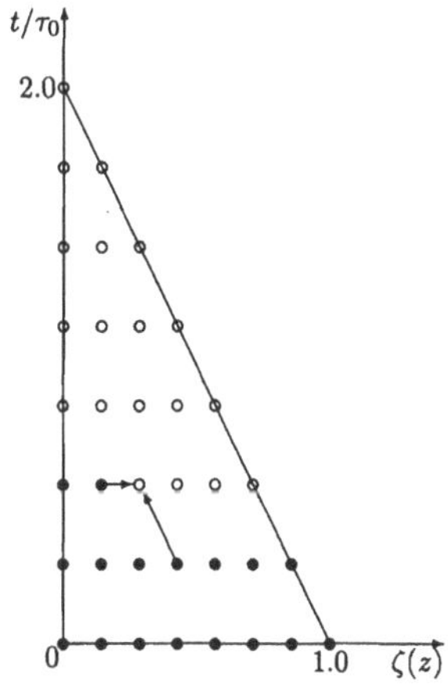

FIG. 2.1. The grid for the direct problem using the Green's function approach.

$$G^-(z,0) = g(z),$$
$$G^+(0,t) = \nu G^-(0,t),$$

where $g(z) = -\frac{c}{2}a\gamma$, and the constant ν is zero if the impedance is matched at $z = 0$ ($\nu \neq 0$ when there is an impedance mismatch at $z = 0$; cf. (2.84) and (2.80) below). Space and time are discretized in a grid of points (z_i, t_j), $i, j = 0, 1, 2, ..., N$, ($z_0 = 0$, $z_N = l$, $t_0 = 0$, $t_N = 2\tau(l)$), where the gridsize in time, Δt, remains constant and the gridsize Δz in the z direction is related to Δt by $\Delta z = c(z)\Delta t/2$. The discretization of the functions is then

$$G^{\pm}_{i,j} = G^{\pm}(z_i, t_j), \qquad \alpha_i = \alpha(z_i), \quad \text{etc.}$$

The numerical algorithm is obtained by integrating (2.29) along the z direction and (2.31) along the characteristic direction $dz = -c(z)dt/2$ using the trapezoid rule approximations. Figure 2.1 shows the grid for the direct problem (the filled circles are the grid points at which the Green's functions have been calculated at certain stage of the calculation; the arrows indicate the direction of integration for G^+ and G^-; in this and the other figures in this chapter we use travel time coordinates, i.e., $\zeta(z) = \frac{1}{\tau_0}\int_0^z \frac{1}{c(z')}\,dz'$, $\tau_0 = \int_0^l \frac{1}{c(z)}\,dz$). The area in Fig. 2.1 is thus built up by adding one horizontal row at a time, starting from the left and moving to the right in each row. The Green's functions are calculated in order of the time, i.e. $j = 0, 1, 2, ..., N-1$, ($i = 0, 1, 2, ..., N-j-1$ for a certain j). The initial values are given by

$$G^-_{i,0} = g_i, \quad i = 0, 1, 2, ..., N,$$

$$
\begin{aligned}
G^+_{0,0} &= \nu g_0, \\
G^+_{i,0} &= \{(1 + \frac{1}{2}\alpha_{i-1}\Delta z_i)G^+_{i-1,0} \\
&\quad + \frac{1}{2}(\beta_i G^-_{i,0} + \beta_{i-1}G^-_{i-1,0})\Delta z_i\}/(1 - \frac{1}{2}\alpha_i \Delta z_i), \quad i = 1, 2, ..., N,
\end{aligned}
$$

where $\Delta z_i = z_i - z_{i-1}$. Assume that we are at the stage of calculating the Green's functions at the time $t = t_j$, $j \geq 1$ (the Green's functions at time $t < t_j$ are assumed to have been calculated already). The boundary values for the Green's functions at time $t = t_j$ are

$$
\begin{aligned}
G^-_{0,j} &= \{G^-_{1,j-1} - \frac{1}{2}(\gamma_1 G^+_{1,j-1} + \vartheta_1 G^-_{1,j-1})\Delta z_1\}/[1 + \frac{1}{2}(\vartheta_0 + \nu\gamma_0)\Delta z_1], \\
G^+_{0,j} &= \nu G^-_{0,j}.
\end{aligned}
$$

The Green's functions at (z_i, t_j), $i \geq 1$, $j \geq 1$, are then given by (cf. (2.29) and (2.31))

$$
\begin{aligned}
G^+_{i,j} &= [h^{(1)}_{i,j} n^{(2)}_{i,j} - h^{(2)}_{i,j} n^{(1)}_{i,j}]/d_{i,j}, \\
G^-_{i,j} &= [h^{(2)}_{i,j} m^{(1)}_{i,j} - h^{(1)}_{i,j} m^{(2)}_{i,j}]/d_{i,j},
\end{aligned}
$$

where

$$
\begin{aligned}
m^{(1)}_{i,j} &= 1 - \frac{1}{2}\alpha_i \Delta z_i, \qquad n^{(1)}_{i,j} = -\frac{1}{2}\beta_i \Delta z_i, \\
h^{(1)}_{i,j} &= (1 + \frac{1}{2}\alpha_{i-1}\Delta z_i)G^+_{i-1,j} + \frac{1}{2}\beta_{i-1}\Delta z_i G^-_{i-1,j}, \\
m^{(2)}_{i,j} &= \frac{1}{2}\gamma_i \Delta z_{i+1}, \qquad n^{(2)}_{i,j} = 1 + \frac{1}{2}\vartheta_i \Delta z_{i+1}, \\
h^{(2)}_{i,j} &= (1 - \frac{1}{2}\vartheta_{i+1}\Delta z_{i+1})G^-_{i+1,j-1} - \frac{1}{2}\gamma_{i+1}\Delta z_{i+1} G^+_{i+1,j-1}, \\
d_{i,j} &= m^{(1)}_{i,j} n^{(2)}_{i,j} - m^{(2)}_{i,j} n^{(1)}_{i,j}.
\end{aligned}
$$

After one round trip, one needs to add the discontinuity in G^- using (2.35) (since σ and ϵ_z may be discontinuous at the surfaces $z = 0$ and $z = l$), and then calculate the Green's functions $G^\pm(z,t)$ for $t > 2\tau(l) - 2\tau(z)$ in an analogous way.

2.1.4 *Numerical implementation of an inverse problem*

In this subsection we give the numerical implementation of the reconstruction of one parameter ($\sigma(z)$ or $\epsilon(z)$; the other parameter is assumed to be known) using only one round trip of reflection data, i.e., $R(t)$ with $t \in [0, 2\tau(l)]$.

In the inverse problem the PDEs are solved starting at the known boundary values of the Green's functions at $z = 0$ (cf. (2.30) and (2.39)). We use the same

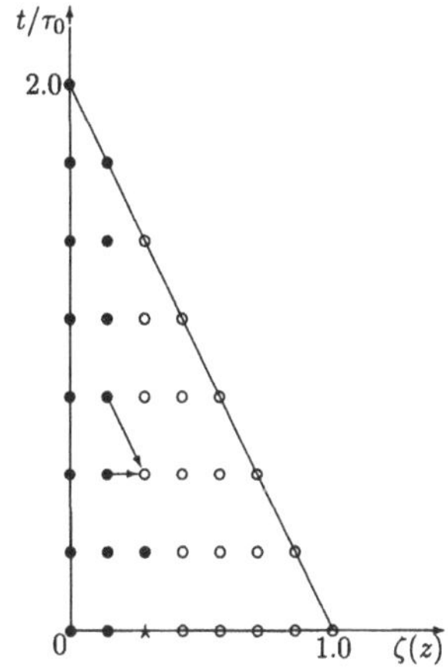

FIG. 2.2. The grid for the inverse problem of one-parameter reconstruction using the Green's function approach.

discretization as for the direct problem (as described in Section 2.1.3). Figure 2.2 shows the grid for the inverse problem. In the reconstruction algorithm the values of σ or ϵ_z (note that in the reconstruction of the permittivity, the logarithmic derivative of the permittivity is determined by the algorithm and the permittivity is obtained by an integration in z) are determined from the left and moving to the right in Fig. 2.2. The triangle in Fig. 2.2 is filled by adding one vertical column at a time moving from the bottom to the top in each column. The star $\star$ in Fig. 2.2 indicates the position at which the value of G^- is used to reconstruct the parameter at this layer. The parameter can be reconstructed according to the following steps:

(1) From the value $G^-(0,0) = R(0)$ one can determine the unknown parameter σ or ϵ_z at the surface $z = 0$ from the condition (2.32). Furthermore one knows both G^+ and G^- at the surface $z = 0$ (cf. (2.30) and (2.39)).

(2) If the Green's functions $G^{\pm}_{i-1,j}$ at the position z_{i-1}, $i \geq 1$, $j = N-(i-1)$, are known, as well as the parameters at this position, one can calculate the Green's functions at the next position z_i with either an explicit or an implicit algorithm.

An explicit algorithm: Using the first-order approximation to integrate the PDEs (2.29) and (2.31) along the characteristics to the point (z_i, t_j), $j = 0, 1, 2, \ldots$, $N - i$, one obtains

$$G^+_{i,j} - G^+_{i-1,j} = \Delta z_i \left[\alpha_{i-1} G^+_{i-1,j}\right] + \Delta z_i \left[\beta_{i-1} G^-_{i-1,j}\right],$$
$$G^-_{i,j} - G^-_{i-1,j+1} = \Delta z_i \left[\gamma_{i-1} G^+_{i-1,j+1}\right] + \Delta z_i \left[\vartheta_{i-1} G^-_{i-1,j+1}\right],$$

which gives the following explicit expressions for the two unknown quantities $G^+_{i,j}$ and $G^-_{i,j}$,

$$G^+_{i,j} = G^+_{i-1,j} + \Delta z_i \left[\alpha_{i-1} G^+_{i-1,j} + \beta_{i-1} G^-_{i-1,j}\right], \tag{2.41}$$

$$G^-_{i,j} = G^-_{i-1,j+1} + \Delta z_i \left[\gamma_{i-1} G^+_{i-1,j+1} + \vartheta_{i-1} G^-_{i-1,j+1}\right]. \tag{2.42}$$

An implicit algorithm: Applying the trapezoidal rule to integrate the PDEs (2.29) and (2.31) along the characteristics to the point (z_i, t_j), $j = 0, 1, 2, ..., N-i$, yields

$$G^+_{i,j} - G^+_{i-1,j} = \frac{1}{2}\Delta z_i \left[\alpha_{i-1} G^+_{i-1,j} + \alpha_i G^+_{i,j}\right] + \frac{1}{2}\Delta z_i \left[\beta_{i-1} G^-_{i-1,j} + \beta_i G^-_{i,j}\right],$$

$$G^-_{i,j} - G^-_{i-1,j+1} = \frac{1}{2}\Delta z_i \left[\gamma_{i-1} G^+_{i-1,j+1} + \gamma_i G^+_{i,j}\right] + \frac{1}{2}\Delta z_i \left[\vartheta_{i-1} G^-_{i-1,j+1} + \vartheta_i G^-_{i,j}\right],$$

which can be rewritten as the following linear system of equations of order two for the two unknown quantities $G^+_{i,j}$ and $G^-_{i,j}$,

$$\mathbf{A}_i \begin{bmatrix} G^+_{i,j} \\ G^-_{i,j} \end{bmatrix} = \mathbf{B}_{i-1} \begin{bmatrix} G^+_{i-1,j} \\ G^-_{i-1,j} \end{bmatrix} + \mathbf{C}_{i-1} \begin{bmatrix} G^+_{i-1,j+1} \\ G^-_{i-1,j+1} \end{bmatrix},$$

where

$$\mathbf{A}_i = \begin{bmatrix} 1 - \frac{1}{2}\Delta z_i \alpha_i & -\frac{1}{2}\Delta z_i \beta_i \\ -\frac{1}{2}\Delta z_i \gamma_i & 1 - \frac{1}{2}\Delta z_i \vartheta_i \end{bmatrix}, \tag{2.43}$$

$$\mathbf{B}_{i-1} = \begin{bmatrix} 1 + \frac{1}{2}\Delta z_i \alpha_{i-1} & +\frac{1}{2}\Delta z_i \beta_{i-1} \\ 0 & 0 \end{bmatrix}, \tag{2.44}$$

$$\mathbf{C}_{i-1} = \begin{bmatrix} 0 & 0 \\ \frac{1}{2}\Delta z_i \gamma_{i-1} & 1 + \frac{1}{2}\Delta z_i \vartheta_{i-1} \end{bmatrix}. \tag{2.45}$$

Since α, β, γ and ϑ are bounded, one can always choose Δz_i to be small enough so that the matrix $\mathbf{A}_i$ is invertible. Therefore, one has

$$\begin{bmatrix} G^+_{i,j} \\ G^-_{i,j} \end{bmatrix} = [\mathbf{A}_i]^{-1} \mathbf{B}_{i-1} \begin{bmatrix} G^+_{i-1,j} \\ G^-_{i-1,j} \end{bmatrix} + [\mathbf{A}_i]^{-1} \mathbf{C}_{i-1} \begin{bmatrix} G^+_{i-1,j+1} \\ G^-_{i-1,j+1} \end{bmatrix}. \tag{2.46}$$

Note that the matrices $\mathbf{A}$, $\mathbf{B}$ and $\mathbf{C}$ are independent of the time t, and thus the same matrices are used in calculating $G^+_{i,j}$ for all different times t_j at the same position z_i. The algorithm (2.46) is implicit since at this stage one does not know α, β, γ and ϑ at the position z_i. One can simply approximate α_i, β_i, γ_i and ϑ_i in $\mathbf{A}_i$ with α_{i-1}, β_{i-1}, γ_{i-1} and ϑ_{i-1}, respectively (the accuracy can be improved by updating α_i, β_i, γ_i and ϑ_i after the parameter is reconstructed at the position z_i).

(3) Using the initial condition (2.32) at $z = z_i$, one can immediately reconstruct the unknown parameter σ or ϵ_z at the position $z = z_i$ from the calculated

value of $G_{i,0}^-$. By repeatedly applying steps (2) and (3) with $i = 1, 2, 3, ..., N$, one thus reconstructs the parameter layer by layer in the whole inhomogeneous slab.

As shown in He et al. (1993c), the parameter reconstructed using the explicit formulas (2.41) and (2.42) converges to the true profile as the discretization stepsize goes to zero. The implicit formula (2.46) gives a more accurate reconstruction than the explicit formulas, if the same discretization stepsize is used. The reconstruction algorithm (using either the explicit or implicit formula) is robust and stable for noise-contaminated $R(t)$ (cf. e.g. the reconstruction of the inhomogeneous permittivity for a coaxial transmission line, using experimental data for the transient reflection in Section 2.5 below).

2.1.5 *An example of an exact and explicit solution*

Exact and explicit solutions for special cases are useful for the insight they provide and as references, e.g., in establishing benchmarks for general numerical algorithms for direct problems. The closed-form solution to the transient reflection and transmission can also be directly put into an inverse code in a time domain inverse problem as an independent check (to avoid any numerical 'marching back effect'). In He (1995b), the equations for the Green's functions $G^{\pm}$ have been explicitly solved for a lossy exponential transmission line (exponential transmission lines are used in microwave applications as, e.g., resonators, transformers, etc.). Through suitable specifications of the parameters of the transmission line problem considered in He (1995b), one arrives at equations corresponding to the present electromagnetic scattering problem (with suitable replacements). Thus, consider an electromagnetic plane wave, normally incident on a stratified slab which has the following z-dependent parameters

$$\mu(z) = \mu_0 e^{2qz}, \quad \epsilon(z) = \epsilon_0 e^{-2qz}, \tag{2.47}$$

$$\sigma(z) = \sigma_0 e^{-2qz}, \tag{2.48}$$

where μ_0, ϵ_0, σ_0 are constants, and the constant q defines the taper of the parameter profiles. Therefore, the speed of the propagation and impedance at position z are given by

$$c = \frac{1}{\sqrt{\mu_0 \epsilon_0}}, \qquad Z(z) = \sqrt{\frac{\mu_0}{\epsilon_0}} e^{2qz},$$

respectively. Note that the speed of the propagation is a constant in this special case. On either side of the slab there is a homogeneous, lossless medium with the permittivity and permeability continuous at $z = 0$ and $z = l$.

Let

$$G^{\pm}(z,t) = e^{(q-k)z-kct} g^{\pm}(z,t), \tag{2.49}$$

where

$$k = \frac{1}{2}\sigma_0\sqrt{\frac{\mu_0}{\epsilon_0}}. \tag{2.50}$$

The PDEs together with the initial and boundary conditions for $g^{\pm}(z,t)$ are as follows (deduced from the corresponding equations for $G^{\pm}$ given in the previous subsection)

$$\frac{\partial}{\partial z}g^{+} = -(q+k)g^{-}, \tag{2.51}$$

$$(\frac{\partial}{\partial z} - \frac{2}{c}\frac{\partial}{\partial t})g^{-} = -(q-k)g^{+}, \tag{2.52}$$

$$g^{-}(z,0) = \frac{c}{2}(q-k), \tag{2.53}$$

$$g^{+}(0,t) = 0, \quad g^{-}(l,t) = 0. \tag{2.54}$$

Equations (2.51)–(2.54) constitute a set of equations from which $g^{\pm}(z,t)$ can be determined for all $t \geq 0$, $z \in [0,l]$. This set of equations can be solved analytically, by means of the Laplace transform.

With notation according to

$$\bar{g}^{\pm}(z,p) = L[g^{\pm}(z,t)] \equiv \int_0^{\infty} g^{\pm}(z,t)e^{-pt}\,dt, \tag{2.55}$$

it then follows from (2.51)–(2.53) that

$$\frac{d}{dz}\begin{bmatrix}\bar{g}^{+}\\ \bar{g}^{-}\end{bmatrix} = \begin{bmatrix}0 & -(q+k)\\ -(q-k) & 2p/c\end{bmatrix}\begin{bmatrix}\bar{g}^{+}\\ \bar{g}^{-}\end{bmatrix} + \begin{bmatrix}0\\ -(q-k)\end{bmatrix}. \tag{2.56}$$

The boundary conditions $\bar{g}^{\pm}$ are (cf. (2.54))

$$\bar{g}^{+}(0,p) = 0, \tag{2.57}$$

$$\bar{g}^{-}(l,p) = 0. \tag{2.58}$$

Note that (2.56) is a linear system of inhomogeneous ordinary differential equations (ODEs) with constant coefficients. The solution for the system (2.56) with the boundary condition (2.57) is (He 1995b)

$$\begin{bmatrix}\bar{g}^{+}\\ \bar{g}^{-}\end{bmatrix}(z,p) = \frac{\bar{g}^{-}(0,p)}{1+(q^2-k^2)/\bar{\lambda}_2^2}\begin{bmatrix}\{(q+k)/\bar{\lambda}_2\}(e^{\bar{\lambda}_1 z} - e^{\bar{\lambda}_2 z})\\ \{(q^2-k^2)/\bar{\lambda}_2^2\}e^{\bar{\lambda}_1 z} + e^{\bar{\lambda}_2 z}\end{bmatrix}$$
$$-\frac{q-k}{1+(q^2-k^2)/\bar{\lambda}_2^2}\begin{bmatrix}\{(q+k)/\bar{\lambda}_2\}\{(e^{\bar{\lambda}_1 z}-1)/\bar{\lambda}_1 - (e^{\bar{\lambda}_2 z}-1)/\bar{\lambda}_2\}\\ \{(q^2-k^2)/\bar{\lambda}_2^2\}(e^{\bar{\lambda}_1 z}-1)/\bar{\lambda}_1 + (e^{\bar{\lambda}_2 z}-1)/\bar{\lambda}_2\end{bmatrix}, \tag{2.59}$$

where

$$\bar{\lambda}_1 = \frac{p}{c} - (\frac{p^2}{c^2} + q^2 - k^2)^{1/2}, \quad \bar{\lambda}_2 = \frac{p}{c} + (\frac{p^2}{c^2} + q^2 - k^2)^{1/2}. \tag{2.60}$$

Since $\bar{g}^-(l,p) = 0$ (cf. (2.58)), it follows from the second component of (2.59) that

$$\bar{g}^-(0,p) = -\frac{q-k}{\bar{\lambda}_2}\,\frac{e^{\bar{\lambda}_1 l} - e^{\bar{\lambda}_2 l}}{[(q^2-k^2)/\bar{\lambda}_2^2]e^{\bar{\lambda}_1 l} + e^{\bar{\lambda}_2 l}}. \tag{2.61}$$

Substituting the above equation into the first component of (2.59) with $z = l$, one obtains the following equation after some rearrangements

$$\bar{g}^+(l,p) = -1 + \frac{(q^2-k^2) + [(\frac{p^2}{c^2}+q^2-k^2)^{1/2} - \frac{p}{c}]^2}{(q^2-k^2) + [(\frac{p^2}{c^2}+q^2-k^2)^{1/2} - \frac{p}{c}]^2 e^{-2(\frac{p^2}{c^2}+q^2-k^2)^{1/2}l}}\, e^{-[(\frac{p^2}{c^2}+q^2-k^2)^{1/2} - \frac{p}{c}]l}. \tag{2.62}$$

With the notation

$$\bar{a} = c(q^2-k^2)^{1/2}, \tag{2.63}$$

it follows from (2.61) that

$$\begin{aligned}
&\frac{\bar{a}}{(q-k)c}\bar{g}^-(0,p) \\
&= \frac{\sqrt{p^2+\bar{a}^2}-p}{\bar{a}}\,\frac{1-e^{-(2l/c)\sqrt{p^2+\bar{a}^2}}}{1+(\frac{\sqrt{p^2+\bar{a}^2}-p}{\bar{a}})^2 e^{-(2l/c)\sqrt{p^2+\bar{a}^2}}} \\
&= \frac{\sqrt{p^2+\bar{a}^2}-p}{\bar{a}}\left[1-e^{-(2l/c)\sqrt{p^2+\bar{a}^2}}\right] \times \\
&\quad \left\{1+\sum_{n=1}^{\infty}(-1)^n[\frac{\sqrt{p^2+\bar{a}^2}-p}{\bar{a}}]^{2n} e^{-n(2l/c)\sqrt{p^2+\bar{a}^2}}\right\} \\
&= \frac{\sqrt{p^2+\bar{a}^2}-p}{\bar{a}} + \sum_{n=1}^{\infty}(-1)^n \bar{f}_n\left(\sqrt{p^2+\bar{a}^2}\right),
\end{aligned} \tag{2.64}$$

where

$$\bar{f}_n(p) = \left[\left(\frac{p-\sqrt{p^2-\bar{a}^2}}{\bar{a}}\right)^{2n-1} + \left(\frac{p-\sqrt{p^2-\bar{a}^2}}{\bar{a}}\right)^{2n+1}\right] e^{-n(2l/c)p},\ n \geq 1. \tag{2.65}$$

Using the following property of the inverse Laplace transform (Oberhettinger and Badii 1973)

$$L^{-1}[\bar{f}(\sqrt{p^2+\bar{a}^2})] = f(t) - \bar{a}\int_0^t f(t')\frac{t'}{\sqrt{t^2-t'^2}}J_1[\bar{a}\sqrt{t^2-t'^2}]\,dt',$$

where $f(t)$ is the inverse Laplace transform of $\bar{f}(p)$, i.e., $f(t) \equiv L^{-1}[\bar{f}(p)]$, one can invert each term on the right-hand side of (2.64) and one obtains (cf. (2.49))

$$G^{-}(0,t) = \frac{(q-k)c}{\bar{a}} e^{-kct} \frac{J_1(\bar{a}t)}{t} + \frac{(q-k)c}{\bar{a}} e^{-kct} \sum_{n=1}^{\infty} (-1)^n F_n(t) \Theta(t - \frac{2nl}{c}), \tag{2.66}$$

where $\Theta(t)$ is the Heaviside unit step function and

$$F_n(t) = K_n(t - \frac{2nl}{c}) - \bar{a} \int_{2nl/c}^{t} K_n(t' - \frac{2nl}{c}) \frac{t'}{\sqrt{t^2 - t'^2}} J_1 \left[\bar{a}\sqrt{t^2 - t'^2} \right] dt', \quad n \geq 1, \tag{2.67}$$

and where

$$K_n(t) = (2n-1) I_{2n-1}(\bar{a}t)/t + (2n+1) I_{2n+1}(\bar{a}t)/t, \quad n \geq 1 \tag{2.68}$$

($J_n(t)$, $I_n(t)$ are the Bessel function and the modified Bessel function of order n, respectively). Note that $G^{-}(0,t)$ for the first round trip (i.e., $0 \leq t \leq 2l/c$) is given by the first term in (2.66), and the term containing F_n, $n \geq 1$, contributes to $G^{-}(0,t)$ only after the n-th round trip time (note that $\Theta(t - \frac{2nl}{c})$ is the Heaviside unit step function which vanishes for $t < \frac{2nl}{c}$). Physically, these discontinuities in the Green's function at $t = 2nl/c$ are due to the discontinuities in the conductivity and the z derivative of the impedance at the two endpoints $z = 0$ and $z = l$ (thus a series summation of the type given in (2.66) is inevitable in every explicit solution for a δ pulse reflection, regardless of the method used in the derivation). For a half-space problem, one obtains the following result by taking the limit $l \to +\infty$ of (2.66)

$$G^{-}(0,t) = \frac{q-k}{(q^2-k^2)^{1/2}} e^{-kct} \frac{J_1(c(q^2-k^2)^{1/2}t)}{t}.$$

The reflected field $E^r(t)$ for an arbitrary incident field $E^i(t)$ is given by $E^r(t) = G^{-}(0,t) * E^i(t)$.

Similarly, one can expand (2.62) as follows

$$\bar{g}^{+}(l,p) = -1 + \left[1 + \bar{a}^{-2}(\sqrt{p^2+\bar{a}^2} - p)^2\right] e^{-(\sqrt{p^2+\bar{a}^2}-p)l/c} + \left[1 + \bar{a}^{-2}(\sqrt{p^2+\bar{a}^2} - p)^2\right] e^{-(\sqrt{p^2+\bar{a}^2}-p)l/c} \times \sum_{n=1}^{\infty} (-1)^n \left[\bar{a}^{-2}(\sqrt{p^2+\bar{a}^2} - p)^2\right]^n e^{-n(2l/c)\sqrt{p^2+\bar{a}^2}}. \tag{2.69}$$

After some algebra, the inversion of the above equation gives the following result

$$G^+(l,t) = e^{(q-k)l-kct}g^+(l,t)$$
$$= e^{-kct}E_0(t) + e^{-kct}\sum_{n=1}^{\infty}(-1)^n E_n(t)H(t-\frac{2nl}{c}), \quad (2.70)$$

where

$$E_0(t) = -\frac{2te^{(q-k)l}}{(t^2+2lt/c)^2}\left[\bar{a}(l/c)^2(t^2+2lt/c)^{1/2}J_1(\bar{a}(t^2+2lt/c)^{1/2}) -t^2J_2(\bar{a}(t^2+2lt/c)^{1/2})\right], \quad (2.71)$$

$$E_n(t) = L_n(t-2nl/c) -2n(l/c)\bar{a}^2\int_{2nl/c}^{t}L_n(t-t')\frac{J_1(\bar{a}(t'^2-4n^2l^2/c^2)^{1/2})}{\bar{a}(t'^2-4n^2l^2/c^2)^{1/2}}dt', \quad n \geq 1, \quad (2.72)$$

and where

$$L_n(t) = -\bar{a}^2e^{(q-k)l} \times\left[\left(\frac{\partial}{\bar{a}\partial u}\right)^{2n}+\left(\frac{\partial}{\bar{a}\partial u}\right)^{2n+2}\right]\left[\frac{J_1(\bar{a}(t^2+2ut)^{1/2})}{\bar{a}(t^2+2ut)^{1/2}}u\right]_{u=2l/c}, \quad n \geq 1. \quad (2.73)$$

From (2.70) one sees that $G^+(l,t)$ for the first round trip (i.e., $0 \leq t \leq 2l/c$) is given by the first term in (2.70), and the term containing E_n, $n \geq 1$, contributes to $G^+(l,t)$ only after the n-th round trip time. Note that the transmitted field $E^t(t)$ for an arbitrary excitation $E^i(t)$ is given by $E^t(t+l/c) = e^{(q-k)l}E^i(t) + G^+(l,t) * E^i(t)$.

2.1.6 *The case with an impedance mismatch at the front surface*

We consider the modifications of the previous treatment that are appropriate when there is an impedance mismatch at the surface $z = 0$. Thus there are instantaneous reflection and transmission coefficients associated with this surface and the boundary conditions provide an expression for these coefficients in terms of the impedance mismatch. In this section the Green's function formalism is extended to this case. The permeability μ may also vary with the depth z in this subsection, and thus the same applies to the impedance

$$Z(z) = \sqrt{\frac{\mu(z)}{\epsilon(z)}}. \quad (2.74)$$

If there is an impedance mismatch at a boundary, one can still introduce Green's functions (including additional coefficients) which directly map the incident electric field outside the inhomogeneous slab to the internal split fields. In He and

Ström (1992), the Green's function approach has been applied to the scattering problem for a non-uniform transmission line with an impedance mismatch at the excitation endpoint (cf. also the discussion in Kristensson and Krueger (1987) concerning a front-end discontinuity in the imbedding approach). The treatment given in He and Ström (1992) is valid for the present electromagnetic scattering problem (with small modifications).

If there is an impedance mismatch at the surface $z = 0$, it is appropriate to define the physical reflection kernel and the Green's functions as follows (He and Ström 1992)

$$E^-(0^-,t) = r_0 E^+(0^-,t) + R^{(\mathrm{ph})}(t) * E^+(0^-,t), \tag{2.75}$$
$$E^+(z,t+\tau(0,z)) = a(z)t_0 E^+(0^-,t) + t_0[G^+(z,t)] * [E^+(0^-,t)], \tag{2.76}$$
$$E^-(z,t+\tau(0,z)) = t_0[G^-(z,t)] * [E^+(0^-,t)], \quad z > 0, \tag{2.77}$$

where the reflection coefficient r_0 and the transmission coefficient t_0 will be determined from the boundary conditions (note that $E^+(0^-,t) = E^i(t)$). One notes that $G^\pm(z,t)$ is defined only for $z > 0$ and thus $G^\pm(0,t)$ stands for the limit $z \to 0^+$ of $G^\pm(z,t)$. From the continuity of the tangential fields, one has the following boundary conditions

$$E(0^-,t) = E(0^+,t), \tag{2.78}$$
$$H(0^-,t) = H(0^+,t). \tag{2.79}$$

Note that $E(0^\pm,t) = E^+(0^\pm,t) + E^-(0^\pm,t)$, $H(0^\pm,t) = [Z(0^\pm)]^{-1}[E^+(0^\pm,t) - E^-(0^\pm,t)]$, (cf. (2.3)). Substituting (2.76), (2.77) (with $z = 0^+$) and (2.75) into the boundary conditions (2.78) and (2.79), one obtains the following equations

$$\begin{cases} 1 + r_0 = t_0, \\ R^{(\mathrm{ph})}(t) = t_0[G^+(0,t) + G^-(0,t)], \end{cases}$$
$$\begin{cases} [Z(0^-)]^{-1}(1 - r_0) = [Z(0^+)]^{-1} t_0, \\ -[Z(0^-)]^{-1} R^{(\mathrm{ph})}(t) = [Z(0^+)]^{-1} t_0 [G^+(0,t) - G^-(0,t)], \end{cases}$$

respectively. Therefore, one has the following values of the coefficients r_0 and t_0

$$r_0 = \frac{Z(0^+) - Z(0^-)}{Z(0^+) + Z(0^-)}, \tag{2.80}$$
$$t_0 = (1 + r_0) = \frac{2Z(0^+)}{Z(0^+) + Z(0^-)}, \tag{2.81}$$

and the following relations between $G^\pm(0,t)$ and $R^{(\mathrm{ph})}(t)$

$$G^-(0,t) = \frac{1}{1 - r_0^2} R^{(\mathrm{ph})}(t), \tag{2.82}$$

$$G^+(0,t) = -\frac{r_0}{1-r_0^2} R^{(\mathrm{ph})}(t), \tag{2.83}$$

$$G^+(0,t) = -r_0 G^-(0,t). \tag{2.84}$$

The PDEs for the Green's functions $G^\pm(z,t)$, the initial and boundary conditions for G^- remain the same as given by (2.29), (2.31), (2.32) and (2.33).

In the direct problem, the two PDEs for $G^\pm$ (cf. (2.29), (2.31)) together with the initial condition for G^- (cf. (2.32)), the boundary condition at $z = l$ for G^- (cf. (2.33)), and the boundary relation (2.84) between G^+ and G^- at $z = 0$, constitute a set of equations from which $G^\pm(z,t)$ can be determined for $t \geq 0$, $z \in [0,l]$ (see Section 2.1.3 for the numerical implementation). The physical reflection kernel $R^{(\mathrm{ph})}(t)$ is then obtained from the boundary relation (2.82).

In the inverse problem, the physical reflection kernel $R^{(\mathrm{ph})}(t)$ is assumed to be known, and the boundary values $G^\pm(0,t)$ are obtained from the boundary relations (2.82) and (2.83). Hence the unknown parameter (permittivity or conductivity) can be reconstructed by propagating the boundary values of the Green's functions to the initial values (cf. (2.32); see Section 2.1.4 for the numerical implementation).

The case with an impedance mismatch at the back boundary $z = l$ can be treated analogously (see He et al. (1994c); see also Section 2.3 for such a treatment in the context of the compact Green's function approach).

2.1.7 *Two-sided excitation and connections with the imbedding kernels*

In some inverse problems of simultaneous reconstruction, one may need to use the scattering data for both the left side and right side excitations (see Kristensson and Krueger (1986b, 1987) for the imbedding approach to such a case; cf. also Exercise 1.3). In this subsection, we consider a situation in which the inhomogeneous slab is excited by incident fields from both sides. Assume that the field inside the slab is identically zero before the time $t = 0$. For the direct problem, if the incident fields $E^+(0,t)$ and $E^-(l,t)$ are known for all times $t > 0$, one can determine the internal fields $E^+(z,t)$ and $E^-(z,t)$ for all z in $[0,l]$ and for all times $t > 0$. This is achieved by solving the well-posed initial-boundary value problem for the system (2.5) with the initial conditions

$$E^+(z,0) = E^-(z,0) = 0, \quad 0 < z < l, \tag{2.85}$$

and the boundary conditions

$$\begin{cases} E^+(0,t) = E^i_+(t), & t > 0, \\ E^-(l,t) = E^i_-(t), & t > 0. \end{cases} \tag{2.86}$$

The solution can be expressed in terms of a Green's tensor $\hat{G}_{ij}$, $i,j = 1,2$, defined as a mapping from the incident fields at $z = 0$ and $z = l$ to the internal split fields according to the following definition which constitutes an extension to two-sided excitations of the treatment given in Section 2.1.2 (see also Krueger and Ochs (1989))

$$\begin{bmatrix} E^+(z,t) \\ E^-(z,t) \end{bmatrix} = \begin{bmatrix} \hat{G}_{11}(z) & \hat{G}_{12}(z) \\ \hat{G}_{21}(z) & \hat{G}_{22}(z) \end{bmatrix} \begin{bmatrix} E^+(0,t) \\ E^-(l,t) \end{bmatrix}$$
$$\equiv \begin{bmatrix} \hat{G}_{11}(z) & 0 \\ \hat{G}_{21}(z) & 0 \end{bmatrix} \begin{bmatrix} E^+(0,t) \\ 0 \end{bmatrix} + \begin{bmatrix} 0 & \hat{G}_{12}(z) \\ 0 & \hat{G}_{22}(z) \end{bmatrix} \begin{bmatrix} 0 \\ E^-(l,t) \end{bmatrix}. \tag{2.87}$$

Note that $\hat{G}_{ij}(z)$, $i,j = 1,2$, are operators (cf. (2.88)–(2.91) below). Here the right-moving and left-moving incident fields are arbitrary (the physical interpretation of the splitting guarantees that $E^+(0,t) \equiv E^i_+(t)$, $E^-(l,t) \equiv E^i_-(t)$ for two-sided incidence and for all times $t > 0$, since the right-moving incident field from the left side of the slab $z \leq 0$ will not generate left-moving fields but only a right-moving field on the other side $z \geq l$, and an analogous statement holds for the left-moving incident field from the right side $z \geq l$). Since $E^+(0,t)$ and $E^-(l,t)$ are arbitrary, it can be seen from (2.87) that the resulting equations for $\hat{G}_{11}$ and $\hat{G}_{21}$ will be coupled, and the same applies to the equations for $\hat{G}_{12}$ and $\hat{G}_{22}$, but there will be no coupling between these pairs of Green's functions.

The operators $\hat{G}_{ij}(z)$, $i,j = 1,2$, have the following representations (cf. (2.22) and (2.23)):

$$[\hat{G}_{11}(z)]E^i_+(t)$$
$$= a^+(0,z)E^i_+(t-\tau(0,z)) + \int_0^{t-\tau(0,z)} \mathcal{G}_{11}(z,t-t')E^i_+(t')\,dt', \tag{2.88}$$

$$[\hat{G}_{21}(z)]E^i_+(t) = \int_0^{t-\tau(0,z)} \mathcal{G}_{21}(z,t-t')E^i_+(t')\,dt', \tag{2.89}$$

$$[\hat{G}_{12}(z)]E^i_-(t) = \int_0^{t-\tau(z,l)} \mathcal{G}_{12}(z,t-t')E^i_-(t')\,dt', \tag{2.90}$$

$$[\hat{G}_{22}(z)]E^i_-(t)$$
$$= a^-(z,l)E^i_-(t-\tau(z,l)) + \int_0^{t-\tau(z,l)} \mathcal{G}_{22}(z,t-t')E^i_-(t')\,dt', \tag{2.91}$$

where $\tau(z_1,z_2)$ is defined by (1.20) and

$$a^{\pm}(z_1,z_2) = \left[\frac{\epsilon(z_1)}{\epsilon(z_2)}\right]^{\pm\frac{1}{4}} \exp\left[-\frac{1}{2}\int_{z_1}^{z_2} \sigma(z')\sqrt{\mu_0/\epsilon(z')}\,dz'\right], \quad z_1 \leq z_2. \tag{2.92}$$

As described in Section 2.1.2, it is convenient to modify the arguments for the Green's functions by a time shift, and consider

$$G_{i1}(z,t) \equiv \mathcal{G}_{i1}(z,t+\tau(0,z)), \qquad i = 1,2, \tag{2.93}$$

$$G_{i2}(z,t) \equiv \mathcal{G}_{i2}(z,t+\tau(z,l)), \qquad i=1,2. \tag{2.94}$$

Then $G_{i1}(z,t)$, $i=1,2$, satisfy all the equations for $G^{\pm}(z,t)$ given in Section 2.1.2, with the replacements $G^{+}(z,t) \to G_{11}(z,t)$, $G^{-}(z,t) \to G_{21}(z,t)$, $a(z) \to a^{+}(0,z)$, $\tau(z) \to \tau(0,z)$, and $E^{i}(t) \to E^{i}_{+}(t)$. Therefore, the Green's functions for the left side excitation, $G_{i1}(z,t)$, $i=1,2$, can be computed for all times $t \geq 0$ as described in Section 2.1.3.

The other pair of Green's functions $G_{i2}(z,t)$, $i=1,2$, satisfy the following equations:

$$\frac{\partial}{\partial z}\begin{bmatrix} G_{12} \\ G_{22} \end{bmatrix} + \frac{2}{c}\frac{\partial}{\partial t}\begin{bmatrix} G_{12} \\ 0 \end{bmatrix} = \begin{bmatrix} \alpha & \beta \\ \gamma & \vartheta \end{bmatrix}\begin{bmatrix} G_{12} \\ G_{22} \end{bmatrix}, \tag{2.95}$$

$$G_{12}(z,0) = \frac{1}{2}ca^{-}(z,l)\beta(z), \tag{2.96}$$

$$G_{12}(0,t) = 0, \quad t > 0, \tag{2.97}$$

$$G_{22}(l,t) = 0, \quad t > 0. \tag{2.98}$$

Assume that σ and ϵ_z are continuous except at a finite number of positions z_m, $m = 1,2,...,M$. The discontinuity in G_{12} along its characteristics $\{\Gamma : \frac{\partial t}{\partial z} = \frac{2}{c(z)}\}$ is

$$\begin{aligned}
[G_{12}]_{\Gamma} &= [G_{12}(z,0)]_{z_m} \cdot \exp\left[\int_{z_m}^{z} \alpha(z')\, dz'\right] \\
&= \frac{1}{2}[ca^{-}(z,l)\beta(z)]_{z=z_m} \cdot \exp\left[\int_{z_m}^{z} \alpha(z')\, dz'\right], \quad z \geq z_m.
\end{aligned} \tag{2.99}$$

The system of equations (2.95)–(2.99) constitute a set of equations from which the Green's functions for the right side excitation, $G_{i2}(z,t)$, $i=1,2$, can be calculated for all times $t \geq 0$, $z \in [0,l]$, in analogy with the Green's functions for the left side excitation (cf. Section 2.1.3).

As shown in Karlsson et al. (1993b), one can construct the Green's tensor for the entire slab from the Green's tensors for its imbedded subslabs (the original slab is partitioned into an arbitrary number of subslabs; each subslab is terminated by homogeneous lossless materials at two endpoints with continuous permittivity, and thus the Green's tensor for each subslab satisfies a set of equations which is analogous to the equations for $G_{ij}(z,t)$, $i,j=1,2$, as given in this subsection). Such a partitioning technique leads to an algorithm suitable for parallel computing, and is particularly useful for wave propagation problems for dispersive media, periodic media, and composite structures (i.e., structures consisting of different types of materials, such as isotropic, chiral and anisotropic materials).

Inserting the definitions for the Green's functions G_{ij}, $i,j=1,2$, into the definitions for the imbedding kernels $R^{\pm}$ and $T^{\pm}$ (introduced in Chapter 1), and using the fact that $E^{+}(0,t)$ and $E^{-}(l,t)$ are arbitrary incident fields, one obtains the following relations between the Green's functions and the imbedding kernels (He 1992a; He and Karlsson 1993; Karlsson et al. 1993b)

$$
\begin{aligned}
G_{21}(z,t) &= a^{+}(0,z)R^{+}(z,l,t) + G_{11}(z,t) * R^{+}(z,l,t), && (2.100)\\
G_{11}(z_2,t) &= a^{+}(z_1,z_2)G_{11}(z_1,t) + a^{+}(0,z_1)T^{+}(z_1,z_2,t) \\
&\quad + T^{+}(z_1,z_2,t) * G_{11}(z_1,t), && (2.101)\\
G_{12}(z,t) &= a^{-}(z,l)R^{-}(0,z,t) + G_{22}(z,t) * R^{-}(0,z,t), && (2.102)\\
G_{22}(z_1,t) &= a^{-}(z_1,z_2)G_{22}(z_2,t) + a^{-}(z_2,l)T^{-}(z_1,z_2,t) \\
&\quad + T^{-}(z_1,z_2,t) * G_{22}(z_2,t). && (2.103)
\end{aligned}
$$

These relations indicate that the Green's tensor contains all the scattering information, and in particular, the boundary values of the Green's functions give the scattering kernels for the physical scattering geometry

$$
\begin{bmatrix} G_{11}(l,t) & G_{12}(l,t) \\ G_{21}(0,t) & G_{22}(0,t) \end{bmatrix} = \begin{bmatrix} T^{+} & R^{-} \\ R^{+} & T^{-} \end{bmatrix} (0,l,t), \qquad (2.104)
$$

as is obvious from their definitions.

In fact, one can derive the imbedding equations (together with the initial conditions) for $R^{\pm}$ and $T^{\pm}$ by differentiating the relations (2.100)–(2.103) with respect to z (or z_1, or z_2) and using the PDEs for the Green's functions (see e.g., He and Karlsson (1993) and Exercise 2.1).

2.2 Mathematical analysis of the Green's function approach

In this section, it will be shown that the system of equations for the Green's functions is well-posed for the inverse problem (the well-posedness of such a system for the direct problem is proved in He and Weston (1992)). An analysis of the stability, existence, uniqueness and local correctness will be given here for the Green's function approach to the inverse problem of a one parameter reconstruction in which the reflection kernel $R(t)$ is taken as input data (note that the reflection kernel $R(t)$ is obtained from measurements of the reflected fields through an ill-posed deconvolution process, see Section 2.5.2). Readers who are not interested in the details of such an analysis may pass over this section and go directly to Section 2.3.

To simplify the analysis, the travel time coordinate $x(z) = \int_0^z \sqrt{\epsilon(z')\mu_0}\, dz'$ is used to replace the depth z (in terms of x the characteristic curves of the PDEs for $\mathcal{G}^{\pm}$ become straight lines, cf. Figs 2.1 and 2.2). Thus, the PDEs for the Green's functions $\mathcal{G}^{\pm}$ (given in Section 2.1.1) may be written

$$
\left(\frac{\partial}{\partial x} + K\frac{\partial}{\partial t} + B\right)\mathcal{G}(x,t) = 0, \qquad (2.105)
$$

where

$$\mathcal{G}(x,t) = \begin{bmatrix} \mathcal{G}_1(x,t) \\ \mathcal{G}_2(x,t) \end{bmatrix} \equiv \begin{bmatrix} \mathcal{G}^+(z(x),t) \\ \mathcal{G}^-(z(x),t) \end{bmatrix}, \tag{2.106}$$

$$K = \mathrm{diag}(1,-1), \qquad B = \begin{bmatrix} b_1 & b_2 \\ -b_1 & -b_2 \end{bmatrix}, \tag{2.107}$$

with

$$b_1 = \frac{1}{4\hat{\epsilon}}\frac{d\hat{\epsilon}}{dx} + \frac{\hat{\sigma}}{2\hat{\epsilon}}, \tag{2.108}$$

$$b_2 = -\frac{1}{4\hat{\epsilon}}\frac{d\hat{\epsilon}}{dx} + \frac{\hat{\sigma}}{2\hat{\epsilon}}, \tag{2.109}$$

$$\hat{\epsilon}(x) = \epsilon(z(x)), \qquad \hat{\sigma}(x) = \sigma(z(x)).$$

Note that $z(x) = \int_0^x 1/\sqrt{\hat{\epsilon}(x')\mu_0}\,dx'$. The initial condition for $\mathcal{G}_2(x,t)$ is (cf. (2.21))

$$\mathcal{G}_2(x,t=x) = \beta_2(x), \tag{2.110}$$

where

$$\beta_2(x) = -\frac{1}{2}b_1\alpha_1, \tag{2.111}$$

$$\alpha_1(x) = \exp(-\int_0^x b_1(x')\,dx')$$

$$= [\frac{\epsilon(0)}{\hat{\epsilon}(x)}]^{\frac{1}{4}} \exp[-\frac{1}{2}\int_0^x (\hat{\sigma}(x')/\hat{\epsilon}(x'))\,dx']. \tag{2.112}$$

In an inverse problem using only reflection data up to a certain time T, the boundary values of the Green's functions are provided by (cf. (2.19) and (2.39))

$$\mathcal{G}_2(0,t) = R(t), \qquad t \in [0,T], \tag{2.113}$$

$$\mathcal{G}_1(0,t) = 0. \tag{2.114}$$

For the analysis of the stability, existence and uniqueness it is useful to rewrite the PDEs for the Green's functions in integral form. Integrating (2.105) along the characteristic lines $t = \pm x + \mathrm{const}$, yields

$$\mathcal{G}_1(x,t) + \int_0^x (B\mathcal{G})_1(x', x'+t-x)\,dx' = 0, \tag{2.115}$$

$$\mathcal{G}_2(x,t) + \int_0^x (B\mathcal{G})_2(x', -x'+t+x)\,dx' = R(x+t) \tag{2.116}$$

($(B\mathcal{G})_1$ denotes the first component of the vector $B\mathcal{G}$, etc). Setting $t = x$ in (2.116) and using the condition (2.110), yields

$$\beta_2(x) = R(2x) - \int_0^x (B\mathcal{G})_2(x', -x'+2x)\,dx'. \tag{2.117}$$

Stability estimates. Stability estimates (i.e., estimates of how much a small perturbation of the scattering data will influence the reconstruction result) will be considered first. Assume that the parameters and their z derivatives are bounded, i.e.

$$0 < m_0 \le \epsilon(z) \le M_1, \qquad |\frac{d\epsilon}{dz}| \le M_2,$$
$$0 \le \sigma(z) \le M_3, \qquad |\frac{d\sigma}{dz}| \le M_4.$$

Then one has the following inequalities (cf. (2.108), (2.109) and (2.111))

$$|b_i| \le \frac{M_2}{4m_0\sqrt{m_0\mu_0}} + \frac{M_3}{2m_0} \equiv M, \qquad i = 1,2,$$
$$|\beta_2| \le \frac{1}{2} M \sqrt[4]{\frac{\epsilon(0)}{m_0}} \equiv P.$$

Integrating (2.105) along the characteristic lines $t = x +$ constant (or $t = -x +$ constant) from (x,t) to $(0, t-x)$ (or $(\frac{x+t}{2}, \frac{x+t}{2})$), yields

$$\mathcal{G}_1(x,t) + \int_{t-x}^{t} (B\mathcal{G})_1(t' - t + x, t')\, dt' = 0, \tag{2.118}$$
$$\mathcal{G}_2(x,t) - \int_{\frac{x+t}{2}}^{t} (B\mathcal{G})_2(x + t - t', t')\, dt' = \beta_2(\frac{x+t}{2}). \tag{2.119}$$

Let

$$\bar{\mathcal{G}}(t) = \max_{k=1,2} \max_{0\le x \le t} |\mathcal{G}_k(x,t)|, \tag{2.120}$$

then (cf. (2.118) and (2.119))

$$\bar{\mathcal{G}}(t) \le P + 2M \int_0^t \bar{\mathcal{G}}(t')\, dt',$$

or,

$$\bar{\mathcal{G}}(t) \le P\exp(2Mt) \le P\exp(2MT) \equiv G_0, \qquad t \le T. \tag{2.121}$$

We give stability estimates for the inverse problem of reconstructing the permittivity when the conductivity is known, using the reflection data $R(t)$, $t \in [0,T]$. Assume that in the domain $D(T) = \{(x,t) | 0 \le x \le T/2, x \le t \le T - x\}$ there exist two solutions $\hat{\epsilon}^{(k)}(x)$, $z^{(k)}(x)$, $k = 1,2$, corresponding to two sets of reflection data $R^{(k)}(x)$, $t \in [0,T]$, $k = 1,2$, respectively. Denote the corresponding other quantities by $\hat{\sigma}^{(k)}(x)$, $\mathcal{G}_1^{(k)}(x)$, $\mathcal{G}_2^{(k)}(x)$, $b_1^{(k)}(x)$, $b_2^{(k)}(x)$, $\beta_2^{(k)}(x)$, etc. Let $\tilde{\epsilon} = \hat{\epsilon}^{(1)} - \hat{\epsilon}^{(2)}$, $\tilde{\sigma} = \hat{\sigma}^{(1)} - \hat{\sigma}^{(2)}$, $\tilde{\mathcal{G}}_1 = \mathcal{G}_1^{(1)} - \mathcal{G}_1^{(2)}$, $\tilde{R} = R^{(1)} - R^{(2)}$, $\tilde{b}_1 = b_1^{(1)} - b_1^{(2)}$, etc. (a stacking symbol ˜ is used to indicate the difference between the two sets of quantities; note that $\hat{\sigma}^{(k)} \equiv \sigma(z^{(k)}(x))$ and thus the difference $\tilde{\sigma}$ is caused by the difference $\tilde{z}$). After straightforward (but lengthy) algebraic estimates for (2.115)–(2.117), one can show that (He et al. 1993c; $C^m(D)$ denotes m times continuously differentiable functions in the domain D, and $C(D)$ continuous functions in the domain D)

$$\|\hat{\epsilon}^{(1)}(x) - \hat{\epsilon}^{(2)}(x)\|_{C[0,T/2]} \leq \bar{C}_1 \|R^{(1)}(t) - R^{(2)}(t)\|_{C[0,T]}, \tag{2.122}$$

$$\|\hat{z}^{(1)}(x) - \hat{z}^{(2)}(x)\|_{C[0,T/2]} \leq \bar{C}_1 \|R^{(1)}(t) - R^{(2)}(t)\|_{C[0,T]}, \tag{2.123}$$

where

$$\bar{C}_1 = 2\exp(\frac{1}{2}TM)\exp(\bar{C}T),$$

$$\bar{C} = \max[(4M + 4G_0 + 2P)\exp(\frac{1}{2}TM) + \frac{M_4}{2m_0^2\sqrt{m_0\mu_0}} + \frac{M_3}{m_0^2}(4M_1 + 4M + 2), \frac{M_4}{2m_0\sqrt{m_0\mu_0}}, \frac{1}{2m_0\sqrt{m_0\mu_0}}].$$

From (2.122) and (2.123), one obtains the following stability estimate,

$$\|\epsilon^{(1)}(z) - \epsilon^{(2)}(z)\|_{C[0,l]} \leq \bar{C}_2 \|R^{(1)}(t) - R^{(2)}(t)\|_{C[0,T]}, \tag{2.124}$$

where $\bar{C}_2 = \bar{C}_1(1 + M_2\sqrt{M_1/m_0})$ and

$$l = \frac{T}{2}\frac{1}{\sqrt{\mu_0 M_1}}. \tag{2.125}$$

Note that the estimate (2.124) also indicates uniqueness of the solution 'in the large' (i.e., T could be large) for the inverse problem using $R(t)$ to determine $\epsilon(z)$ when $\sigma(z)$ is known, i.e., if $R^{(1)}(t) = R^{(2)}(t)$ for $t \in [0,T]$, then $\epsilon^{(1)}(z) - \epsilon^{(2)}(z)$ for $z \in [0,l]$.

Using the contraction mapping principle (cf. e.g. Krasnoselskii et al. (1972)), one can further prove the following theorem of existence and uniqueness (see He et al. (1993c) for a detailed proof):

Theorem 2.1 *If $R(t) \in C[0,T_1]$ for some $T_1 > 0$ is given, if $\sigma(z) \in C^1$ is a known function and if*

$$0 \leq \sigma(z) \leq M_3, \quad |\frac{d\sigma}{dz}| \leq M_4,$$

then there exists a $T^* = T^*(\mu_0, \epsilon(0), M_3, M_4, \|R(t)\|_{C[0,T_1]})$, such that for all $T \in (0,T^*)$, the inverse problem for the system of equations (2.115)–(2.117) has one and only one solution, i.e. there exists a unique pair of functions $\hat{\epsilon}(x)$, $z(x) \in C^1[0,T/2]$ with $\hat{\epsilon}(x) > 0$, $\frac{d}{dx}z(x) > 0$.

Correctness in the neighbourhood of the exact solution means that if the solution to the inverse problem exists for the data $R(t) \in C[0,T]$, then there exists a solution to the inverse problem for any data in a neighbourhood of $R(t)$. Again, using the contraction mapping principle one can show the following theorem of local correctness (see He et al. (1993c) for a detailed proof):

Theorem 2.2 *Suppose that for any $T > 0$ (which could be large) and $R^{(1)}(t) \in C[0, T]$ there exists a solution to the system of (2.115)–(2.117) (where the conductivity $\sigma(z)$ is a known function) in $D(T)$ and $\hat{\epsilon}(x) \geq m_0 > 0$. Then there exists a $\delta > 0$ such that for every $R^{(2)}(t) \in C[0, T]$ which satisfies*

$$\|R^{(1)}(t) - R^{(2)}(t)\|_{C[0,T]} \leq \delta,$$

there exists a unique solution to the system of (2.115)–(2.117), continuous in $D(T)$, and one has the following estimates

$$\|\hat{\epsilon}^{(1)}(x) - \hat{\epsilon}^{(2)}(x)\|_{C[0,T/2]} \leq C_1\delta,$$
$$\|\hat{z}^{(1)}(x) - \hat{z}^{(2)}(x)\|_{C[0,T/2]} \leq C_1\delta.$$

For the inverse problem of reconstructing the conductivity when the permittivity is a known function, one has corresponding theorems on stability, existence, uniqueness and local correctness.

In a similar way, one can give theorems of existence, uniqueness and stability for the Green's function approach to inverse problems of simultaneous reconstruction of $\epsilon(z)$ and $\sigma(z)$ using the following types of scattering data: reflection and transmission data for one round trip, two-sided reflection data for one round trip, and reflection data for two round trips (see He et al. (1994c) for detailed proofs).

2.3 The compact Green's function approach

In Sections 2.1 and 2.2 we have considered the use of Green's functions defined as a map from the incident field to the internal split fields and this concept is useful both for inverse problems on a finite interval (a slab) or on a semi-infinite interval (a half-space). In the derivation of the equations for the Green's functions one then uses the fact that the incident field is arbitrary (cf. Section 2.1.2). However, if we consider specifically inverse problems for a slab, we have both reflected and transmitted fields and these are also 'arbitrary'. Thus one may define alternative Green's functions, as mappings from the reflected or transmitted fields to the internal split fields. Of these, the Green's functions defined with respect to the transmitted fields turn out to be particularly useful, especially for the simultaneous reconstruction of several parameters (He 1993a). There are two major reasons for this.

(i) The Green's functions that map the transmitted field back to the internal split fields have compact support in time for non-dispersive media, i.e., they differ from zero only for a certain finite period of time (see the theorem below; the usual Green's functions have non-zero values for infinite times). Because of their compact support in time, they are referred to as *compact Green's functions* in order to distinguish them from the usual Green's functions. Such a compactness indicates that the compact Green's functions contain a priori more information than the previously defined Green's functions, and one has an additional boundary condition that is very useful for a simultaneous reconstruction of more than one parameter (see (2.136) below).

(ii) The data that may be introduced into the usual Green's function approach for the reconstruction in an inverse problem are $G^-(0,t) = R(t)$ and $G^+(l,t) = T(t)$. Thus, an inverse problem formulated by means of the usual Green's function approach leads to the problem of solving a system of PDEs with two boundary conditions at two endpoints $z = 0$ and $z = l$. Such a two-endpoint problem can only be solved in an iterative way (one then loses some of the advantage of the time domain Green's function approach as an exact and non-iterative method). The compact Green's function approach overcomes the problems associated with two endpoints, by converting the two-endpoint problem to a one-endpoint problem (see (2.139) below).

To introduce the compact Green's function approach, we consider a situation in which the impedance $Z = \sqrt{\mu_0/\epsilon}$ may have a mismatch at the back boundary $z = l$. In this case one has the following boundary relations between the split fields at $z = l$ (using the continuity of E and H)

$$\begin{cases} E^+(l^-, t+\tau_0) = \nu_1 E^t(t), \\ E^-(l^-, t+\tau_0) = r_1 \nu_1 E^t(t), \end{cases} \tag{2.126}$$

where the transmitted field $E^t(t) = E^+(l^+, t+\tau_0)$, $\tau_0 = \tau(0,l)$, and

$$\nu_1 = \frac{Z(l^+) + Z(l^-)}{2Z(l^+)}, \tag{2.127}$$

$$r_1 = \frac{Z(l^+) - Z(l^-)}{Z(l^+) + Z(l^-)}, \tag{2.128}$$

Consider the following initial-boundary value problem for the system (2.5),

$$\begin{cases} E^+(z,0) = E^-(z,0) = 0, \quad 0 < z < l, \\ \\ E^-(l^+, t+\tau_0) = 0, \\ \\ E^+(l^+, t+\tau_0) = E^t(t), \quad t > 0 \end{cases} \tag{2.129}$$

(note that there is a time τ_0 shift in the last boundary condition). The solution to this initial-boundary value problem can be expressed in terms of two *compact*

Green's functions $G^{c\pm}$ defined as mappings from the transmitted field at $z = l$ to the internal split fields according to (He 1993a)

$$\begin{bmatrix} E^+ \\ E^- \end{bmatrix}(z, t + \tau(0,z)) = \begin{bmatrix} \nu_1[a^+(z,l)]^{-1}E^t(t) \\ r_1\nu_1 a^-(z,l)E^t(t - 2\tau(z,l)) \end{bmatrix} + \begin{bmatrix} G^{c+}(z,t) * E^t(t) \\ G^{c-}(z,t) * E^t(t) \end{bmatrix}. \tag{2.130}$$

Since the relation between the transmitted field $E^t(t)$ and the incident field $E^i(t)$ is a Volterra integral equation of the second kind, the existence of such a map from E^t to $E^\pm$ can easily be deduced from (2.24) (see He (1993a)).

Following a derivation similar to the one given in Section 2.1.2 (and noting that $E^t(t)$ is an arbitrary field), one can show that $G^{c\pm}(z,t)$ satisfy the following system of PDEs,

$$\frac{\partial}{\partial z}\begin{bmatrix} G^{c+} \\ G^{c-} \end{bmatrix} - \frac{2}{c}\frac{\partial}{\partial t}\begin{bmatrix} 0 \\ G^{c-} \end{bmatrix} = \begin{bmatrix} \alpha & \beta \\ \gamma & \vartheta \end{bmatrix}\begin{bmatrix} G^{c+} \\ G^{c-} \end{bmatrix}, \tag{2.131}$$

with α, β, γ and ϑ given by (2.6). Note that the above system of PDEs is linear and homogeneous in $G^{c\pm}$.

The boundary conditions for $G^{c\pm}$ at $z = l$ are

$$G^{c+}(l,t) = 0, \qquad t > 0, \tag{2.132}$$

$$G^{c-}(l,t) = 0, \qquad t > 0. \tag{2.133}$$

The above two boundary conditions are trivial consequences of the definition (2.130) and equation (2.126).

The Green's functions $G^{c\pm}$ have the property mentioned above in (i) of having compact support in the time variable. More precisely, one has the following theorem (He 1993a):

Theorem 2.3 *$G^{c\pm}(z,t)$ defined by (2.130) have compact support in the time variable t, and one has*

$$G^{c\pm}(z,t) = 0, \qquad t > 2\tau(z,l). \tag{2.134}$$

The proof is straightforward since the system (2.131) of PDEs for $G^{c\pm}(z,t)$ is linear and *homogeneous*, and $G^{c\pm}$ in the domain of dependence of any point $P(z,t) \in \Omega \equiv \{(z,t) : 0 \le z \le l, t > 2\tau(z,l)\}$ are identically zero (see Fig. 2.3, the domain of dependence for the point P is the segment AB on the line $z = l$ where both G^{c+} and G^{c-} vanish, cf. (2.132) and (2.133)).

Through a derivation similar to the one given in Section 2.1.2, one obtains the following initial condition for G^{c-}

$$G^{c-}(z,0) = -\frac{c(z)}{2}\nu_1[a^+(z,l)]^{-1}\gamma(z)$$

$$= -\frac{c}{2}\nu_1[a^+(z,l)]^{-1}[\frac{1}{4}(\frac{\epsilon_z}{\epsilon}) + \frac{1}{2}\sqrt{\mu_0/\epsilon}\sigma]. \tag{2.135}$$

Similarly, $G^{c+}(z,t)$ at the time $t = 2\tau(z,l)^-$ has the following value

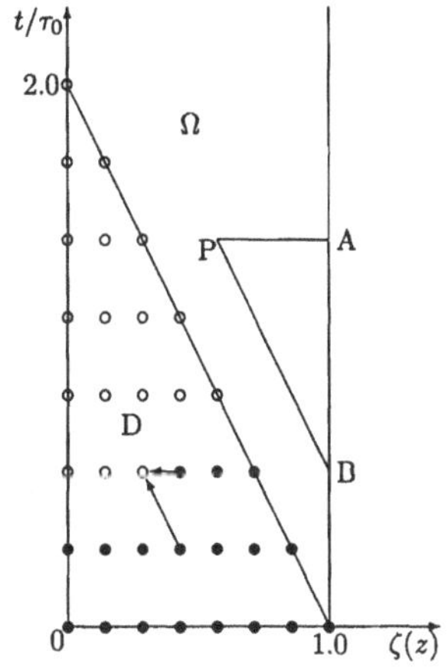

FIG. 2.3. The grid for the direct problem of calculating the compact Green's functions.

$$G^{c+}(z, 2\tau(z,l)^-) = -\frac{c(z)}{2} r_1 \nu_1 a^-(z,l)\beta(z)$$

$$= -\frac{c}{2} r_1 \nu_1 a^-(z,l)[\frac{1}{4}(\frac{\epsilon_z}{\epsilon}) - \frac{1}{2}\sqrt{\mu_0/\epsilon\sigma}]. \qquad (2.136)$$

Equations (2.131), (2.135) and (2.136) constitute a set of equations from which $G^{c+}(z,t)$ and $G^{c-}(z,t)$ can be determined for all $(z,t) \in D = \{(z,t) : 0 \leq z \leq l, t \leq 2\tau(z,l)\}$, where $G^{c\pm}(z,t)$ have non-vanishing values.

If σ and ϵ_z are continuous except at a finite number of positions z_m, $m = 1, 2, ..., M$, $G^{c-}(z,t)$ is continuous everywhere except on its characteristic curves $t = 2\int_z^{z_m} \frac{dz'}{c(z')}$, $z \leq z_m$, $m = 1, 2, ..., M$. The discontinuity in G^{c-} along its characteristics $\{\Gamma\colon \frac{\partial t}{\partial z} = -\frac{2}{c(z)}\}$ is

$$[G^{c-}]_\Gamma = [G^{c-}(z,0)]_{z_m} \cdot \exp\left[-\int_z^{z_m} \vartheta(z')\,dz'\right]$$

$$= -\frac{1}{2}\nu_1 \left[c\left(a^+(z,L)\right)^{-1}\gamma\right]_{z=z_m} \cdot \exp\left[-\int_z^{z_m} \vartheta(z')\,dz'\right], \; z \leq z_m. \qquad (2.137)$$

Similarly, the discontinuity in G^{c+} along its characteristics $t = 2\tau(z_m, l)$ (a constant) is

$$\begin{aligned}
[G^{c+}]_{t=2\tau(z_m,l)} &= [G^{c+}(z,2\tau(z,l))]_{z_m} \cdot \exp\left[-\int_z^{z_m} \alpha(z')\,dz'\right] \\
&= -\frac{1}{2}r_1\nu_1 \left[ca^-(z,l)\beta(z)\right]_{z=z_m} \cdot \exp\left[-\int_z^{z_m} \alpha(z')\,dz'\right], \ z \le z_m.
\end{aligned} \tag{2.138}$$

Note that if the impedance is matched at the back boundary $z = l$, i.e., $r_1 = 0$, then there is no discontinuity in G^{c+}.

The two independent conditions (2.135) and (2.136) can be used for a simultaneous reconstruction of the permittivity and conductivity, when $r_1 \neq 0$ (He 1993a) (in fact, also when $r_1 = 0$, i.e., if the impedance is matched at $z = l$, one can use the compact Green's function approach to reconstruct the permittivity and conductivity simultaneously, see Section 3.1.1). Note that the condition (2.136) is a consequence of the compactness of $G^{c\pm}(z,t)$. In the usual Green's function approach, there is no similar condition.

From the definition (2.130) for the compact Green's functions with $z = 0$ one sees that the first round trip data of the reflection and transmission give the boundary values of the compact Green's functions $G^{c\pm}(0,t)$, $0 \le t \le 2\tau_0$, which will be used as the input for the simultaneous reconstruction in the inverse problem. In fact, one has

$$\begin{bmatrix} G^{c+} \\ G^{c-} \end{bmatrix}(0,t) = \begin{bmatrix} W \\ V \end{bmatrix}(t), \tag{2.139}$$

where $V(t)$ is the propagator kernel, and $W(t)$ is the resolvent of the transmission kernel (see Kristensson and Krueger (1986a) and Section 1.7). Therefore, as a special case of the theorem of compactness, it follows immediately that $W(t)$ and $V(t)$ have compact support, i.e., $W(t)$ and $V(t)$ vanish after the first round trip time $2\tau_0$. Note that this property was shown in Kristensson and Krueger (1986a) by considering the Riemann function for the wave equation, and played a key role in the reconstruction algorithm described in Kristensson and Krueger (1986b) for the simultaneous reconstruction of the permittivity and conductivity. The present formalism gives new insight into this property and allows it to be derived in a simple and straightforward way.

The numerical algorithm. The numerical implementation for the direct problem is obtained by integrating (2.131) along the $-z$ direction for G^{c+} and along the characteristic direction $dz = -c(z)dt/2$ for G^{c-}. The compact Green's functions are calculated in order of the time. The area in Fig. 2.3 is thus built up by adding one horizontal row at a time, starting from the right and going to the left in each row. The numerical algorithm for the compact Green's function approach to the inverse problem is analogous to the one for the usual Green's function approach. In the inverse problem, the values $G^{c\pm}(0,t)$ can be obtained from measurable quantities (cf. the definition (2.130) of the compact Green's

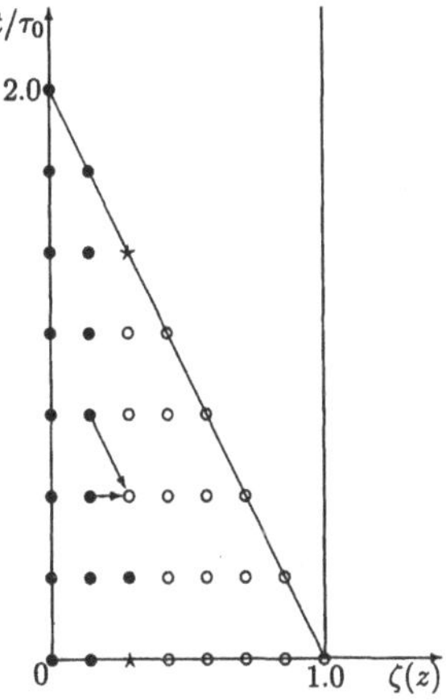

FIG. 2.4. The grid for the inverse problem of two-parameter reconstruction using the compact Green's function approach.

functions). One then propagates the boundary values $G^{c\pm}(0,t_j)$, $0 \leq j \leq N$, to $G^{c\pm}(z_i,t_j)$ at the internal layer $z = z_i$ in the order of $i = 1, 2, ..., N$ using the system of PDEs for $G^{c\pm}$. The triangle in Fig. 2.4 is filled by adding one vertical column at a time going from the bottom to the top in each column. At the layer $z = z_i$ one uses the two values $G^{c-}(z_i, 0)$ and $G^{c+}(z_i, t_{N-i})$ for the simultaneous reconstruction of the permittivity and conductivity (cf. (2.135) and (2.136)). The two stars $\star$ in Fig. 2.4 indicate the positions at which the values of $G^{c\pm}$ are used to reconstruct the permittivity and the conductivity at this layer.

Figure 2.5 shows an example of a numerical reconstruction. The circles are the simultaneous reconstruction of the permittivity $\epsilon(z)$ and the conductivity $\sigma(z)$ (the true profiles are given by the solid lines; the impedance has a discontinuity at $z = l$ with $r_1 = 1/3$). To check the stability of the reconstruction algorithm, data with Gaussian random noise with a root mean square signal to noise ratio (rms S/N) of 3.7 (i.e., a considerable noise) were considered. The noisy data can be used directly as input in the simultaneous reconstruction without being smoothed, and the reconstruction is shown by the dashed lines in Fig. 2.5. Note that even though the noisy data are used directly, the reconstruction of the permittivity comes out smooth, while this is not the case for the reconstruction of the conductivity. The reason is that the permittivity is determined indirectly by the algorithm (i.e. the z derivative of the permittivity is determined by the algorithm and the permittivity is obtained by an integration in z) but the conductivity is directly determined by the algorithm. One can also smooth the noisy input data first and then use the smoothed data as the input for the reconstruction. The triangle markers in Fig. 2.5 indicate the reconstruction results using the smoothed noisy data by means of a five point linear least squares

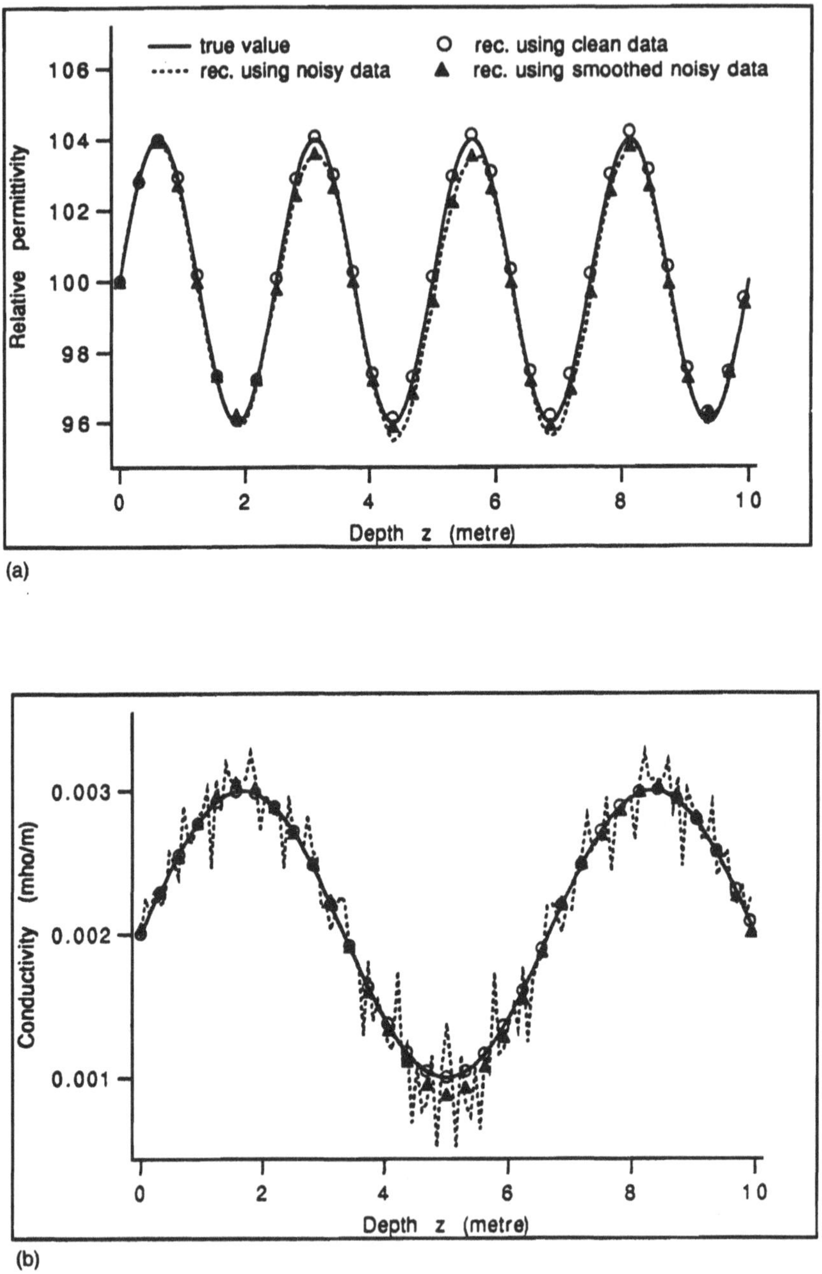

FIG. 2.5. Simultaneous reconstruction of the permittivity and conductivity using the compact Green's function approach.

smoother. The numerical results show that the reconstruction algorithm is robust and stable under noise.

2.4 The propagator approach

As shown in Section 2.1.7, the imbedding kernels and Green's functions are closely related to each other (cf. (2.100)–(2.103)). In fact all the three approaches that have been described so far, namely, the invariant imbedding approach (Chapter 1), the usual Green's function approach (Section 2.1), and the compact Green's function approach (Section 2.3) have close connections. This is easily seen in the so-called propagator approach, which was developed recently (Karlsson and Stewart 1995; Karlsson and Kreider 1996; Karlsson 1996).

Consider the same one-dimensional scattering problem (with the left side excitation) as in Section 2.1.1. The wave propagators $\mathcal{P}^{\pm}(z, z')$ are linear operators that map a right-moving component $E^{+}(z', t)$ at one position z' to the split components at another position z, i.e.,

$$\begin{bmatrix} E^{+} \\ E^{-} \end{bmatrix} \left(z, t + \int_{z'}^{z} \frac{dz''}{c(z'')} \right) = \begin{bmatrix} \mathcal{P}^{+} \\ \mathcal{P}^{-} \end{bmatrix} (z, z') E^{+}(z', t). \tag{2.140}$$

Note that there is no restriction on the relative values of z' and z in the above definition. When $z > z'$ $(z < z')$, $\mathcal{P}^{\pm}(z, z')$ propagate the field $E^{+}(z', t)$ forward (backward) in the positive (negative) z direction.

From the above definition one sees immediately that the propagators have the following properties:

$$\mathcal{P}^{+}(z, z) = \mathcal{I} \text{ (identity operator)}\,, \tag{2.141}$$

$$\mathcal{P}^{+}(z'', z') = \mathcal{P}^{+}(z'', z)\mathcal{P}^{+}(z, z'), \tag{2.142}$$

$$\mathcal{P}^{-}(z, z') = \mathcal{P}^{-}(z, z)\mathcal{P}^{+}(z, z'). \tag{2.143}$$

From (2.141) and (2.142) one obtains

$$\mathcal{P}^{+}(z', z)\mathcal{P}^{+}(z, z') = \mathcal{I}, \tag{2.144}$$

i.e., the inverse operator of the propagator $\mathcal{P}^{+}(z, z')$ is given by

$$[\mathcal{P}^{+}(z, z')]^{-1} = \mathcal{P}^{+}(z', z). \tag{2.145}$$

From the dynamic system (2.5) one obtains the following differential equations for the propagators $\mathcal{P}^{\pm}(z, z')$,

$$\frac{\partial}{\partial z} \begin{bmatrix} \mathcal{P}^{+}(z, z') \\ \mathcal{P}^{-}(z, z') \end{bmatrix} - \frac{2}{c(z)} \frac{\partial}{\partial t} \begin{bmatrix} 0 \\ \mathcal{P}^{-}(z, z') \end{bmatrix} = \begin{bmatrix} \alpha(z) & \beta(z) \\ \gamma(z) & \vartheta(z) \end{bmatrix} \begin{bmatrix} \mathcal{P}^{+}(z, z') \\ \mathcal{P}^{-}(z, z') \end{bmatrix}. \tag{2.146}$$

From (2.143) and the first equation of the above system, one obtains

$$\frac{\partial}{\partial z}\mathcal{P}^{+}(z,z') = \mathcal{N}(z)\mathcal{P}^{+}(z,z'), \tag{2.147}$$

where

$$\mathcal{N}(z) = \alpha(z) + \beta(z)\mathcal{P}^{-}(z,z). \tag{2.148}$$

Equation (2.147) has the following solution (cf. (2.141))

$$\mathcal{P}^{+}(z,z') = \exp\left[\int_{z'}^{z} \mathcal{N}(z'')\,dz''\right], \tag{2.149}$$

where the exponential function of the operator is defined by the following expansion,

$$\exp\left[\int_{z'}^{z} \mathcal{N}(z'')\,dz''\right] = \sum_{n=0}^{\infty} \frac{\left[\int_{z'}^{z} \mathcal{N}(z'')\,dz''\right]^{n}}{n!}. \tag{2.150}$$

If the impedance is continuous everywhere, the propagators $\mathcal{P}^{\pm}(z,z')$ have the following explicit representations,

$$\mathcal{P}^{+}(z,z')E^{+}(z',t) = a(z,z')E^{+}(z',t) + P^{+}(z,z',t) * E^{+}(z',t), \tag{2.151}$$

$$\mathcal{P}^{-}(z,z')E^{+}(z',t) = P^{-}(z,z',t) * E^{+}(z',t), \tag{2.152}$$

where $a(z,z') \equiv a^{+}(z,z')$ is defined by (2.92) when $z \le z'$. Note that the attenuation factor has the following property:

$$a(z',z) = \frac{1}{a(z,z')}. \tag{2.153}$$

Substituting (2.151) and (2.152) into (2.146) and following a derivation similar to the one given in Section 2.1.2, one obtains the following system of PDEs for the propagator kernels $P^{\pm}(z,z',t)$,

$$\frac{\partial}{\partial z}\begin{bmatrix} P^{+}(z,z',t) \\ P^{-}(z,z',t) \end{bmatrix} - \frac{2}{c(z)}\frac{\partial}{\partial t}\begin{bmatrix} 0 \\ P^{-}(z,z',t) \end{bmatrix} = \begin{bmatrix} \alpha(z) & \beta(z) \\ \gamma(z) & \vartheta(z) \end{bmatrix}\begin{bmatrix} P^{+}(z,z',t) \\ P^{-}(z,z',t) \end{bmatrix}, \tag{2.154}$$

together with the initial condition

$$P^{-}(z,z',0) = -\frac{c(z)}{2}a(z,z')\gamma(z). \tag{2.155}$$

From (2.141) one has

$$P^{+}(z',z',t) = 0. \tag{2.156}$$

In the case of a slab, there is no left-moving wave at the backwall $z = l$, and thus one has

$$P^{-}(l,z',t) = 0. \tag{2.157}$$

Equations (2.154)–(2.157) constitute a set of equations from which $P^{\pm}(z,z',t)$ can be determined for any $z' \in [0,l]$. One first computes $P^{\pm}(z,z',t)$ for $z \in [z',l]$

with an algorithm similar to the one for the usual Green's functions approach as described in Section 2.1.3. Using the computed $P^-(z', z', t)$, one can then compute $P^{\pm}(z, z', t)$ for $z \in [0, z']$ with an algorithm similar to the one for the compact Green's functions as described in Section 2.3.

Substituting (2.151) into (2.144), one obtains

$$a(z, z')P^+(z, z', t) + a(z', z)P^+(z', z, t) + P^+(z, z', t) * P^+(z', z, t) = 0. \tag{2.158}$$

The above equation is a Volterra equation of the second kind for $P^+(z, z', t)$ (or $P^+(z', z, t)$), which has a unique solution when $P^+(z', z, t)$ (or $P^+(z, z', t)$) is known.

If σ, ϵ_z are continuous except at a finite number of positions $z = z_m$, $m = 1, 2, ..., M$, $P^-(z, z, t)$ is continuous everywhere except on its characteristic curves $t = 2\int_z^{z_m} \frac{dz''}{c(z'')}$, $z \leq z_m$, $m = 1, 2, ..., M$ ($P^+(z, z', t)$ is continuous in the whole domain). The discontinuity in $P^-(z, z', t)$ along its characteristics $\{\Gamma\colon \frac{\partial t}{\partial z} = -\frac{2}{c(z)}\}$ is

$$\begin{aligned}
&[P^-(z, z', t)]_\Gamma \\
&= [P^-(z, z', 0)]_{z=z_m} \cdot \exp\left[-\int_z^{z_m} \vartheta(z'')\, dz''\right] \\
&= -\frac{1}{2}\left[ca(z', z)\gamma\right]_{z=z_m} \cdot \exp\left[-\int_z^{z_m} \vartheta(z'')\, dz''\right], \quad z \leq z_m.
\end{aligned} \tag{2.159}$$

From the definition of the propagators, one immediately sees that the propagator kernels are related to the imbedding kernels, the Green's functions, and the compact Green's functions by the following relations:

$$R^+(z, l, t) = P^-(z, z, t), \tag{2.160}$$
$$T^+(z, l, t) = P^+(l, z, t), \tag{2.161}$$
$$G^{\pm}(z, t) = P^{\pm}(z, 0, t), \tag{2.162}$$
$$G^{c\pm}(z, t) = P^{\pm}(z, l, t). \tag{2.163}$$

This propagator approach is very suitable for use for wave propagation in periodic structures (periodic along the propagation direction) (Karlsson and Stewart 1995) and in structures with internal sources (Åberg and Karlsson 1997).

2.5 Measurements in a coaxial cable and the reconstruction of the permittivity

In many contexts it is appropriate to model the interface between a solid dielectric and the surrounding air by a step function for the permittivity. However,

Table 2.1 *Specifications of the pulse generator PSPL Model 3050A*

Pulse width (FWHM, 50%)	60 ps
Pulse width (10%)	115 ps
Rise & falltime (10–90%)	45 ps
Amplitude	4 V
Residual ringing	±3%
Precursor	±1%
VSWR	3:1
Jitter	±2.5 ps
Repetition rate	1 MHz
Trigger risetime	0.5 ns

Table 2.2 *Specifications of the Tektronix TDR/Sampler 7S12 with S-6 head installed in a 7854 frame*

Risetime	30 ps
Displayed noise	5 mV
Jitter	10 ps
Maximum input voltage	1 V p-p
Input resistance	10 kOhm
Deflection factor accuracy	3%
Time/Div accuracy	3%

the wave-splitting and imbedding treatment is based on the assumption that the impedance is continuous and piecewise differentiable. Therefore we model the interface by means of a continuous function with a very steep slope. This situation can be handled by the reconstruction algorithms of the wave-splitting methods, as evidenced by the reconstruction results based on experimental data in this subsection.

2.5.1 *Experimental set-up*

To meet the assumptions made in the time domain one-dimensional wave-splitting approach and to have an easily controlled environment for the experiment, a coaxial component set-up is adopted. The one-dimensional variation in the permittivity is then easy to realize and the dispersion free wave propagation in the waveguide is guaranteed by the use of a TEM wave mode (such an experimental set-up is thus a mimic of the scattering phenomena due to a normally incident plane wave). A schematic experimental set-up is given in Fig. 2.6. The permittivity profile of the sample (including its thickness) can be reproduced from the measured time trace of the reflected wave with the wave-splitting inversion

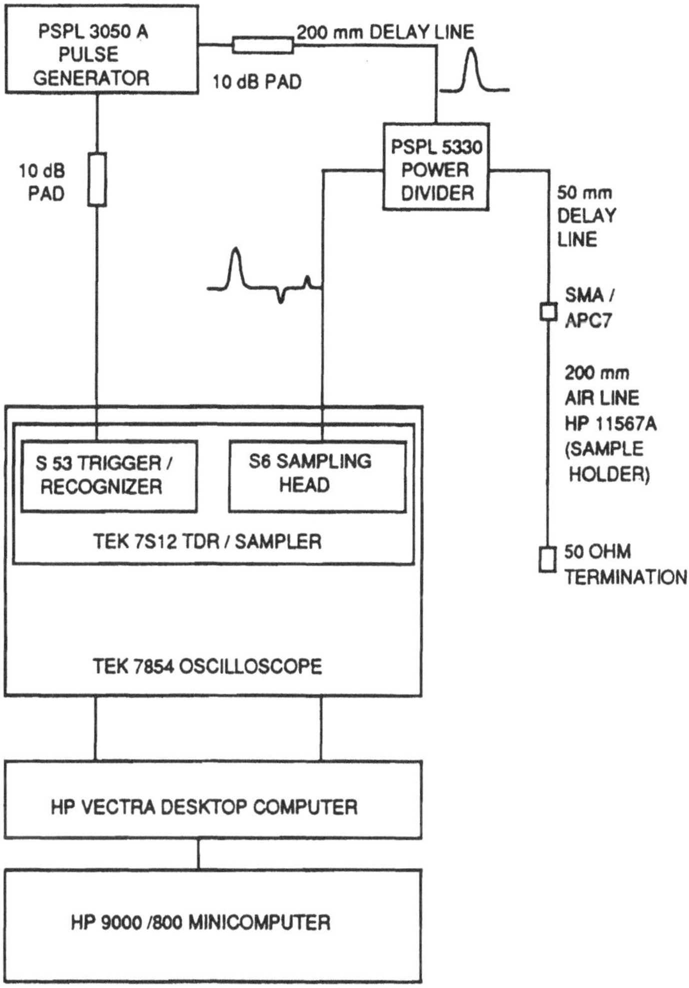

FIG. 2.6. Schematic experimental set-up.

treatment. The measurement system can be described as a pulsed radar. A short pulse is generated and sent towards the dielectric sample to be investigated. The reflected wave is recorded. The set-up is similar to a time domain reflectometer (TDR), however, a short duration pulse is used as excitation instead of a step. The longest air line available was only 200 mm long which requires short samples and, consequently, pulses with very short rise time have to be used in order to obtain adequate relative resolution.

The two main components of the system are the pulse generator (model

3050A, Picosecond Pulse Labs, USA) and the sampling oscilloscope (Tektronix TDR/Sampler 7S12 with a S-6 head installed in a 7854 frame). The specifications of these components are summarized in Tables 2.1 and 2.2, respectively. The signal from the generator is connected to the sample cell and to the oscilloscope through a power divider optimized for broadband pulses. The generator is also a source of trigger pulses. The reflection from the sample is continuously monitored and can be sampled, averaged, and stored under the control of a desktop computer. The oscilloscope averages a number of time traces to improve the signal to noise ratio. Data are then transferred to a mainframe computer for additional processing. Immediately after recording of the reflection data from an unknown sample a short is placed in its place and a new recording is made. This wave is an inverted copy of the incident wave (including all imperfections of the measurement system) and is used as a reference for the deconvolution. We refer to Fuks et al. (1990) for a more detailed description of the experiment.

One of the prominent advantages with the wave-splitting approaches is that they only require a finite time trace (one round trip) of reflection data as input data for the inversion of the permittivity profile of a stratified dielectric sample of finite length. The design of the experiment can therefore be made such that all unwanted reflections from the experimental set-up (e.g., feeder connections, waveguide crosses, sampling heads, etc.) arrive after one round trip. In this way these reflections do not affect the measurements and it is possible to avoid unwanted interference with the experimental set-up. This is not possible in measurements of the reflection from a dielectric sample in the frequency domain.

2.5.2 *Deconvolution*

The reconstruction algorithm requires the reflection kernel $R(t)$ defined as

$$V^r(t) = R(t) * V^i(t).$$

$R(t)$ has to be computed by means of a deconvolution. Although it is possible to perform the deconvolution directly in the time domain (see e.g. Rothwell and Sun (1990)), the deconvolution code used here is based on data processing in the frequency domain.

The signal reflected from the dielectric sample is sampled in the time domain and then transformed to the frequency domain through a discrete Fourier transform (DFT). Assume that the recorded time interval $[0, T]$ is partitioned into N equidistant points $t_k = k\Delta t$, $\Delta t = \frac{T}{N}$, $k = 0, 1, 2, \ldots, N-1$. Then the DFT of $R_k = R(t_k)$ is defined as

$$r_p = \sum_{k=0}^{N-1} R_k e^{-j\omega_p t_k} \equiv \sum_{k=0}^{N-1} R_k W^{pk}, \quad p = 0, 1, 2, \ldots, N-1, \qquad (2.164)$$

where $\omega_p = \frac{2\pi p}{T}, p = 0, 1, 2, \ldots, N-1$ and $W = \exp\left\{\frac{-j2\pi}{N}\right\}$. This can be computed in an efficient way using any of the FFT algorithms that are available. The inverse transform is

$$R_k = \frac{1}{N}\sum_{p=0}^{N-1} r_p e^{j\omega_p t_k}, \quad k = 0, 1, 2, \ldots, N-1.$$

The limitation of this ordinary DFT method is that the sample points $z_k = \exp\left\{\frac{i2\pi k}{N}\right\}$ are equally spaced around the unit circle in the complex z-plane, and $\Delta t \Delta f = \frac{T}{N}\frac{1}{T} = \frac{1}{N}$. Thus, the resolution in the time domain and the number of sampling points determine the resolution in the frequency domain. This is often undesirable, and the limited resolution in the frequency domain is unacceptable. This problem may be solved by using 'zero padding', which means increasing the N with a number of dummy zeros, but consequently the computing time will increase drastically. A more efficient way to remedy this problem, and to be free from the limitations inherent in the ordinary DFT, is to use the chirp z-transform (CZT) (Rabiner and Schafer 1969). The CZT has the advantage that it is possible to choose the number of time-sampling points different from the number of frequency-sampling points (the spacing and the starting point of the argument of r_p are arbitrary) and it is possible to calculate the frequency-sampling points only in the frequency interval of interest. The reference signal is also transformed to the frequency domain using the CZT algorithm. The CZT algorithm is based on the identity $pk = \frac{1}{2}[p^2 + k^2 - (k-p)^2]$. The DFT defined in (2.164) then becomes

$$r_p = W^{\frac{p^2}{2}} \sum_{k=0}^{N-1} g_k W^{-\frac{(k-p)^2}{2}}, \quad p = 0, 1, 2, \ldots, M-1, \tag{2.165}$$

where $g_k = R_k W^{\frac{k^2}{2}}, k = 0, 1, 2, \ldots, N-1$. This can be viewed as a three-step process consisting of two multiplications and one convolution, which can be performed using FFT.

Before the CZT transformation, some windowing is sometimes necessary (Harris 1978). Depending on the shape of the signals the window function is chosen so that a good compromise between frequency resolution and amplitude accuracy is obtained. The Kaiser–Bessel weighting function, defined as

$$w(t) = 1 - 1.24\cos\frac{2\pi t}{T} + 0.244\cos\frac{4\pi t}{T} - 0.00305\cos\frac{6\pi t}{T}, \tag{2.166}$$

is excellent for separation of closely spaced components with widely different levels.

A formal inverse DFT of the ration between the DFT of the reflected wave and the DFT of the incident wave is not possible because of noise at the higher frequencies (above 12 GHz). The deconvolution problem is an ill-posed problem

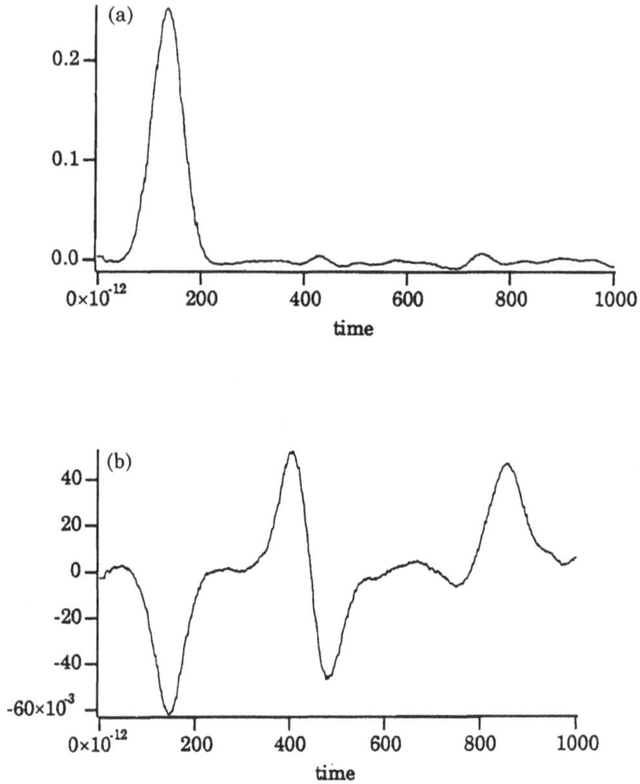

FIG. 2.7. (a) The recorded incident wave; (b) the recorded reflected wave.

and filtering is necessary to yield a stable and physically consistent result. The following regularization filter is used here to stabilize these ill-posedness effects (Parruck and Riad 1984; Nahman 1986; Bennia and Riad 1990)

$$\hat{R}(\omega) = \left[1 + \frac{\lambda\omega^4}{|\hat{V}^i(\omega)|^2}\right]^{-1} \frac{\hat{V}^r(\omega)}{\hat{V}^i(\omega)},$$

where $\hat{V}^r(\omega)$ and $\hat{V}^i(\omega)$ are the Fourier transformed reflected and incident waves, respectively. As suggested by others, the optimum value of the parameter λ can be obtained by minimizing the imaginary part of the inverse Fourier transform of $\hat{R}(\omega)$ (Nahman 1986), or by minimizing the energy in the unknown part of the inverse Fourier transform of $\hat{R}(\omega)$ (Bennia and Riad 1990). In the experiments,

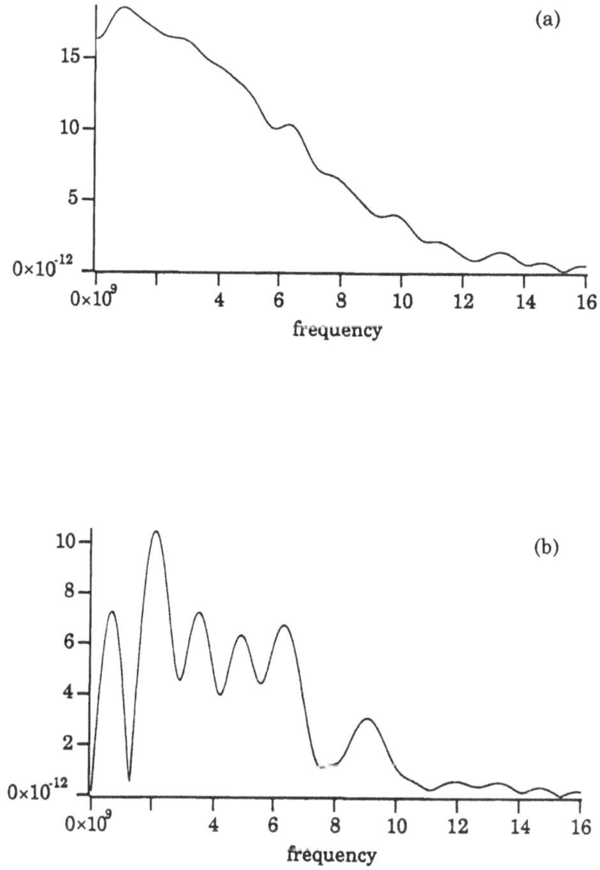

FIG. 2.8. (a) The DFT of the incident wave; (b) the DFT of the reflected wave.

to determine the optimum value of λ, two recordings of the same incident signal are measured and deconvolved, using one of them as a reference. The constant λ is then adjusted and chosen so that a smooth transition from one to zero at the high frequency end of the spectrum is obtained. The Kaiser–Bessel window (2.166) is used on the data before it is transformed into the time domain. The transformation to the time domain is made by using an inverse CZT algorithm.

Note that sometimes one can obtain an approximate scattering kernel by a simple scaling if the system is excited by a pulse of very short duration. For example, to obtain the transmission kernel $T(t)$ defined as (cf. (1.19))

$$V^t(t) = a(l)V^i(t) + T(t) * V^i(t),$$

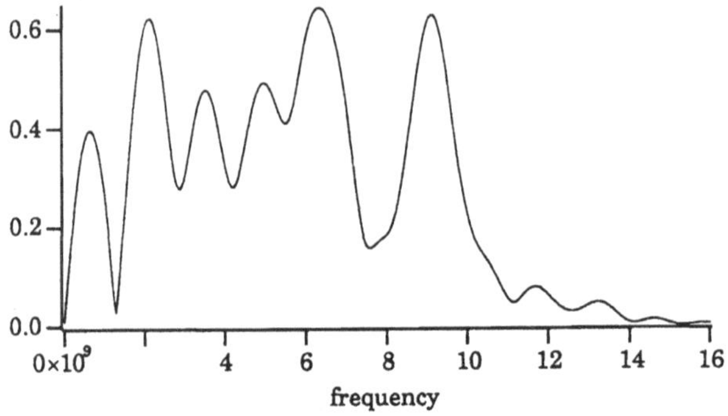

FIG. 2.9. Spectrum of the deconvolved reflected wave.

(note that there is a time shift in the notation of $V^t(t)$, i.e., $V^t(t) = V^+(l, t + \tau(0, l))$, one can perform the following scaling

$$T(t) \approx \frac{V^t(t) - a(l)V^i(t)}{\int_0^{+\infty} V^i(t')dt'}, \tag{2.167}$$

instead of a deconvolution. This approximation is valid when the duration of the incident pulse $V^i(t)$ is much shorter than the durations of 'changes' in $V^t(t) - a(l)V^i(t)$ (which is true, in particular, when the scattering is mainly due to the dispersion of the medium; note that unlike the transmitted pulse, the reflected pulse usually has a duration of the same order as the incident pulse).

2.5.3 *Reconstruction results*

The general scheme for processing the recorded data to obtain the inversion of the permittivity profile is illustrated in Figs 2.7–2.11. The recorded incident and reflected waves are shown in Fig. 2.7, and their Fourier transformations are given in Fig. 2.8. Figure 2.9 shows the deconvolved reflection kernel in the frequency domain. The time domain behaviour of the reflection kernel is obtained from the inverse CZT transform of Fig. 2.9 with the Kaiser–Bessel window (2.166), and is showed in Fig. 2.10. Figure 2.11(a) is the wave-splitting inversion of the permittivity profile from the reflection kernel (shown in Fig. 2.10). If a rectangular window is used to obtain the reflection kernel instead of the Kaiser–Bessel window, then the reconstructed permittivity profile is as shown in Fig. 2.11(b). Figure 2.11(c) shows the original dielectric sample, which consists of the two dielectric samples (of different lengths, but the same permittivity) with an air-gap in between. The length of this air-gap can be varied. The smallest possible air-gap that still gives a fair reconstruction of the vacuum value of the permittivity

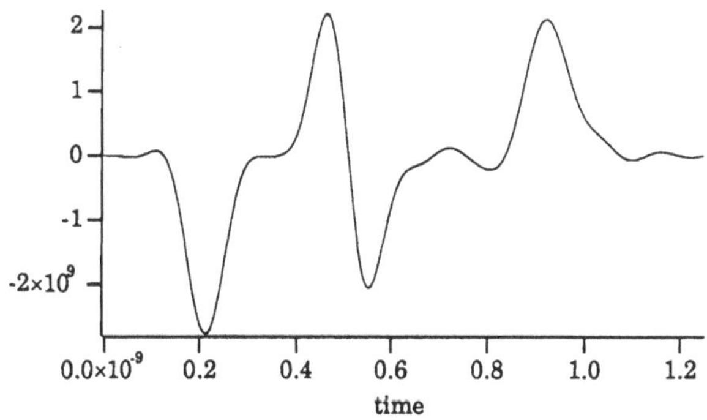

FIG. 2.10. The reflection kernel obtained from the inverse CZT transform with the Kaiser–Bessel window.

is a measure of the resolution of the data processing. Figures 2.11(a) and (b) show the trade-off between the resolution and amplitude accuracy when different windows are used. The dielectric sample has been turned round and new measurements have been made for comparison. The two reconstructions should be mirror images of each other. Such a comparison has shown that the accuracy decreases slightly when the length of the dielectric sample increases. This is due to the losses. If the losses are known a priori, then these losses can be compensated for (however, no such compensation is made here).

Figure 2.12 shows the reconstruction of another permittivity profile. The original sample is depicted in Fig. 2.12(a), and the inversion of the permittivity profile is shown in Fig. 2.12(b).

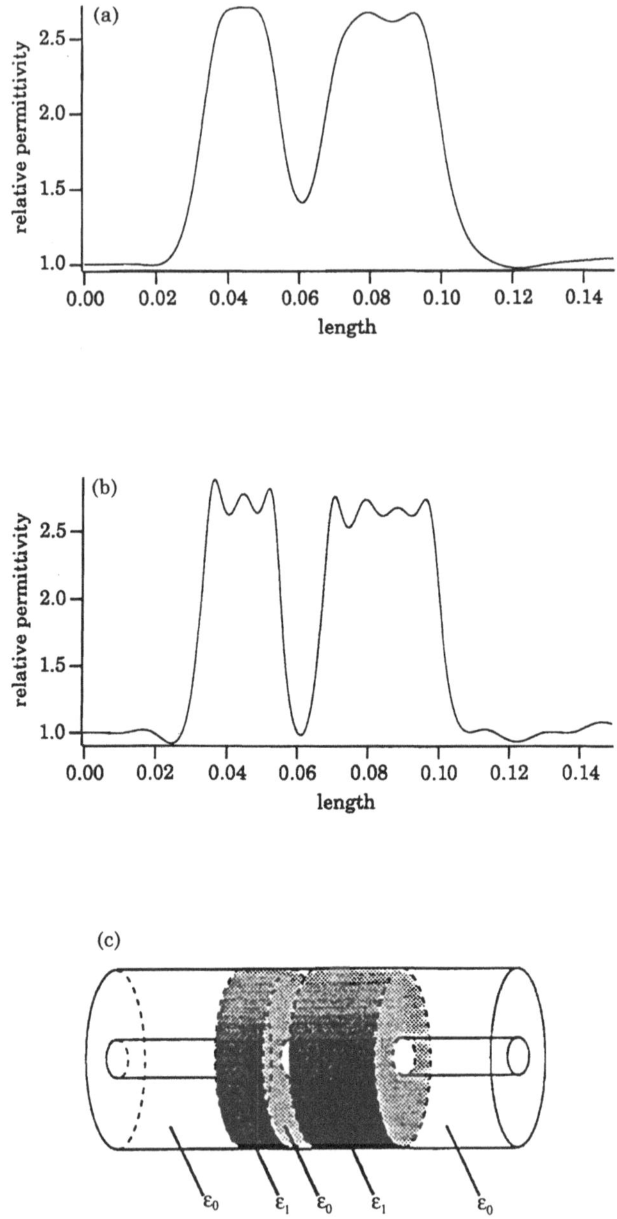

FIG. 2.11. Reconstructed permittivity profile using the reflection kernel obtained from the inverse CZT transform (a) with the Kaiser–Bessel window, or (b) with a rectangular window. The original dielectric sample is depicted in (c).

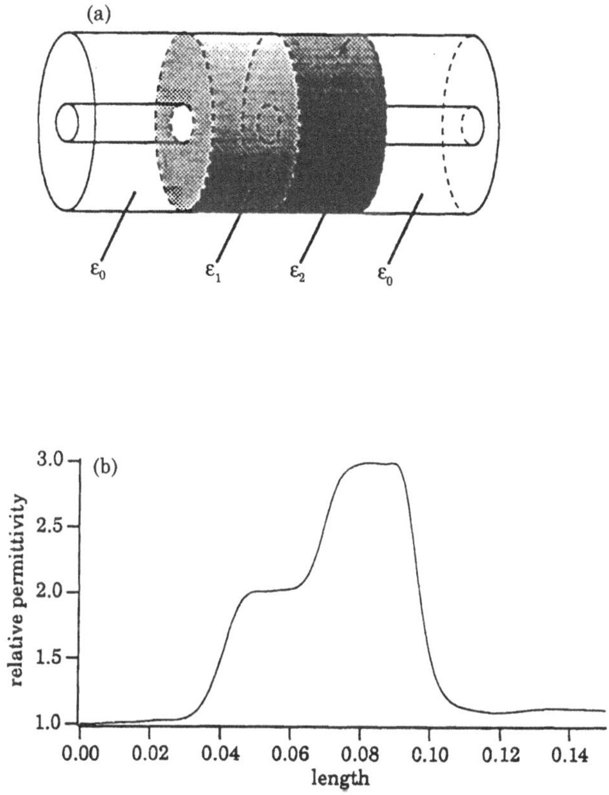

FIG. 2.12. (a) Another dielectric sample. (b) The corresponding permittivity profile reconstructed from the experimental reflection data.

Exercises for Chapter 2

Exercise 2.1. Derive the imbedding equation for the reflection kernel $R^+(z,l)$ and its initial condition by differentiating the relation (2.100) with respect to z and using the system of equations for the Green's functions.

Exercise 2.2. Write a computer program according to the inversion algorithm for the Green's function approach (Section 2.1.4) to reconstruct the conductivity with the reflection data when the permittivity and permeability are known. Test the code with the explicit solution (2.66) (as the given reflection data) for the profiles (2.47) and (2.48).

3

EXTENSIONS OF THE ONE-DIMENSIONAL WAVE-SPLITTING APPROACHES

In this chapter, the one-dimensional wave-splitting approaches described in the previous two chapters are extended and applied to structures which lead to one-dimensional problems such as propagation along non-uniform LCRG transmission lines, cases with obliquely incident plane waves, and propagation of plane electromagnetic waves in dispersive media.

3.1 Non-uniform LCRG transmission lines

A transmission line usually consists of two conductors (e.g., wires). Transmission lines are widely used in applications as, e.g., filters, impedance matching sections, couplers, as well as communication lines. Consider a lossy non-uniform transmission line of length l between $x = 0$ and $x = l$. The line is assumed to be non-dispersive, i.e., it is characterized by time independent electrical properties. Denote by $L(x)$, $C(x)$, $R(x)$ and $G(x)$ the inductance, capacitance, resistance and shunt conductance respectively per unit length at the position x. Thus the following equations (telegrapher's equations) for the voltage $V(x,t)$ and current $I(x,t)$ are obtained:

$$\frac{\partial V}{\partial x} + L(x)\frac{\partial I}{\partial t} + R(x)I = 0, \tag{3.1}$$

$$\frac{\partial I}{\partial x} + C(x)\frac{\partial V}{\partial t} + G(x)V = 0. \tag{3.2}$$

The condition for the above telegrapher's equations to be valid is that the electromagnetic field surrounding the two conductors is approximately a TEM mode, i.e., both the electric and magnetic fields are orthogonal to the propagation direction. This is expected to be true if the distance between the two conductors is much less than the characteristic distance over which the signal varies significantly. Note that the above system of first-order PDEs does not reduce to a second-order hyperbolic equation for a single function $V(x,t)$ or $I(x,t)$ unless G and C are constants or R and L are constants. The coefficients depend here only on the space variable, but we note that extensions to systems where the coefficients depend both on the space and time variables have been studied recently by the present approach (He 1993b; Åberg et al. 1995). In this section the general non-uniform LCRG transmission line problem is considered, i.e. all four parameters L, C, R and G are x-dependent. One notes that with a suitable reinterpretation of the model quantities, a special case of the above telegrapher's equations corresponds to Maxwell's equations for a normally incident plane wave propagating in a dielectric medium where the permittivity, conductivity and permeability vary with the depth (He and Ström 1992).

The split components are defined by (cf. (2.3))

$$V^{\pm}(x,t) = (1/2)[V \pm ZI], \tag{3.3}$$

where the impedance Z is given by

$$Z = \sqrt{L/C} = 1/Y, \tag{3.4}$$

where Y is the admittance. V^+ is the right-moving voltage, while V^- is the left-moving voltage on a uniform LC line. As in the previously studied cases the split components $V^{\pm}$ no longer have a direct physical interpretation as right- and left-moving waves on a non-uniform LCRG line. Equation (3.3) is nevertheless a useful change of basis for the hyperbolic equations (3.1) and (3.2) from the variables $V(x,t)$, $I(x,t)$ to $V^{\pm}(x,t)$. One notes that the splitting (3.3) is different from the splitting which is often used for a second-order hyperbolic equation (denoted $V_1^{\pm}$, see Chapter 1), namely

$$\begin{aligned} V_1^{\pm}(x,t) &= (1/2)[V \mp c\int_{-\infty}^{t} \partial_x V(x,t')\,dt'] \\ &= V^{\pm} \pm (1/2)cR\int_{-\infty}^{t} I(x,t')\,dt', \end{aligned} \tag{3.5}$$

where $c = 1/\sqrt{LC}$ is the wavefront velocity. On a lossless part of the line the splitting (3.3) is identical to the splitting (3.5), while on a lossy part (when $R \neq 0$) they are different. Therefore, in general the dynamic system for the split components will not be the same in the two cases. As shown in He and Ström (1992), the splitting (3.3) is more convenient to use for a non-uniform LCRG line than the splitting (3.5). For instance, the imbedding equation for the reflection kernel associated with the splitting (3.3) has a simpler form than the one for the reflection kernel associated with the splitting (3.5).

In this section we consider how the wave-splitting approach applies to the problems of electrical parameter reconstruction, signal restoration, and internal sources reconstruction for a non-uniform transmission line.

3.1.1 *Reconstruction of the electrical parameters*

It is assumed that the non-uniform LCRG line is connected to two uniform LC lines at the two ends, and one has

$$R = G = 0, \quad x < 0, \text{ or } x > l. \tag{3.6}$$

The incident signal is assumed to impinge on the left endpoint $x = 0$ at $t = 0$ from the left side. This leads to the initial conditions

$$I(x,0) = 0, \quad V(x,0) = 0, \quad \text{for } x > 0. \tag{3.7}$$

For simplicity of analysis, it is assumed in this subsection that the impedance $Z(x)$ is continuous everywhere except at $z = l$ and piecewise continuously differentiable on the portion $0 < z < l$ of the line. From (3.1)–(3.3) one obtains the following dynamic equations for the split voltages

$$\frac{\partial}{\partial x}\begin{bmatrix} V^+ \\ V^- \end{bmatrix} + \frac{1}{c(x)}\frac{\partial}{\partial t}\begin{bmatrix} V^+ \\ -V^- \end{bmatrix} = \begin{bmatrix} \alpha & \beta \\ \gamma & \vartheta \end{bmatrix}\begin{bmatrix} V^+ \\ V^- \end{bmatrix}, \tag{3.8}$$

where

$$\alpha = -\frac{1}{2}(GZ + RY) + \frac{1}{2}\frac{d}{dx}(\ln Z), \tag{3.9}$$

$$\beta = -\frac{1}{2}(GZ - RY) - \frac{1}{2}\frac{d}{dx}(\ln Z), \tag{3.10}$$

$$\gamma = \frac{1}{2}(GZ - RY) - \frac{1}{2}\frac{d}{dx}(\ln Z), \tag{3.11}$$

$$\vartheta = \frac{1}{2}(GZ + RY) + \frac{1}{2}\frac{d}{dx}(\ln Z). \tag{3.12}$$

In the region $x < 0$, V^+ and V^- correspond to the incident voltage and the reflected voltage, respectively. In the region $x > l$, V^+ is the transmitted voltage. From the dynamic equations, one sees that the interactions between the split voltages are through the factors $\beta(x)$ and $\gamma(x)$, and furthermore $\gamma(x) = 0$ is the distortionless condition for a non-uniform line, i.e., signals propagate without distortion, however, with possible attenuation, along the line (Lundstedt 1995a).

The compact Green's functions are defined by (cf. Section 2.3)

$$V^+(x, t + \tau(0,x)) = a^+(x)V^t(t) + G^{c+}(x,t) * V^t(t), \tag{3.13}$$

$$V^-(x, t + \tau(0,x)) = a^-(x)V^t(t - 2\tau(x,l)) + G^{c-}(x,t) * V^t(t), \tag{3.14}$$

where

$$\begin{aligned} a^+(x) &= a^+(l)\exp\left(\int_l^x \alpha(x')\,dx'\right) \\ &= \frac{Z(l^+) + Z(l^-)}{2Z(l^+)}\sqrt{\frac{Z(x)}{Z(l)}}\exp\left(\tfrac{1}{2}\int_x^l (GZ + RY)\,dx'\right), \end{aligned} \tag{3.15}$$

$$a^-(x) = a^-(l)\exp\left(\int_l^x \vartheta(x')\,dx'\right)$$

$$= \frac{Z(l^+) - Z(l^-)}{2Z(l^+)}\sqrt{\frac{Z(x)}{Z(l)}}\exp\left(-\tfrac{1}{2}\int_x^l (GZ + RY)\,dx'\right). \quad (3.16)$$

The PDEs for the compact Green's functions $G^{c\pm}(x,t)$ are obtained by substituting the definitions (3.13) and (3.14) into the dynamic system (3.8). One obtains

$$\frac{\partial}{\partial z}\begin{bmatrix} G^{c+}(z,t) \\ G^{c-}(z,t) \end{bmatrix} - \frac{2}{c(z)}\frac{\partial}{\partial t}\begin{bmatrix} 0 \\ G^{c-}(z,t) \end{bmatrix} = \begin{bmatrix} \alpha(z) & \beta(z) \\ \gamma(z) & \vartheta(z) \end{bmatrix}\begin{bmatrix} G^{c+}(z,t) \\ G^{c-}(z,t) \end{bmatrix}. \quad (3.17)$$

For the inverse problem, it is assumed that one knows the boundary values $G^{c\pm}(0,t)$ (cf. (2.139)), as well as the factors $a^{\pm}(0)$. One then propagates the boundary values into the internal values of the compact Green's functions along the characteristic curves using the system (3.17) of PDEs. When the impedance has a mismatch at $x = l$, any two of the parameters $Z(x)$, $c(x)$, $R(x)$, and $G(x)$ can then be reconstructed from the following two conditions (see Section 2.3):

$$G^{c+}(x, 2\tau(x,l))/a^-(x) = -\tfrac{1}{4}c\left[-GZ + RY - \left(\frac{d}{dx}Z\right)Y\right], \quad (3.18)$$

$$G^{c-}(x,0)/a^+(x) = -\tfrac{1}{4}c\left[+GZ - RY - \left(\frac{d}{dx}Z\right)Y\right]. \quad (3.19)$$

Equation (3.18) is useful only when $a^-(x) \neq 0$, i.e., when there is a discontinuity of the impedance at the back (cf. (3.16)). When the impedance is matched at $x = l$, one can reconstruct two parameters from the following two conditions (Lundstedt and Ström 1996):

$$G_t^{c+}(x, 2\tau(x,l))/G^{c-}(x, 2\tau(x,l))$$

$$= -\tfrac{1}{4}c\left[-GZ + RY - \left(\frac{d}{dx}Z\right)Y\right], \quad (3.20)$$

$$G^{c-}(x,0)/a^+(x) = -\tfrac{1}{4}c\left[+GZ - RY - \left(\frac{d}{dx}Z\right)Y\right] \quad (3.21)$$

($G_t^{c+} \equiv \partial_t G^{c+}$). Time-differentiation will be used to obtain useful conditions for simultaneous reconstruction of several parameters in several other places of the book (cf. Section 4.1; for noisy scattering data a regularization technique is needed to handle the second or higher order time-differentiations). Equation

Table 3.1 *Conditions for a unique simultaneous reconstruction of two parameters*

Parameters	Condition for uniqueness
Z and R (or Z and G)	no restriction
c and R (or c and G)	$\frac{d}{dx}Z(x) \neq 0$
c and Z	$GZ - RY \neq 0$
R and G	$\frac{d}{dx}Z(x) \neq 0$ or $GZ - RY \neq 0$

(3.20) is less restrictive, but it still requires $G^{c-}(x, 2\tau(x,l)) \neq 0$, which is equivalent to

$$\gamma(l^-) = \frac{1}{2}[GZ - RY - \frac{d}{dx}(\ln Z)]|_{x=l^-} \neq 0. \tag{3.22}$$

For one-parameter reconstruction, there is no restriction in the reconstruction of $Z(x)$, $R(x)$, or $G(x)$. However, the condition $\frac{d}{dx}Z(x) \neq 0$ or $G(x)Z(x) - R(x)Y(x) \neq 0$ is necessary for a unique reconstruction of c at the position x. Table 3.1 gives the conditions for a unique simultaneous reconstruction of two parameters. Note that the simultaneous reconstruction of the two dissipation parameters $R(x)$ and $G(x)$ is more difficult and necessitates a special treatment that includes numerical differentiation (Lundstedt and Ström 1996).

It may be shown that all the boundary values of the compact Green's functions for both left- and right-sided incidences (which are all possible inputs for an inverse problem) can be uniquely determined if one knows the following two functions

$$c(s)\beta(s)e^{\int_0^s (GZ+RY)c\,ds'} \quad \text{and} \quad c(s)\gamma(s)e^{-\int_0^s (GZ+RY)c\,ds'}$$

(where $s(x) = \tau(0,x)$ and $c(s) \equiv c(x(s))$, etc.), as well as the two constants $\tau(0,l)$ and $e^{\int_0^l (GZ+RY)\,dx}$, cf. Lundstedt (1995b). Therefore, one can only reconstruct up to two unknown electrical parameters of a non-uniform transmission line with any possible measurement of the scattering data at the two ends.

Numerical illustrations. The profiles are chosen as shown by the solid lines in Figs 3.1 and 3.2. The derivative of Z is zero at $x \approx 0.75$ and thus one expects problems in reconstructing c together with R (or G) there. $GZ - RY$ is zero at $x \approx 0.85$ and similarly that causes the reconstruction of c and Z to fail there. In order to illustrate the convergence, reconstructions with 25, 50 and 100 discretization steps are plotted. The dotted, dashed and dot-dashed lines refer to reconstructions with 25, 50 and 100 discretization steps, respectively. Figure 3.1

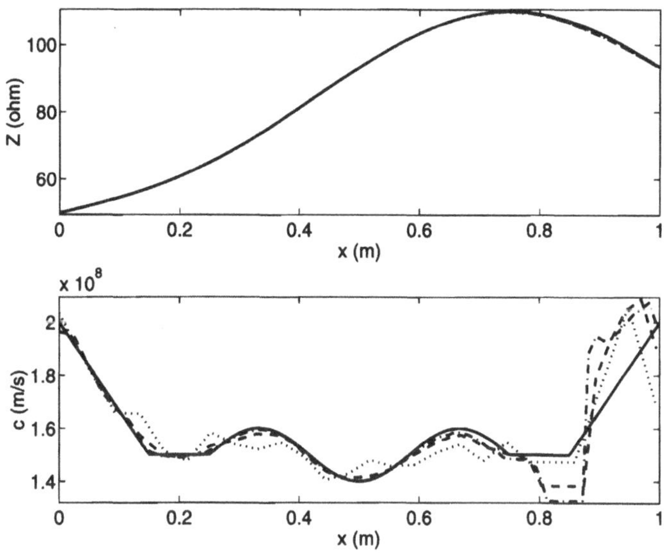

FIG. 3.1. Simultaneous reconstruction of $Z(x)$ and $c(x)$.

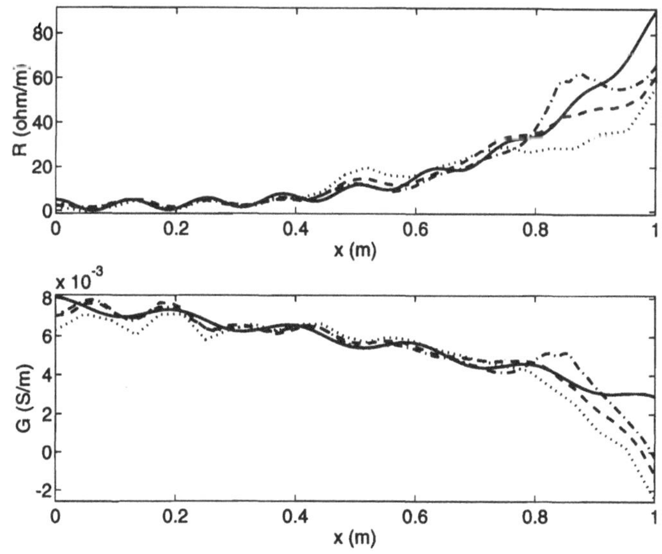

FIG. 3.2. Simultaneous reconstruction of $R(x)$ and $G(x)$.

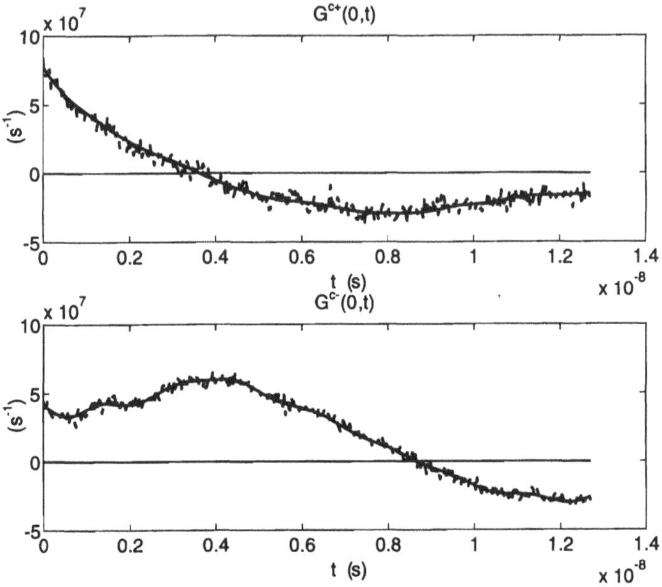

FIG. 3.3. $G^{c\pm}(0,t)$ with Gaussian noise for the discontinuous line. Solid line: true values. Dashed line: noisy data.

shows the simultaneous reconstruction of $Z(x)$ and $c(x)$, while Fig. 3.2 shows the simultaneous reconstruction of $R(x)$ and $G(x)$. The failure of the reconstructions around points which violate the conditions in Table 3.1 is evident in these figures.

In order to indicate how sensitive the reconstruction is to noise, Z and R are reconstructed from noisy $G^{c\pm}(0,t)$. The noise is of Gaussian type with standard deviations equal to 5% of the peak values of $G^{c+}(0,t)$ and $G^{c-}(0,t)$, respectively. The reconstruction is made both with a line that is continuous and discontinuous, respectively, at the endpoint $x = l$. The reconstructions are performed with 400 discretization steps. $G^{c\pm}(0,t)$ are smoothed before the reconstruction with a Gaussian distribution of width 0.05. In Fig. 3.3, the clean and noisy Green's functions are illustrated for the discontinuous line. Figure 3.4 shows the simultaneous reconstruction of the parameters $Z(x)$ and $R(x)$. In the continuous case, the reconstructed parameter $R(x)$ is smoothed with a Gaussian distribution of width 0.12. It is clearly seen that the presence of a discontinuity at the end stabilizes the reconstruction considerably (for the clean data, the reconstructed parameters are indistinguishable from the true values on the scale of Fig. 3.3).

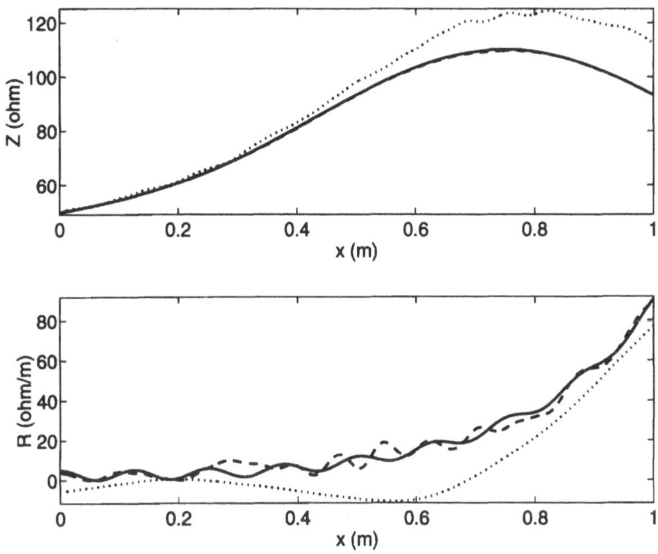

FIG. 3.4. Simultaneous reconstruction of $Z(x)$ and $R(x)$ with noisy data. Dashed line: reconstruction with discontinuous impedance, $Z(l^+) = \frac{1}{2}Z(l^-)$. Dotted line: reconstruction with continuous impedance.

3.1.2 *Signal restoration after transmission through a non-uniform LCRG line*

In this subsection, the problem of restoring an incident signal after transmission through a known non-uniform LCRG line is considered, and it is shown that the compact Green's function approach provides the solution to this problem (Lundstedt and He 1994).

As described in Section 2.3 and the previous subsection, the compact Green's functions $G^{c\pm}(x,t)$ and the factor $a^+(0)$ can be calculated if the electrical parameters of the non-uniform transmission line are known. Thus, the incident signal $V^i(t)$ at $x = 0$ can be restored from the received signal $V^t(t)$ at $x = l$ by the following formula (obtained by putting $x = 0$ in (3.13))

$$V^i(t) = a^+(0)V^t(t) + G^{c+}(0,t) * V^t(t). \tag{3.23}$$

In order to illustrate some numerical features of this solution, the parameters of the non-uniform line are chosen as shown in Fig. 3.5. The length of the non-uniform line is $l = 10$ metres. Note that the resistance R and the conductance G have jumps at the two endpoints as well as at the middle point in this example. The corresponding compact Green's functions $G^{c+}(x,t)$ and $G^{c-}(x,t)$ are shown in Fig. 3.6. The factor $a^+(0)$ is 12.44. The incident signal at the left endpoint $x = 0$ is chosen arbitrarily to be a combination of a few rectangular waves

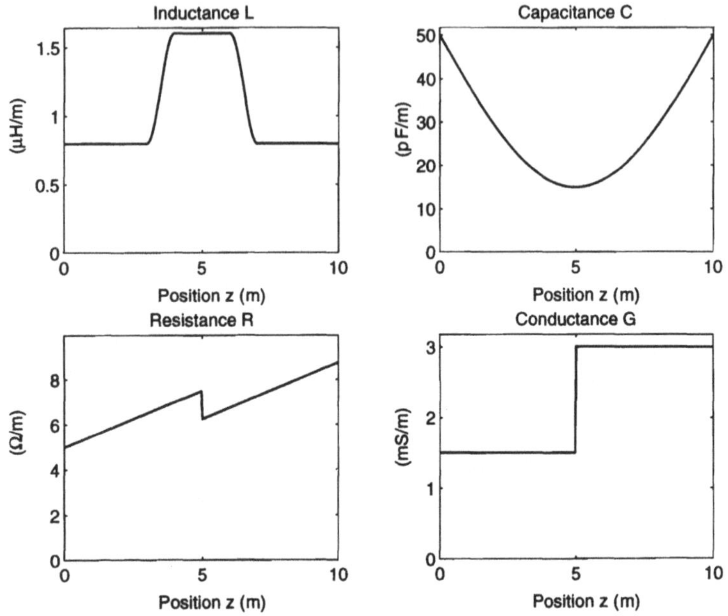

FIG. 3.5. The profile for a non-uniform LCRG line.

followed by one period of a sine wave, as shown by the solid line in Fig. 3.7(a). After transmission through this non-uniform LCRG line, the signal is distorted and is received at the right endpoint $x = l$. The received signal at $x = l$ is given by the solid line in Fig. 3.7(b), which is calculated using the usual Green's function approach (see Section 2.1 for the usual Green's function approach). To show the effect of noise on the restoration, a random noise is added to the received signal as shown by the dashed line in Fig. 3.7(b). Using (3.23), the noisy distorted signal is restored and the restoration result is shown by the dashed line in Fig. 3.7(a). In this restoration, 256 data points are used to calculate $G^{c\pm}$.

One may wish to design a circuit to restore the signal instantaneously. If $h(t)$ represents the impulse response of a linear, time-invariant circuit, then the desired output signal V^{out} is given by the following convolution integral (Nilsson 1990)

$$V^{\text{out}}(t) = h(t) * V^{\text{input}}(t), \tag{3.24}$$

where $V^{\text{input}}(t)$ is the input signal which is zero for $t < 0$. Comparing (3.23) with (3.24), one see that if

$$h(t) = G^{c+}(0, t), \tag{3.25}$$

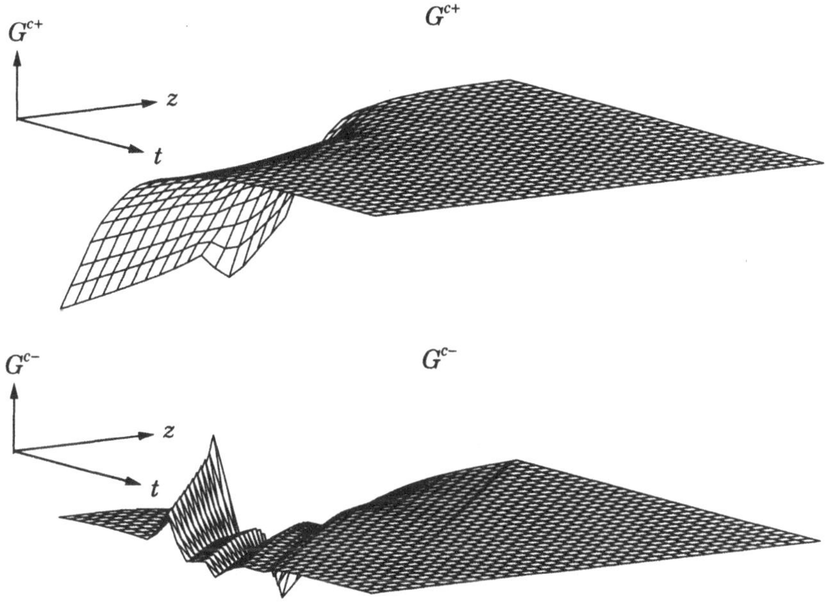

FIG. 3.6. The compact Green's functions corresponding to the non-uniform line in Fig. 3.5.

$$V^{\text{input}}(t) = V^t(t), \tag{3.26}$$

then the original signal impinging on the transmission line at $x = 0$ can be restored by the following formula (cf. (3.23))

$$V^i(t) = V^{\text{out}}(t) + a^+(0)V^{\text{input}}(t). \tag{3.27}$$

Therefore, the remaining task is to find a circuit whose impulse response behaves approximately like $G^{c+}(0,t)$.

As indicated earlier, $G^{c+}(x,t)$ is continuous everywhere in its domain of definition. For the signal restoration, only the boundary value $G^{c+}(0,t)$ is used. The values of $G^{c+}(0,t)$ corresponding to the profiles in Fig. 3.5 are plotted with the solid line in Fig. 3.8(a). Due to the compactness of $G^{c+}(0,t)$, the amplitude of $G^{c+}(0,t)$ goes smoothly to zero at the time for one round trip (i.e., for $t = 2\tau(0,l)$). In this example $G^{c+}(0,t)$ behaves approximately like an exponentially decreasing function (cf. the solid line in Fig. 3.8(a)). Using a standard root mean square fit to the first half of the solid line ($t \leq \tau_0$) in Fig. 3.8(a), the $G^{c+}(0,t)$ corresponding to the profiles in Fig. 3.5 is approximated by the following function (in SI units)

$$h(t) = -5.95 \times 10^8 e^{-t/(1.43\times 10^{-8})},$$

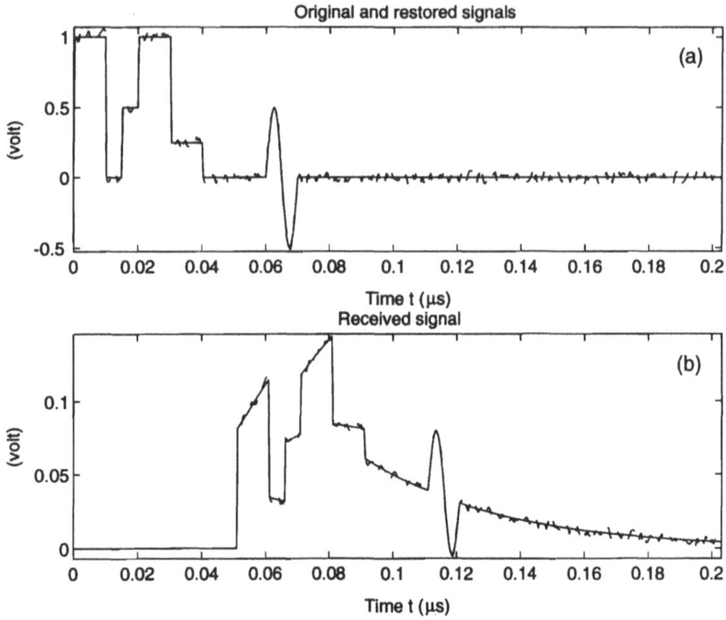

FIG. 3.7. (a) The original signal (the solid line) and the restored signal (the dashed line) using the received noisy signal (as shown in (b)). (b) The received clean signal (the solid line) and the received noisy signal (the dashed line).

which is shown by the dashed line in Fig. 3.8(a). This result may be compared with that for the voltage on the capacitance in a RC circuit, for which one has the following impulse response

$$h_0(t) = \frac{1}{R_0 C_0} e^{-t/(R_0 C_0)},$$

where R_0 and C_0 are the resistance and capacitance of the circuit, respectively. Therefore, any signal after transmission through the non-uniform line with parameters given by Fig. 3.5 can be restored approximately by the circuit shown in Fig. 3.9, where $R_0 C_0 = 1.43 \times 10^{-8}$/second and the amplification factor $A = 8.51$. Such a restoration is shown by the crosses in Fig. 3.8(b), where one can see that the approximation gives a good restoration of the original signal.

Numerical results indicate that if the loss of the non-uniform line is reasonably large and is dominated by either R or G (in the example of Fig. 3.8(a), the loss is dominated by G since $GZ >> RY$), $G^{c+}(0,t)$ behaves approximately like an exponentially decreasing function. Otherwise, $G^{c+}(0,t)$ may not behave like an exponentially decreasing function and one then needs to use some other

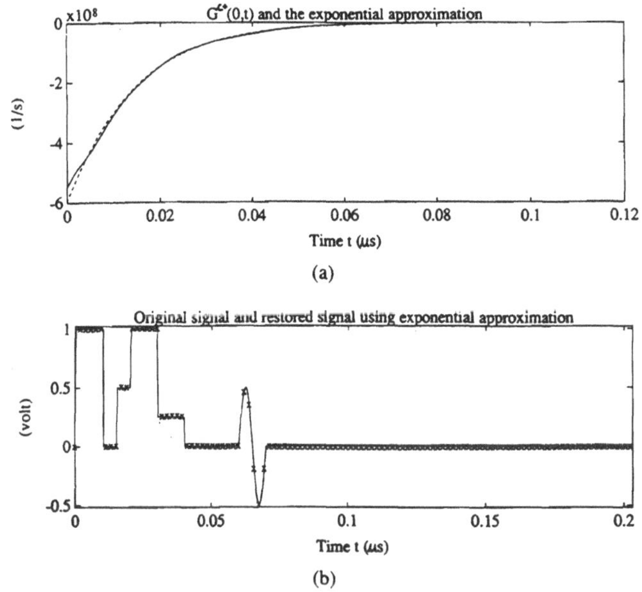

FIG. 3.8. (a) The compact Green's function $G^{c+}(0,t)$ (the solid line) and its exponential approximation (the dashed line). (b) The restored signal (crosses) using the exponential approximation (the solid line is for the original signal).

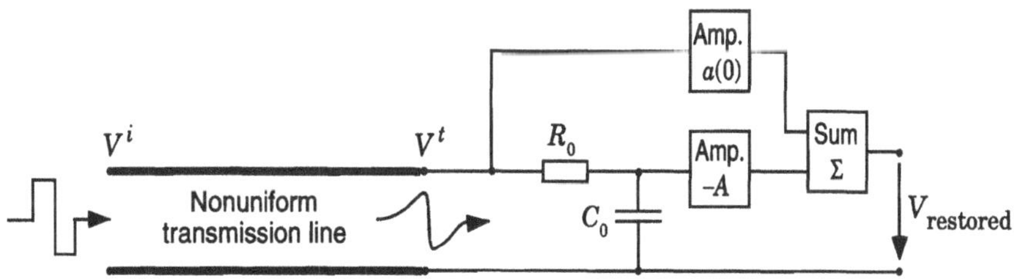

FIG. 3.9. The diagram for a restoration circuit.

circuits for instantaneous restoration, e.g. a digital filter.

3.1.3 *Reconstruction of a transient source on a transmission line*

When dealing with transmission lines one often encounters problems associated with the appearance of disturbances (such as a stroke of lightning) along the line. The waves generated by such disturbances are then usually measured at the endpoints of the line and one is interested in the problem of using the results of these measurements to infer the nature of the disturbance. The disturbance appears as a source term in the PDEs and this type of problem is therefore referred to as a source reconstruction problem. In this subsection it is shown how the compact Green's function approach can be used to solve a class of such problems in which the time dependence of the voltage (and current) source is determined, assuming that its position is known (Lundstedt and He 1997).

Consider the same non-uniform $LCRG$ transmission line of length as before in Section 3.1. A disturbance is assumed to appear on the line at the position x_0, and induces a distributed series voltage drop $V^s(t)v^s(x)$ and a distributed shunt current $I^s(t)i^s(x)$, where $v^s(x)$ and $i^s(x)$ represent the spatial distributions of the induced voltage and current, respectively. If the induced voltage and current are distributed in a narrow region around the position x_0, i.e. if

$$v^s(x) = i^s(x) = 0, \qquad \text{for } |x - x_0| > \Delta x,$$

with $\Delta x << l$, one can model the problem approximately with the following partial differential equations for the voltage $V(x,t)$ and the current $I(x,t)$,

$$\frac{\partial}{\partial x}V(x,t) + L(x)\frac{\partial}{\partial t}I(x,t) + R(x)I(x,t) = V^i(t)\delta(x - x_0), \tag{3.28}$$

$$\frac{\partial}{\partial x}I(x,t) + C(x)\frac{\partial}{\partial t}V(x,t) + G(x)V(x,t) = I^i(t)\delta(x - x_0), \tag{3.29}$$

where

$$V^i(t) = V^s(t)\int_{x_0-\Delta x}^{x_0+\Delta x} v^s(x)\,dx, \qquad I^i(t) = I^s(t)\int_{x_0-\Delta x}^{x_0+\Delta x} i^s(x)\,dx.$$

The right hand sides of (3.28) and (3.29) represent a voltage source $V^i(t)$ and a current source $I^i(t)$, respectively, at the position $x = x_0$. As before it is assumed that the non-uniform line is connected to two uniform LC lines (with reflectionless terminations) at the two ends, so that (3.6) is valid. The induced voltage and current are assumed to be zero before the time $t = 0$, i.e.,

$$V^i(t) = 0, \quad I^i(t) = 0, \qquad \text{for } t < 0. \tag{3.30}$$

Consider the inverse problem of reconstructing the voltage source $V^i(t)$ and the current source $I^i(t)$ if the received signals $V(0,t)$ and $V(l,t)$, as well as the position x_0, are known. The compact Green's function approach can conveniently be used to solve such an inverse source problem. The compact Green's function approach is then applied separately to the two parts $[0, x_0)$ and $(x_0, l]$ respectively of the line and by combining the results at x_0^- and x_0^+, the source functions can be reconstructed.

Consider first the following initial boundary value problem for the system (3.8) for the right portion $x \in (x_0, l]$ of the line

$$\begin{cases} \text{(i) } V^+(x,0) = V^-(x,0) = 0, \quad x_0 < x \le l, \\ \text{(ii) } V^-(l,t) = 0, \\ \text{(iii) } V^+(l, t + \tau(x_0, l)) = V_l(t), \quad t > 0, \end{cases} \tag{3.31}$$

where $V_l(t)$ is the measured voltage at the endpoint $x = l$ with the time shift $\tau(x_0, l)$. The solution to the above initial-boundary value problem can be expressed in terms of two compact Green's functions $G_R^{c\pm}$ defined as mappings from the measured voltage at $x = l$ to the internal split components according to (where the split fields $V^\pm$ are defined as before, cf. (3.3))

$$\begin{bmatrix} V^+(x, t + \tau(x_0, x)) \\ V^-(x, t + \tau(x_0, x)) \end{bmatrix} = \begin{bmatrix} a_R(x) V_l(t) \\ 0 \end{bmatrix} + \begin{bmatrix} G_R^{c+}(x,t) * V_l(t) \\ G_R^{c-}(x,t) * V_l(t) \end{bmatrix}, \quad x_0 < x \le l, \tag{3.32}$$

where

$$a_R(x) = \sqrt{\frac{Z(x)}{Z(l)}} \exp\left(\frac{1}{2} \int_x^l (GZ + RY)\, dx' \right), \quad x_0 < x \le l. \tag{3.33}$$

The system of equations for the compact Green's functions $G_R^{c\pm}$ are, as before, obtained by inserting the definition of $G_R^{c\pm}$ into the dynamic system (3.8) of the split voltages. One finds

$$\begin{cases} \frac{\partial}{\partial x} G_R^{c+}(x,t) = \alpha(x) G_R^{c+}(x,t) + \beta(x) G_R^{c-}(x,t), \\ \frac{\partial}{\partial x} G_R^{c-}(x,t) - \frac{2}{c(x)} \frac{\partial}{\partial t} G_R^{c-}(x,t) = \gamma(x) G_R^{c+}(x,t) + \vartheta(x) G_R^{c-}(x,t), \\ G_R^{c-}(x,0) = -\frac{1}{2} c(x)\, a_R(x)\, \gamma(x), \\ G_R^{c-}(l,t) = 0, \\ G_R^{c+}(l,t) = 0. \end{cases} \tag{3.34}$$

As shown in Section 2.3, $G_R^{c\pm}(x,t)$ can be determined in their domain of definition from the above set of equations. Furthermore, $G_R^{c\pm}(x,t)$ have compact support in the time variable t, i.e., they vanish after a finite period of time and one has

$$G_R^{c\pm}(x,t) = 0, \qquad t > 2\tau(x,l), \; x_0 < x \le l. \tag{3.35}$$

Similarly, one can consider the following initial boundary value problem for the system (3.8) for the left portion $x \in [0, x_0)$ of the line

$$\begin{cases} \text{(i) } V^+(x,0) = V^-(x,0) = 0, \quad 0 \le x < x_0, \\ \text{(ii) } V^+(0,t) = 0, \\ \text{(iii) } V^-(0,t+\tau(0,x_0)) = V_0(t), \quad t > 0, \end{cases} \tag{3.36}$$

where $V_0(t)$ is the measured voltage at the left endpoint $x = 0$ with the time shift $\tau(0,x_0)$. The solution to the above initial-boundary value problem can be expressed in terms of two compact Green's functions $G_L^{c\pm}$ defined as mappings from the measured voltage at $x = 0$ to the internal split components according to

$$\begin{bmatrix} V^+(x,t+\tau(x,x_0)) \\ V^-(x,t+\tau(x,x_0)) \end{bmatrix} = \begin{bmatrix} 0 \\ a_L(x)V_0(t) \end{bmatrix} + \begin{bmatrix} G_L^{c+}(x,t) * V_0(t) \\ G_L^{c-}(x,t) * V_0(t) \end{bmatrix}, \quad 0 \le x < x_0, \tag{3.37}$$

where

$$a_L(x) = \sqrt{\frac{Z(x)}{Z(0)}} \exp\left(\frac{1}{2}\int_0^x (GZ + RY)\, dx'\right), 0 \le x < x_0. \tag{3.38}$$

The system of equations for $G_L^{c\pm}$ is obtained by repeating the process used in deriving the system (3.34) for $G_R^{c\pm}$:

$$\begin{cases} \frac{\partial}{\partial x} G_L^{c+}(x,t) + \frac{2}{c(x)}\frac{\partial}{\partial t} G_L^{c+}(x,t) = \alpha(x)G_L^{c+}(x,t) + \beta(x)G_L^{c-}(x,t), \\ \frac{\partial}{\partial x} G_L^{c-}(x,t) = \gamma(x)G_L^{c+}(x,t) + \vartheta(x)G_L^{c-}(x,t), \\ G_L^{c+}(x,0) = \frac{1}{2}c(x)\, a_L(x)\, \beta(x), \\ G_L^{c-}(0,t) = 0, \\ G_L^{c+}(0,t) = 0. \end{cases} \tag{3.39}$$

From the above set of equations $G_L^{c\pm}$ can be determined in their domain of definition. Similarly, $G_L^{c\pm}$ have compact support in the time variable t, i.e.,

$$G_L^{c+}(x,t) = 0, \qquad t > 2\tau(0,x),\ 0 \le x < x_0. \tag{3.40}$$

Note that if the location of the sources is not known, then x_0 has to be determined from the arrival times of the signals at the two endpoints of the line. The arrival time is the first moment the signal is observed. From the knowledge of the arrival times and the wavefront velocity $c(x)$ on the line, one can uniquely determine x_0 (see Fig. 3.16 below for the influence of an error in the value of x_0 on the reconstruction). From the location x_0 and the known parameters of the non-uniform transmission line, one can calculate $G_R^{c\pm}(x_0,t)$ and $G_L^{c\pm}(x_0,t)$. The voltage source $V^i(t)$ can then be reconstructed from the received voltages at the two endpoints by means of the following relations

$$V(x_0^+,t) = V^+(x_0^+,t) + V^-(x_0^+,t)$$

$$= a_R(x_0)V_l(t) + \left[G_R^{c+}(x_0,t) + G_R^{c-}(x_0,t)\right] * V_l(t), \qquad (3.41)$$

$$V(x_0^-,t) = V^+(x_0^-,t) + V^-(x_0^-,t)$$

$$= a_L(x_0)V_0(t) + \left[G_L^{c+}(x_0,t) + G_L^{c-}(x_0,t)\right] * V_0(t), \qquad (3.42)$$

$$V^i(t) \quad = V(x_0^+,t) - V(x_0^-,t). \qquad (3.43)$$

The current source $I^i(t)$ is calculated as follows,

$$I(x_0^+,t) \quad = I^+(x_0^+,t) + I^-(x_0^+,t) = Y(x_0)\left(V^+(x_0^+,t) - V^-(x_0^+,t)\right)$$

$$= Y(x_0)\left(a_R(x_0)V_l(t) + \left[G_R^{c+}(x_0,t) - G_R^{c-}(x_0,t)\right] * V_l(t)\right), \quad (3.44)$$

$$I(x_0^-,t) \quad = I^+(x_0^-,t) + I^-(x_0^-,t) = Y(x_0)\left(V^+(x_0^-,t) - V^-(x_0^-,t)\right)$$

$$= Y(x_0)\left(-a_L(x_0)V_0(t) + \left[G_L^{c+}(x_0,t) - G_L^{c-}(x_0,t)\right] * V_0(t)\right), (3.45)$$

$$I^i(t) \quad = I(x_0^+,t) - I(x_0^-,t). \qquad (3.46)$$

Since the compact Green's functions have compact support in the time variable, one only needs to calculate $G_R^{c\pm}(x_0,t)$ for $t \in [0, 2\tau(0,x_0)]$, and $G_L^{c\pm}(x_0,t)$ for $t \in [0, 2\tau(x_0,l)]$. The voltage source $V^i(t)$ and the current source $I^i(t)$ can then be reconstructed up to an arbitrary time if the received signals $V_0(t)$ and $V_l(t)$ are known up to that time. From the present approach, one sees clearly that a non-radiating *point* source does not exist in one dimension (since if $V_l(t) = V_0(t) \equiv 0$, it then follows immediately from (3.41)–(3.46) that $V^i(t) = I^i(t) \equiv 0$). Thus it is seen that using the concept of the compact Green's functions for the two parts $[0, x_0)$ and $(x_0, l]$ respectively of the non-uniform transmission line, we arrive at a scheme which allows us to determine $V^i(t)$ and $I^i(t)$ in a straightforward way (and a simplification is achieved as compared to schemes based on the usual Green's function approach (Sun 1992)).

As a numerical example, the parameter profile of the non-uniform transmission line is chosen as shown in Fig. 3.10. The length of the line is $l = 1000$ m and the internal voltage and current sources are located at the position $x_0 = 400$ m.

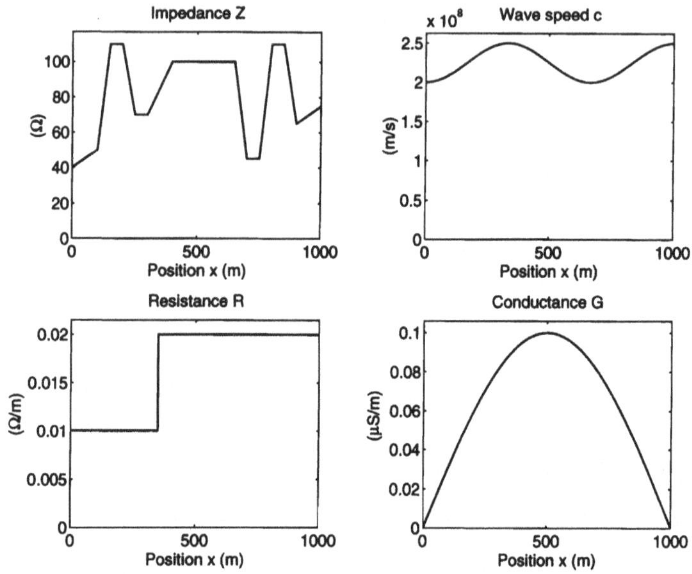

FIG. 3.10. The parameter profile for the non-uniform transmission line.

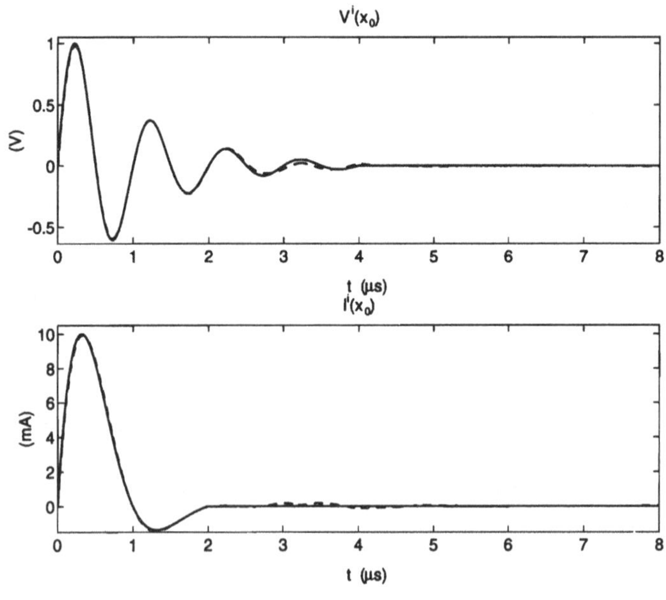

FIG. 3.11. The reconstructed voltage and current sources.

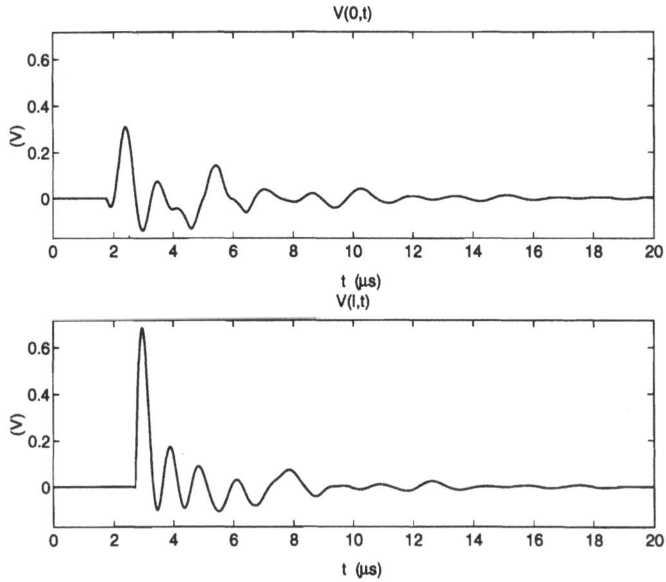

FIG. 3.12. The voltages, $V(0,t)$ and $V(l,t)$, that are received at the ends of the line.

The internal sources tested in the numerical example are shown by the solid lines in Fig. 3.11.

A finite difference algorithm (see Lundstedt and He (1997), Richtmyer and Morton (1967)) is used to compute the transient signals $V(0,t)$ and $V(l,t)$, which are then used as inputs by the inverse algorithm to reconstruct the internal sources. Thus, the direct and inverse problems are solved by means of fundamentally different methods. A discretization grid with 400 intervals in the x variable and 2400 intervals in the t variable is used in the finite difference scheme for the direct problem. The resulting total voltages at the two endpoints $x = 0$ and $x = l$ are shown in Fig. 3.12, where one sees that the received signals at the two endpoints are distorted considerably (compared to the source voltage and current) due to the non-uniformity and losses of the line.

The compact Green's functions $G_R^{c\pm}$ and $G_L^{c\pm}$ are solved from the systems (3.34) and (3.39) by integration along the characteristics, and the numerical results are shown in Figs. 3.13 and 3.14 (256 discretization intervals were used in both the space and time variables). Note that the compact Green's functions $G_R^{c\pm}(x_0,\, t)$ are zero for $t > 2\tau(x_0, l) \approx 5.4$ μs and that $G_L^{c\pm}(x_0, t)$ are zero for $t > 2\tau(0, x_0) \approx 3.5$ μs. The reconstruction of the source voltage and current is carried out through (3.41)–(3.46) using the received voltages (at the endpoints) shown in Fig. 3.12. The reconstructed voltage and current sources are given by

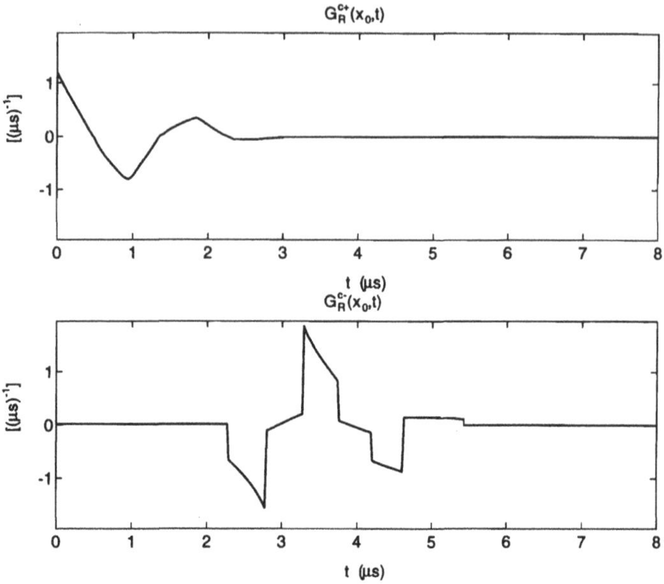

FIG. 3.13. The compact Green's functions, $G_R^{c\pm}(x_0, t)$, for the right part of the line ($x_0 = 400$ m).

the dashed lines in Fig. 3.11.

In order to see how noise influences the reconstruction, noise with a normal distribution is added to the received voltages $V(0,t)$ and $V(l,t)$ shown in Fig. 3.12 (with different samples of noise for the two signals). The standard deviation is $\sigma = 0.04$ for the noise distributions in both $V(0,t)$ and $V(l,t)$. This corresponds to a signal to noise ratio (SNR) of 6.8 for $V(0,t)$ and 18.2 for $V(l,t)$. Note that SNR has no unique definition in the time domain, but to conform reasonably well with its definition in the frequency domain, the following definition for SNR is used,

$$\mathrm{SNR} = \frac{\|S\|^2}{\|N\|^2}, \quad \|S\|^2 = \int_{t_0}^{t_0+T} [S(t)]^2 \, dt, \quad \|N\|^2 = \int_{t_0}^{t_0+T} [N(t)]^2 \, dt,$$

where $S(t)$ and $N(t)$ denote the signal and noise, respectively, t_0 is the arrival time for the signal, and T is the time period of interest, which is chosen to be twice the duration time of the voltage source ($T = 8\mu$s). Thus $\|S\|^2$ is proportional to the energy content of the signal between t_0 and $t_0 + T$. The result of the reconstruction is shown by the dashed lines in Fig. 3.15. The error in the reconstruction due to the noise is proportional to the noise level of the measured voltages, as expected (the reconstruction map is linear, cf. (3.41)–(3.46)). With

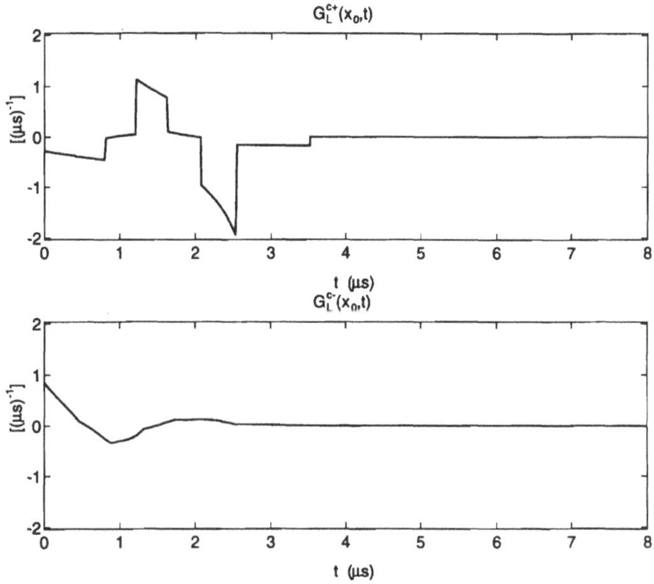

FIG. 3.14. The compact Green's functions, $G_L^{c\pm}(x_0, t)$, for the left part of the line ($x_0 = 400$ m).

a definition that the noise in the reconstruction is the difference between reconstructions using clean data and noisy data, i.e. $N(t) = S_{\text{noisy}}(t) - S_{\text{clean}}(t)$, the SNRs for the reconstruction of $V^i(t)$ and $I^i(t)$ from noisy data are 10.5 and 11.9, respectively. The SNRs of the reconstructions are averages of the SNRs of the given data, which implies that the reconstruction is stable with respect to noise.

Another type of error may be introduced in the estimate of the location x_0 of the sources. This type of error is tested by setting x_0 to an incorrect value instead of the correct value 400 m. Numerical results show that if the offset is less than 10 m the reconstruction is reasonably good. However, as shown in Fig. 3.16, the error in the reconstruction is significant for an offset of 20 m, i.e., $x_0 = 420$ m, and the reconstruction fails when the offset is 50 m. Using the same definition of SNR as above, one obtains SNRs equal to 3.0 and 6.9 for $V^i(t)$ and $I^i(t)$, respectively, when $x_0 = 420$ m, and 0.9 and 2.5 when $x_0 = 450$ m. The error of the reconstruction due to an erroneous source position arises for two reasons. Firstly, $G_R^{c\pm}$ and $G_L^{c\pm}$ are computed for two parts of the line which have erroneous lengths. Secondly, from the definitions of $V_0(t)$ and $V_l(t)$ in (3.31) and (3.36), one has $V_0(t) = V(0, t - \tau(0, x_0))$ and $V_l(t) = V(l, t - \tau(x_0, l))$, from which one sees that the important timing of $V_0(t)$ and $V_l(t)$ becomes erroneous. In fact,

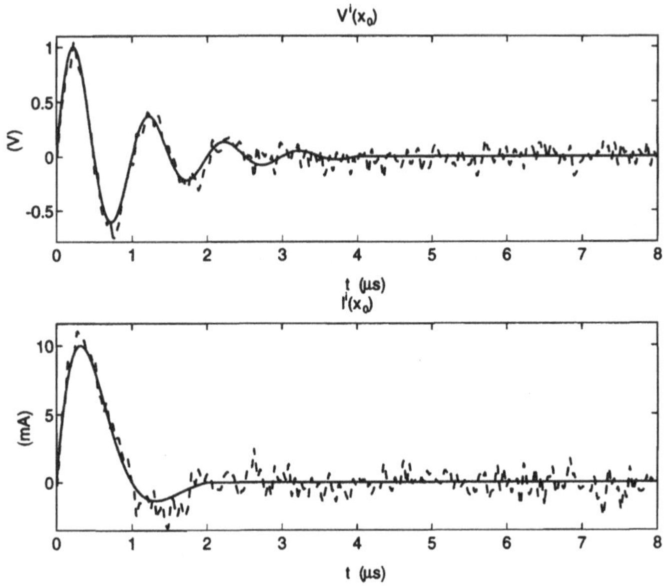

FIG. 3.15. The influence of noise on the reconstruction. Solid lines: true sources; dashed lines: reconstruction using $V_0(t)$ and $V_l(t)$ corrupted with Gaussian noise.

the timing error causes larger errors in the reconstruction than the errors in the compact Green's functions do (at least for rapidly varying signals).

Note that the compact Green's functions are continuous functions of the parameters $\frac{d}{dx}Z$, c, R and G, i.e., a small error in the parameters of the line causes a small error in the compact Green's function, which results in a small error in the reconstruction of the internal sources. Thus, if the source position is known, the inverse problem is well-conditioned with respect to errors in the parameters of the line. However, a successful reconstruction is, in general, heavily dependent on a successful determination of the source position.

The present approach to the inverse source problem can be generalized in a straightforward way to the case with dispersive parameters. If, e.g., the dispersion in the resistance has to be considered in order to obtain an accurate description of the signal propagation on the transmission line, one should replace (3.28) with the following equation (cf. He (1993b)),

$$\frac{\partial}{\partial x}V(x,t) + L(x)\frac{\partial}{\partial t}I(x,t) + R_0(x)I(x,t) + R(x,t) * I(x,t) = V^i(t)\delta(x - x_0).$$

Note that in the above equation $R_0(x)I(x,t) + R(x,t) * I(x,t)$ comes from the inverse Fourier transform of $\tilde{R}(x,\omega)\tilde{I}(x,\omega)$ (here $\tilde{R}(x,\omega)$, $\tilde{I}(x,\omega)$ denote the re-

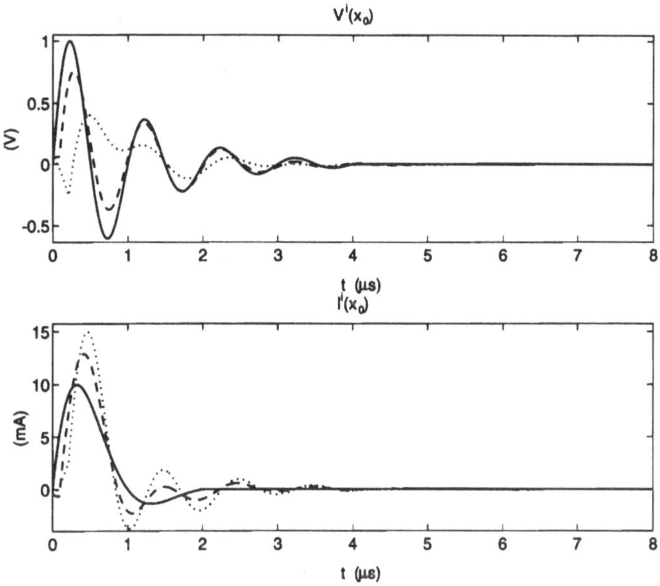

FIG. 3.16. Reconstructions with an erroneous estimation of the source position x_0. Solid lines: true sources at $x_0 = 400$ m; dashed lines: reconstruction with $x_0 = 420$ m; dotted lines: reconstruction with $x_0 = 450$ m.

sistance and current, respectively, at the frequency ω). $R_0(x)$ denotes the instantaneous response of the resistance, i.e., the resistance at the highest frequencies of the signals. The inclusion of the dispersion model does not change the inverse source algorithm described in this subsection. The only difference is that several additional convolution terms appear in the PDEs for the compact Green's functions, which make the Green's functions $G_R^{c\pm}$ and $G_L^{c\pm}$, in the dispersive case, lose the compact support property in the time variable. However, we note that when one considers inverse problems in dispersive media which involve reconstructing the dispersive properties, then the influence on the method of solution is more profound (cf. Section 3.3).

Finally, it is noted that if the disturbance gives rise to a source distribution in a wide region (instead of in a narrow region as considered in this subsection) on a non-uniform LCRG line, then the spatial source distribution functions on the line can be reconstructed by other wave-splitting methods (see Section 5.1.1). Other types of transient source problems have also been solved by wave-splitting methods (see Corones and Sun (1993), Åberg and Karlsson (1997)).

3.2 Propagation and scattering of obliquely incident electromagnetic plane waves

In the present section we extend the 1D wave-splitting approach to the case of plane waves which are obliquely incident on a half-space. The parameter reconstruction for a lossy half-space characterized by depth-dependent permittivity, permeability, and conductivity is considered. This reconstruction utilizes a wave-splitting which is physical in a homogeneous lossless medium. As a basis for the study of propagation and scattering of electromagnetic waves in lossy media, the construction of an exact wave-splitting in a homogeneous lossy medium is carried out. It is also shown how such a (physical) wave-splitting provides a natural framework for the study of the reflection of obliquely incident plane waves from a homogeneous lossy half-space.

Consider an electromagnetic plane wave obliquely impinging on a finitely conducting half-space from the left side $z < 0$. The medium on the left side $z < 0$ is a vacuum. Thus Maxwell's equations give

$$\nabla \times \vec{\mathbf{E}} = -\mu \partial_t \vec{\mathbf{H}}, \tag{3.47}$$

$$\nabla \times \vec{\mathbf{H}} = \epsilon \partial_t \vec{\mathbf{E}} + \sigma \vec{\mathbf{E}}, \tag{3.48}$$

where $\vec{\mathbf{E}} = (E_1, E_2, E_3)^T$, etc.

If the plane of incidence is the xz plane, then the fields have the following forms

$$\vec{\mathbf{E}} = \vec{\mathbf{E}}(z, t - \frac{\sin\theta}{c_0}x), \qquad \vec{\mathbf{H}} = \vec{\mathbf{H}}(z, t - \frac{\sin\theta}{c_0}x), \tag{3.49}$$

i.e.,

$$\partial_x = -\frac{\sin\theta}{c_0}\partial_t, \qquad \partial_y = 0, \tag{3.50}$$

where θ is the incident angle, and $c_0 = 1/\sqrt{\epsilon_0\mu_0}$. Thus, one only needs to consider the fields at those points with $x = 0$, and the condition $x = 0$ will be suppressed hereafter in all the equations in this section. The fields are assumed to be identically zero in the conducting half-space $z > 0$ (with $x = 0$) for the time $t \leq 0$, i.e.

$$\vec{\mathbf{E}}(z,t)|_{t\leq 0} = 0, \qquad \vec{\mathbf{H}}(z,t)|_{t\leq 0} = 0, \quad \text{for } z > 0. \tag{3.51}$$

3.2.1 *Wave-splitting in a homogeneous lossless half-space*

First we determine the right- and left-going waves in a homogeneous lossless half-space (with a permittivity ϵ), i.e. $\sigma = 0$. From the third components of

Maxwell's equations, one expresses the third components of the electromagnetic fields in terms of the tangential fields as

$$H_3 = \frac{\sin\theta}{\mu c_0} E_2, \tag{3.52}$$

$$E_3 = -\frac{\sin\theta}{\epsilon c_0} H_2. \tag{3.53}$$

Using the above two equations, one rewrites the first two components of Maxwell's equations in the following form

$$\partial_z \begin{bmatrix} \mathbf{E} \\ \mathbf{H} \end{bmatrix} = W \begin{bmatrix} \mathbf{E} \\ \mathbf{H} \end{bmatrix}, \tag{3.54}$$

where

$$\mathbf{E} = \begin{bmatrix} E_1 \\ E_2 \end{bmatrix}, \qquad \mathbf{H} = \begin{bmatrix} H_1 \\ H_2 \end{bmatrix},$$

$$W = \begin{bmatrix} 0 & 0 & 0 & -\mu(1 - \frac{\sin^2\theta}{c_0^2 \epsilon\mu}) \\ 0 & 0 & \mu & 0 \\ 0 & \epsilon(1 - \frac{\sin^2\theta}{c_0^2 \epsilon\mu}) & 0 & 0 \\ -\epsilon & 0 & 0 & 0 \end{bmatrix} \partial_t. \tag{3.55}$$

The right-going waves $\mathbf{E}^+$ and left-going waves $\mathbf{E}^-$ can then be identified from (3.54). Let

$$\begin{bmatrix} \mathbf{E}^+ \\ \mathbf{E}^- \end{bmatrix} = T \begin{bmatrix} \mathbf{E} \\ \mathbf{H} \end{bmatrix}, \tag{3.56}$$

where

$$\mathbf{E}^\pm = \begin{bmatrix} E_1^\pm \\ E_2^\pm \end{bmatrix}.$$

Differentiating (3.56) with respect to z and using (3.54), yields

$$\partial_z \begin{bmatrix} \mathbf{E}^+ \\ \mathbf{E}^- \end{bmatrix} = TWT^{-1} \begin{bmatrix} \mathbf{E}^+ \\ \mathbf{E}^- \end{bmatrix}, \tag{3.57}$$

where T^{-1} is the inverse of T. In order to allow a physical interpretation of $\mathbf{E}^+$ and $\mathbf{E}^-$ as right- and left-moving waves, TWT^{-1} should be diagonal. From the usual process of diagonalizing a matrix, it follows that the diagonal elements are the eigenvalues of the matrix W and the matrix T^{-1} consists of the eigenvectors of the matrix W (the eigenvectors should be ordered appropriately so that

$\mathbf{E}^+$ and $\mathbf{E}^-$ have the physical interpretation as right- and left-moving waves, respectively). Performing the indicated derivation, one obtains

$$T = \frac{1}{2}\begin{bmatrix} I & Z \\ I & -Z \end{bmatrix}, \tag{3.58}$$

where I is the 2×2 identity matrix and

$$Z = \begin{bmatrix} 0 & \sqrt{\frac{\mu}{\epsilon}}\sqrt{1 - \frac{\sin^2\theta}{c_0^2\epsilon\mu}} \\ -\sqrt{\frac{\mu}{\epsilon}}/\sqrt{1 - \frac{\sin^2\theta}{c_0^2\epsilon\mu}} & 0 \end{bmatrix}. \tag{3.59}$$

Note that

$$TWT^{-1} = \mathrm{diag}[-1,\ -1,\ 1,\ 1]\,\sqrt{\mu\epsilon}\partial_t. \tag{3.60}$$

As a special case, put $\epsilon = \epsilon_0$, $\mu = \mu_0$ in (3.58) to obtain the right- and left-going waves in vacuum

$$\begin{bmatrix} \mathbf{E}^+ \\ \mathbf{E}^- \end{bmatrix}(z,t) = \frac{1}{2}\begin{bmatrix} I & Z_0 \\ I & -Z_0 \end{bmatrix}\begin{bmatrix} \mathbf{E} \\ \mathbf{H} \end{bmatrix}(z,t) \equiv T_0\begin{bmatrix} \mathbf{E} \\ \mathbf{H} \end{bmatrix}(z,t), \quad z < 0, \tag{3.61}$$

where

$$Z_0 = \sqrt{\frac{\mu_0}{\epsilon_0}}\begin{bmatrix} 0 & \cos\theta \\ -\dfrac{1}{\cos\theta} & 0 \end{bmatrix}. \tag{3.62}$$

As can easily be checked (Norgren and He 1994), in the vacuum region $z < 0$ the E_1^+, E_1^- waves are right-going (i.e., propagating along the $+z$ direction) and left-going (i.e., propagating along the $-z$ direction) TM modes, respectively, and the E_2^+, E_2^- waves are right- and left-going TE modes, respectively ($E_i^\pm$, $i = 1, 2$, give the amplitudes of the *tangential* components of the electric field for these modes).

3.2.2 *Parameter reconstruction for a stratified half-space*

For the inverse problem, we consider the case when the incident field has a transverse electric (TE) polarization, i.e., $E_3 = 0$. Consequently, one has $E_1 = H_2 = 0$. Thus, from Maxwell's equations, one has

$$\partial_z E_2 = \mu\partial_t H_1, \tag{3.63}$$

$$\epsilon\left(1 - \frac{\sin^2\theta}{c_0^2\epsilon\mu}\right)\partial_t E_2 + \sigma E_2 = \partial_z H_1. \tag{3.64}$$

Note that if an incident field with transverse magnetic (TM) polarization is used, i.e., $H_3 = 0$, then the equation corresponding to (3.63) will contain a convolution term (He 1994b). The method for solving the inverse problem is then the same as the method developed in Section 3.3 below for dispersive media.

The wave-splitting that is convenient for use for the inverse problem is (cf. (3.58) and (3.59))

$$E^{\pm} = \frac{1}{2}\left[E_2 \mp \sqrt{\frac{\mu}{\epsilon(1-\frac{\sin^2\theta}{c_0^2\epsilon\mu})}}H_1\right]. \tag{3.65}$$

Note that E^+ and E^- have physical interpretation as right- and left-going electric fields, respectively, only in a lossless region, e.g., the exterior region $z<0$. In the lossy region, the split fields $E^{\pm}$ no longer have a direct physical interpretation as right- and left-moving waves (the physical splitting in the lossy region is given by (3.72)–(3.74) below). In this subsection both the permittivity ϵ and the permeability μ are assumed to be continuous and differentiable.

From (3.63)–(3.65), one obtains the following dynamic equation for the split TE fields,

$$\begin{aligned}\partial_z\begin{bmatrix}E^+\\E^-\end{bmatrix} + \frac{1}{c(z;\theta)}\partial_t\begin{bmatrix}E^+\\-E^-\end{bmatrix}\\ = \left(\frac{1}{2}\sqrt{\frac{\mu}{\epsilon q(z;\theta)}}\,\sigma\begin{bmatrix}-1&-1\\1&1\end{bmatrix} + \frac{1}{4}\frac{d}{dz}\ln\frac{\mu}{\epsilon q(z;\theta)}\begin{bmatrix}1&-1\\-1&1\end{bmatrix}\right)\begin{bmatrix}E^+\\E^-\end{bmatrix},\end{aligned} \tag{3.66}$$

where $q(z;\theta) = 1 - \frac{\sin^2\theta}{c_0^2\epsilon(z)\mu(z)}$, and $c(z;\theta) = 1/\sqrt{\epsilon(z)\mu(z)q(z;\theta)}$.

The system of equations for the Green's functions are (cf. Section 2.1 for the derivation)

$$\partial_z G^+ = -\frac{1}{2}\sqrt{\frac{\mu}{\epsilon q(z;\theta)}}\,\sigma(G^+ + G^-) + \frac{1}{4}\frac{d}{dz}\ln\frac{\mu}{\epsilon q(z;\theta)}(G^+ - G^-), \tag{3.67}$$

$$\partial_z G^- - \frac{2}{c(z;\theta)}\partial_t G^- = \frac{1}{2}\sqrt{\frac{\mu}{\epsilon q(z;\theta)}}\,\sigma(G^+ + G^-) - \frac{1}{4}\frac{d}{dz}\ln\frac{\mu}{\epsilon q(z;\theta)}(G^+ - G^-), \tag{3.68}$$

$$G^-(z,0;\theta) = -\frac{c(z;\theta)}{4q(z;\theta)}a(z;\theta)\left[\sqrt{\frac{\mu q(z;\theta)}{\epsilon}}\sigma - \frac{1}{2\epsilon}\frac{d\epsilon}{dz} - \frac{1-2q(z;\theta)}{2\mu}\frac{d\mu}{dz}\right], \tag{3.69}$$

$$G^+(0,t;\theta) = 0, \tag{3.70}$$

$$G^-(0,t;\theta) = R(t;\theta), \tag{3.71}$$

where the attenuation factor

$$a(z;\theta) = \left[\frac{\epsilon_0\mu(z)q(0;\theta)}{\epsilon(z)\mu_0 q(z;\theta)}\right]^{1/4} \exp\left\{-\frac{1}{2}\int_0^z \sqrt{\frac{\mu(z')}{\epsilon(z')q(z';\theta)}}\,\sigma(z')\,dz'\right\}.$$

Using the known reflection kernel $R(t;\theta)$ for three different incident angles, one can reconstruct all the three parameters $\epsilon(z)$, $\mu(z)$, and $\sigma(z)$ simultaneously. In the inverse problem, one propagates the known boundary values (cf. (3.70) and (3.71)) to the initial values using the PDEs for the Green's functions for three different incident angles $\theta = \theta_1, \theta_2, \theta_3$. The parameters can then be reconstructed according to the initial condition (3.69), which can be rewritten as

$$a_1(z;\theta_j)\sigma + a_2(z;\theta_j)\frac{d\epsilon}{dz} + a_3(z;\theta_j)\frac{d\mu}{dz} = G^-(z,0;\theta_j), \quad j = 1,2,3$$

(expressions for the coefficients a_1, a_2, and a_3 can be found from (3.69)). The above system of equations has a unique solution for σ, $\frac{d\epsilon}{dz}$ and $\frac{d\mu}{dz}$ when the determinant of the coefficient matrix is not zero. The parameters $\epsilon(z)$ and $\mu(z)$ are then obtained by integrations in z of their z derivatives. Since one needs to solve the system of equations (3.67)–(3.71) for several different incident angles simultaneously, the grid in the z direction should be the same for all incident angles. The grid in the t direction can be chosen as $\Delta t = 2\Delta z/\tilde{c}(z;\theta_1, 0)$, and thus an interpolation is needed for the other two incident angles in propagating $G^{\pm}$ to the next layer along the characteristic directions. Note that the incident angles need to be separated well enough for the simultaneous reconstruction to be stable under noise. We refer to Cao and He (1997) and He and Ström (1995b) for numerical reconstructions using different incident angles in similar acoustic inverse scattering problems.

3.2.3 *Wave-splitting in a finitely conducting medium*

In this subsection we consider plane waves obliquely incident on a conducting half-space. We concentrate on the direct propagation problem and the solution is based on identifying time domain right- and left-going waves in a homogeneous finitely conducting medium. In the rest of this section, it is assumed that $\mu = \mu_0$. The wave-splitting can be obtained by means of a Laplace transform. In the Laplace transform space, ϵ in (3.58) and (3.59) is replaced by $(\epsilon + \frac{\sigma}{s})$ for a conducting half-space, i.e.

$$T(s) = \frac{1}{2}\begin{bmatrix} I & Z(s) \\ I & -Z(s) \end{bmatrix},$$

where

$$Z(s) = \begin{bmatrix} 0 & \frac{\sqrt{\mu_0(\epsilon+\frac{\sigma}{s}-\epsilon_0\sin^2\theta)}}{\epsilon+\frac{\sigma}{s}} \\ -\sqrt{\frac{\mu_0}{\epsilon+\frac{\sigma}{s}-\epsilon_0\sin^2\theta}} & 0 \end{bmatrix}.$$

After an inverse Laplace transform, one obtains the following time domain right- and left-going waves in a conducting half-space (He and Ström 1995a)

$$\begin{bmatrix} \mathcal{E}^+ \\ \mathcal{E}^- \end{bmatrix}(z,t) = \frac{1}{2}\begin{bmatrix} I & (Z_1 + Z_2*) \\ I & -(Z_1 + Z_2*) \end{bmatrix}\begin{bmatrix} \mathbf{E} \\ \mathbf{H} \end{bmatrix}(z,t) \equiv \mathcal{T}\begin{bmatrix} \mathbf{E} \\ \mathbf{H} \end{bmatrix}(z,t), \quad (3.72)$$

where

$$Z_1 = \begin{bmatrix} 0 & \frac{\sqrt{\mu_0(\epsilon-\epsilon_0\sin^2\theta)}}{\epsilon} \\ -\sqrt{\frac{\mu_0}{\epsilon-\epsilon_0\sin^2\theta}} & 0 \end{bmatrix}, \quad (3.73)$$

$$Z_2 = \sqrt{\frac{\mu_0}{\epsilon_0}}\begin{bmatrix} 0 & -\eta_M \\ \eta_E & 0 \end{bmatrix}, \quad (3.74)$$

and where

$$\begin{aligned} \eta_E(t) &= -\sqrt{\frac{\epsilon_0}{\mu_0}}\mathcal{L}^{-1}[\sqrt{\frac{\mu_0}{\epsilon + \frac{\sigma}{s} - \epsilon_0\sin^2\theta}} - \sqrt{\frac{\mu_0}{\epsilon - \epsilon_0\sin^2\theta}}] \\ &= -\sqrt{\frac{\epsilon_0}{\epsilon_1}}\mathcal{L}^{-1}[\sqrt{\frac{s}{s+\sigma/\epsilon_1}} - 1] \\ &= \frac{\sigma}{2\epsilon_1}\sqrt{\frac{\epsilon_0}{\epsilon_1}}[I_0(\frac{\sigma}{2\epsilon_1}t) - I_1(\frac{\sigma}{2\epsilon_1}t)]e^{-\frac{\sigma}{2\epsilon_1}t}, \end{aligned} \quad (3.75)$$

$$\begin{aligned} \eta_M(t) &= -\frac{\sqrt{\epsilon_0\epsilon_1}}{\epsilon}\mathcal{L}^{-1}[(\frac{\sqrt{s}\sqrt{s+\sigma/\epsilon_1}}{s+\sigma/\epsilon} - \frac{\sqrt{s}}{\sqrt{s+\sigma/\epsilon_1}}) + (\frac{\sqrt{s}}{\sqrt{s+\sigma/\epsilon_1}} - 1)] \\ &= -\frac{\sqrt{\epsilon_0\epsilon_1}}{\epsilon}(\frac{1}{\epsilon_1} - \frac{1}{\epsilon})\sigma\mathcal{L}^{-1}[\frac{\sqrt{s}}{(s+\sigma/\epsilon)\sqrt{s+\sigma/\epsilon_1}}] + \frac{\epsilon_1}{\epsilon}\eta_E \\ &= \frac{\epsilon_1}{\epsilon}\eta_E - \frac{\sqrt{\epsilon_0\epsilon_1}}{\epsilon}(\frac{1}{\epsilon_1} - \frac{1}{\epsilon})\sigma e^{-\frac{\sigma}{\epsilon}t} \\ &\quad \times \left\{1 - \frac{\sigma}{2\epsilon_1}\int_0^t e^{(\frac{\sigma}{\epsilon} - \frac{\sigma}{2\epsilon_1})t'}[I_0(\frac{\sigma}{2\epsilon_1}t') - I_1(\frac{\sigma}{2\epsilon_1}t')]\,dt'\right\}, \end{aligned} \quad (3.76)$$

$$\epsilon_1 = \epsilon - \epsilon_0\sin^2\theta. \quad (3.77)$$

Note that the convolution splitting kernels η_E and η_M are due to the conductivity, i.e., $\eta_M = \eta_E = 0$ if $\sigma = 0$.

Equation (3.72) shows that the condition for having only right-going waves (i.e. no left-going wave) in a homogeneous conducting half-space is

$$\mathbf{E} - (Z_1 + Z_2*)\mathbf{H} = 0, \qquad z \geq 0. \quad (3.78)$$

This condition is suitable for use as an absorbing boundary condition for, e.g. the finite-difference time domain (FDTD) modelling of electromagnetic wave propagation in a stratified slab over a conducting half-space.

3.2.4 *Calculation of the transient reflection from a conducting half-space*

Transient reflection from the earth-air interface needs to be considered in many electromagnetic coupling and scattering problems. One can sometimes simplify complicated scattering problems by removing the earth and model the transient reflection from the earth by means of additional sources (see e.g. the extensive recent developments concerning the image theory by Lindell and co-workers as can be traced from Lindell (1993)). A conventional way to obtain the transient reflection from a finitely conducting earth is to perform a numerical Fourier transform of the frequency domain reflection spectrum. However, it is desirable in many applications to compute the transient reflection directly in the time domain and then use the result in electromagnetic coupling problems. Next we show how this can be achieved by means of the physical splitting derived in Section 3.2.3.

At $z = 0^-$, i.e., in the vacuum just above $z = 0$, the left-going (reflected) waves $\mathbf{E}^-(0^-, t) = \mathbf{E}^r(t)$ are related to the right-going (incident) waves $\mathbf{E}^+(0^-, t)$ $= \mathbf{E}^i(t)$ as follows

$$\mathbf{E}^r(t) = \bar{\bar{r}}\,\mathbf{E}^i(t) + \bar{\bar{R}}\,(t) * \mathbf{E}^i(t), \tag{3.79}$$

where

$$\bar{\bar{r}} = \begin{bmatrix} r_{11} & r_{12} \\ r_{21} & r_{22} \end{bmatrix}, \quad \bar{\bar{R}} = \begin{bmatrix} R_{11} & R_{12} \\ R_{21} & R_{22} \end{bmatrix}. \tag{3.80}$$

The reflection coefficient matrix $\bar{\bar{r}}$ and the reflection kernel matrix $\bar{\bar{R}}$ relate the amplitudes of the *tangential* components of the reflected and incident electric fields to each other for TM and TE modes.

By the continuity of the tangential fields $\mathbf{E}$ and $\mathbf{H}$ across the interface $z = 0$, one has (cf. (3.72))

$$\begin{bmatrix} \mathcal{E}^+ \\ \mathcal{E}^- \end{bmatrix}(0^+, t) = \mathcal{T}\begin{bmatrix} \mathbf{E} \\ \mathbf{H} \end{bmatrix}(0^+, t) = \mathcal{T}\begin{bmatrix} \mathbf{E} \\ \mathbf{H} \end{bmatrix}(0^-, t)$$

$$= \mathcal{T}\mathcal{T}_0^{-1}\begin{bmatrix} \mathbf{E}^+ \\ \mathbf{E}^- \end{bmatrix}(0^-, t) = \mathcal{T}\mathcal{T}_0^{-1}\begin{bmatrix} \mathbf{E}^+(0^-, t) \\ (\bar{\bar{r}} + \bar{\bar{R}}\,*)\mathbf{E}^+(0^-, t) \end{bmatrix}$$

$$= \begin{bmatrix} \bar{\bar{n}} & \bar{\bar{m}} \\ \bar{\bar{m}} & \bar{\bar{n}} \end{bmatrix}\begin{bmatrix} \mathbf{E}^i(t) \\ (\bar{\bar{r}} + \bar{\bar{R}}\,*)\mathbf{E}^i(t) \end{bmatrix}, \tag{3.81}$$

where

$$\overline{\overline{m}} = \frac{1}{2}\left[I - (Z_1 + Z_2 *) Z_0^{-1}\right], \tag{3.82}$$

$$\overline{\overline{n}} = \frac{1}{2}\left[I + (Z_1 + Z_2 *) Z_0^{-1}\right]. \tag{3.83}$$

Since $\boldsymbol{\mathcal{E}}^-(0^+, t) = 0$ (which is due to the fact that there is no left-going wave in the conducting half-space $z > 0$), from (3.81) it follows that

$$\overline{\overline{m}}\, \mathbf{E}^i(t) + \overline{\overline{n}}\, (\overline{\overline{r}} + \overline{\overline{R}} *) \mathbf{E}^i(t) = 0,$$

i.e.,

$$\left[(I + Z_1 Z_0^{-1})\, \overline{\overline{r}} + (I - Z_1 Z_0^{-1})\right] \mathbf{E}^i(t)$$

$$+ \left[Z_2 Z_0^{-1}\, \overline{\overline{r}} + \overline{\overline{R}} + Z_1 Z_0^{-1}\, \overline{\overline{R}} + Z_2 Z_0^{-1} *\, \overline{\overline{R}} - Z_2 Z_0^{-1}\right] * \mathbf{E}^i(t) = 0. \tag{3.84}$$

Since the incident field $\mathbf{E}^+(0^-, t)$ is arbitrary, the following equations must hold

$$(I + Z_1 Z_0^{-1})\, \overline{\overline{r}} + (I - Z_1 Z_0^{-1}) = 0,$$

$$Z_2 Z_0^{-1}\, \overline{\overline{r}} + \overline{\overline{R}} + Z_1 Z_0^{-1}\, \overline{\overline{R}} + Z_2 Z_0^{-1} *\, \overline{\overline{R}} - Z_2 Z_0^{-1} = 0,$$

which give, respectively,

$$\overline{\overline{r}} = -[I + Z_1 Z_0^{-1}]^{-1}[I - Z_1 Z_0^{-1}]$$

$$= \begin{bmatrix} -\dfrac{1 - \dfrac{\sqrt{\epsilon_0}\sqrt{\epsilon - \epsilon_0 \sin^2\theta}}{\epsilon\cos\theta}}{1 + \dfrac{\sqrt{\epsilon_0}\sqrt{\epsilon - \epsilon_0 \sin^2\theta}}{\epsilon\cos\theta}} & 0 \\ 0 & \dfrac{1 - \dfrac{\sqrt{\epsilon - \epsilon_0 \sin^2\theta}}{\sqrt{\epsilon_0}\cos\theta}}{1 + \dfrac{\sqrt{\epsilon - \epsilon_0 \sin^2\theta}}{\sqrt{\epsilon_0}\cos\theta}} \end{bmatrix}, \tag{3.85}$$

$$[I + Z_1 Z_0^{-1}]\, \overline{\overline{R}} + [Z_2 Z_0^{-1}] *\, \overline{\overline{R}} = Z_2 Z_0^{-1}[I - \overline{\overline{r}}]. \tag{3.86}$$

As is customary, the reflection coefficients for TM and TE modes are defined as

$$E_{TE}^{ref} = r_{TE} E_{TE}^{inc} + R_{TE}(t) * E_{TE}^{inc}, \tag{3.87}$$

$$E_{TM}^{ref} = r_{TM} E_{TM}^{inc} + R_{TM}(t) * E_{TM}^{inc}, \tag{3.88}$$

where E_{TE}^{ref} (E_{TE}^{inc}) is the amplitude of the reflected (incident) electric field for the TE mode, etc., and r_{TE}, R_{TE} (r_{TM}, R_{TM}) are the reflection coefficient and reflection kernel for TE (TM) mode incidence, respectively. It is easy to see that

$$r_{TM} = -r_{11}, \quad r_{TE} = r_{22}, \tag{3.89}$$
$$R_{TM} = -R_{11}, \quad R_{TE} = R_{22} \tag{3.90}$$

(note that $r_{12} = r_{21} = R_{12} = R_{21} = 0$ for the present problem since the scattering medium is isotropic). For the TM mode, r_{11} (R_{11}) has a sign difference as compared to r_{TM} (R_{TM}) because the tangential reflected electric field is in the opposite direction to the tangential incident electric field for the TM mode, which contributes a minus sign in the definition of r_{11} and R_{11}. Therefore, it follows from (3.85), (3.86), (3.89) and (3.90) that

$$r_{TM} = \frac{1 - \dfrac{\sqrt{\epsilon_0}\sqrt{\epsilon - \epsilon_0 \sin^2\theta}}{\epsilon\cos\theta}}{1 + \dfrac{\sqrt{\epsilon_0}\sqrt{\epsilon - \epsilon_0 \sin^2\theta}}{\epsilon\cos\theta}}, \qquad r_{TE} = \frac{1 - \dfrac{\sqrt{\epsilon - \epsilon_0 \sin^2\theta}}{\sqrt{\epsilon_0}\cos\theta}}{1 + \dfrac{\sqrt{\epsilon - \epsilon_0 \sin^2\theta}}{\sqrt{\epsilon_0}\cos\theta}}, \tag{3.91}$$

$$R_{TM} - \eta_{TM} * R_{TM} = (1 + r_{TM})\eta_{TM} \tag{3.92}$$
$$R_{TE} - \eta_{TE} * R_{TE} = -(1 - r_{TE})\eta_{TE}, \tag{3.93}$$

where

$$\eta_{TM} = \frac{\epsilon}{\epsilon\cos\theta + \sqrt{\epsilon_0\epsilon_1}}\eta_M, \qquad \eta_{TE} = \frac{\sqrt{\epsilon_1}\cos\theta}{\sqrt{\epsilon_1} + \sqrt{\epsilon_0}\cos\theta}\eta_E \tag{3.94}$$

(ϵ_1 is given by (3.77)). Equation (3.91) gives the reflection coefficients for TM and TE modes, which are identical to the well-known reflection coefficients for a lossless half-space (Kong 1986).

Equations (3.92) and (3.93) are Volterra equations of the second kind, from which the reflection kernels are easily obtained numerically (see (3.95) and (3.96) below). The time is discretized with a constant stepsize Δt, and we use the notation $R_{TM}(j\Delta t) = R_{TM}^{(j)}$, etc. The reflection kernel R_{TM} can be easily obtained from the Volterra equation (3.92) as follows (using the trapezoidal rule):

$$R_{TM}^{(0)} = (1 + r_{TM})\eta_{TM}^{(0)}, \tag{3.95}$$

$$R_{TM}^{(i)} = \frac{(1+r_{TM})\eta_{TM}^{(i)} + [\sum_{j=1}^{i-1} \eta_{TE}^{(i-j)} R_{TM}^{(j)} + 0.5\eta_{TE}^{(i)} R_{TM}^{(0)}]\Delta t}{(1-0.5\eta_{TM}^{(0)}\Delta t)}, i = 1, 2, 3, ... \tag{3.96}$$

(with $\sum_{j=1}^{-1} = 0$ when $i = 1$). Similarly the Volterra equation (3.93) can be solved numerically for R_{TE}.

Numerical results concerning transient reflection from a conducting half-space treated by means of frequency-domain methods have been given in Barnes and Tesche (1991), from which earlier work on this problem can be traced. In order to compare with the numerical results given in Barnes and Tesche (1991), we choose the same parameters as used there for the conducting half-space, i.e., $\epsilon = 10\epsilon_0$, $\sigma = 0.01$ mho/m. We also choose the same Bell Laboratory waveform incident field:

$$E^{inc} = A_0(e^{-\alpha t} - e^{-\beta t}), \tag{3.97}$$

with $A_0 = 5.25 \times 10^4$ (V/m), $\alpha = 4 \times 10^6$ (1/s) and $\beta = 4.76 \times 10^8$ (1/s). This incident field is given by the dotted line in Fig. 3.17(a) (note that E^{inc} is positive for all times). The reflected electric fields given by (3.92) and (3.93) are calculated for the incident angles $\theta = 5°$, $45°$, $85°$ and are plotted in Fig. 3.17(a) and Fig. 3.17(b) for the TM and TE modes, respectively. The reflected electric fields for $\theta = 45°$ shown in Fig. 3.17 may be compared with the results given in Fig. 3 of Barnes and Tesche (1991), which are indicated by the circles in Fig. 3.17. For the TE mode, the reflected electric field always has negative values for all incident angles (note that the reflection coefficient r_{TE} is negative for all incident angles). For the TM mode, the reflected electric field has positive values for all incident angles that are smaller than the Brewster angle. The Brewster angle is the incident angle at which the reflection coefficient r_{TM} becomes zero, i.e. $\theta_B = \tan^{-1}\sqrt{\frac{\epsilon}{\epsilon_0}}$ (cf. (3.91) and Kong (1986)). Note that $r_{TM} > 0$ for $\theta < \theta_B$, and $r_{TM} < 0$ for $\theta > \theta_B$ (in the limit, $r_{TM} \longrightarrow -1$, when $\theta \longrightarrow 90°$). As a consequence, the reflected electric field for a TM mode may become negative at some early times for an incident angle larger than θ_B, and then become positive (it certainly goes to zero eventually). The dot-dashed line in Fig. 3.17(a) for $\theta = 85°$ shows such a case clearly (note that $\theta_B = 72.45°$ for this profile).

3.3 Electromagnetically dispersive media

For any real material, the permittivity is frequency dependent (i.e., dispersive), and this frequency dependence cannot be ignored in the study of propagation of a broadband pulse. In a time domain treatment, the frequency dependence gives

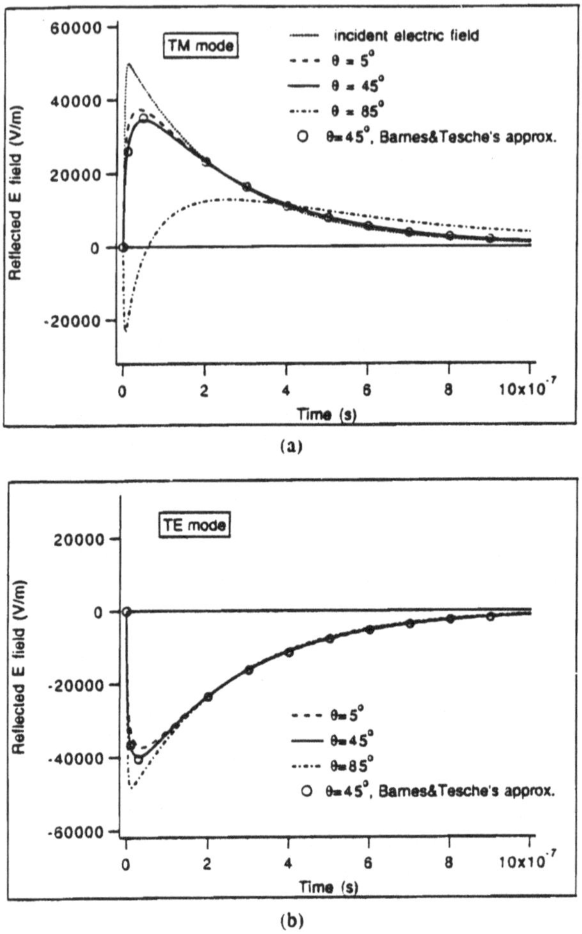

FIG. 3.17. The reflected electric field from a conducting half-space with $\epsilon = 10\epsilon_0$, $\sigma = 0.01$ mho/m for the Bell Laboratory waveform incident field: (a) TM mode; (b) TE mode.

rise to a convolution with dispersion kernel (i.e. the susceptibility kernel) in the constitutive relations (see e.g. Jackson (1975), Bohren and Huffman (1983)). A recent analysis of the time domain formulation of the model for general dispersive media (media characterized by one or more memory functions) is found in Karlsson and Kristensson (1992).

In this section, we consider a stratified dispersive medium characterized by the following constitutive relations,

$$D(z,t) = \epsilon_0 \left[\epsilon_r(z)E(z,t) + \chi_e(z,t) * E(z,t)\right], \tag{3.98}$$

$$B(z,t) = \mu_0 \left[\mu_r(z)H(z,t) + \chi_m(z,t) * H(z,t)\right], \tag{3.99}$$

where χ_e, χ_m are the electric and magnetic susceptibility kernels, respectively, which describes the dispersion. Here $\epsilon_r(z)$, $\mu_r(z)$ are the usual relative static permittivity and permeability, respectively, and ϵ_0, μ_0 are the permittivity and permeability in vacuum, respectively.

In Section 3.3.1–3.3.3 we apply an approach, which is analogous to the one used in Section 3.2 for a conducting half-space, to study propagation in a homogeneous dispersive medium. The key point is again that of finding a splitting which defines exact forward and backward propagating modes in a homogeneous medium. Two often used models for dispersive media are the Lorentz and Debye models and we note in particular that for these models the convolution kernels in the definition of the wave-splitting can be given in a useful explicit form. The problem of the *first precursor* for dispersive media is a classic one and it is shown in Section 3.3.4 that splitting and propagator kernels at early times for the Lorentz and Debye models are obtained in explicit form. This allows for a simple and direct computation of these precursors. In analogy with what was presented in Section 3.2.3, it is shown in Section 3.3.5 how the exact splitting for a homogeneous dispersive medium provides a natural framework for treating reflection from a half-space filled by such a medium.

In Sections 3.3.6 and 3.3.7 we consider several examples of inverse problems of reconstructing the temporal behaviour of a dispersive medium (note that if the temporal behaviour is known, then the spatial dependence of the parameters can be reconstructed in the same way as described in Section 2.1). The measurements are assumed to be made in an exterior non-dispersive medium, and the treatment there is based on a simple non-dispersive splitting. Extensive work based on the time domain wave-splitting approach has been done concerning dispersive media, as exemplified by Beezley and Krueger (1985), Karlsson (1989), He and Ström (1996), and Rikte (in press).

3.3.1 *Time domain models for dispersive media*

Consider a plane wave propagating in a homogeneous dispersive medium. Assume that the electric field $\vec{\mathbf{E}}$ is polarized in the $\hat{\mathbf{e}}_x$ direction, the magnetic field $\vec{\mathbf{H}}$ is polarized in the $\hat{\mathbf{e}}_y$ direction, and the plane wave propagates in the $\hat{\mathbf{e}}_z$ direction. It is assumed that the incident wave will not reach the position $z = 0$ until time $t = 0$, i.e., one has the following initial conditions in the region $z \in (0, +\infty)$

$$E(z,t) = 0, \quad H(z,t) = 0, \quad \text{for } t \leq 0. \tag{3.100}$$

In this subsection, only the region $z \in (0, +\infty)$ is of interest, and thus in all the formulas below in this subsection z should be understood to have non-negative values.

From Maxwell's equations, one has

$$\partial_z E = -\mu_0 \partial_t H, \tag{3.101}$$

$$\partial_z H = -\partial_t D. \tag{3.102}$$

At a fixed frequency ω, the displacement field $\hat{D}(z,\omega)$ (the Fourier transform of $D(z,t)$) is related to the electric field $\hat{E}(z,\omega)$ (the Fourier transform of $E(z,t)$) by the following linear relation

$$\hat{D}(z,\omega) = \hat{\epsilon}(\omega)\hat{E}(z,\omega), \tag{3.103}$$

where $\hat{\epsilon}(\omega)$ is the frequency-dependent permittivity. In a time domain treatment, the frequency dependence of the permittivity gives rise to a convolution with a dispersion kernel (i.e. the susceptibility kernel) in the following time domain constitutive relation

$$D(z,t) = \epsilon[E(z,t) + \chi(t) * E(z,t)], \tag{3.104}$$

where ϵ is a constant, and the susceptibility kernel $\chi(t)$ is given by

$$\chi(t) = \frac{1}{2\pi}\int_{-\infty}^{+\infty} [\hat{\epsilon}(\omega)/\epsilon - 1]e^{-j\omega t}\, d\omega. \tag{3.105}$$

Note that one has $\hat{\epsilon}(\omega) = \epsilon\left[1 + \int_0^{+\infty}\chi(t)e^{j\omega t}\, dt\right]$. Also note that due to the causality, $\chi(t)$ vanishes for negative times, and this should be understood in all the relevant formulas below in this section.

In particular, the following two dispersion models are often used as models for dispersive solid materials and fluids respectively and they will be used for illustrations of the general results and in the numerical calculations.

The Lorentz (single resonance) model. For many solid non-magnetic materials of relatively low density, it is appropriate to assume that each harmonically bound electron (with mass m and charge $-e$) in the medium is affected by three forces, namely, an electric force, a restoring harmonic force, and a friction force proportional to the velocity of the electron (caused by the collisions between the electron and other molecules). From the equation of motion for the electron, one can easily obtain the following expression for the permittivity (Jackson 1975; Bohren and Huffman 1983)

$$\hat{\epsilon}(\omega) = \epsilon_0\left[1 + \frac{\omega_p^2}{\omega_0^2 - \omega^2 - i\gamma\omega}\right], \tag{3.106}$$

where ω_0 is the resonant frequency (i.e., the harmonic frequency of the restoring force), $\omega_p = \sqrt{Ne^2/(\epsilon_0 m)}$ is the plasma frequency of the medium (N is the electron density), and $\gamma \geq 0$ is a damping constant. In the Lorentz model, $\epsilon = \epsilon_0$ (the permittivity in vacuum). The corresponding susceptibility kernel is (for $t \geq 0$) (Jackson 1975)

$$\chi(t) = \omega_p^2 e^{-\gamma t/2} \frac{\sin(\nu_0 t)}{\nu_0}, \tag{3.107}$$

where $\nu_0 = \sqrt{\omega_0^2 - \gamma^2/4}$. Note that $\chi(t)$ vanishes for $t < 0$.

The Debye (relaxation) model. Polar liquids can be described accurately by relaxation models. The molecules of these materials have permanent dipole moments, and they tend to align the polarization with the electric field, however, with a relaxation time τ (due to the tendency toward random orientation caused by thermal buffeting). Thus, one obtains the following expression for the permittivity (Bohren and Huffman 1983)

$$\hat{\epsilon}(\omega) = \epsilon_\infty + \frac{\epsilon_s - \epsilon_\infty}{1 - i\omega\tau}, \tag{3.108}$$

where ϵ_s is the static dielectric constant as $\omega \to 0$, and ϵ_∞ is the high frequency limit as $\omega \to \infty$. In the Debye model, $\epsilon = \epsilon_\infty$ (which is usually greater than the permittivity in vacuum). The corresponding susceptibility kernel is (for $t \geq 0$)

$$\chi(t) = \frac{q}{\tau} e^{-t/\tau}, \tag{3.109}$$

where $q = \frac{\epsilon_s - \epsilon_\infty}{\epsilon_\infty}$.

3.3.2 *Wave-splitting in dispersive media*

Maxwell's equations for a dispersive medium characterized by the constitutive relation (3.104) are rewritten in the following matrix form

$$\partial_z \begin{bmatrix} E \\ H \end{bmatrix} = \begin{bmatrix} 0 & -\mu_0 \\ -\epsilon(1+\chi*) & 0 \end{bmatrix} \partial_t \begin{bmatrix} E \\ H \end{bmatrix} \equiv W \begin{bmatrix} E \\ H \end{bmatrix}. \tag{3.110}$$

In the non-dispersive case (i.e., $\chi \equiv 0$), the total electric field can be decomposed into a right-moving component $E^+(z,t)$ and a left-moving component $E^-(z,t)$ by $E^\pm = \frac{1}{2}(E + \sqrt{\mu_0/\epsilon}\, H)$. In the dispersive case, the expressions for the split components are modified and the following ansatz is used (cf. (3.72))

$$\begin{bmatrix} \mathcal{E}^+ \\ \mathcal{E}^- \end{bmatrix}(z,t) = \frac{1}{2} \begin{bmatrix} 1 & \sqrt{\mu_0/\epsilon}\,[1+\eta(t)*] \\ 1 & -\sqrt{\mu_0/\epsilon}\,[1+\eta(t)*] \end{bmatrix} \begin{bmatrix} E \\ H \end{bmatrix} \equiv \mathcal{T} \begin{bmatrix} E \\ H \end{bmatrix}, \tag{3.111}$$

where $\eta(t)$ will be determined later. Note that $\mathcal{E}^+ + \mathcal{E}^- = E$. The inverse of $\mathcal{T}$ is given by

$$\mathcal{T}^{-1} = \left[\frac{1}{\sqrt{\epsilon/\mu_0}\,[1+\xi(t)*]} \quad \frac{1}{-\sqrt{\epsilon/\mu_0}\,[1+\xi(t)*]} \right], \tag{3.112}$$

where $\xi(t)$ satisfies the following Volterra equation of the second kind,

$$\xi + \eta + \eta * \xi = 0. \tag{3.113}$$

From the above Volterra equation, one can compute either $\eta(t)$ or $\xi(t)$ if the other is known.

In order to allow a physical interpretation of $\mathcal{E}^+$ and $\mathcal{E}^-$ as right- and left-moving waves, $\mathcal{T}W\mathcal{T}^{-1}$ should be diagonal. Diagonalizing the matrix W, one obtains

$$\partial_z \begin{bmatrix} \mathcal{E}^+ \\ \mathcal{E}^- \end{bmatrix} = \frac{1}{c} \begin{bmatrix} -(1+\xi*) & 0 \\ 0 & (1+\xi*) \end{bmatrix} \partial_t \begin{bmatrix} \mathcal{E}^+ \\ \mathcal{E}^- \end{bmatrix}, \tag{3.114}$$

where the wavefront propagation speed $c = 1/\sqrt{\epsilon\mu_0}$, and

$$\xi(t) = \mathcal{L}^{-1}\left\{\sqrt{1+\bar{\chi}} - 1\right\}, \tag{3.115}$$

$$\eta(t) = \mathcal{L}^{-1}\left\{\frac{1}{\sqrt{1+\bar{\chi}}} - 1\right\}, \tag{3.116}$$

and where $\bar{\chi}$ is the Laplace transform of $\chi(t)$, i.e., $\bar{\chi}(p) = \mathcal{L}\{\chi(t)\} \equiv \int_0^\infty \chi(t)e^{-pt}\,dt$. Therefore, one sees that the condition for the right-moving wave in a dispersive medium is (cf. (3.111) with $\mathcal{E}^- = 0$)

$$E(z,t) - \sqrt{\mu_0/\epsilon}[1+\eta(t)*]H(z,t) = 0, \tag{3.117}$$

or, equivalently (cf. (3.114))

$$[\partial_z + \frac{1}{c}(1+\xi*)\partial_t]E = 0. \tag{3.118}$$

Similarly, the condition for the left-moving wave is

$$E(z,t) + \sqrt{\mu_0/\epsilon}[1+\eta(t)*]H(z,t) = 0, \tag{3.119}$$

or equivalently,

$$[\partial_z - \frac{1}{c}(1+\xi*)\partial_t]E = 0. \tag{3.120}$$

Explicit forms of right- and left-moving wave conditions can thus be found if the inverse Laplace transform of $\{\sqrt{1+\bar{\chi}} - 1\}$ or $\left\{\frac{1}{\sqrt{1+\bar{\chi}}} - 1\right\}$ can be found explicitly (cf. (3.115) and (3.116)). In particular, one finds (He and Ström 1996)

$$\eta(t) = \begin{cases} -\frac{1}{2}\omega_p^2 e^{-\gamma t/2} \int_0^t [I_0(\omega_p t') - I_2(\omega_p t')] J_0[\sqrt{\nu_0^2 + \omega_p^2}\,\sqrt{t^2 - t'^2}]\,dt', \\ \qquad\qquad\qquad\qquad \text{for the Lorentz model,} \\ -\frac{q}{2\tau}[I_0(\frac{q}{2\tau}t) - I_1(\frac{q}{2\tau}t)]e^{-\frac{(1+q/2)}{\tau}t}, \qquad \text{for the Debye model.} \end{cases} \tag{3.121}$$

$$\xi(t) = \begin{cases} \frac{1}{2}\omega_p^2 e^{-\gamma t/2} \int_0^t [J_0(\omega_p t') + J_2(\omega_p t')] J_0[\nu_0 \sqrt{t^2 - t'^2}]\, dt', & \\ \qquad\qquad\qquad\qquad\qquad\qquad \text{for the Lorentz model,} & \\ \frac{q}{2\tau}[I_0(\frac{q}{2\tau}t) + I_1(\frac{q}{2\tau}t)] e^{-\frac{(1+q/2)}{\tau}t}, \quad \text{for the Debye model.} & \end{cases} \tag{3.122}$$

In the general case when none of the inverse Laplace transforms of $\{\sqrt{1+\bar{\chi}} - 1\}$ and $\left\{\frac{1}{\sqrt{1+\bar{\chi}}} - 1\right\}$ can be found explicitly, one can compute $\xi(t)$ by solving numerically the following second-kind Volterra integral equation for $\xi(t)$

$$2\xi(t) + \xi(t) * \xi(t) = \chi(t), \tag{3.123}$$

which is obtained in a straightforward way from (3.115). $\eta(t)$ is then obtained by solving numerically the second-kind Volterra integral equation (3.113).

3.3.3 *The propagator kernel*

The one-way propagation conditions (3.118) and (3.120), respectively, can be utilized to calculate the transient fields propagating in a homogeneous dispersive medium.

Specifically, we assume that in such a medium there is only a right-moving wave (i.e., there is no left-moving reflected wave) and thus one has

$$\mathcal{E}^+(z,t) = E(z,t), \tag{3.124}$$

$$\mathcal{E}^-(z,t) = 0. \tag{3.125}$$

If the field at $z = 0$ is known and given by $f(t)$, the transient field at any position $z > 0$ can be determined. The solution can be expressed in terms of the forward propagator $\mathcal{P}^+(z,0)$ (the other propagator $\mathcal{P}^-(z,z')$ is identically zero in this case; cf. Section 2.4), i.e.,

$$E(z,t+z/c) = [\mathcal{P}^+(z,0)]f(t) \equiv a(z)f(t) + G(z,t) * f(t), \tag{3.126}$$

where $G(z,t)$ is the propagator kernel, and $a(z)$ is an attenuation factor which will be determined below. According to the above definition, the propagator kernel $G(z,t)$ is the regular part of the electric field due to an incident delta-function pulse at $z = 0$.

By differentiating (3.126) with respect to z, the following equation is obtained

$$(\partial_z E)(z,t+z/c) + \frac{1}{c}(\partial_t E)(z,t+z/c) = (\partial_z a)f(t) + (\partial_z G) * f(t).$$

From the right-moving wave condition (3.118), one has

$$(\partial_z E)(z,t+z/c) = -\frac{1}{c}(1+\xi*)(\partial_t E)(z,t+z/c).$$

Therefore, from the above two equations it follows that

$$-\frac{1}{c}\xi * (\partial_t E)(z, t + z/c) = (\partial_z a) f(t) + (\partial_z G) * f(t).$$

Since

$$\xi * (\partial_t E) = \partial_t(\xi * E) = \xi_t * E + \xi(0) E$$

($\xi_t \equiv \partial_t \xi$), one obtains the following equation (cf. (3.126))

$$\left\{\partial_z G + \frac{1}{c} a \xi_t + \frac{1}{c}\xi(0) G + \frac{1}{c}\xi_t * G\right\} * f(t) + \left\{\partial_z a + \frac{1}{c}\xi(0) a\right\} f(t) = 0.$$

Since the incident field $f(t)$ is arbitrary, the above equation implies that the following two equations must hold, namely

$$\partial_z G(z, t) = -\frac{1}{c} a(z) \xi_t - \frac{1}{c}\xi(0) G - \frac{1}{c}\xi_t * G, \tag{3.127}$$

and

$$\partial_z a + \frac{1}{c}\xi(0) a = 0.$$

The latter gives (note that $a(0) = 1$)

$$a(z) = e^{-\xi(0) z/c}. \tag{3.128}$$

From the definition (3.126), it follows that

$$G(0, t) = 0. \tag{3.129}$$

In general, one cannot obtain an explicit solution for $G(z, t)$ from (3.127) and (3.129). However, the propagator kernel $G(z, t)$ can be easily calculated numerically by integrating (3.127) along the $+z$ direction starting from $z = 0$ (cf. (3.129)). Such an integration may be time-consuming if one wants to calculate $G(z, t)$ at a large propagation distance. Since the medium is homogeneous, the propagator $\mathcal{G}(z)$ has the following property (cf. (2.142))

$$\mathcal{P}^+(z_1 + z_2, 0) = \mathcal{P}^+(z_1, 0)\mathcal{P}^+(z_2, 0), \qquad z_1, z_2 \geq 0. \tag{3.130}$$

Taking $z_1 = z_2 = z$ in the above equation, one obtains the following useful formula (cf. the definition (3.126))

$$G(2z, t) = 2a(z) G(z, t) + G(z, t) * G(z, t). \tag{3.131}$$

Using (3.131), one only needs to compute $G(z, t)$, $t \in [0, t_0]$, at a small distance $z = z_0$, and then quickly builds up $G(z, t)$, $t \in [0, t_0]$, at a large distance $z = 2^n z_0$,

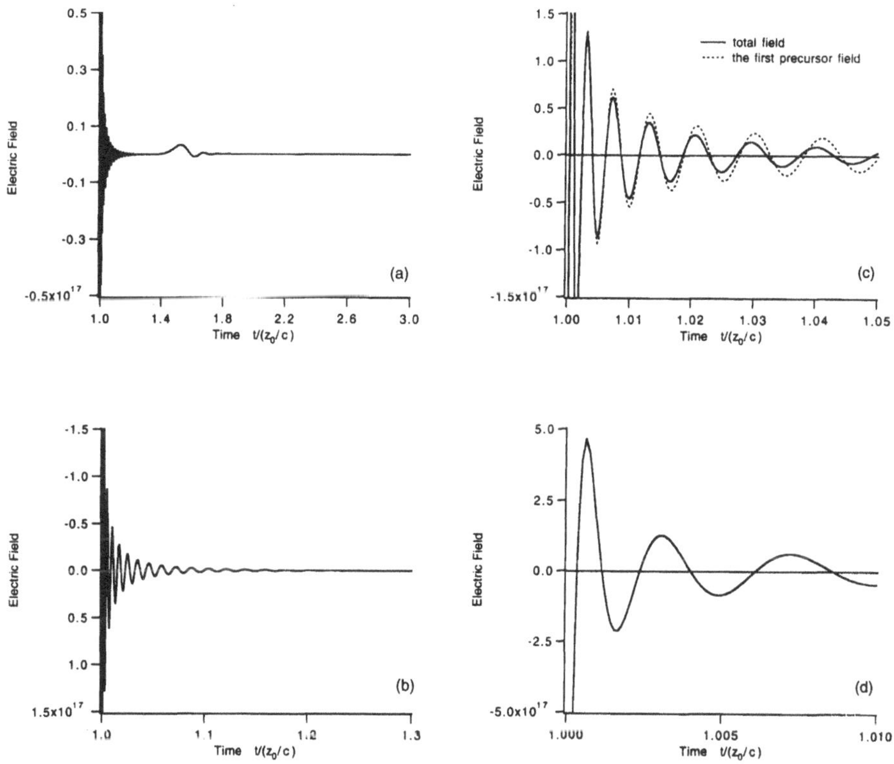

FIG. 3.18. The electric field $E(z_0, t)$ (due to an incident electric field $\delta(t)$ at $z = 0$) at a propagation distance $z = z_0 = 1.0 \times 10^6$ m in a Lorentz medium with $\omega_p = \sqrt{20} \times 10^{16}\,\mathrm{s}^{-1}$, $\omega_0 = 4.0 \times 10^{16}\,\mathrm{s}^{-1}$, $\gamma = 5.6 \times 10^{15}\,\mathrm{s}^{-1}$. (b)–(d) give expanded views for three different ranges of time. The dotted line in (c) shows the first precursor field obtained from the formula (3.133).

$n = 1, 2, 3, \ldots$, by convoluting n times.

Two numerical examples

In the first example, the Lorentz model is considered and the parameters are chosen to have the same values as the parameters used in Wyns et al. (1989), i.e.,

$$\omega_p = \sqrt{20} \times 10^{16}\,\mathrm{s}^{-1}, \qquad \omega_0 = 4.0 \times 10^{16}\,\mathrm{s}^{-1},$$
$$\gamma = 5.6 \times 10^{15}\,\mathrm{s}^{-1}.$$

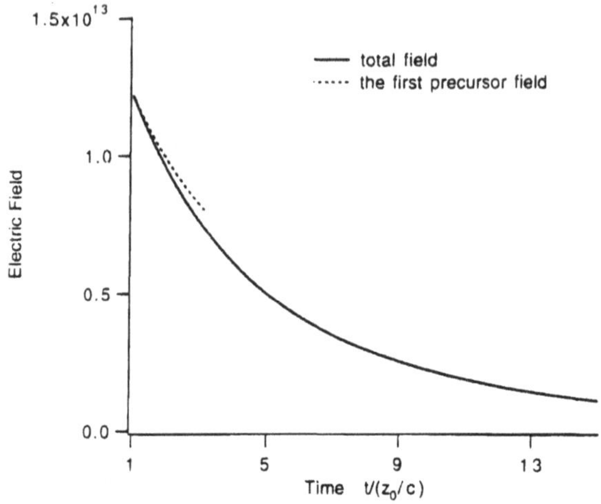

FIG. 3.19. The electric field $E(z_0, t)$ (due to an incident electric field $\delta(t)$ at $z = 0$) at a propagation distance $z = z_0 = 1.0 \times 10^6$ m in a Debye medium with $q = 6.0$, $\tau = 4.0 \times 10^{14}$ s. The dotted line shows the first precursor field obtained from the formula (3.133).

The propagator kernel $G(z, t)$ at a propagation distance $z = z_0 = 1.0 \times 10^6$ m is computed, and the electric field $E(z_0, t) = \delta(t - z_0/c) + G(z, t - z_0/c)$ (due to an incident electric field $\delta(t)$ at $z = 0$) is shown in Fig. 3.18(a). In this computation, 32 points in z and 8192 points in t are used. One sees clearly the arrival of the second (Brillouin) precursor around $t = 1.5z_0/c$ in Fig. 3.18(a). Expanded views of the electric field $E(z_0, t)$ for three different ranges of time are given in Fig. 3.18(b)–(d). Note that the time is scaled by the arrival time z_0/c in these figures. The numerical results shown in Fig. 3.18 are in good agreement with the corresponding numerical results shown in Wyns et al. (1989) (obtained by an approximate numerical method of evaluation of the Laplace integral representation, cf. Figs 1 and 2 given there).

In the second example, the Debye model is considered and the parameters are chosen to be

$$q = 6.0, \qquad \tau = 4.0 \times 10^{14}\,\text{s}.$$

The propagator kernel $G(z, t)$ at a propagation distance $z = z_0 = 1.0 \times 10^6$ m is computed, and the electric field due to an incident electric field $\delta(t)$ is shown in Fig. 3.19. In this computation, 32 points in z and 256 points in t are used. Unlike the Lorentz model, the Debye model does not require so many discretization points in t in the numerical calculation (this is because the splitting kernel

$\xi(t)$ for the Lorentz model is given by Bessel functions which have an oscillating asymptotic behaviour, while $\xi(t)$ for the Debye model is given by modified Bessel functions which do not have such behaviour; that is also the reason why the propagator kernel for the Lorentz model is oscillating, while the propagator kernel for the Debye model at a fixed propagation distance is a smoothly decreasing function of time).

3.3.4 *The first precursor*

In general, one cannot solve the system of equations (3.127) and (3.129) for $G(z,t)$ explicitly. Applying a Laplace transform to the system, one obtains

$$G(z,t) = a(z)\mathcal{L}^{-1}\left\{e^{-\frac{z}{c}[p\bar{\xi}-\xi(0)]} - 1\right\}, \tag{3.132}$$

where $\bar{\xi}$ is the Laplace transform of $\xi(t)$. Only for some special cases can one find an explicit expression for the above inverse Laplace transform. The first precursors for the Lorentz and Debye models are such cases (He and Ström 1996). The first (Sommerfeld) precursor describes the early arrival of the signal (Brillouin 1960). The short-time behaviour of $\chi(t)$ is

$$\chi(t) \approx \begin{cases} \omega_p^2 t, & \text{for the Lorentz model,} \\ \frac{q}{\tau}(1-\frac{t}{\tau}), & \text{for the Debye model,} \end{cases}$$

It then follows from (3.115) that

$$\xi(t) \approx \begin{cases} \omega_p \left[\int_0^t \frac{J_1(\omega_p t')}{t'}\, dt'\right], & \text{for the Lorentz model,} \\ -\frac{\sqrt{q(1+q/4)}}{\tau}\left[\int_0^t \frac{I_1(\sqrt{q(1+q/4)}\, t'/\tau)}{t'}\, e^{-\frac{q}{2\tau}t'}\, dt'\right] + \frac{q}{2\tau}, & \text{for the Debye model.} \end{cases}$$

Therefore, (3.132) gives

$$G(z,t) \approx \begin{cases} -\frac{\omega_p}{c} z \left[\frac{J_1\left(\omega_p\sqrt{t^2+2zt/c}\right)}{\sqrt{t^2+2zt/c}}\right], & \text{for the Lorentz model,} \\ \frac{\sqrt{q(1+q/4)}}{\tau c} z \left[\frac{I_1\left(\frac{\sqrt{q(1+q/4)}}{\tau}\sqrt{t^2+2zt/c}\right)}{\sqrt{t^2+2zt/c}}\right] \exp[-q(z+ct)/(2c\tau)], & \text{for the Debye model.} \end{cases} \tag{3.133}$$

The dotted lines in Fig. 3.18(c) and Fig. 3.19 are the corresponding first precursors for the electric field in the Lorentz and Debye models, respectively, using the above explicit expressions. One sees from Fig. 3.18(c) and Fig. 3.19 that at the early arrival time, the first precursor is a good approximation for the total field. A further discussion of propagation in dispersive media and of precursors is also found in Karlsson and Rikte (1997).

3.3.5 *Transient reflection from a dispersive half-space*

The reflection of a transient pulse at an interface between the vacuum and a homogeneous dispersive medium can be treated in a way that is analogous to the one described in Section 3.2.4 for a conducting half-space.

When $\chi(t)$ is known the reflection kernel $R(t)$ can be calculated from the following Volterra equation of the second kind (He and Ström 1996)

$$R(t) + \frac{\sqrt{\frac{\epsilon_0}{\epsilon}}}{\sqrt{\frac{\epsilon_0}{\epsilon}}+1}\,\eta(t) * R(t) = \frac{2\sqrt{\frac{\epsilon_0}{\epsilon}}}{\left(\sqrt{\frac{\epsilon_0}{\epsilon}}+1\right)^2}\,\eta(t), \tag{3.134}$$

where $\eta(t)$ is the splitting kernel defined by (3.116). In particular, the Volterra equation (3.134) can be solved explicitly (through a Laplace transform) for the Lorentz and Debye models. The explicit form of the reflection kernels are

$$R(t) = \begin{cases} -2e^{-\gamma t/2}\frac{J_2(\omega_p t)}{t} + 2\nu_0 e^{-\gamma t/2}\int_0^t \frac{J_2(\omega_p t')J_1(\nu_0\sqrt{t^2-t'^2})}{\sqrt{t^2-t'^2}}\,dt', \\ \qquad\qquad \text{for the Lorentz model,} \\ -\frac{q\sqrt{\epsilon_0/\epsilon}}{\tau}e^{-\left[1+\frac{q}{1-\epsilon_0/\epsilon}\right]\frac{t}{\tau}}\left[\frac{1}{(1+\sqrt{\epsilon_0/\epsilon})^2}\right. \\ \left. \quad + \frac{1}{1-\epsilon_0/\epsilon}\int_0^t \frac{I_1(\frac{qt'}{2\tau})}{t'}e^{\frac{q(1+\epsilon_0/\epsilon)}{2\tau(1-\epsilon_0/\epsilon)}t'}\,dt'\right], \\ \qquad\qquad \text{for the Debye model.} \end{cases} \tag{3.135}$$

Note that for the Lorentz model one has $\epsilon = \epsilon_0$, while for the Debye model one usually has $\epsilon > \epsilon_0$. If $\epsilon = \epsilon_0$ for the Debye model, then one has

$$R(t) = -\frac{I_1(\frac{qt}{2\tau})}{t}e^{-(1+q/2)\frac{t}{\tau}}, \tag{3.136}$$

instead of the one given in the above equation.

3.3.6 *Reconstruction of the temporal behaviour for a stratified dispersive slab*

It is assumed that $\epsilon_r(z)$ and $\mu_r(z)$ in (3.98) and (3.103) are continuous and piecewise differentiable within the slab $(0, l)$. It is furthermore assumed that the medium outside the dispersive slab is non-dispersive. Thus, for the description of the reflected and transmitted fields outside the dispersive slab we use the simple splitting available for non-dispersive media. This splitting is then used as a basis for the (usual) Green's function formalism for the dispersive medium. We also derive the equations of the invariant imbedding approach and use them in the solution of some of the inverse problems considered (we refer to Romanov and He (1995) for theorems on existence and uniqueness, as well as stability estimates for the inverse problems of determining the susceptibility kernel using reflection and/or transmission data). In the inverse problems studied in the rest of this section, we assume that the exciting field is a normally incident plane wave. The approaches can be easily generalized to the case of oblique incidence (cf. e.g. Hellberg (1995)).

From Maxwell's equations for this case, one has

$$\partial_z \begin{bmatrix} E \\ H \end{bmatrix} = \begin{bmatrix} 0 & -\mu_0\mu_r\partial_t - \mu_0\chi_m * \partial_t \\ -\epsilon_0\epsilon_r\partial_t - \epsilon_0\chi_e * \partial_t & 0 \end{bmatrix} \begin{bmatrix} E \\ H \end{bmatrix}. \quad (3.137)$$

Introduce the following non-dispersive wave-splitting (cf. (3.58) and (3.59))

$$\begin{bmatrix} E^+ \\ E^- \end{bmatrix}(z,t) = \frac{1}{2}\begin{bmatrix} 1 & Z(z) \\ 1 & -Z(z) \end{bmatrix} \begin{bmatrix} E \\ H \end{bmatrix}(z,t), \quad (3.138)$$

where the impedance $Z(z) = \sqrt{\frac{\mu_0\mu_r(z)}{\epsilon_0\epsilon_r(z)}}$. Note that E^+ and E^- have a direct physical interpretation as right- and left-moving electric fields, respectively, only in a homogeneous *non-dispersive* region (the use of a splitting that gives physical right- and left-moving waves in a homogeneous dispersive medium was illustrated in Sections 3.3.2–3.3.5). Differentiating (3.138) with respect to z and using (3.137), one obtains the following system of dynamic equations for the split fields

$$\frac{\partial}{\partial z}\begin{bmatrix} E^+(z,t) \\ E^-(z,t) \end{bmatrix} + \frac{1}{c(z)}\frac{\partial}{\partial t}\begin{bmatrix} E^+(z,t) \\ -E^-(z,t) \end{bmatrix} = \begin{bmatrix} \alpha & \beta \\ \gamma & \vartheta \end{bmatrix} \begin{bmatrix} E^+(z,t) \\ E^-(z,t) \end{bmatrix}, \quad (3.139)$$

where the wave velocity $c = 1/\sqrt{\epsilon_0\mu_0\epsilon_r\mu_r}$ and

$$\begin{cases} \alpha = \frac{1}{2}\frac{d}{dz}\ln Z - A_1 * \partial_t, \\ \beta = -\frac{1}{2}\frac{d}{dz}\ln Z - A_2 * \partial_t, \\ \gamma = -\frac{1}{2}\frac{d}{dz}\ln Z + A_2 * \partial_t, \\ \vartheta = \frac{1}{2}\frac{d}{dz}\ln Z + A_1 * \partial_t, \end{cases} \quad (3.140)$$

and where

$$A_1(z,t) = \frac{1}{2}\left[\epsilon_0 Z(z)\chi_e(z,t) + \mu_0 Z^{-1}(z)\chi_m(z,t)\right], \tag{3.141}$$

$$A_2(z,t) = \frac{1}{2}\left[\epsilon_0 Z(z)\chi_e(z,t) - \mu_0 Z^{-1}(z)\chi_m(z,t)\right]. \tag{3.142}$$

As described in Section 2.1, the Green's functions in a stratified dispersive slab are defined through

$$\begin{bmatrix} E^+ \\ E^- \end{bmatrix}(z, t+\tau(0,z)) = a(0,z)\begin{bmatrix} E^+(0,t) \\ 0 \end{bmatrix} + \begin{bmatrix} G^+(z,t) * E^+(0,t) \\ G^-(z,t) * E^+(0,t) \end{bmatrix}, \tag{3.143}$$

where $\tau(z_1,z_2) = \int_{z_1}^{z_2} \frac{dz'}{c(z')}$, and

$$a(z_1,z_2) = \sqrt{\frac{Z(z_2)}{Z(z_1)}}\exp\left[-\int_{z_1}^{z_2} A_1(z',0)\,dz'\right]. \tag{3.144}$$

After a derivation similar to the one given in Section 2.1, one obtains the following system of equations for the Green's functions

$$\begin{cases} \frac{\partial}{\partial z}G^+ = -a(0,z)\frac{\partial}{\partial t}A_1 + \frac{1}{2}\left(\frac{d}{dz}\ln Z\right)(G^+ - G^-) - \frac{\partial}{\partial t}(A_1 * G^+ + A_2 * G^-), \\ \left(\frac{\partial}{\partial z} - \frac{2}{c}\frac{\partial}{\partial t}\right)G^- = a(0,z)\frac{\partial}{\partial t}A_2 - \frac{1}{2}\left(\frac{d}{dz}\ln Z\right)(G^+ - G^-) \\ \qquad\qquad + \frac{\partial}{\partial t}(A_2 * G^+ + A_1 * G^-), \\ G^-(z,0) = -\frac{1}{2}c\,a(0,z)[A_2(z,0) - \frac{1}{2}\frac{d}{dz}\ln Z], \\ G^+(0,t) = 0, \\ G^-(l,t) = 0. \end{cases} \tag{3.145}$$

As shown in Section 2.1, $G^{\pm}(z,t)$ can be determined in their domain of definition from the above set of equations in the direct problem. The numerical forward algorithm for the dispersive case is completely analogous to the one for the non-dispersive case (for both the Green's function approach and the invariant imbedding approach).

For the inverse problem, it is assumed that one has the following information

$$G^-(0,t) = R(t), \tag{3.146}$$

$$G^+(l,t) = T(t), \tag{3.147}$$

where $R(t)$, $T(t)$ are the reflection and transmission kernels, respectively, for the dispersive slab.

In the above framework we now consider the inverse problem of reconstructing the temporal behaviour of the dispersion. From a physical point of view, the following assumption can often be expected to be relevant

$$\chi_e(z,t) = \tilde{\chi}_e(z)\bar{\chi}_e(t), \qquad \chi_m(z,t) = \tilde{\chi}_m(z)\bar{\chi}_m(t), \tag{3.148}$$

where $\tilde{\chi}_e(z) > 0$ and $\tilde{\chi}_m(z) > 0$ describe the spatial distribution of the dispersion (cf. e.g. the Lorentz and Debye media; the assumption (3.148) then means that the atomic level processes which define $\bar{\chi}_e(t)$ are the same at all depths z, although the dispersion concentration may be different). Then one can reconstruct the functions $\bar{\chi}_e(t)$ and $\bar{\chi}_m(t)$ (the temporal behaviour of the dispersion) simultaneously from the scattering data $R(t)$ and $T(t)$ for a dispersive slab if the other parameters are known, as well as $\bar{\chi}_e(0)$ and $\bar{\chi}_m(0)$. The reconstruction algorithm (Karlsson 1989) is based upon the forward algorithm for the direct problem. The forward algorithm is the same as the one described in Section 2.1.3. Note that $\bar{\chi}_e'(t) \equiv \frac{d}{dt}\bar{\chi}_e$ and $\bar{\chi}_m'(t) \equiv \frac{d}{dt}\bar{\chi}_m$ are reconstructed by the algorithm, and $\bar{\chi}_e(t)$ and $\bar{\chi}_m(t)$ are then obtained by integration in t. If one knows $\bar{\chi}_e'(t_j)$ and $\bar{\chi}_m'(t_j)$ for $j = 0, 1, 2, ..., J$, then one can calculate $R(t_j)$ and $T(t_j)$ up to $t = t_J$ using the forward algorithm. The parameters $\bar{\chi}_e'(t)$ and $\bar{\chi}_m'(t)$ can be reconstructed simultaneously as follows:

(1) Initial values. Letting $t \to 0^+$ in the PDE for $G^+(z,t)$ (cf. (3.145)), and integrating from $z = 0$ to $z = l$, yields

$$\begin{aligned}\epsilon_0 &\left[\int_0^l Z(z)\tilde{\chi}_e(z)\,dz\right]\bar{\chi}_e'(0) + \mu_0\left[\int_0^l Z^{-1}(z)\tilde{\chi}_m(z)\,dz\right]\bar{\chi}_m'(0)\\ &= \int_0^l c(z)\left[p_2^2(z) - \frac{1}{4}(\frac{d}{dz}\ln Z)^2\right]dz - 2[a(0,l)]^{-1}T(0).\end{aligned}$$

Similarly, letting $t \to 0^+$ and $z \to 0^+$ in the PDE for $G^-(z,t)$, and using the initial condition for $G^-(z,t)$, yields

$$\begin{aligned}&\epsilon_0 Z(0)\tilde{\chi}_e(0)\bar{\chi}_e'(0) - \mu_0 Z^{-1}(0)\tilde{\chi}_m(0)\bar{\chi}_m'(0)\\ &= -2R(0)\left[A_1(z,0) + \frac{1}{2}\frac{d}{dz}\ln Z\right]_{z=0+}\\ &-\frac{d}{dz}\left[ca(0,z)\left(A_2(z,0) - \frac{1}{2}\frac{d}{dz}\ln Z\right)\right]_{z=0+} - \frac{4}{c(0)}(\frac{d}{dt}R)(0),\end{aligned}$$

From the above two equations, one can uniquely determine $\bar{\chi}_e'(0)$ and $\bar{\chi}_m'(0)$.

(2) Assume that $\bar{\chi}_e'(t_j)$ and $\bar{\chi}_m'(t_j)$ are known for $j = 0, 1, 2, ..., J-1$, where $J \geq 1$. Discretization of the PDEs for $G^{\pm}(z,t)$ in the direct problem (using the trapezoidal rule) indicates that

$$R(t_J) = B_1(t_J)\bar{\chi}_e'(t_J) + B_2(t_J)\bar{\chi}_e'(t_J) + B_3(t_J),$$

$$T(t_J) = C_1(t_J)\bar{\chi}_e'(t_J) + C_2(t_J)\bar{\chi}_e'(t_J) + C_3(t_J),$$

where $B_i(t_J)$, $C_i(t_J)$, $i = 1, 2, 3$, depend only on the values of $\bar{\chi}_e'(t_j)$ and $\bar{\chi}_m'(t_j)$ for $j < J$. Now choose three different sets of dummy values for $(\bar{\chi}_e'(t_J), \bar{\chi}_m'(t_J))$, and calculate the corresponding dummy values of $(R(t_J), T(t_J))$ using the forward algorithm. Thus, $B_i(t_J)$, $C_i(t_J)$, $i = 1, 2, 3$, are determined from a linear system of six algebraic equations. The true values for $\bar{\chi}_e'(t_J)$ and $\bar{\chi}_m'(t_J)$ are then determined from the true values of $R(t_J)$ and $T(t_J)$ using the above two equations. Applying this procedure repeatedly, one can determine $\bar{\chi}_e'(t)$ and $\bar{\chi}_m'(t)$ for the time period of interest.

Such a reconstruction algorithm can be easily generalized to the bi-isotropic case (Kristensson and Rikte 1993).

We note that the above inverse problem can also be solved by using the invariant imbedding method. One then considers the imbedding geometry, i.e. a subslab $[z, l]$ of the original dispersive slab $[0, l]$, and assumes that at the left side of the subslab $\chi_e = \chi_m = 0$, and ϵ_r, μ_r are temporarily redefined to be constants, continuously terminated to the subslab at the left endpoint z (cf. Chapter 1). For this imbedding geometry, the reflection kernel (denoted $R(z,t)$) and the transmission kernel (denoted $T(z,t)$) are defined as follows

$$E^-(z,t) = R(z,t) * E^+(z,t), \tag{3.149}$$
$$E^+(l, t + \tau(z,l)) = a(z,l)E^+(z,t) + T(z,t) * E^+(z,t). \tag{3.150}$$

After a derivation along the lines given in Chapter 1, one obtains the following imbedding equations for $R(z,t)$ and $T(z,t)$ together with suitable initial and boundary conditions

$$\begin{cases} \left(\frac{\partial}{\partial z} - \frac{2}{c}\frac{\partial}{\partial t}\right) R = \frac{\partial}{\partial t} A_2 + \frac{1}{2}\left(\frac{d}{dz}\ln Z\right) R * R + \frac{\partial}{\partial t}(2A_1 * R + A_2 * R * R), \\ R(z,0) = -\frac{1}{2}c\left[A_2(z,0) - \frac{1}{2}\frac{d}{dz}\ln Z\right], \\ R(l,t) = 0, \\ \left[R(z,t)\right]_{t=2(l-z)/c} = \frac{1}{2}c(l)\left[A_2(z,0) - \frac{1}{2}\frac{d}{dz}\ln Z\right]_{z=l}\exp\left[-2\int_z^l A_1(z',0)\,dz'\right], \end{cases} \tag{3.151}$$

$$\begin{cases} \frac{\partial}{\partial z} T = \frac{1}{2}\left(\frac{d}{dz}\ln Z\right)\left[a(z,l)R - T + R * T\right] \\ \qquad + a(z,l)\frac{\partial}{\partial t}[A_1 + A_2 * R] + \frac{\partial}{\partial t}[A_1 * T + A_2 * R * T], \\ T(l,t) = 0. \end{cases} \tag{3.152}$$

The above system of equations can be used to solve inverse problems of reconstructing the temporal behaviour in a way that is analogous to the Green's function approach.

3.3.7 *Reconstruction of the electric susceptibility kernel for a homogeneous dispersive medium*

In this subsection, we consider the inverse problem of reconstructing the susceptibility kernel using the reflection or transmission data from a homogeneous electrically dispersive slab, i.e., $\chi_m \equiv 0$, $\mu_r = 1$, $\chi_e = \chi(t)$, and $\epsilon_r = 1$. For a *homogeneous* medium, the invariant imbedding approach is more convenient to use. Thus, the system (3.151) reduces to

$$\left(2c\frac{\partial}{\partial z} - 4\frac{\partial}{\partial t}\right) R = \chi' + \chi(0)[2R + R * R] + \chi' * [2R + R * R], \quad (3.153)$$

$$R(z,0) = -\frac{1}{4}\chi(0), \quad (3.154)$$

$$R(l,t) = 0, \quad (3.155)$$

$$[R(z,t)]_{t=2(l-z)/c} = \frac{1}{4}\chi(0)\exp\left[-(l-z)\chi(0)\,dz'\right] \quad (3.156)$$

(where $\chi'(t) = \frac{d}{dt}\chi(t)$).

3.3.7.1 *Reconstruction using the reflection data from a homogeneous half-space*

When the dispersive medium fills a half-space, i.e., when $l = \infty$, the reflection produced by the portion $[z, \infty)$ is independent of z, i.e., $R(z,t) = R(t)$ for a homogeneous dispersive half space. Therefore, it follows from (3.153) and (3.154) that

$$4R(t) + \chi(t) + [\chi * (2R + R * R)](t) = 0. \quad (3.157)$$

For the inverse problem of reconstructing the susceptibility kernel $\chi(t)$, (3.157) is a Volterra equation of the second kind for $\chi(t)$, which is well-posed and can be easily solved by the following scheme (using the trapezoidal rule):

$$\chi_0 = -4R_0, \quad (3.158)$$

$$\chi_J = -\frac{4R_J + [\sum_{j=1}^{J-1} S_{J-j}\,\chi_j + 0.5S_J\,\chi_0]\Delta t}{1 + 0.5S_0\Delta t}, \qquad J = 1, 2, 3, \ldots \quad (3.159)$$

(with $\sum_{j=1}^{J-1} = 0$ when $J = 1$), where $S(t) = (2R + R * R)(t)$.

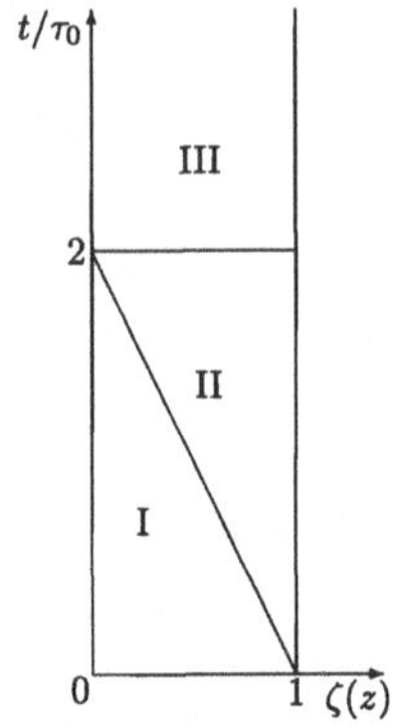

FIG. 3.20. Computation domain for reconstruction of the susceptibility kernel using the reflection data.

3.3.7.2 *Reconstruction using the reflection data from a homogeneous slab*
For a homogeneous dispersive slab between two non-dispersive half-spaces, one expects to be able to use either reflection or transmission data to reconstruct the susceptibility kernel $\chi(t)$ ($\equiv \chi_e(t)$). Consider first the use of reflection data in the equations of the imbedding approach, i.e. (3.153)–(3.156).

Introduce a rectangular grid of points in the computation domain with $\Delta z = l/N$ being the stepsize in z, and $\Delta t = 2\Delta z/c$ being the stepsize in t. Denote $R_{ij} = R(i\Delta z, j\Delta t)$, $\chi_j = \chi(j\Delta t)$. The $\chi'(t)$ can be reconstructed as follows (the susceptibility kernel $\chi(t)$ is then obtained by an integration in t) (Beezley and Krueger 1985):

(1) Solve the Volterra equation (3.157) to determine χ_j for $j = 0, 1, 2, ..., N$. By a difference formula, one immediately obtains χ'_j for $j = 0, 1, 2, ..., N$.

(2) Evaluate $R(z,t)$ in region I (see Fig. 3.20 for the computation domain) by the following formula,

$$R_{i,j} = R_{0,j}, \qquad j = 0, 1, 2, ..., N, \quad i = 0, 1, ..., N - j, \tag{3.160}$$

i.e., R is constant on the horizontal lines t = constant in region I, which can be explained by the fact that the subslab $[z, l]$ for the imbedding geometry gives a reflection that is identical to the one from the original homogeneous slab before the reflections from the back wall return and make a contribution to R. Furthermore, adding the discontinuity, one obtains

$$\begin{aligned} &R_{i,(N-i)+} \\ &= R_{i,(N-i)-} + \frac{1}{8}\chi_0 \exp[-\chi_0 (N-i)\Delta z/c], \quad i = 0, 1, ..., N-1. \end{aligned} \tag{3.161}$$

(3) Calculate $R_{i,j}$ in region II by integrating the imbedding equation (3.153) along the characteristics, from the right to left, bottom to top (starting from $R_{N,j} = 0$, cf. (3.155)). Integration of the imbedding equation (3.153) along the characteristics leads to the following discretized equation (using the trapezoidal rule):

$$\begin{aligned}&2c[R_{i+1,j-1} - R_{i,j}] \\ &= \Delta z[\chi'_j + \chi'_{j-1}]/2 + \Delta z \chi_0 [R_{i+1,j-1} + R_{i,j}] + (\Delta z)^2 \chi_0 [F_{i+1,j-1} + F_{i,j}]/c \\ &\quad +2(\Delta z)^2 [B_{i+1,j-1} + B_{i,j}]/c + 2(\Delta z)^3 [C_{i+1,j-1} + C_{i,j}]/c, \qquad (3.162)\end{aligned}$$

where $F_{i,j} = \sum_{k=1}^{j} R_{i,j-k} R_{i,k}$, $B_{i,j} = (\chi'_j R_{i,0} + \chi'_0 R_{i,j})/2 + \sum_{k=1}^{j} \chi'_{i,j-k} R_{i,k}$, and $C_{i,j} = \chi'_0 F_{i,j}/2 + \sum_{k=1}^{j-1} \chi'_{i,j-k} F_{i,k}$.

(4) Finally in region III, assuming χ'_j and $R_{i,j}$ are known for $j = 1, 2, ..., J-1$, $i = 1, 2, ..., N$, with $J > N$, one can determine χ'_J from (3.162) with $i = 0$ since $R_{0,J}$ is known. $R_{i,J}$, $i = 1, 2, ..., N$, is then calculated from (3.162) after χ'_J is determined. Applying this procedure repeatedly, one can reconstruct χ'_j for $j = N+1, N+2, ...$

The results from the above reconstruction algorithm are illustrated in Fig. 3.21. The following two-resonance (Lorentz-type) model is chosen for the susceptibility kernel

$$\chi(t) = \sin(1.6\pi t)\exp(-0.2t) + 0.5\sin(6\pi t)\exp(-0.5t), \qquad (3.163)$$

as indicated by the solid line in Fig. 3.21. The thickness of the slab is $l = 0.8$, and the wave velocity is chosen to be $c = 1$. The circles in Fig. 3.21 show the reconstruction using the transmission data with 64 points in z. If more than 128 points in z are used the reconstructed $\chi(t)$ is indistinguishable from the true $\chi(t)$. The reconstruction algorithm has been tested for noisy data, and the results are reasonably stable.

3.3.7.3 *Reconstruction using the transmission data from a homogeneous slab*

For a homogeneous dispersive slab, the system (3.152) of equations reduces to

$$2c\frac{\partial}{\partial z}T = \exp\left[-\frac{1}{2c}(l-z)\chi(0)\right]\frac{\partial}{\partial t}[\chi + \chi * R] + \frac{\partial}{\partial t}[\chi * T + \chi * R * T], \qquad (3.164)$$

$$T(l,t) = 0. \qquad (3.165)$$

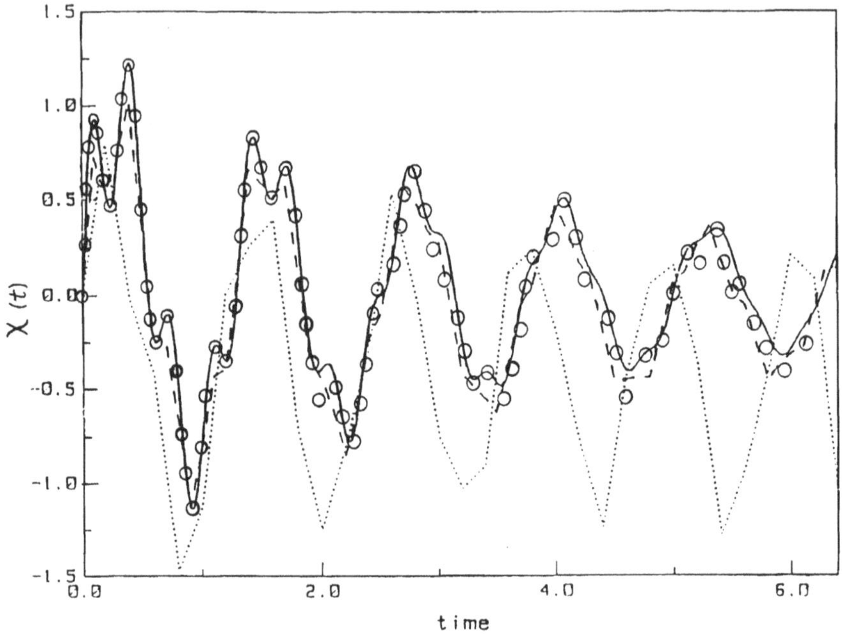

FIG. 3.21. Reconstruction of the susceptibility kernel $\chi(t)$ for a homogeneous dispersive slab. The solid line shows the true values for $\chi(t)$. The circles show the reconstruction using the transmission data with 64 points in z. The dotted line is the reconstruction result using the reflection data with 32 points in z, and the dashed line is the reconstruction result using the reflection data with 64 points in z.

Using the transmission data $T(0,t)$, one can reconstruct $\chi'(t)$ with the following algorithm (if $\chi(0)$ is assumed to be known from the knowledge of, e.g., the attenuation factor $a(0,l)$) (Karlsson 1989):

(1) Initial value. Letting $t \to 0^+$ in the imbedding equation for $T(z,t)$, and integrating from $z=0$ to $z=l$, yields $\chi_0' = -2cT(0,0)/l + \frac{1}{4}\chi_0^2$.

(2) Assume that χ_j' are known for $j = 0,1,2,...,J-1$, where $J \geq 1$. Thus, $R_{i,j}$ can be calculated for $j = 0,1,2,...,J-1$, where $J \geq 1$. Numerical integration of the imbedding equation (3.164) for $T(z,t)$ from $z=l$ to $z=0$ gives

$$T_{0,J} = C_1(t_J)\chi_J' + C_2(t_J), \tag{3.166}$$

where $C_i(t_J)$, $i=1,2$, depend only on the values of χ_j' for $j<J$. Now choose a trivial dummy value for χ_J', i.e., $\chi_J' = 0$, and calculate the corresponding

$T_{0,J}$ using a forward algorithm for the direct problem (first calculate $R_{i,j}$ for $j = 0, 1, 2, ..., J$, $i = 1, 2, ..., N$ using the system of equations (3.153)–(3.156)). This dummy value for $T_{0,J}$ is equal to C_2. Then choose a non-zero dummy value for χ'_J, calculate again the corresponding $T_{0,J}$ using a forward algorithm. Thus C_1 is determined from (3.166). The true values for $\chi'_J = 0$ are finally determined from the true values of $T_{0,J}$ using (3.166). Applying such a procedure repeatedly, one can determine $\chi'(t)$ for the time period of interest ($\chi(t)$ is then obtained by a straightforward integration in t).

To test the reconstruction algorithm, we use the same two-resonance model as used in Section 3.3.7.2, and the reconstruction results are given in the same figure (Fig. 3.21). In Fig. 3.21, the dotted line is the reconstruction result using the reflection data with 32 points in z, and the dashed line is the reconstruction result using the reflection data with 64 points in z (the solid line is the true $\chi(t)$). If more than 128 points in z are used the reconstructed $\chi(t)$ is virtually indistinguishable from the true $\chi(t)$. The reconstruction of the susceptibility kernel $\chi(t)$ has also been carried out using experimental data as described in Section 5.3.

These reconstruction algorithms can be generalized to the case when the impedance is discontinuous at the two interfaces $z = 0$ and $z = l$ (cf. e.g. Karlsson (1989)).

Exercises for Chapter 3

Exercise 3.1. *Transmission lines with frequency-dependent parameters.* Consider a time domain model for a non-uniform line with frequency-dependent parameters. The model system of differential equations for the transient voltage $V(z,t)$ and current $I(z,t)$ is taken to be

$$\begin{aligned}\partial_z I(z,t) = & -\bar{G}(z)V(z,t) - \bar{\bar{G}}(z,t) * V(z,t) \\ & -\bar{C}(z)\partial_t V(z,t) - \partial_t[\bar{\bar{C}}(z,t) * V(z,t)], \\ \partial_z V(z,t) = & -\bar{R}(z)I(z,t) - \bar{\bar{R}}(z,t) * I(z,t) \\ & -\bar{L}(z)\partial_t I(z,t) - \partial_t[\bar{\bar{L}}(z,t) * I(z,t)].\end{aligned}$$

Derive the corresponding system of equations for the Green's functions for the case of a dispersive line terminated continuously by two homogeneous non-dispersive LC lines at the two ends $z = 0$ and $z = l$.

Exercise 3.2. *Reconstruction of the electron density of an inhomogeneous plasma.* Consider a neutral plasma characterized by an electron density n, an effective collision frequency ν (a non-negative constant related to the collisions between the electron and other molecules). If the positive charges are heavy and the restoring force between the electrons and the positive charges is neglected, the polarization of the plasma is governed by the following equation of motion for an electron

$$m\left[\partial_t^2\mathbf{r} + \nu\partial_t\mathbf{r}\right] = q\vec{\mathbf{E}}(\mathbf{r},t),$$

where m, q are the mass and charge of the electron, respectively. By introducing the following definition of the current density $\mathbf{J}$

$$\mathbf{J}(\mathbf{r},t) = nq\partial_t\mathbf{r},$$

the equation of motion becomes

$$\partial_t\mathbf{J} + \nu\mathbf{J} = \frac{nq^2}{m}\vec{\mathbf{E}},$$

which has the following solution

$$\mathbf{J}(\mathbf{r},t) = \frac{q^2n}{m}\int_{-\infty}^{t} e^{-\nu(t-t')}\vec{\mathbf{E}}(\mathbf{r},t')\,dt'.$$

From the above equation, one sees that a plasma can be characterized by a frequency-dependent conductivity in the frequency domain, or, equivalently, a conductivity kernel in the time domain. Based upon the above physical model, we consider the following Maxwell's equation for a plasma

$$\begin{aligned}\nabla\times\vec{\mathbf{H}} &= \epsilon\partial_t\vec{\mathbf{E}} + \int_{-\infty}^{t}\Sigma(\cdot,t-t')\vec{\mathbf{E}}(\cdot,t')\,dt',\\ \nabla\times\vec{\mathbf{E}} &= -\mu\partial_t\vec{\mathbf{H}},\end{aligned}$$

where the permittivity ϵ, and the permeability μ are constant parameters (ϵ and μ for most plasmas are close to the values of the permittivity and permeability in vacuum, respectively), and the conductivity kernel $\Sigma(x,t)$ is given by

$$\Sigma(x,t) = \frac{q^2n}{m}e^{-\nu t} \equiv Qe^{-\nu t}.$$

Consider a plasma half-space with a stratified electron density $n(z)$, which is quiescent before the time $t=0$.

(1) Derive the corresponding system of equations for the Green's functions for normally incident plane waves.

(2) Write a computer program to reconstruct the electron density $n(z)$ using the reflection data.

Exercise 3.3. *Impulsive reflection from a homogeneous dispersive half-space.* Show that the impulsive reflection $R(t)$ from a homogeneous dispersive half-space satisfies the Volterra equation (3.134).

Exercise 3.4. *Design of a reflectionless profile.* If the susceptibility kernels $\chi_e(z,t)$ and $\chi_m(z,t)$ are known, then the permittivity $\epsilon_r(z)$ or permeability $\mu_r(z)$ can be reconstructed using the reflection data $R(t)$ for one round trip period of time. The reconstruction algorithm is the same as the one described in Section 2.1.4. The design problem for a reflectionless half-space is a special type of inverse problem. For a non-dispersive medium, the PDEs for the Green's functions are homogeneous, and thus a sufficient and necessary condition for a reflectionless half-space is that $\gamma = 0$ for all $z \in [0, l]$ (cf. (2.32)), which guarantees that there is no local reflection at any position z (i.e., $G^-(z,t) = 0$ for all z). For a dispersive medium, the PDEs for the Green's functions are not homogeneous, and consequently the condition $\gamma = 0$ is not a necessary condition for a reflectionless half-space. In other words, the half-space can be reflectionless even though internal local reflections exist (i.e., $G^-(0,t) \equiv 0$ even though $G^-(z,t)$ may be non-zero for $z > 0$). Thus, setting

$$R(t) = 0$$

as prescribed values in the code for the reconstruction, one can reconstruct a reflectionless profile for the permittivity $\epsilon_r(z)$ or permeability $\mu_r(z)$ (the other parameters are assumed to be known). Write a numerical code to obtain the reflectionless permittivity profile $\epsilon_r(z)$ for an electrically dispersive half-space ($\mu_r(z) = 1$, $\chi_m(z,t) \equiv 0$; to the left of the stratified half-space, it is a homogeneous non-dispersive dielectric medium with $\epsilon_r = 2$) for (a) a Lorentz model with $\chi_e = \omega_p^2 \frac{\sin\sqrt{\omega_0^2-\gamma_0^2}\,t}{\sqrt{\omega_0^2-\gamma_0^2}} \exp[-\gamma_0 t]$, with $\omega_p = 5\times 10^9$ s^{-1}, $\omega_0 = 2\times 10^{10}$ s^{-1}, and $\gamma_0 = 5\times 10^9$ s^{-1}; (b) a Debye model with $\chi_e = \alpha_0 \exp[-t/\tau_0]$ with $\alpha_0 = 5 \times 10^8$ s^{-1} and $\tau_0 = 10^{-9}$ s. Note that such a design problem is well-posed. Also note that if the dispersive medium is highly lossy, then one can cut off the designed profile at a certain position $z = z_0 > 0$, and coat the designed slab $[0, z_0]$ on a planar surface to reduce the reflection from this surface. (Time domain wave-splitting treatments of non-reflecting or low-reflecting media are given in e.g. Hellberg et al. (1992), Hellberg (1995).)

4

INVERSE PROBLEMS RELATED TO FIELDS FROM LOCALIZED SOURCES OVER A STRATIFIED HALF-SPACE

Inverse problems involving a localized source over a stratified half-space are relevant in many different contexts, such as non-invasive investigations of subsurface physical properties by means of acoustic, electromagnetic or elastodynamic waves. In the present chapter the previous treatment is generalized inasmuch as instead of normally or obliquely incident plane waves, we now consider more general incident fields generated within a bounded region. The main assumption about the half-space to be investigated is the same as before, i.e., it is assumed to be plane-stratified. For these more general incident fields it is shown how the use of suitably defined moments of the fields and the reflection kernel can be used to solve inverse problems. In particular it is shown how the use of the zeroth, second and fourth moments of the three-dimensional fields reduce the inverse problem to a set of one-dimensional problems. When one has additional symmetry properties of the incident field, such as axial symmetry, it is shown how the inverse problem is reduced to a one-dimensional problem through the use of a Hankel transform. In this case an explicit form of the wave-splitting operator is found in the Hankel transformed space. The Green's function formalism for this case is developed. The explicit form of the wave-splitting is also used to treat wave propagation in homogeneous plasmas, dispersive media and waveguides.

The following model equation will be considered

$$\partial_t^2 u = \frac{1}{r}\nabla \cdot (p\nabla u) - b\partial_t u + qu, \tag{4.1}$$

where the coefficients r, p, b and q are functions of the depth z in the lower half-space $z > 0$, with p, r being positive and twice differentiable with bounded derivatives, and

$$\frac{p}{r} = c^2 > 0. \tag{4.2}$$

The scalar field u may stand for, e.g. an acoustic pressure field generated by an explosive source, the Hertz potential or the azimuthal component of the electric (magnetic) field of a vertical magnetic (electric) dipole, etc. Equation (4.1) is sometimes referred to as the telegraph equation. All sources are assumed to be located in the upper half-space $z \leq 0$ which is assumed to be free space, i.e., it is characterized by

$$c = c_0, \quad p = 1, \quad b = q = 0, \quad z \leq 0. \tag{4.3}$$

Besides reconstruction problems associated with (4.1), a permittivity tensor reconstruction for an anisotropic medium is considered in a treatment which is based on Maxwell's equations in vector form.

4.1 Parameter reconstruction using moments of the fields

4.1.1 *Equations for the split moments*

In the upper half-space $z \leq 0$, (4.1) reduces to the wave equation

$$\left(\frac{1}{c^2}\partial_t^2 - \nabla^2\right) u = 0, \tag{4.4}$$

and the wave-splitting of this equation which gives up- and downwards travelling waves has been given by Weston (1987) and the reflection operator is referred to this splitting. We refer to Chapter 7 for a more detailed discussion of this splitting and in the present section we shall confine ourselves to quoting some of its main properties.

Since the medium is plane-stratified, (4.1) reduces to the following equation

$$\frac{1}{c^2}\partial_t^2 u - \nabla^2 u + \frac{b}{c^2}\partial_t u - (\partial_z \ln p)\partial_z u \quad \frac{q}{c^2} u = 0. \tag{4.5}$$

The sources, with compact spatial support, are located in the homogeneous upper half-space $z < 0$. The sources are turned on at a certain time and the incident fields will not reach the surface $z = 0$ until the time $t = 0$. This leads to the initial conditions

$$u(x, y, z, 0) = \partial_t u(x, y, z, 0) = 0, \quad \text{for } z > 0. \tag{4.6}$$

The scattering configuration is shown in Fig. 4.1.

In the homogeneous upper half-space we have the wave equation (4.5), which is also the principal part of (4.5), and we introduce the following wave-splitting (see Section 7.1)

$$u^{\pm} = (1/2)[u \mp \mathbf{K}u_z] \tag{4.7}$$

($u_z \equiv \partial_z u$), where $\mathbf{K}$ and its inverse $\mathbf{K}^{-1}$ satisfy

$$\mathbf{K}^{-1} = (\frac{1}{c^2}\partial_t^2 - \partial_x^2 - \partial_y^2)\mathbf{K}. \tag{4.8}$$

Down-going waves at plane $z = z_0$ are defined as those generated by sources in the region $z < z_0$, with the other half-space $z > z_0$ being source-free. Up-going

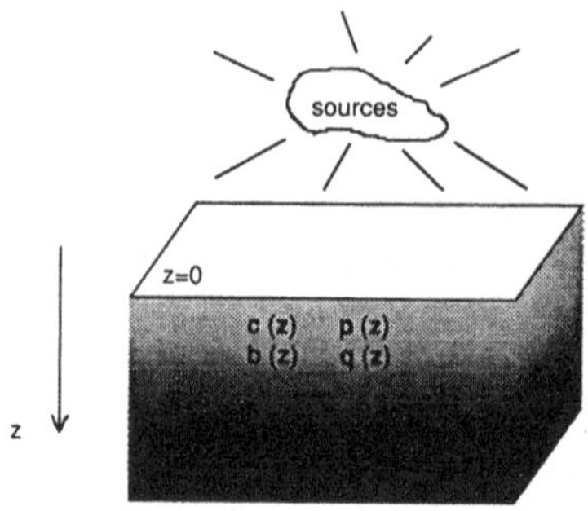

FIG. 4.1. The scattering configuration.

waves are defined correspondingly. The operator $\mathbf{K}$ has the following explicit form (see Section 7.1; cf. (7.26))

$$\mathbf{K}f(x,y,z,t) = \frac{1}{2\pi}\int\int_{\mathbb{R}^2} \frac{1}{\rho} f(x',y',z,\tau)\Theta(\tau)\,dx'\,dy', \tag{4.9}$$

where $\rho = \sqrt{(x-x')^2 + (y-y')^2}$, $\tau = t - \frac{\rho}{c(z)}$, $\Theta(\tau)$ is the Heaviside unit step function defined by (2.12), and $\mathbb{R}^2$ denotes the whole (x, y) plane. In the homogeneous non-dissipative upper half-space, the splitting (4.7) splits the total field $u(x, y, z, t)$ into two components, a down-going field $u^+(x, y, z, t)$ (incident field) and an up-going field $u^-(x, y, z, t)$ (reflected field). However, as in the one-dimensional cases treated in the previous chapters, this splitting has no physical significance in the inhomogeneous lower half-space but is merely a useful change of variables from $u(x, y, z, t)$, $u_z(x, y, z, t)$ to $u^\pm(x, y, z, t)$.

The up-going wave u^- is linearly related to the down-going wave u^+ through the reflection operator $\mathcal{R}$ defined by (Weston 1988b)

$$\begin{aligned} u^-(x,y,z,t) &= \mathcal{R}u^+(x,y,z,t) \\ &\equiv \int_{\mathbb{R}^2}\int_0^t R(x-x',y-y',z,t-s)u(x',y',z,s)\,dx'\,dy'\,ds, \end{aligned} \tag{4.10}$$

where $R(x-x', y-y', z, t-s)$ is the reflection kernel. As shown in Buzdin (1971), Weston (1988b), He and Weston (1992), for the inverse problem in a stratified medium, one does not need to know the precise form of the reflection operator in order to be able to reconstruct the parameters; it suffices to know several of its moments.

We shall consider the zeroth, second and fourth moments of the field u, defined by

$$u_0(z,t) = \int\int_{\mathbb{R}^2} u(x,y,z,t)\,dx\,dy, \tag{4.11}$$

$$u_2(z,t) = \int\int_{\mathbb{R}^2} (x^2+y^2)u(x,y,z,t)\,dx\,dy, \tag{4.12}$$

$$u_4(z,t) = \int\int_{\mathbb{R}^2} (x^2+y^2)^2 u(x,y,z,t)\,dx\,dy, \tag{4.13}$$

respectively. Since the sources have compact support in space, the field $u(x,y,z,t)$ is non-zero only within a finite region $\sqrt{x^2+y^2} < \rho_0$ on a plane $z = z_0 \geq 0$ at any finite time $t = t_0$. Therefore, a transverse spatial transform, which reduces the three-dimensional problem to a set of one-dimensional problems, does not have any convergence problems in a time domain method. In other words, all the moments converge for all finite times.

For a plane-stratified medium, the reflection kernel $R(x-x', y-y', z, t-s)$ in (4.10) has the form

$$R(x-x', y-y', z, t-s) = R(\sqrt{(x-x')^2+(y-y')^2}, z, t-s).$$

Setting $x - x' = \rho\cos\phi$, $y - y' = \rho\sin\phi$, $x' = \rho'\cos\phi'$, $y' = \rho'\sin\phi'$, one has $x^2+y^2 = \rho^2+\rho'^2+2\rho\rho'\cos(\phi-\phi')$, and the nth moments ($n = 0, 2, 4$) of the reflection operator are

$$\begin{aligned}
&(\mathbf{R}u^+)_n \\
&= \int_{\mathbb{R}^2}\int_{\mathbb{R}^2}\int_0^t (x^2+y^2)^n R(x-x', y-y', z, t-s)u^+(x', y', z, s)\,dx'\,dy'\,dx\,dy\,ds \\
&= \int_{\mathbb{R}^2}\int_{\mathbb{R}^2}\int_0^t [\rho^2+\rho'^2+2\rho\rho'\cos(\phi-\phi')]^n R(\rho, z, t-s)u^+(\rho', \phi', z, s)\,d\sigma'\,d\sigma\,ds,
\end{aligned}$$

where $d\sigma = dx\,dy$ and $d\sigma' = dx'\,dy'$. From (4.10) and the above equation, one can show that the split moments are related to each other by the moments of the reflection operator (Weston 1988b; He et al. 1993a)

$$u_0^-(z,t) = R_0(z,t) * u_0^+(z,t), \tag{4.14}$$

$$u_2^-(z,t) = R_2(z,t) * u_0^+(z,t) + R_0(z,t) * u_2^+(z,t), \tag{4.15}$$

$$\begin{aligned}
&u_4^-(z,t) \\
&= R_4(z,t) * u_0^+(z,t) + 2R_2(z,t) * u_2^+(z,t) + R_0(z,t) * u_4^+(z,t),
\end{aligned} \tag{4.16}$$

where R_0, R_2 and R_4 are the zeroth, second and fourth moments of the reflection kernel R, respectively (cf. Exercise 4.1). Note that $R_0(0,t)$, $R_2(0,t)$ and $R_4(0,t)$ are obtained from measurable quantities and later they will be used as given data for the reconstruction.

Similarly, taking the zeroth, second and fourth moments of (4.7), one obtains the split moments in the following forms (cf. Exercise 4.2)

$$u_0^\pm = (1/2)[u_0 \mp k_0 * u_{0z}], \tag{4.17}$$

$$u_2^\pm = (1/2)[u_2 \mp (k_0 * u_{2z} + k_2 * u_{0z})], \tag{4.18}$$

$$u_4^{\pm} = (1/2)[u_4 \mp (k_0 * u_{4z} + 2k_2 * u_{2z} + k_4 * u_{0z})] \tag{4.19}$$

($u_{0z} \equiv \partial_z u_0$, etc.). Using the explicit form (4.9) of the **K** operator, one obtains

$$k_n = c^{n+1} t^n, \quad n = 0, 2, 4. \tag{4.20}$$

For convenience of calculation, the following notation and property are used,

$$\partial_t^{-n} f(z,t) \equiv \underbrace{1 * 1 * \ldots * 1 *}_{n} f(z,t) = \frac{1}{(n-1)!} t^{n-1} * f(z,t), \qquad n = 1, 2, 3, \ldots. \tag{4.21}$$

Therefore, it follows from (4.17)–(4.19) that

$$\begin{bmatrix} u_0^+ \\ u_0^- \\ u_2^+ \\ u_2^- \\ u_4^+ \\ u_4^- \end{bmatrix} = \begin{bmatrix} T_0 & 0 & 0 \\ T_1 & T_0 & 0 \\ T_2 & 2T_1 & T_0 \end{bmatrix} \begin{bmatrix} u_0 \\ u_{0z} \\ u_2 \\ u_{2z} \\ u_4 \\ u_{4z} \end{bmatrix} \equiv T \begin{bmatrix} u_0 \\ u_{0z} \\ u_2 \\ u_{2z} \\ u_4 \\ u_{4z} \end{bmatrix}, \tag{4.22}$$

where

$$T_0 = \frac{1}{2} \begin{bmatrix} 1 & -c\partial_t^{-1} \\ 1 & c\partial_t^{-1} \end{bmatrix}, \quad T_1 = \begin{bmatrix} 0 & -c^3 \partial_t^{-3} \\ 0 & c^3 \partial_t^{-3} \end{bmatrix}, \quad T_2 = \begin{bmatrix} 0 & -12c^5 \partial_t^{-5} \\ 0 & 12c^5 \partial_t^{-5} \end{bmatrix}.$$

Note that $u_i = u_i^+ + u_i^-$, $i = 0, 2, 4$.

Taking moments of the telegraph equation (4.5), one obtains the following equations

$$\frac{1}{c^2}\partial_t^2 u_0 - \partial_z^2 u_0 + \frac{b}{c^2}\partial_t u_0 - (\partial_z \ln p)\partial_z u_0 - \frac{q}{c^2} u_0 = 0, \tag{4.23}$$

$$\frac{1}{c^2}\partial_t^2 u_2 - \partial_z^2 u_2 + \frac{b}{c^2}\partial_t u_2 - (\partial_z \ln p)\partial_z u_2 - \frac{q}{c^2} u_2 - 4u_0 = 0, \tag{4.24}$$

$$\frac{1}{c^2}\partial_t^2 u_4 - \partial_z^2 u_4 + \frac{b}{c^2}\partial_t u_4 - (\partial_z \ln p)\partial_z u_4 - \frac{q}{c^2} u_4 - 16u_2 = 0. \tag{4.25}$$

Note that the reduced wave equation (4.24) for u_2 is coupled to u_0 and the reduced wave equation (4.25) for u_4 is coupled to u_2. Equations (4.23)–(4.25) can be rewritten in matrix form (by adding three identities) as

$$\partial_z \begin{bmatrix} u_0 \\ u_{0z} \\ u_2 \\ u_{2z} \\ u_4 \\ u_{4z} \end{bmatrix} = \begin{bmatrix} D_0 & 0 & 0 \\ D_1 & D_0 & 0 \\ 0 & D_2 & D_0 \end{bmatrix} \begin{bmatrix} u_0 \\ u_{0z} \\ u_2 \\ u_{2z} \\ u_4 \\ u_{4z} \end{bmatrix} \equiv D \begin{bmatrix} u_0 \\ u_{0z} \\ u_2 \\ u_{2z} \\ u_4 \\ u_{4z} \end{bmatrix}, \tag{4.26}$$

where

$$D_0 = \begin{bmatrix} 0 & 1 \\ \frac{1}{c^2}\partial_t^2 + \frac{b}{c^2}\partial_t - \frac{q}{c^2} & -\partial_z \ln p \end{bmatrix}, \quad D_1 = \begin{bmatrix} 0 & 0 \\ -4 & 0 \end{bmatrix}, \quad D_2 = \begin{bmatrix} 0 & 0 \\ -16 & 0 \end{bmatrix}.$$

Differentiating (4.76) with respect to z and using (4.75), yields

$$\partial_z \begin{bmatrix} u_0^+ \\ u_0^- \\ u_2^+ \\ u_2^- \\ u_4^+ \\ u_4^- \end{bmatrix} = TDT^{-1} \begin{bmatrix} u_0^+ \\ u_0^- \\ u_2^+ \\ u_2^- \\ u_4^+ \\ u_4^- \end{bmatrix} + \frac{\partial T}{\partial z} T^{-1} \begin{bmatrix} u_0^+ \\ u_0^- \\ u_2^+ \\ u_2^- \\ u_4^+ \\ u_4^- \end{bmatrix}. \tag{4.27}$$

After a matrix calculation, one has (He et al. 1993a)

$$\partial_z \begin{bmatrix} u_0^+ \\ u_0^- \\ u_2^+ \\ u_2^- \\ u_4^+ \\ u_4^- \end{bmatrix} = \begin{bmatrix} A & 0 & 0 \\ P & A & 0 \\ Q & 2P & A \end{bmatrix} \begin{bmatrix} u_0^+ \\ u_0^- \\ u_2^+ \\ u_2^- \\ u_4^+ \\ u_4^- \end{bmatrix}, \tag{4.28}$$

where

$$A = \begin{bmatrix} \alpha & \beta \\ \gamma & \vartheta \end{bmatrix}, \quad P = \begin{bmatrix} p_{11} & p_{12} \\ p_{21} & p_{22} \end{bmatrix}, \quad Q = \begin{bmatrix} q_{11} & q_{12} \\ q_{21} & q_{22} \end{bmatrix},$$

with

$$\begin{cases} \alpha = -\frac{1}{c}\partial_t - \frac{b}{2c} + \frac{q}{2c}\partial_t^{-1} - \frac{1}{2}(\partial_z \ln p) + \frac{c_z}{2c}, \\ \beta = -\frac{b}{2c} + \frac{q}{2c}\partial_t^{-1} + \frac{1}{2}(\partial_z \ln p) - \frac{c_z}{2c}, \\ \gamma = \frac{b}{2c} - \frac{q}{2c}\partial_t^{-1} + \frac{1}{2}(\partial_z \ln p) - \frac{c_z}{2c}, \\ \vartheta = \frac{1}{c}\partial_t + \frac{b}{2c} - \frac{q}{2c}\partial_t^{-1} - \frac{1}{2}(\partial_z \ln p) + \frac{c_z}{2c}, \end{cases} \tag{4.29}$$

$$\begin{cases} p_{11} = 2c\partial_t^{-1} - bc\partial_t^{-2} + qc\partial_t^{-3} + 2cc_z\partial_t^{-2}, \\ p_{12} = -bc\partial_t^{-2} + qc\partial_t^{-3} - 2cc_z\partial_t^{-2}, \\ p_{21} = bc\partial_t^{-2} - qc\partial_t^{-3} - 2cc_z\partial_t^{-2}, \\ p_{22} = -2c\partial_t^{-1} + bc\partial_t^{-2} - qc\partial_t^{-3} + 2cc_z\partial_t^{-2}, \end{cases} \tag{4.30}$$

$$\begin{cases} q_{11} = 4c^3\partial_t^{-3} - 12c^3(b\partial_t^{-4} - q\partial_t^{-5}) + 40c^3c_z\partial_t^{-4}, \\ q_{12} = -12c^3\partial_t^{-3} - 12c^3(b\partial_t^{-4} - q\partial_t^{-5}) - 40c^3c_z\partial_t^{-4}, \\ q_{21} = 12c^3\partial_t^{-3} + 12c^3(b\partial_t^{-4} - q\partial_t^{-5}) - 40c^3c_z\partial_t^{-4}, \\ q_{22} = -4c^3\partial_t^{-3} + 12c^3(b\partial_t^{-4} - q\partial_t^{-5}) + 40c^3c_z\partial_t^{-4}. \end{cases} \tag{4.31}$$

The set of equations in the system (4.28) expresses the dynamics for the split moments.

4.1.2 *Imbedding equations and parameter reconstruction*

The equations for the zeroth moment R_0 can be derived in a completely analogous way as the one shown in Section 1.6 for the reflection kernel. The imbedding equation for the zeroth moment R_0 is found to be

$$\begin{aligned} \partial_z R_0 = (\frac{2}{c}\partial_t + \frac{b}{c})R_0 + \frac{1}{2}(\frac{b}{c} - \partial_z \ln p + \frac{c_z}{c})R_0 * R_0 \\ -\frac{q}{2c}(1 + 2 * R_0 + 1 * R_0 * R_0), \end{aligned} \tag{4.32}$$

and the initial condition is

$$R_0(z,0) = \frac{1}{4}(c_z - b - c\partial_z \ln p). \tag{4.33}$$

In the inverse problem, the initial condition (4.33) can be used to reconstruct c_z (c is then obtained by an integration in z), or b, or p, by propagating the boundary data $R_0(0,t)$ to the initial data $R_0(z,0)$ using the imbedding equation (4.32) when the other three parameters are known. If the reconstruction is required only up to a depth $z = z_{max}$, it suffices to know the boundary data $R_0(0,t)$ for a finite period of time, i.e., $0 \le t \le 2\int_0^{z_{max}} \frac{1}{c(z')}\,dz'$.

Substituting (4.15) into the dynamics equation (4.28) yields

$$\begin{aligned} &R_{2z} * u_0^+ + R_2 * (\alpha u_0^+ + \beta R_0 * u_0^+) + R_{0z} * u_2^+ \\ &+R_0 * [p_{11}u_0^+ + p_{12}(R_0 * u_0^+) + \alpha u_2^+ + \beta(R_2 * u_0^+ + R_0 * u_2^+)] \\ &= p_{21}u_0^+ + p_{22}(R_0 * u_0^+) + \gamma u_2^+ + \vartheta(R_2 * u_0^+ + R_0 * u_2^+)\,(\equiv \partial_z u_2^-). \end{aligned} \tag{4.34}$$

Since the kernels R_i, $i = 0,2,4$, depend only on the parameters of the medium, but not on the form of the incident field, any relation between these kernels (including the imbedding equations) should also be independent of the form of the incident field u^+. Therefore, the easiest way to obtain the imbedding equation for R_2 is to set the arbitrary function $u_2^+ \equiv 0$ in (4.34), which gives

$$\begin{aligned} &R_{2z} * u_0^+ + R_2 * (\alpha u_0^+ + \beta R_0 * u_0^+) + R_0 * [p_{11}u_0^+ + p_{12}(R_0 * u_0^+) + \beta(R_2 * u_0^+)] \\ &= p_{21}u_0^+ + p_{22}(R_0 * u_0^+) + \vartheta(R_2 * u_0^+). \end{aligned} \tag{4.35}$$

Substituting (4.29) and (4.30) into (4.35), yields

$$\begin{aligned} &\Big\{-R_{2z} + \frac{2}{c}R_{2t} + [(b - 2c_z)ct - \frac{1}{2}qct^2] + 2[(bt-2)c - \frac{1}{2}qct^2] * R_0 \\ &+[(b+2c_z)ct - \frac{1}{2}qct^2] * R_0 * R_0 + \frac{b}{c}R_2 - \frac{q}{c} * R_2 + (\frac{b}{c} - \partial_z \ln p + \frac{c_z}{c})R_0 * R_2 \\ &-\frac{q}{c} * R_0 * R_2\Big\} * u_0^+ + \frac{2}{c}R_2(z,0)u_0^+ = 0, \end{aligned} \tag{4.36}$$

which gives the following imbedding equation and initial condition for the second moment R_2,

$$\partial_z R_2 = \left(\frac{2}{c}\partial_t + \frac{b}{c}\right) R_2 + F_1 * R_2 + F_2, \tag{4.37}$$

$$R_2(z,0) = 0, \tag{4.38}$$

where

$$F_1(z,t) = \left(\frac{b}{c} - \partial_z \ln p + \frac{c_z}{c}\right) R_0 - \frac{q}{c} * R_0 - \frac{q}{c}, \tag{4.39}$$

$$F_2(z,t) = [(b+2c_z)ct - \frac{1}{2}qct^2] * R_0 * R_0 + 2[(bt-2)c - \frac{1}{2}qct^2] * R_0$$
$$+[(b-2c_z)ct - \frac{1}{2}qct^2]. \tag{4.40}$$

Equations (4.37) and (4.38) constitute the sought-for equations for the second moment R_2. However, in the inverse problem the initial condition (4.38) does not give any information for the reconstruction. Thus one needs to find a non-zero initial condition. Letting $t = 0$ in (4.37), one obtains the initial value of the first time derivative of the kernel R_2 (denoted $\dot{R}_2(z,t)$)

$$\dot{R}_2(z,0) = 0. \tag{4.41}$$

Differentiating (4.37) with respect to t, the following PDE for the kernel $\dot{R}_2(z,t)$ is obtained

$$\partial_z \dot{R}_2 = (\frac{2}{c}\partial_t + \frac{b}{c})\dot{R}_2 + F_1 * \dot{R}_2 + \dot{F}_2. \tag{4.42}$$

Setting $t = 0$ in (4.42), we obtain the initial value of the second time derivative of the R_2 kernel (denoted $\ddot{R}_2(z,t)$)

$$\ddot{R}_2(z,0) = \frac{1}{2}c^2(3c_z - 2b - c\partial_z \ln p). \tag{4.43}$$

The PDE for $\ddot{R}_2(z,t)$ follows from (4.37)

$$\partial_z \ddot{R}_2 = (\frac{2}{c}\partial_t + \frac{b}{c})\ddot{R}_2 + F_1 * \ddot{R}_2 + \ddot{F}_2. \tag{4.44}$$

After a derivation similar to the one that led to (4.37) and (4.38), one obtains the following imbedding equation and initial condition for the fourth moment R_4

$$\partial_z R_4 = (\frac{2}{c}\partial_t + \frac{b}{c})R_4 + F_1 * R_4 + F_4, \tag{4.45}$$
$$R_4(z,0) = 0, \tag{4.46}$$

where

$$F_4(z,t) = E_0 + B_0 * R_0 + B_2 * R_2 + C_0 * R_0 * R_0 + C_2 * R_0 * R_2$$
$$+(\frac{b}{c} - \partial_z \ln p + \frac{c_z}{c})R_2 * R_2 - \frac{q}{c} * R_2 * R_2, \tag{4.47}$$

$$\begin{cases} E_0 = 6c^3t^2 + \frac{1}{2}c^3[4bt^3 - qt^4] - \frac{20}{3}c^3c_zt^3, \\ B_0 = -4c^3t^2 + c^3(4bt^3 - qt^4), \\ B_2 = -8c + 2c(2bt - qt^2), \\ C_0 = 6c^3t^2 + \frac{1}{2}c^3[4bt^3 - qt^4] + \frac{20}{3}c^3c_zt^3, \\ C_2 = 2c(2bt - qt^2) + 8cc_zt. \end{cases} \tag{4.48}$$

Through a derivation analogous to the one that led to (4.41) and (4.43), one obtains

$$\begin{aligned}
\dot{R}_4(z,0) &= 0,\\
\ddot{R}_4(z,0) &= 0,\\
\dddot{R}_4\,(z,0) &= -6c^4,\\
\ddddot{R}_4\,(z,0) &= c^4(15c_z - 11b - 3c\partial_z \ln p).
\end{aligned} \tag{4.49}$$

The imbedding equation for $\ddddot{R}_4$ is as follows

$$\partial_z\, \ddddot{R}_4 = (\frac{2}{c}\partial_t + \frac{b}{c})\, \ddddot{R}_4 + F_1 *\, \ddddot{R}_4 - 6c^4 F_1 + \ddddot{F}_4\,. \tag{4.50}$$

We note that although the moment approach is not the only way to reduce the three-dimensional problem to a set of one-dimensional problems, it seems intuitively attractive. Furthermore, it has the effect of averaging the noise contained in the data. Note also that time-differentiations of reflection kernels are required in an inverse problem of parameter reconstruction using higher order moments. In realistic applications to parameter reconstructions, the data are likely to be noisy and some regularization technique will be required in the implementation of the time differentiations (see, e.g. the convolution smoothing process with a Fejer kernel in Section 7.1). In applications to design problems, the given data are free of noise (and often prescribed in terms of explicit formulas, see e.g. Jordan and Lakshmanasamy (1989)). Therefore, the numerical performance shown in this section will be characteristic also of design problems treated by means of the present method. If the present moment approach is combined with an optimization approach, one does not need to use time-differentiation of the reflection kernel (see Chapter 5).

Numerical example for the reconstruction of q using R_0. In this example it is assumed that $c = c_0$ (constant), $p = 1$, $b = 0$ and thus one has

$$\begin{aligned}
\partial_z R_0 &= \frac{2}{c_0}\partial_t R_0 - \frac{q}{2c_0}(1 + 2 * R_0 + 1 * R_0 * R_0),\\
R_0(z,0) &= 0.
\end{aligned}$$

To reconstruct $q(z)$, one needs the following initial condition

$$\dot{R}_0(z,0) = \frac{1}{4}q.$$

The PDE for $\dot{R}$ is

$$\partial_z \dot{R}_0 = \frac{2}{c_0}\partial_t \dot{R}_0 - \frac{q}{2c_0}(2R_0 + R_0 * R_0).$$

The reconstruction of q is shown in Fig. 4.2. The number of the data points used in the reconstruction is 32.

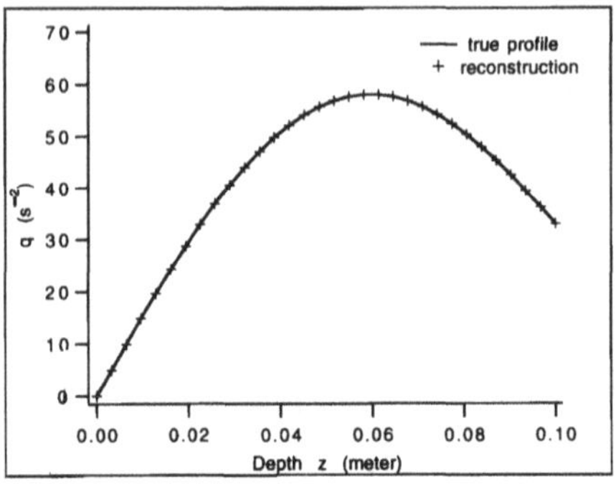

FIG. 4.2. Reconstruction of $q(z)$ using $R_0(0,t)$.

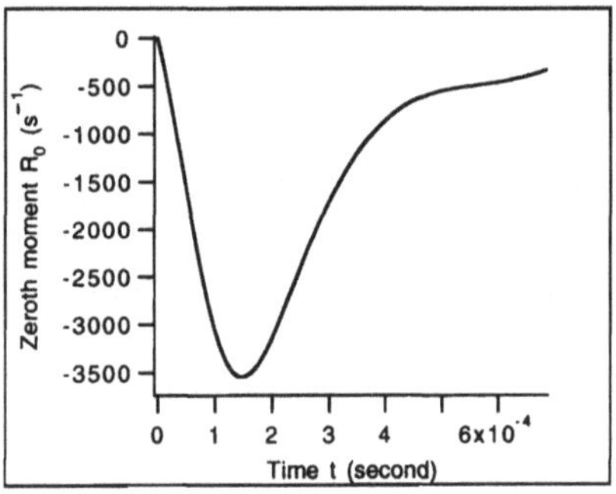

(a)

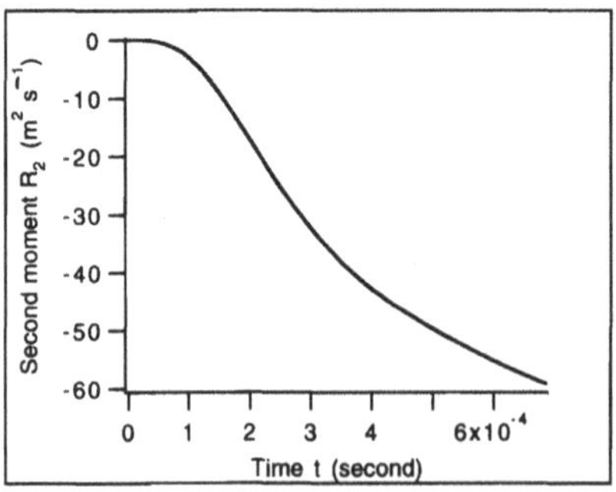

(b)

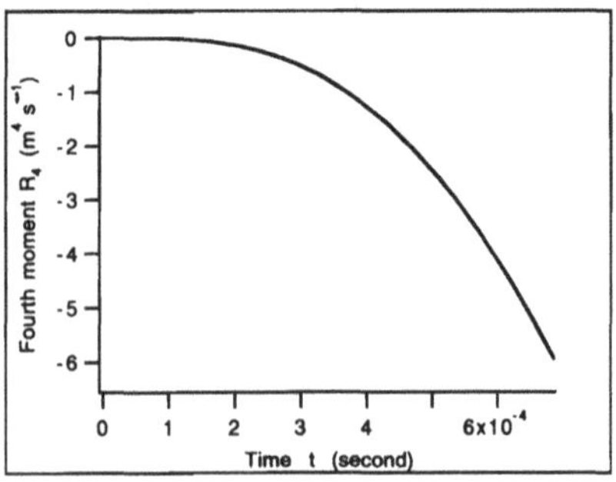

(c)

FIG. 4.3. The zeroth, second and fourth moments of the reflection operator corresponding to the profiles in Fig. 4.4 (with $q = 0$).

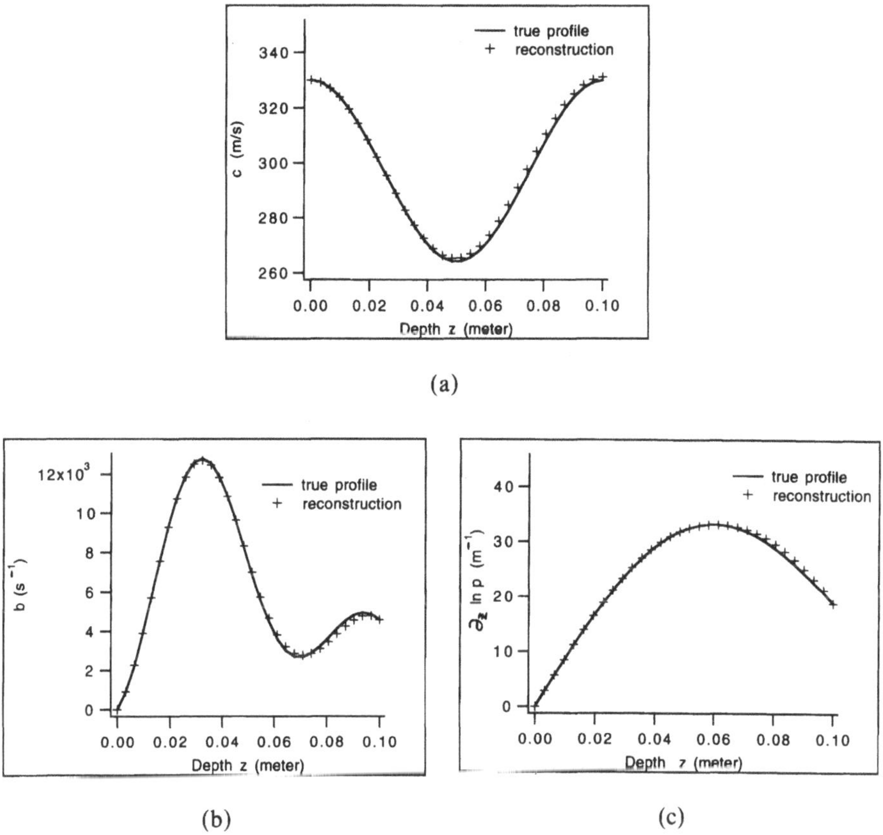

FIG. 4.4. Simultaneous reconstruction of c, b and p using $R_0(0,t)$, $R_2(0,t)$ and $R_4(0,t)$.

Numerical example for the simultaneous reconstruction of c, b and p using R_0, R_2 and R_4. In this example it is assumed that $q = 0$. In the reconstruction algorithm, one propagates the boundary values $R_0(0,t)$, $R_2(0,t)$, $R_4(0,t)$ and their time derivatives to the initial values of R_0, R_2, R_4 and their time derivatives using the imbedding equations for the corresponding kernels. The parameters c, b and p can then be reconstructed simultaneously by combinations of these initial values according to the conditions (4.33), (4.43) and (4.49). The zeroth, second and fourth moments corresponding to the profiles in Fig. 4.4 are given in Fig. 4.3. The simultaneous reconstruction of c, b and p using 32 data points is shown in Fig. 4.4. If the number of data points is greater than 128, the reconstruction and the true values essentially coincide on the scale of Fig. 4.4.

If the upper half-space $z \leq 0$ is a homogeneous lossy medium with a dissipation coefficient b_0 (constant), then one should use a wave-splitting which factorizes the operator $\left[\frac{1}{c^2}\partial_t^2 - \nabla^2 + \frac{b_0}{c^2}\partial_t\right]$. The corresponding split moments are used to solve the inverse problem for such a case in He and Weston (1992). The approach using the split moments can also be applied to anisotropic wave equations (cf. Exercise 4.3 and He (1994a)).

4.1.3 *Reconstruction of the permittivity tensor*

In the previous subsections, we have considered the second-order hyperbolic equation (4.1) for a scalar field u. As noted above, u may be the Hertz potential in an electromagnetic problem. However, in general, in electromagnetic problems we wish to refer directly to Maxwell's equations. In this subsection we show how moments of the fields and reflection kernel can be used in the reconstruction of the material parameters in a vector field formalism, though the detailed derivations are not given here. Time domain wave-splitting of the three-dimensional Maxwell's equations for the isotropic case was introduced in Weston (1993) (see Section 8.1). In this subsection, the following Maxwell's equations for an anisotropic medium are considered,

$$\nabla \times \vec{\mathbf{E}} = -\partial_t \vec{\mathbf{B}}, \qquad \nabla \times \vec{\mathbf{H}} = \partial_t \vec{\mathbf{D}}, \tag{4.51}$$

with the constitutive relations

$$\vec{\mathbf{D}} = \overline{\overline{\epsilon}}\, \vec{\mathbf{E}}, \qquad \vec{\mathbf{B}} = \mu \vec{\mathbf{H}}, \tag{4.52}$$

where

$$\overline{\overline{\epsilon}} = \begin{bmatrix} \epsilon_{11} & \epsilon_{12} & \epsilon_{13} \\ \epsilon_{21} & \epsilon_{22} & \epsilon_{23} \\ \epsilon_{31} & \epsilon_{32} & \epsilon_{33} \end{bmatrix}$$

is the permittivity tensor for the anisotropic medium, and μ is the (scalar) permeability. Energy conservation requires the permittivity tensor to be symmetric, i.e. $\epsilon_{ij} = \epsilon_{ji}$, $i, j = 1, 2, 3$ (cf. e.g. Born and Wolf (1980)). There always exists a coordinate transformation that transforms this symmetric matrix into a diagonal matrix. The three coordinate axes are referred to as the principal axes of the anisotropic medium (Kong 1986). It is assumed that the anisotropic medium is stratified along one of the principal axes. This principal axis is chosen as the z axis and the other two principal axes as the x axis and y axis, respectively. Thus one has

$$\overline{\overline{\epsilon}} = \begin{bmatrix} \epsilon_1(z) & 0 & 0 \\ 0 & \epsilon_2(z) & 0 \\ 0 & 0 & \epsilon_3(z) \end{bmatrix} \tag{4.53}$$

(the permeability μ is assumed to be a constant). The sources, with compact spatial support, are located in the upper half-space $z < 0$, which is homogeneous but possibly anisotropic (a one-dimensional inverse problem for a plane wave normally incident on an anisotropic slab has been treated in Stewart (1989)). The parameters ϵ_i $(i = 1, 2, 3)$ are assumed to be continuous everywhere and continuously differentiable in the inhomogeneous half-space $z > 0$. It is assumed that the incident fields will not reach the surface $z = 0$ until the time $t = 0$, which leads to the initial conditions

$$\vec{\mathbf{E}}(x, y, z, 0) = 0, \qquad \vec{\mathbf{H}}(x, y, z, 0) = 0, \quad \text{for } z > 0. \tag{4.54}$$

The third component of Maxwell's equations gives

$$H_3 = -\frac{1}{\mu}\partial_t^{-1}[\partial_x E_2 - \partial_y E_1], \tag{4.55}$$

$$E_3 = \frac{1}{\epsilon_3}\partial_t^{-1}[\partial_x H_2 - \partial_y H_1], \tag{4.56}$$

which express the third component of the fields in terms of the tangential components. From the above two equations, it follows that

$$\nabla_T E_3 \equiv \begin{bmatrix} \partial_x E_3 \\ \partial_y E_3 \end{bmatrix} = \frac{1}{\epsilon_3}\partial_t^{-1} J\mathbf{H}, \tag{4.57}$$

$$\nabla_T H_3 \equiv \begin{bmatrix} \partial_x H_3 \\ \partial_y H_3 \end{bmatrix} = -\frac{1}{\mu}\partial_t^{-1} J\mathbf{E}, \tag{4.58}$$

where

$$\mathbf{E} = \begin{bmatrix} E_1 \\ E_2 \end{bmatrix}, \qquad \mathbf{H} = \begin{bmatrix} H_1 \\ H_2 \end{bmatrix},$$

$$J = \begin{bmatrix} -\partial_{xy} & \partial_{xx} \\ -\partial_{yy} & \partial_{xy} \end{bmatrix}. \tag{4.59}$$

The first two components of Maxwell's equations can be written in the following form

$$\begin{aligned} \partial_z \begin{bmatrix} \mathbf{E} \\ \mathbf{H} \end{bmatrix} &= \begin{bmatrix} 0 & -\mu I_2 \partial_t \\ I_2\, \bar{\epsilon}\, \partial_t & 0 \end{bmatrix} \begin{bmatrix} \mathbf{E} \\ \mathbf{H} \end{bmatrix} + \begin{bmatrix} \nabla_T E_3 \\ \nabla_T H_3 \end{bmatrix} \\ &= \begin{bmatrix} 0 & -\mu I_2 \partial_t + \frac{1}{\epsilon_3}\partial_t^{-1} J \\ I_2\, \bar{\epsilon}\, \partial_t - \frac{1}{\mu}\partial_t^{-1} J & 0 \end{bmatrix} \begin{bmatrix} \mathbf{E} \\ \mathbf{H} \end{bmatrix} \\ &\equiv \hat{D} \begin{bmatrix} \mathbf{E} \\ \mathbf{H} \end{bmatrix}, \end{aligned} \tag{4.60}$$

where

$$\bar{\epsilon} = \begin{bmatrix} \epsilon_1 & 0 \\ 0 & \epsilon_2 \end{bmatrix}, \qquad I_2 = \begin{bmatrix} 0 & 1 \\ -1 & 0 \end{bmatrix}. \tag{4.61}$$

Eigenmodes of Maxwell's equations for an anisotropic medium can then be identified by diagonalizing the matrix $\hat{D}$. The eigenmodes consist of two pairs and the two components of each pair correspond to down-going and up-going waves, respectively. The down-going modes $\mathbf{F}^+ \equiv \begin{bmatrix} F_1^+ \\ F_2^+ \end{bmatrix}$ and up-going modes $\mathbf{F}^- \equiv \begin{bmatrix} F_1^- \\ F_2^- \end{bmatrix}$ are determined to be (He 1994c)

$$\begin{bmatrix} \mathbf{F}^+ \\ \mathbf{F}^- \end{bmatrix} = \frac{1}{2} \begin{bmatrix} I_\alpha & N_\alpha \\ I_\alpha & -N_\alpha \end{bmatrix} \begin{bmatrix} \mathbf{E} \\ \mathbf{H} \end{bmatrix} \equiv \hat{T} \begin{bmatrix} \mathbf{E} \\ \mathbf{H} \end{bmatrix}, \tag{4.62}$$

where the splitting operator N_α and its inverse N_α^{-1} satisfy

$$N_\alpha D_2 D_1 N_\alpha^{-1} = \mathcal{K}^{-2} \tag{4.63}$$

(i.e. N_α^{-1} consists of the two eigenvectors of the 2×2 matrix $D_2 D_1$ corresponding to the eigenvalues $\mathcal{K}_1^{-2}$ and $\mathcal{K}_2^{-2}$), with

$$D_1 = -\mu I_2 \partial_t + \frac{1}{\epsilon_3} \partial_t^{-1} J, \tag{4.64}$$

$$D_2 = I_2 \, \bar{\epsilon} \, \partial_t - \frac{1}{\mu} \partial_t^{-1} J, \tag{4.65}$$

$$\mathcal{K} = \begin{bmatrix} \mathcal{K}_1 & 0 \\ 0 & \mathcal{K}_2 \end{bmatrix}. \tag{4.66}$$

The operator $\mathcal{K}_i$ and its inverse $\mathcal{K}_i^{-1}$ $(i = 1, 2)$ satisfy

$$\mathcal{K}_1^{-1} = \left(K_o^{-2} + D_e^{-1} \right) \mathcal{K}_1, \tag{4.67}$$

$$\mathcal{K}_2^{-1} = \left(K_o^{-2} - D_e^{-1} \right) \mathcal{K}_2, \tag{4.68}$$

where the operators K_o, D_e and their inverses satisfy

$$K_o^{-1} = \left[\frac{\mu(\epsilon_1 + \epsilon_2)}{2} \partial_t^2 - \frac{\epsilon_1 + \epsilon_3}{2\epsilon_3} \partial_x^2 - \frac{\epsilon_2 + \epsilon_3}{2\epsilon_3} \partial_y^2 \right] K_o, \tag{4.69}$$

$$D_e^{-1} = \frac{1}{4} \left[\mu^2 (\epsilon_1 - \epsilon_2)^2 \partial_t^4 + 2\mu \left(\frac{\epsilon_3 - \epsilon_1}{\epsilon_3} \partial_x^2 - \frac{\epsilon_3 - \epsilon_2}{\epsilon_3} \partial_y^2 \right) (\epsilon_1 - \epsilon_2) \partial_t^2 \right.$$
$$\left. + \left(\frac{\epsilon_3 - \epsilon_1}{\epsilon_3} \partial_x^2 + \frac{\epsilon_3 - \epsilon_2}{\epsilon_3} \partial_y^2 \right)^2 \right] D_e. \tag{4.70}$$

The matrix I_α is given by

$$I_\alpha = -\mathcal{K} N_\alpha D_2. \tag{4.71}$$

The matrix $\hat{D}$ is thus diagonalized as follows,

$$\hat{T}\hat{D}\hat{T}^{-1} = \mathrm{diag}[\,-\mathcal{K}_1^{-1}, -\mathcal{K}_2^{-1}, \mathcal{K}_1^{-1}, \mathcal{K}_2^{-1}\,]. \tag{4.72}$$

For a uniaxially anisotropic medium, i.e. $\epsilon_2 = \epsilon_1$, (4.67) and (4.68) reduce to

$$\mathcal{K}_1^{-1} = [\,\mu\epsilon_1\,\partial_t^2 - \frac{\epsilon_1}{\epsilon_3}(\partial_x^2 + \partial_y^2)\,]\mathcal{K}_1,$$

$$\mathcal{K}_2^{-1} = [\,\mu\epsilon_1\,\partial_t^2 - (\partial_x^2 + \partial_y^2)\,]\mathcal{K}_2,$$

which give the following results (He 1994a,c)

$$\mathcal{K}_1 f(x,y,z,t) = \frac{1}{2\pi}(\frac{\sqrt{\epsilon_3}}{\sqrt{\epsilon_1}})\int\int_{\mathcal{R}^2}\frac{1}{r}f(x',y',z,\tau_1)\Theta(\tau_1)\,dx'\,dy',$$

$$\mathcal{K}_2 f(x,y,z,t) = \frac{1}{2\pi}\int\int_{\mathcal{R}^2}\frac{1}{r}f(x',y',z,\tau_2)\Theta(\tau_2)\,dx'\,dy',$$

where $r = \sqrt{(x-x')^2 + (y-y')^2}$, $\tau_1 = t - \sqrt{\mu\epsilon_3}\,r$, $\tau_2 = t - \sqrt{\mu\epsilon_1}\,r$ and $\Theta(\tau)$ is the Heaviside unit step function. For an isotropic medium, $\mathcal{K}_1 = \mathcal{K}_2$ and all the results reduce to the ones given in Weston (1993) (see Section 8.1).

The zeroth and second moments of the tangential fields are defined by

$$\mathbf{E}^{(0)}(z,t) = \int\int_{\mathbb{R}^2}\mathbf{E}\,dx\,dy, \tag{4.73}$$

$$\mathbf{E}^{(2)}(z,t) = \int\int_{\mathbb{R}^2}(x^2+y^2)\mathbf{E}\,dx\,dy, \tag{4.74}$$

respectively, with similar definitions for $\mathbf{H}^{(0)}$, $\mathbf{H}^{(2)}$, etc. Taking zeroth and second moments of (4.60) and (4.62), one obtains

$$\partial_z\begin{bmatrix}\mathbf{E}^{(0)}\\ \mathbf{H}^{(0)}\\ \mathbf{E}^{(2)}\\ \mathbf{H}^{(2)}\end{bmatrix} = \begin{bmatrix}D & 0\\ D_0 & D\end{bmatrix}\begin{bmatrix}\mathbf{E}^{(0)}\\ \mathbf{H}^{(0)}\\ \mathbf{E}^{(2)}\\ \mathbf{H}^{(2)}\end{bmatrix} \equiv \tilde{D}\begin{bmatrix}\mathbf{E}^{(0)}\\ \mathbf{H}^{(0)}\\ \mathbf{E}^{(2)}\\ \mathbf{H}^{(2)}\end{bmatrix}, \tag{4.75}$$

$$\begin{bmatrix}\mathbf{F}^{(0)+}\\ \mathbf{F}^{(0)-}\\ \mathbf{F}^{(2)+}\\ \mathbf{F}^{(2)-}\end{bmatrix} = \begin{bmatrix}T & 0\\ T_0 & T\end{bmatrix}\begin{bmatrix}\mathbf{E}^{(0)}\\ \mathbf{H}^{(0)}\\ \mathbf{E}^{(2)}\\ \mathbf{H}^{(2)}\end{bmatrix} \equiv \tilde{T}\begin{bmatrix}\mathbf{E}^{(0)}\\ \mathbf{H}^{(0)}\\ \mathbf{E}^{(2)}\\ \mathbf{H}^{(2)}\end{bmatrix} \tag{4.76}$$

where

$$D = \begin{bmatrix}0 & -\mu I_2\partial_t\\ I_2\,\bar{\epsilon}\,\partial_t & 0\end{bmatrix}, \qquad D_0 = \begin{bmatrix}0 & \frac{2}{\epsilon_3}I_2\partial_t^{-1}\\ -\frac{2}{\mu}I_2\partial_t^{-1} & 0\end{bmatrix}, \tag{4.77}$$

$$T = \frac{1}{2}\begin{bmatrix} I & N \\ I & -N \end{bmatrix}, \qquad T_0 = \frac{1}{2}\begin{bmatrix} 0 & N_0 \\ 0 & -N_0 \end{bmatrix}, \tag{4.78}$$

$$N = \begin{bmatrix} 0 & \sqrt{\mu/\epsilon_1} \\ -\sqrt{\mu/\epsilon_2} & 0 \end{bmatrix}, \qquad N_0 = \begin{bmatrix} 0 & \xi_1 \\ -\xi_2 & 0 \end{bmatrix} \partial_t^{-2}, \tag{4.79}$$

$$\xi_i = \frac{1}{\sqrt{\mu\epsilon_i}}\left(\frac{1}{\epsilon_i} - \frac{1}{\epsilon_3}\right), \quad i = 1, 2. \tag{4.80}$$

Therefore, one has the following dynamic equations for the split moments,

$$\partial_z \begin{bmatrix} \mathbf{F}^{(0)+} \\ \mathbf{F}^{(0)-} \\ \mathbf{F}^{(2)+} \\ \mathbf{F}^{(2)-} \end{bmatrix} = \tilde{T}\tilde{D}\tilde{T}^{-1} \begin{bmatrix} \mathbf{F}^{(0)+} \\ \mathbf{F}^{(0)-} \\ \mathbf{F}^{(2)+} \\ \mathbf{F}^{(2)-} \end{bmatrix} + (\partial_z \tilde{T})\tilde{T}^{-1} \begin{bmatrix} \mathbf{F}^{(0)+} \\ \mathbf{F}^{(0)-} \\ \mathbf{F}^{(2)+} \\ \mathbf{F}^{(2)-} \end{bmatrix}$$

$$= \begin{bmatrix} \boldsymbol{\alpha} & \boldsymbol{\beta} & 0 & 0 \\ \boldsymbol{\gamma} & \boldsymbol{\vartheta} & 0 & 0 \\ \mathbf{p_1} & \mathbf{p_2} & \boldsymbol{\alpha} & \boldsymbol{\beta} \\ \mathbf{p_3} & \mathbf{p_4} & \boldsymbol{\gamma} & \boldsymbol{\vartheta} \end{bmatrix} \begin{bmatrix} \mathbf{F}^{(0)+} \\ \mathbf{F}^{(0)-} \\ \mathbf{F}^{(2)+} \\ \mathbf{F}^{(2)-} \end{bmatrix}, \tag{4.81}$$

where

$$\boldsymbol{\alpha} = -\begin{bmatrix} \sqrt{\mu\epsilon_1} & 0 \\ 0 & \sqrt{\mu\epsilon_2} \end{bmatrix} \partial_t - \frac{1}{4}\begin{bmatrix} \frac{\epsilon_{1z}}{\epsilon_1} & 0 \\ 0 & \frac{\epsilon_{2z}}{\epsilon_2} \end{bmatrix}, \tag{4.82}$$

$$\boldsymbol{\beta} = \boldsymbol{\gamma} = \frac{1}{4}\begin{bmatrix} \frac{\epsilon_{1z}}{\epsilon_1} & 0 \\ 0 & \frac{\epsilon_{2z}}{\epsilon_2} \end{bmatrix}, \tag{4.83}$$

$$\boldsymbol{\vartheta} = \begin{bmatrix} \sqrt{\mu\epsilon_1} & 0 \\ 0 & \sqrt{\mu\epsilon_2} \end{bmatrix} \partial_t - \frac{1}{4}\begin{bmatrix} \frac{\epsilon_{1z}}{\epsilon_1} & 0 \\ 0 & \frac{\epsilon_{2z}}{\epsilon_2} \end{bmatrix}, \tag{4.84}$$

$$\mathbf{p_1} = \begin{bmatrix} \frac{\epsilon_1 + \epsilon_3}{\epsilon_3\sqrt{\mu\epsilon_1}}, & 0 \\ 0 & \frac{\epsilon_2 + \epsilon_3}{\epsilon_3\sqrt{\mu\epsilon_2}} \end{bmatrix} \partial_t^{-1} + \frac{1}{2}\begin{bmatrix} \eta_1 & 0 \\ 0 & \eta_2 \end{bmatrix} \partial_t^{-2}, \tag{4.85}$$

$$\mathbf{p_2} = \mathbf{p_3} = -\frac{1}{2}\begin{bmatrix} \eta_1 & 0 \\ 0 & \eta_2 \end{bmatrix} \partial_t^{-2}, \tag{4.86}$$

$$\mathbf{p_4} = -\begin{bmatrix} \frac{\epsilon_1 + \epsilon_3}{\epsilon_3\sqrt{\mu\epsilon_1}}, & 0 \\ 0 & \frac{\epsilon_2 + \epsilon_3}{\epsilon_3\sqrt{\mu\epsilon_2}} \end{bmatrix} \partial_t^{-1} + \frac{1}{2}\begin{bmatrix} \eta_1 & 0 \\ 0 & \eta_2 \end{bmatrix} \partial_t^{-2}, \tag{4.87}$$

$$\eta_i = \sqrt{\epsilon_i/\mu}\,\xi_{iz} + \frac{1}{2}\frac{\epsilon_3 - \epsilon_i}{\mu\epsilon_i^3}\epsilon_{iz}, \quad i = 1, 2. \tag{4.88}$$

In (4.81) one notes that the split second moments $\mathbf{F}^{(2)\pm}$ are not involved in the dynamic equations for the split zeroth moments $\mathbf{F}^{(0)\pm}$, but the split zeroth moments $\mathbf{F}^{(0)\pm}$ appear in the dynamic equations for split second moments $\mathbf{F}^{(2)\pm}$.

For an imbedding geometry, the split moments are related to each other by the moments of the reflection operator matrix according to

$$\mathbf{F}^{(0)-}(z,t) = \begin{bmatrix} R_{11}^{(0)} & R_{12}^{(0)} \\ R_{21}^{(0)} & R_{22}^{(0)} \end{bmatrix} (z,t) * \mathbf{F}^{(0)+}(z,t), \tag{4.89}$$

$$\mathbf{F}^{(2)-}(z,t) = \begin{bmatrix} R_{11}^{(2)} & R_{12}^{(2)} \\ R_{21}^{(2)} & R_{22}^{(2)} \end{bmatrix} (z,t) * \mathbf{F}^{(0)+}(z,t) + \begin{bmatrix} R_{11}^{(0)} & R_{12}^{(0)} \\ R_{21}^{(0)} & R_{22}^{(0)} \end{bmatrix} (z,t) * \mathbf{F}^{(2)+}(z,t), \tag{4.90}$$

After a derivation analogous to the one given in Section 4.1.2, one obtains the following imbedding equations (He 1994c)

$$(\partial_z - 2\sqrt{\mu\epsilon_i}\partial_t)R_{ii}^{(0)} = -\frac{1}{4}\frac{\epsilon_{iz}}{\epsilon_i}R_{ii}^{(0)} * R_{ii}^{(0)}, \qquad i = 1, 2, \tag{4.91}$$

$$\begin{aligned}(\partial_z - 2\sqrt{\mu\epsilon_i}\partial_t)R_{ii}^{(2)} &= -\frac{1}{2}\eta_i t - \frac{2(\epsilon_i + \epsilon_3)}{\epsilon_3\sqrt{\mu\epsilon_i}} * R_{ii}^{(0)} \\ &+\frac{1}{2}\eta_i t * R_{ii}^{(0)} * R_{ii}^{(0)} - \frac{1}{2}\frac{\epsilon_{iz}}{\epsilon_i}R_{ii}^{(0)} * R_{ii}^{(2)}, \qquad i = 1, 2,\end{aligned} \tag{4.92}$$

$$R_{ij}^{(0)} = 0, \qquad R_{ij}^{(2)} = 0, \qquad i, j = 1, 2, \ i \neq j, \tag{4.93}$$

and the initial conditions

$$R_{ii}^{(0)}(z, 0) = -\frac{1}{8\sqrt{\mu\epsilon_i}}\frac{\epsilon_{iz}}{\epsilon_i}, \qquad i = 1, 2, \tag{4.94}$$

$$R_{ii}^{(2)}(z, 0) = 0, \qquad i = 1, 2. \tag{4.95}$$

For the inverse problem, one needs the following initial conditions for the second time derivative of $R_{ii}^{(2)}$ (denoted $\ddot{R}_{ii}^{(2)}$),

$$\ddot{R}_{ii}^{(2)}(z, 0) = \frac{1}{8\sqrt{\mu\epsilon_i}}[2\eta_i - \frac{\epsilon_{iz}}{\mu\epsilon_i}(\frac{1}{\epsilon_i} + \frac{1}{\epsilon_3})], \qquad i = 1, 2. \tag{4.96}$$

The imbedding equations for the $\ddot{R}^{(2)}(z,t)$ are

$$(\partial_z - 2\sqrt{\mu\epsilon_i}\partial_t)\ddot{R}_{ii}^{(2)} = -\frac{2(\epsilon_i + \epsilon_3)}{\epsilon_3\sqrt{\mu\epsilon_i}} R_{ii}^{(0)}$$

$$+\frac{1}{2}\eta_i R_{ii}^{(0)} * R_{ii}^{(0)} - \frac{1}{2}\frac{\epsilon_{iz}}{\epsilon_i} R_{ii}^{(0)} * \ddot{R}_{ii}^{(2)}, \qquad i = 1, 2. \tag{4.97}$$

In the inverse problem, the boundary values of the moments of the reflection operator matrix are obtained from the measurement of the fields at the surface $z = 0$ through deconvolution (cf. the definitions (4.89) and (4.90)). One then propagates the boundary values $R_{ii}^{(0)}(0,t)$ $(i = 1, 2)$ and $\ddot{R}_{ii}^{(2)}(0,t)$ $(i = 1, 2)$ along the characteristic curves to the initial values $R_{ii}^{(0)}(z, 0)$ and $\ddot{R}_{ii}^{(2)}(z, 0)$ using the imbedding equations for $R_{ii}^{(0)}(z,t)$ and $\ddot{R}_{ii}^{(2)}(z,t)$, respectively. The permittivity tensor is thus reconstructed from the initial conditions (4.94) and (4.96). Note that (4.94) and (4.96) contain four initial conditions. One only needs to choose three of them, e.g. two from (4.94) and one from (4.96).

4.2 Linearization of the imbedding equations

The imbedding equation (4.32) for the zeroth moment R_0 is non-linear in R_0 (cf. also the imbedding equation (1.80) for the reflection kernel). To develop a faster numerical algorithm for solving the non-linear imbedding equation (4.32), one can replace it by an equivalent linear system. The basic idea behind this resides in the Green's function approach (cf. the relation (2.100)). However, the reduction to a linear system is not unique. The following linearization result will be presented in terms of a generalized form of the imbedding equation (4.32), rather than (4.32) itself. With $\tau(z)$ defined by $\tau(z) \equiv \tau(0, z) = \int_0^z \frac{1}{c(z')}dz'$ (cf. (1.20)), let $R(z,t)$ be the solution of the well-posed problem

$$\left(\partial_z - \frac{2}{c}\partial_t\right) R = f(R), \quad 0 \le t \le 2[\tau(L) - \tau(z)], \quad 0 \le z \le L, \tag{4.98}$$

$$R(z, 0) = r(z), \quad 0 \le z \le L, \tag{4.99}$$

where

$$f(R) = p(z)R + q(z)R * R + A(z,t) + B(z,t) * R + C(z,t) * R * R. \tag{4.100}$$

The coefficients p, q, A, B, C and initial data r are continuous functions of their arguments. Let $G_1(z,t)$, $G_2(z,t)$ be the solution of the linear system

$$\partial_z G_1 = f_1(G_1, G_2), \tag{4.101}$$

$$\left(\partial_z - \frac{2}{c}\partial_t - p(z)\right) G_2 = f_2(G_1, G_2), \tag{4.102}$$
$$\text{for} \qquad 0 \le t \le \xi(L) - \xi(z), \qquad 0 \le z \le L,$$

$$G_1(0,t) = 0, \qquad\qquad 0 \le t \le \xi(L), \tag{4.103}$$

$$G_2(z,0) = r(z), \qquad\qquad 0 \le z \le L, \tag{4.104}$$

where

$$f_1(G_1, G_2) = -q(z)G_2 + A_1(z,t) + A_1(z,t) * G_1 - C(z,t) * G_2, \tag{4.105}$$

$$f_2(G_1, G_2) = -\frac{2}{c}r(z)G_1 + A(z,t) + A(z,t) * G_1 + (B(z,t) + A_1(z,t)) * G_2. \tag{4.106}$$

The coefficient $A_1(z,t)$ is an arbitrary continuous function (the simplest choice is to set A_1 identically zero) which indicates the non-uniqueness of the linearized system. This system is well-posed, as is shown in He and Weston (1992). The connection between the two systems is given by the following linearization theorem (He and Weston 1992):

Theorem 4.1 *The system of (4.98), (4.99) for R is equivalent to the system (4.101)–(4.104) for G_1, G_2 with*

$$G_2(z,t) = R(z,t) + R(z,t) * G_1(z,t). \tag{4.107}$$

Furthermore, one has

$$G_2(z,0) = R(z,0), \tag{4.108}$$

$$G_2(0,t) = R(0,t). \tag{4.109}$$

The above linearization theorem can be utilized to solve both the direct and inverse problems, and improve the speed by an order of N, where N is the number of the discretization points in t.

4.3 Parameter reconstruction using the Hankel transform of a point source field

In the previous section we made only very general assumptions concerning the nature of the incoming field. If more specific types of sources are considered, more detailed and explicit results can be obtained. An important canonical case is that of a point source, or more generally, axially symmetric sources. It will be shown that if the axial symmetry is exploited by means of a Hankel transform, explicit representations of the splitting operator and its inverse can be obtained. The considerations and derivations which are typical for an axially symmetric source are illustrated for the case of an acoustic point source above a stratified half-space.

Consider wave propagation in an inhomogeneous acoustic medium which is described by the following stress-strain equations for fluids (see e.g. Friedlander (1958))

$$p = -\rho c^2 \nabla \cdot \mathbf{u}, \tag{4.110}$$

$$\rho \partial_t^2 \mathbf{u} = -\nabla p, \tag{4.111}$$

where p is the pressure, $\mathbf{u}$ is the displacement, ρ is the fluid density and c is the velocity of the acoustic wave in the medium. Equations (4.110) and (4.111) can be reduced to a scalar equation for the pressure p,

$$\frac{1}{c^2}\partial_t^2 p - \rho \nabla \cdot \left(\frac{1}{\rho} \nabla p \right) = 0. \tag{4.112}$$

Note that the above equation is a special case of the telegraph equation (4.1) considered in the previous section. In this section, a point source over a stratified acoustic half-space is considered. The velocity and the density are functions of the depth z in the half-space $z \geq 0$. The upper half-space $z < 0$ is homogeneous and the point source is located at the position $(0, 0, z_0)$ with $z_0 < 0$. Taking the axial symmetry of the scattering configuration into account, one obtains the following equations for $p = p(r, z, t)$ from (4.112) in the cylindrical coordinates (r, ϕ, z),

$$\frac{1}{c^2}\partial_t^2 p - \frac{1}{r}\partial_r \left(r \partial_r p \right) - \partial_z^2 p + \frac{d}{dz}(\ln \rho)\partial_z p = 0, \quad z > 0, \tag{4.113}$$

$$\frac{1}{c_0^2}\partial_t^2 p - \frac{1}{r}\partial_r \left(r \partial_r p \right) - \partial_z^2 p = 0, \quad z < 0, \quad (z - z_0)^2 + r^2 > 0, \tag{4.114}$$

where $c_0 = c(0)$ is the wave speed in the homogeneous upper half-space. It is assumed that the incident wave will not reach the interface $z = 0$ until the time $t = 0$, i.e. one has the initial conditions

$$p(r, z, 0) = \partial_t p(r, z, 0) = 0, \quad \text{for } z > 0. \tag{4.115}$$

In view of the axial symmetry of the scattering problem, the following Hankel transform is introduced

$$\bar{p}(z,t;\kappa) = \int_0^\infty p(r,z,t) J_0(\kappa r) r \, dr, \tag{4.116}$$

$$p(r,z,t) = \int_0^\infty \bar{p}(z,t;\kappa) J_0(\kappa r) \kappa \, d\kappa, \tag{4.117}$$

where J_0 is the zeroth order Bessel function of the first kind.

Thus from the transformation of (4.113)–(4.115) one obtains

$$\partial_z^2 \bar{p} - \frac{1}{c^2}\partial_t^2 \bar{p} - \kappa^2 \bar{p} - \frac{d}{dz}(\ln \rho)\partial_z \bar{p} = 0, \quad z > 0, \tag{4.118}$$

$$\partial_z^2 \bar{p} - \frac{1}{c_0^2}\partial_t^2 \bar{p} - \kappa^2 \bar{p} = 0, \quad z < 0, \quad (z - z_0)^2 + r^2 > 0, \tag{4.119}$$

$$\bar{p}(z,0;\kappa) = \partial_t \bar{p}(z,0;\kappa) = 0, \quad \text{for } z > 0. \tag{4.120}$$

Note that if the field is not symmetric, one can obtain similar equations by first applying a Fourier series expansion with respect to the azimuth angle and then applying the nth order Hankel transform for the nth Fourier component (cf. He and Karlsson (1993)).

4.3.1 *Explicit form of the splitting*

In the Hankel transformed space, a wave-splitting that decomposes the total pressure $\bar{p}$ into a down-going pressure $\bar{p}^+$ and an up-going pressure $\bar{p}^-$, is defined by

$$\begin{bmatrix} \bar{p}^+ \\ \bar{p}^- \end{bmatrix} = \frac{1}{2}\begin{bmatrix} 1 & -K \\ 1 & K \end{bmatrix}\begin{bmatrix} \bar{p} \\ \partial_z \bar{p} \end{bmatrix} \equiv T \begin{bmatrix} \bar{p} \\ \partial_z \bar{p} \end{bmatrix}, \tag{4.121}$$

where the splitting operator K and its inverse K^{-1} satisfy the following equation (cf. (4.8))

$$K^{-1} = (\frac{1}{c^2}\partial_t^2 + \kappa^2)K. \tag{4.122}$$

Note that $\bar{p}^+$ and $\bar{p}^-$ have a physical interpretation as the transformed incident pressure and the transformed reflected pressure, respectively, in the region $z_0 < z \leq 0$.

Equation (4.118) can be written in a matrix form as

$$\partial_z \begin{bmatrix} \bar{p} \\ \partial_z \bar{p} \end{bmatrix} = \begin{bmatrix} 0 & 1 \\ \frac{1}{c^2}\partial_t^2 + \kappa^2 & \frac{d}{dz}(\ln \rho) \end{bmatrix}\begin{bmatrix} \bar{p} \\ \partial_z \bar{p} \end{bmatrix} \equiv D \begin{bmatrix} \bar{p} \\ \partial_z \bar{p} \end{bmatrix}. \tag{4.123}$$

Differentiating (4.121) with respect to z and using (4.123), yields

$$\partial_z \begin{bmatrix} \bar{p}^+ \\ \bar{p}^- \end{bmatrix} = TDT^{-1} \begin{bmatrix} \bar{p}^+ \\ \bar{p}^- \end{bmatrix} + (\partial_z T)T^{-1} \begin{bmatrix} \bar{p}^+ \\ \bar{p}^- \end{bmatrix}$$

$$= \begin{bmatrix} -1 & 0 \\ 0 & 1 \end{bmatrix} K^{-1} \begin{bmatrix} \bar{p}^+ \\ \bar{p}^- \end{bmatrix} + \begin{bmatrix} 1 & -1 \\ -1 & 1 \end{bmatrix} \left(\frac{\rho_z}{2\rho} + \frac{c_z}{2c}(1 - \kappa^2 K^2) \right) \begin{bmatrix} \bar{p}^+ \\ \bar{p}^- \end{bmatrix}, \tag{4.124}$$

where $c_z = \frac{d}{dz}c$, $\rho_z = \frac{d}{dz}\rho$. Note that in the above derivation, the following relation is used (cf. Exercise 4.4),

$$K\partial_z K^{-1} = -\frac{c_z}{c}(1 - \kappa^2 K^2). \tag{4.125}$$

The system (4.124) expresses the dynamics for the split pressures in the Hankel transformed space.

Explicit forms of the wave-splitting operator K and its inverse K^{-1} were first found in He and Karlsson (1993). The results are conveniently described by the Laplace transform and we use the following notation

$$\bar{f}(s) = \int_0^\infty f(t)e^{-st}\, dt.$$

Thus, the Laplace transform of (4.122) gives

$$\bar{K}(s) = (s^2/c^2 + \kappa^2)^{-1/2}. \tag{4.126}$$

The inverse Laplace transform of $(s^2/c^2 + \kappa^2)^{-1/2}$ is $cJ_0(\kappa ct)$, and thus one obtains the following explicit expression for the wave-splitting operator K,

$$Kf(z,t) = cJ_0(\kappa ct) * f(z,t) \equiv A_0(z,t;\kappa) * f(z,t). \tag{4.127}$$

Similarly, the Laplace transform of the operator K^2 reads

$$\bar{K}^2(s) = \frac{1}{(s^2/c^2 + \kappa^2)}, \tag{4.128}$$

which in the time domain corresponds to

$$K^2 f(z,t) = \frac{c(z)}{\kappa} \sin(\kappa c(z)t) * f(z,t) \equiv C(z,t;\kappa) * f(z,t). \tag{4.129}$$

From (4.122) and (4.127), the explicit expression of K^{-1} follows,

$$K^{-1}f(z,t) = (\frac{1}{c^2}\partial_t^2 + \kappa^2)c \int_0^t J_0(\kappa c(t-t'))f(z,t')\, dt'$$

$$= \frac{1}{c}\partial_t \left\{ f - c\kappa \int_0^t J_1(\kappa c(t-t'))f(z,t')\,dt' \right\} + \kappa^2 c \int_0^t J_0(\kappa c(t-t'))f(z,t')\,dt'$$

$$= \frac{1}{c}\partial_t f + \kappa \frac{J_1(\kappa ct)}{t} * f(z,t) \equiv \frac{1}{c}\partial_t f + A(z,t;\kappa) * f(z,t). \tag{4.130}$$

The explicit expressions (4.127) and (4.130) have many important applications in wave propagation problems (see Section 4.4).

4.3.2 *Equations for the Green's functions*

For simplicity of analysis, it is assumed that the velocity and the density are continuous everywhere (the discontinuous case can be treated in an analogous way, with appropriate modifications of the scattering coefficients, see Section 4.3.4.1). As before, the Green's functions $G^{\pm}(z,t;\kappa)$ are defined as mappings from the incident pressure at $z=0$ to the internal split pressures in the Hankel transformed space according to

$$\begin{bmatrix} \bar{p}^+ \\ \bar{p}^- \end{bmatrix} (z, t+\tau(z);\kappa) = a(z) \begin{bmatrix} \bar{p}^+(0,t;\kappa) \\ 0 \end{bmatrix} + \begin{bmatrix} G^+(z,t;\kappa) * \bar{p}^+(0,t;\kappa) \\ G^-(z,t;\kappa) * \bar{p}^+(0,t;\kappa) \end{bmatrix}, \tag{4.131}$$

where $\tau(z) = \int_0^z \frac{dz'}{c(z')} (= \tau(0,z))$ is the vertical travel time from 0 to z, and $a(z)$ is the attenuation factor

$$a(z) = \sqrt{\frac{\rho(z)c(z)}{\rho(0)c(0)}}. \tag{4.132}$$

From a derivation similar to the one given in Section 2.1.2, one obtains the following imbedding equations for the Green's functions

$$\begin{aligned} \partial_z G^+ &= -aA - \frac{ac_z}{2c}\kappa^2 C - A * G^+ \\ &+ \frac{1}{2}[\frac{c_z}{c} + \frac{\rho_z}{\rho}](G^+ - G^-) - \frac{c_z}{2c}\kappa^2 C * (G^+ - G^-), \end{aligned} \tag{4.133}$$

$$\begin{aligned} \partial_z G^- &= \frac{2}{c}\partial_t G^- + \frac{ac_z}{2c}\kappa^2 C + A * G^- \\ &- \frac{1}{2}[\frac{c_z}{c} + \frac{\rho_z}{\rho}](G^+ - G^-) + \frac{c_z}{2c}\kappa^2 C * (G^+ - G^-) \end{aligned} \tag{4.134}$$

(the coefficients $C(z,t;\kappa)$, $A(z,t;\kappa)$ are defined by (4.129) and (4.130), respectively), together with the following initial condition for G^-

$$G^-(z,0;\kappa) = \frac{1}{4}a(c_z + \frac{c\rho_z}{\rho}). \tag{4.135}$$

The boundary conditions are

$$G^+(0,t;\kappa) = 0, \tag{4.136}$$

$$G^-(0,t;\kappa) = R(t;\kappa), \tag{4.137}$$

where $R(t;\kappa)$ is the reflection kernel defined by

$$\bar{p}^r(0,t;\kappa) = R(t;\kappa) * \bar{p}^i(0,t;\kappa) \tag{4.138}$$

($\bar{p}^r(0,t;\kappa)$ and $\bar{p}^i(0,t;\kappa)$ are the reflected and incident pressures, respectively, along any radial direction at the surface $z = 0$ in the Hankel transformed space). In the inverse problem, the reflection kernel $R(t;\kappa)$ is obtained from the measured transient reflection on the surface $z = 0$.

If the reflection kernel $R(t;\kappa)$ is known for a fixed value of κ, one propagates the known boundary values (cf. (4.136) and (4.137)) along the characteristics to the initial values using the imbedding equations for the Green's functions. The velocity $c(z)$ or the density $\rho(z)$ is then reconstructed according to the initial condition (4.135).

4.3.3 *Parameter reconstruction*

In this subsection, the reflected pressure is computed in physical space with a finite difference method, and the transverse Hankel transform (with a fixed value of κ) is performed for the reflected pressure at the surface as well as for the incident pressure. The reflection kernel $R(t;\kappa)$ is then obtained from a deconvolution according to equation (4.138). The velocity $c(z)$ and the density $\rho(z)$ can be reconstructed simultaneously using the reflection kernel $R(t;\kappa)$ for two different values of κ.

4.3.3.1 *Generating reflection data in physical space by means of a finite difference method*

A finite difference method is used to generate the reflected pressure for a short-pulse point source (the time-varying function $f(t)$ is non-zero only for a short period of time) over a half-space with a plane-stratified density (the velocity is taken to be a constant c_0 in this example in order to make it possible to check with an explicit and exact result).

In the xyz coordinate system, the incident pressure $p^i(x,y,z,t)$ satisfies the following equation

$$\frac{1}{c_0^2}\frac{\partial^2 p^i}{\partial t^2} - \nabla^2 p^i = \delta(\mathbf{r} - \mathbf{r}_0)f(t), \tag{4.139}$$

where $\mathbf{r}_0 = (0,0,z_0)$. Obviously the solution to the above equation is

$$p^i = \frac{f(t - |\mathbf{r} - \mathbf{r}_0|/c_0)}{4\pi|\mathbf{r} - \mathbf{r}_0|}. \tag{4.140}$$

The total pressure $p(x, y, z, t)$ satisfies

$$\frac{1}{c_0^2}\frac{\partial^2 p}{\partial t^2} - \rho\nabla\cdot\left(\frac{1}{\rho}\nabla p\right) = \delta(\mathbf{r}-\mathbf{r}_0)f(t). \tag{4.141}$$

From (4.139) and (4.141), one obtains for the scattered pressure $p^s(x, y, z, t) \equiv p(x, y, z, t) - p^i(x, y, z, t)$

$$\frac{1}{c_0^2}\frac{\partial^2 p^s}{\partial t^2} - \nabla^2 p^s + \frac{\rho_z}{\rho}\partial_z p^s = -\frac{\rho_z}{\rho}\partial_z p^i. \tag{4.142}$$

Using central difference formulas to approximate the above partial differential equation, one obtains the following discretized equation

$$\begin{aligned}
p^s(i,j,k,n+1) &= 2p^s(i,j,k,n) - p^s(i,j,k,n-1)\\
&+\left(\frac{c_0\Delta t}{\Delta z}\right)^2\{[p^s(i+1,j,k,n) - 2p^s(i,j,k,n) + p^s(i-1,j,k,n)]\\
&+[p^s(i,j+1,k,n) - 2p^s(i,j,k,n) + p^s(i,j-1,k,n)]\\
&+[p^s(i,j,k+1,n) - 2p^s(i,j,k,n) + p^s(i,j,k-1,n)]\\
&-\frac{1}{2}\Delta z\left(\frac{\rho_z}{\rho}\right)(i,j,k)[p^s(i,j,k+1,n) - p^s(i,j,k-1,n)\\
&+p^i(i,j,k+1,n) - p^i(i,j,k-1,n)]\},
\end{aligned} \tag{4.143}$$

where $p^s(i, j, k, n) = p^s(i\Delta x, j\Delta y, k\Delta z, n\Delta t)$ with $\Delta x = \Delta y = \Delta z$, etc. The initial values are

$$p^s(i,j,k,-1) = p^s(i,j,k,0) = 0, \tag{4.144}$$

due to the fact that there is no scattering before the time $t = 0$. This algorithm is stable when the following conventional Courant condition is fulfilled,

$$\Delta t \le \frac{\Delta z}{\sqrt{3}c_0}. \tag{4.145}$$

As a numerical example, the short pulse is chosen to have the following time-dependence

$$f(t) = \begin{cases} \sin\left(\frac{t+|z_0|/c_0}{t_0}\pi\right), & \text{when } -|z_0|/c_0 \le t \le -|z_0|/c_0 + t_0,\\ 0, & \text{otherwise.}\end{cases} \tag{4.146}$$

Thus it has non-zero values only within a time width t_0, and reaches the interface $z = 0$ at the time $t = 0$ (note that the position of the point source is $(0, 0, z_0)$). The density profile used in this numerical example is

$$\rho(z) = \rho_0\frac{e^{-2\beta_1 z}(1+\alpha_1)^2}{(e^{-2\beta_1 z}+\alpha_1)^2}, \quad z > 0, \tag{4.147}$$

with $\rho_0 = 1$, $\beta_1 = 1$ and $\alpha_1 = 0.1$ (the reflection kernel for such a density profile can be given in terms of an explicit expression, see (4.148) below). The velocity

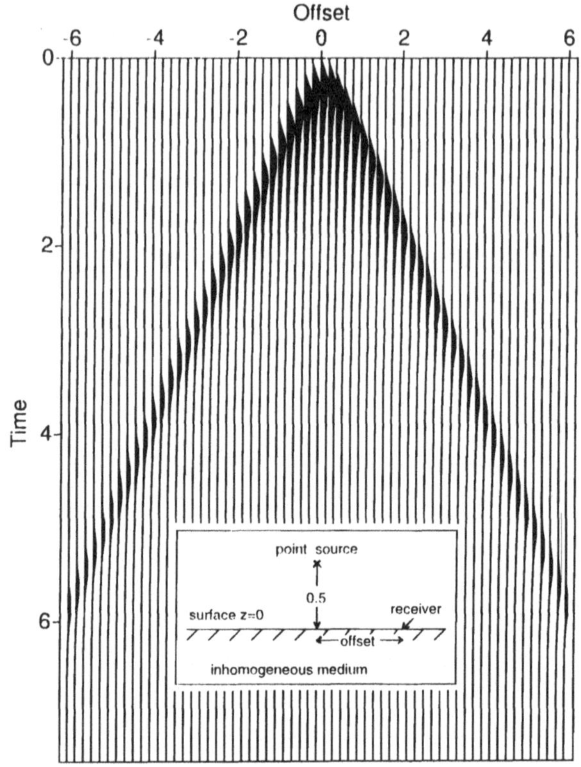

FIG. 4.5. The transient reflected pressure along a line on the surface $z = 0$, for the density profile given by the solid line in Fig. 4.8 (the velocity $c = c_0 = 1$).

is $c = c_0 = 1$. $z_0 = -0.5$ is chosen for the position of the point source, and $t_0 = 0.4$ in (4.146) for the width of the short-pulse in this example. Computation domains have to be finite. In this example, the boundary surfaces of the computation domain are taken to be far enough away so that putting p^s to zero at the boundary surfaces will have no effect on p^s at the receiving points and during the time periods of interest (the source point can be in the computation domain in this example, since p^i is zero (non-singular) at the source point for $t > 0$; note that $t_0 < |z_0|/c_0$ in this example). On the interface $z = 0$, p^s is identical to the reflected pressure p^r. The numerical results for the reflected pressure p^r along a line on the interface $z = 0$ are plotted in Fig. 4.5 (with $\Delta z = \frac{1}{32}$, $\Delta t = 0.2(\Delta z)/c_0$).

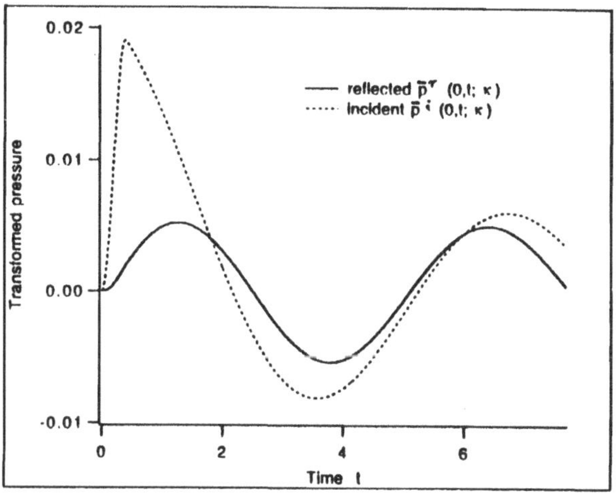

FIG. 4.6. The corresponding Hankel transformed reflected and incident pressures at $z = 0$. The value for the Hankel transform parameter is $\kappa = 1$.

4.3.3.2 *Obtaining the reflection kernel from the reflection data on the surface*

Hankel transform. On the surface $z = 0$, we consider a Hankel transform (4.116) (for a fixed value of κ) of the reflected pressure p^r calculated by the finite difference method in the previous subsection, as well as of the incident pressure p^i (cf. (4.140)). Since a time domain method is used and the pressure $p(r, z, t)$ is non-zero only within a finite region $r < r_1$ on the surface $z = 0$ at any finite time $t = t_1$, the radial Hankel transform (4.116) does not lead to any convergence problems. The numerical results are plotted in Fig. 4.6, where the solid and dashed lines are for the Hankel transformed (with $\kappa = 1$) reflected and incident pressures, respectively, at the surface $z = 0$.

Deconvolution. After the Hankel transform of the reflected and incident pressures on the surface $z = 0$ have been computed, the reflection kernel $R(t; \kappa)$ is obtained from a deconvolution according to (4.138). The deconvolution code used here is based on the fast Fourier transform of (4.138). In a deconvolution process one has to introduce a cutoff frequency ω_{cutoff}, and the problem is ill-posed so that a regularization filter is necessary to yield a stable and physically consistent result (see e.g. Nahman and Guillaume (1981) for a discussion of deconvolutions in the presence of noise). We refer to Section 2.5.2 for details about numerical aspects of the deconvolution. The reflection kernel $R(t; \kappa = 1)$ obtained from a deconvolution with the cutoff frequency $\omega_{cutoff} = 4.5$ is presented in Fig. 4.7 by the solid line. As shown in Cao and He (1994), the reflection kernel for the density profile (4.147) (with $\alpha_1 \neq 1$) has the following exact explicit form

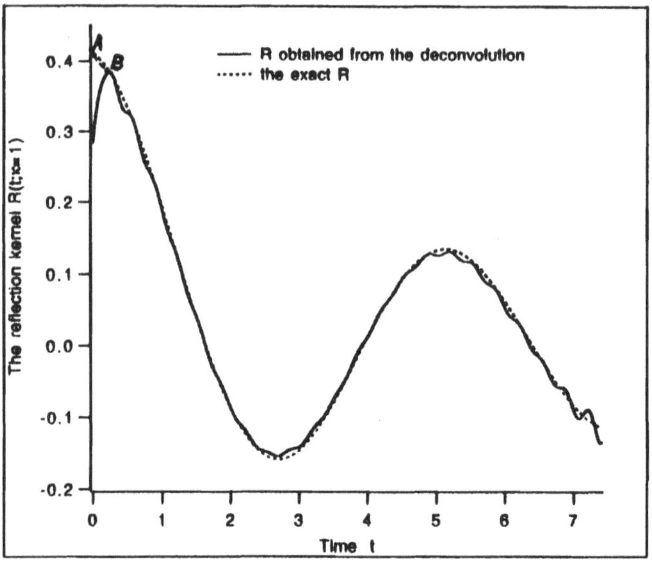

FIG. 4.7. The corresponding reflection kernel $R(t;\kappa = 1)$ (the solid line) obtained from a deconvolution with the cutoff frequency $\omega_{cutoff} = 4.5$. The dotted line is from the exact expression (4.148). The dashed line segment AB is the correction to the deconvolution for small t (using the prior knowledge of $R(t = 0; \kappa = 1)$).

$$R(t;\kappa) = \frac{1}{a_1}\left[\frac{\sqrt{\kappa^2+\beta_1{}^2}}{\beta_1}\cdot\frac{J_1(c_0 t\sqrt{\kappa^2+\beta_1{}^2})}{t} - \frac{\kappa}{\beta_1}\cdot\frac{J_1(c_0\kappa t)}{t}\right] + \frac{a_1^2-1}{a_1}\cdot c_0\beta_1 f_2(t;\kappa), \tag{4.148}$$

where $a_1 = \frac{1-\alpha_1}{1+\alpha_1}$ and

$$f_2(t;\kappa) = \frac{1}{2}f_1(t;\kappa) + \frac{1}{2a_1}f_1(t;\kappa) * \left[\frac{\sqrt{\kappa^2+\beta_1{}^2}}{\beta_1}\cdot\frac{J_1(c_0 t\sqrt{\kappa^2+\beta_1{}^2})}{t} - \frac{\kappa}{\beta_1}\cdot\frac{J_1(c_0\kappa t)}{t}\right],$$

$$f_1(t;\kappa) = J_0(c_0\kappa t) - \frac{(1-a_1^2)c_0\beta_1}{2a_1}\int_0^t J_0(c_0\kappa\sqrt{t^2-t'^2})\, e^{-\frac{(1-a_1^2)c_0\beta_1}{2a_1}t'}\,dt'.$$

The exact $R(t;\kappa)$ given by (4.148) for $\beta_1 = 1$, $\alpha_1 = 0.1$ and $\kappa = 1$ is plotted by the dotted line in Fig. 4.7. As seen from Fig. 4.7, the reflection kernel $R(t;\kappa)$ obtained from the deconvolution is consistent with the exact $R(t;\kappa)$ except for small t. The main reason for the error at small t is the use of the cutoff frequency in the deconvolution process (i.e., one only uses the information with frequencies lower than ω_{cutoff} in the spectra of the reflected and incident pressures; note that $R(t;\kappa)|_{t=0} = \int_{-\infty}^{\infty} \tilde{R}(\omega,\kappa)\, d\omega$). If the parameters are known at the surface $z = 0$, one can obtain the correct value of $R(t = 0; \kappa)$ from the initial condition (4.135) (and thus one knows the position of the point A in Fig. 4.7), and then replace the beginning portion of the $R(t;\kappa)$ curve for small t with a short straight line segment connecting point A to a point with a reasonably reliable value of R (e.g. the point B in Fig. 4.7), as shown by the dashed short line segment AB in Fig. 4.7 (this will be referred to as $R(t;\kappa)$ obtained from a corrected deconvolution, which will also be used in the reconstruction below). One can also set the surface of the inhomogeneous medium a bit lower to the plane $z = z_1 > 0$ so that the parameters in the region $0 \leq z \leq z_1$ are known (corresponding, e.g. to those of air), which means that $R(t;\kappa)$ is known for a short period of time $0 \leq t \leq t_1$. In this way one can remove the error in $R(t;\kappa)$ (obtained from a deconvolution) for small $t \in (0, t_1)$. However, the errors in $R(t;\kappa)$ for small t are not critical for the Green's function approach to the reconstruction, as we will see below.

Assume that in a real experiment, one measures the total transient pressure along a line on the surface $z = 0$ for a finite period of time, and obtains the reflected pressure by subtracting the incident pressure (generated by a point source with short time-varying pulse). One then chooses a specific value of the Hankel transform parameter κ and one performs the transform of the surface (incident and reflected) data. Then one performs the deconvolution to obtain the reflection kernel $R(t;\kappa)$ for this chosen κ, as described above. Once the reflection kernel $R(t;\kappa)$ has been obtained, one can use the Green's function approach described in the previous subsection to reconstruct the parameters in the stratified half-space.

4.3.3.3 *Numerical reconstruction*

Numerical example for reconstruction of the density using the reflection kernel obtained from the deconvolution. The reflection kernel $R(t;\kappa = 1)$ (the solid line in Fig. 4.7) obtained from the deconvolution in the previous section is now used as the boundary value of the Green's function G^- (cf. (4.137)) in the Green's function approach to the reconstruction of the density $\rho(z)$. After propagating the known boundary values (cf. (4.136) and (4.137)) along the characteristics to the initial values using the imbedding equations for the Green's functions, the density $\rho(z)$ is reconstructed using the initial condition (4.135) (see Section 2.1.4 for detailed description of the reconstruction algorithm for the Green's function approach). The reconstructed density $\rho(z)$ is plotted by the dashed line in Fig. 4.8 (the solid line is the true profile). One notes that even though $R(t;\kappa = 1)$ used here is quite different from the true value for small t (cf. Fig. 4.7), the reconstruction is not affected seriously. This indicates that the values of the reflection kernel $R(t;\kappa)$ for small t are not critical for the

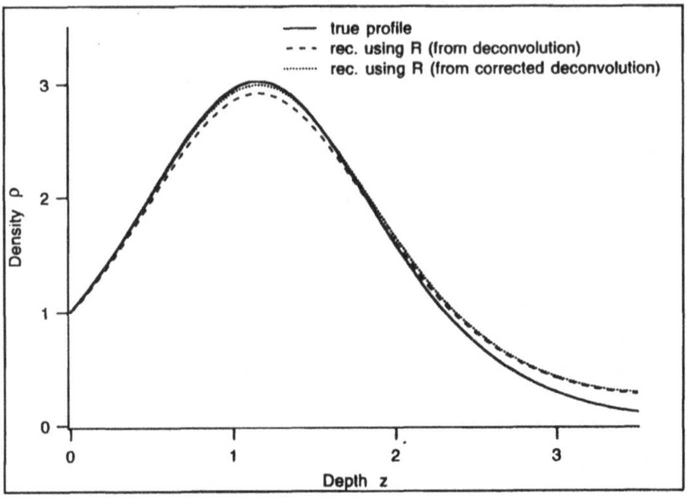

FIG. 4.8. Reconstruction of the density using the reflection kernel obtained from the deconvolution. The solid line is the true profile. The dashed line is the reconstruction using $R(t;\kappa=1)$ directly obtained from the deconvolution (the solid line in Fig. 4.7). The dotted line is the reconstruction using $R(t;\kappa=1)$ obtained from the corrected deconvolution.

reconstruction in the present Green's function approach. The reconstruction can be improved by using the reflection kernel $R(t;\kappa=1)$ obtained from a corrected deconvolution (the dashed line segment AB and the remaining part of the solid curve in Fig. 4.7), as shown by the dotted line in Fig. 4.8. Note that the error of the reconstruction at the deep positions in Fig. 4.8 is due to the computation errors introduced in the process of the finite difference calculation, Hankel transform and deconvolution.

Numerical example for a simultaneous reconstruction of $c(z)$ and $\rho(z)$ using two different values of κ. In this numerical example, the reflection kernel $R(t;\kappa)$ for two different values of κ is used to reconstruct the velocity $c(z)$ and the density $\rho(z)$ simultaneously. Introduce a function $R_T(z,t;\kappa)$ through the following Volterra equation of the second kind

$$aR_T + G^+ * R_T = \partial_t G^- - \frac{1}{4}(c_z + \frac{c\rho_z}{\rho})G^+ \tag{4.149}$$

(in fact $R_T(z,t;\kappa)$ is the time derivative of the reflection kernel), which has a unique solution for $R_T(z,t;\kappa)$ when the Green's functions $G^\pm$ are known at the depth z. Using the imbedding equations for $G^\pm$ and the initial condition (4.135) for G^-, one obtains the following useful condition

$$[(\partial_t R_T)(z,0;\kappa_1) - (\partial_t R_T)(z,0;\kappa_2)]/(\kappa_1^2 - \kappa_2^2) = -\frac{1}{8}c^2(3c_z + \frac{c\rho_z}{\rho}). \qquad (4.150)$$

Equations (4.135) (with $\kappa = \kappa_1$) and (4.150) can be used as two independent conditions for a simultaneous reconstruction. $(\partial_t R_T)(z,0;\kappa)$ in (4.150) can be calculated numerically as follows

$$(\partial_t R_T)(z,0;\kappa) = \frac{R_T(z,\Delta t;\kappa) - R_T(z,0;\kappa)}{\Delta t},$$

where (cf. (4.149))

$$R_T(z,0;\kappa) = \frac{1}{a}[\partial_t G^- - \frac{1}{4}(c_z + \frac{c\rho_z}{\rho})G^+](z,0;\kappa),$$

$$R_T(z,\Delta t;\kappa) =$$

$$\frac{(\partial_t G^-)(z,\Delta t;\kappa) - \frac{1}{4}(c_z + \frac{c\rho_z}{\rho})G^+(z,\Delta t;\kappa) - \frac{1}{2}\Delta t G^+(z,\Delta t;\kappa)R_T(z,0;\kappa)}{a + \frac{1}{2}\Delta t G^+(z,0;\kappa)}.$$

We refer to Cao and He (1997) for details. The simultaneous reconstruction of the velocity $c(z)$ and the density $\rho(z)$ using $R(t;\kappa = 0.01)$ and $R(t;\kappa = 3)$ are given by the circles (output for every fourth point for graphical clarity) in Fig. 4.9. The gridsize used here is $\Delta t = 2\tau(z_{max})/N$, $\Delta z = c(z)\Delta t/2$, with $N = 512$, $z_{max} = 1.1$. Obviously, due to the time-differentiation, this algorithm for the simultaneous reconstruction is sensitive to noise (for noisy reflection data a regularization technique is needed to handle the time-differentiations, see the convolution smoothing process with a Fejer kernel in Section 7.1).

4.3.4 *A dissipative half-space with a velocity mismatch at the surface*

4.3.4.1 *General treatment for a stratified dissipative half-space*

Consider the following dissipative wave equation in a stratified half-space $z > 0$,

$$\frac{1}{c^2}\partial_t^2 p - \nabla^2 p + b\partial_t p = 0, \qquad (4.151)$$

where b is the dissipation coefficient. It is assumed that both $c(z)$ and $b(z)$ are continuous and the velocity $c(z)$ is continuously differentiable in the half-space $z > 0$. However, the phase velocity and the dissipation coefficient are in general assumed to have discontinuities at the surface $z = 0$, which result in the following boundary conditions at the surface

$$\begin{cases} (\partial_z p)(x,y,0^+,t) = s_0\,(\partial_z p)(x,y,0^-,t), \\ p(x,y,0^+,t) = q_0\,p(x,y,0^-,t), \end{cases} \qquad (4.152)$$

where s_0 and q_0 are constants depending on the parameters at $z = 0^+$, $z = 0^-$ and the specific physical model considered (for example, if p is the Hertz potential

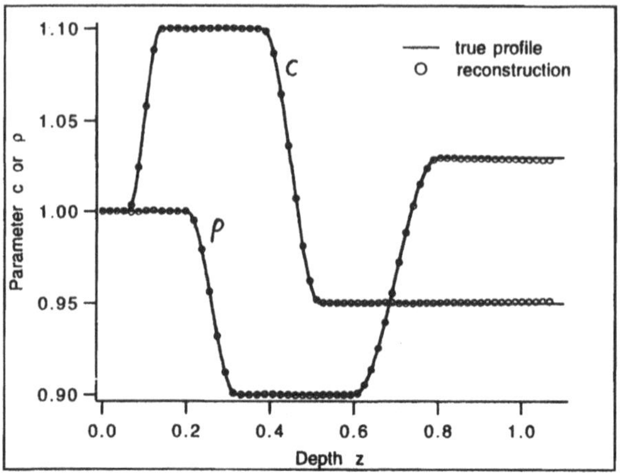

FIG. 4.9. Simultaneous reconstruction of the velocity and density using the reflection kernel $R(t;\kappa)$ for two different values of the Hankel transform parameter κ, $\kappa_1 = 0.01$, and $\kappa_2 = 3$.

of a vertical electric dipole, then $s_0 = 1$ and $q_0 = \epsilon(0^-)/\epsilon(0^+)$, where ϵ is the permittivity). The mismatch of the velocity, i.e. $c(0^-) = c_0$, $c(0^+) = c_1$, where in general $c_1 \neq c_0$, will also affect the splitting across the surface $z = 0$.

For a point source excitation, the same wave-splitting is used in the Hankel transformed space as described in the previous subsections. The derivation for the dynamic equations for the split fields follows the same lines which led to (4.124). One obtains

$$\partial_z \begin{bmatrix} \bar{p}^+ \\ \bar{p}^- \end{bmatrix} = \begin{bmatrix} -1 & 0 \\ 0 & 1 \end{bmatrix} K^{-1} \begin{bmatrix} \bar{p}^+ \\ \bar{p}^- \end{bmatrix} + \begin{bmatrix} 1 & -1 \\ -1 & 1 \end{bmatrix} \left(\frac{c_z}{2c}(1 - \kappa^2 K^2) \right) \begin{bmatrix} \bar{p}^+ \\ \bar{p}^- \end{bmatrix}$$
$$+ \begin{bmatrix} -1 & -1 \\ 1 & 1 \end{bmatrix} \left(\frac{1}{2} bK\partial_t \right) \begin{bmatrix} \bar{p}^+ \\ \bar{p}^- \end{bmatrix}. \tag{4.153}$$

In view of the velocity mismatch at the surface $z = 0$, the physical reflection kernel and the Green's functions are defined as

$$p^-(0^-, t; \kappa) = rp^+(0^-, t; \kappa) + R(t; \kappa) * p^+(0^-, t; \kappa), \tag{4.154}$$

$$\bar{p}^+(z, t + \tau; \kappa) = a(z)t_1\bar{p}^+(0^-, t; \kappa) + a(z)t_1 G^+(z, t; \kappa) * \bar{p}^+(0^-, t; \kappa), \quad z > 0, \tag{4.155}$$

$$\bar{p}^-(z, t + \tau; \kappa) = a(z)t_1 \int_0^t G^-(z, t; \kappa) * \bar{p}^+(0^-, t; \kappa), \quad z > 0, \tag{4.156}$$

where $a(z)$ is the attenuation factor

$$a(z) = \sqrt{\frac{c(z)}{c_1}} \exp(-\frac{1}{2}\int_0^z b(z')c(z')\,dz').$$

Substituting the above definitions for the reflection kernel and the Green's functions into the boundary condition (4.152), one obtains (see Section 2.1.6 for a more detailed derivation for a similar case; cf. also He et al. (1993b))

$$r = \frac{(\frac{c_1}{c_0})s_0 - q_0}{(\frac{c_1}{c_0})s_0 + q_0}, \tag{4.157}$$

$$t_1 = q_0(1+r) = \frac{2(\frac{c_1}{c_0})s_0 q_0}{(\frac{c_1}{c_0})s_0 + q_0}, \tag{4.158}$$

and the following relations between $G^{\pm}(0,t;\kappa)$ and $R(t;\kappa)$

$$G^+(0,t;\kappa) = (q_0/t_1)R(t;\kappa) - G^-(0,t;\kappa), \tag{4.159}$$

$$t_1 J_0(\kappa c_0 t) * G^-(0,t;\kappa) - \frac{1}{2}[q_0 J_0(\kappa c_0 t) + s_0(\frac{c_1}{c_0})J_0(\kappa c_1 t)] * R(t;\kappa) = \frac{t_1}{2}[J_0(\kappa c_0 t) - J_0(\kappa c_1 t)]. \tag{4.160}$$

Equation (4.160) is a Volterra equation of the first kind for $R(t;\kappa)$ (or $G^-(0,t;\kappa)$) if $G^-(0,t;\kappa)$ (or $R(t;\kappa)$) is known. It can be converted to the following Volterra equation of the second kind by differentiating both sides with respect to t,

$$G^-(0,t;\kappa) + A_1 * G^-(0,t;\kappa) + t_1^{-1} d_1 R(t;\kappa) + t_1^{-1} C_1 * R(t;\kappa) + D_1 = 0, \tag{4.161}$$

where

$$A_1 = -\kappa c_0 J_1(\kappa c_0 t), \quad C_1 = \frac{1}{2}\kappa c_0[q_0 J_1(\kappa c_0 t) + s_0 \frac{c_1^2}{c_0^2} J_1(\kappa c_1 t)], \tag{4.162}$$

$$d_1 = -\frac{1}{2}(q_0 + s_0 \frac{c_1}{c_0}), \quad D_1 = \frac{1}{2}\kappa c_0[J_1(\kappa c_0 t) - \frac{c_1}{c_0} J_1(\kappa c_1 t)]. \tag{4.163}$$

From (4.159) and (4.161) one obtains the following boundary relation between G^+ and G^-

$$d_2 G^-(0,t;\kappa) + A_2 * G^-(0,t;\kappa) + d_1 G^+(0,t;\kappa) + C_1 * G^+(0,t;\kappa) + q_0 D_1 = 0, \tag{4.164}$$

where

$$A_2 = \frac{1}{2}(\kappa c_0)[-q_0 J_1(\kappa c_0 t) + s_0 \frac{c_1^2}{c_0^2} J_1(\kappa c_1 t)], \qquad d_2 = \frac{1}{2}(q_0 - s_0 \frac{c_1}{c_0}). \tag{4.165}$$

If there is no parameter mismatch at the surface $z = 0$, i.e. $c_1 = c_0$, $s_0 = q_0 = 1$, then $r = 0$, $t_1 = 1$, and the boundary relations (4.161) and (4.159) reduce to $G^-(0,t;\kappa) = R(t;\kappa)$ and $G^+(0,t;\kappa) = 0$, respectively.

Using (4.124) and the definitions of the Green's functions (4.155) and (4.156), one obtains the following equations for $G^{\pm}$ (cf. He and Karlsson (1993), He et al. (1993b))

$$\begin{aligned}\partial_z G^+ &= -A + \frac{1}{2}bB - \frac{c_z}{2c}\kappa^2 C - A * G^+ - \frac{1}{2}(bc + \frac{c_z}{c})G^- \\ &\quad + \frac{1}{2}bB * (G^+ + G^-) - \frac{c_z}{2c}\kappa^2 C * (G^+ - G^-), \end{aligned} \tag{4.166}$$

$$\begin{aligned}\partial_z G^- - \frac{2}{c}\partial_t G^- &= -\frac{1}{2}bB + \frac{c_z}{2c}\kappa^2 C + A * G^- + \frac{1}{2}(bc - \frac{c_z}{c})G^+ + bcG^- \\ &\quad - \frac{1}{2}bB * (G^+ + G^-) + \frac{c_z}{2c}\kappa^2 C * (G^+ - G^-), \end{aligned} \tag{4.167}$$

$$G^-(z, 0; \kappa) = \frac{1}{4}(c_z - bc^2), \tag{4.168}$$

where $B(z,t;\kappa) = \kappa c^2(z) J_1(\kappa c(z) t)$, and the coefficients $C(z,t;\kappa)$, $A(z,t;\kappa)$ are defined by (4.129) and (4.130).

In the direct problem, the two PDEs (4.166) and (4.167) together with the initial condition for G^-, (4.168), and the boundary relation between $G^{\pm}$, (4.164), constitute a set of equations from which G^+ and G^- can be determined. The reflection kernel $R(t;\kappa)$ is then obtained from the boundary relation (4.159). In the inverse problem, the reflection kernel $R(t;\kappa)$ is assumed to be known and $G^{\pm}(0,t;\kappa)$ are obtained from the boundary relations (4.159) and (4.161). Hence one of the two parameters $c(z)$ and $b(z)$ can be reconstructed by propagating the boundary values along the characteristic curves to the initial values (4.168) (see He et al. (1993b) for a simultaneous reconstruction of $c(z)$ and $b(z)$).

4.3.4.2 *A point source above a homogeneous non-dissipative half-space*

Time domain analysis of a point source over a homogeneous half-space is a classic problem. The simplifications which occur within the present approach when the lower half-space is homogeneous are described below. The fact that for a homogeneous half-space the splitting gives the physical division into up-going and down-going waves can be expected to lead to a simplified equation which determines the reflection kernel (cf. also Section 3.2.4 for the treatment in the case of a conducting half-space).

For a point source located above a homogeneous non-dissipative half-space, one has $b = 0$, $a = 1$, $c_z = 0$. Therefore, from (4.167) and (4.168) it follows that

$$\begin{aligned} G_z^- &= \frac{2}{c}G_t^- + A * G^-, \\ G^-(z, 0) &= 0, \end{aligned}$$

which give the trivial solution

$$G^{-}(z,t;\kappa) = 0, \tag{4.169}$$

as expected, since in the homogeneous non-dissipative half-space $z > 0$ there is only a down-going wave. Substituting (4.169) into (4.161), it follows that

$$R(t;\kappa) + C_3 * R(t;\kappa) + D_3 = 0, \tag{4.170}$$

where

$$C_3 = -\frac{(\kappa c_0)}{(\frac{c_1}{c_0})s_0 + q_0}[q_0 J_1(\kappa c_0 t) + s_0\frac{c_1^2}{c_0^2}J_1(\kappa c_1 t)], \tag{4.171}$$

$$D_3 = -\frac{2(\frac{c_1}{c_0})s_0 q_0}{(\frac{c_1}{c_0})s_0 + q_0}[\frac{(\kappa c_0)}{(\frac{c_1}{c_0})s_0 + q_0}][J_1(\kappa c_0 t) - \frac{c_1}{c_0}J_1(\kappa c_1 t)]. \tag{4.172}$$

Equation (4.170) is a Volterra equation of the second kind which gives the solution for the reflection kernel for the field from a point source over a homogeneous non-dissipative half-space in the Hankel transformed space. Note that a Volterra equation of the second kind is well-posed and has one and only one bounded solution (see e.g. Pogorzelski 1966).

In the limiting case $t \to 0$, one has the following asymptotic solution

$$R(t;\kappa) \sim \frac{a_0}{\nu}\sinh(\nu t), \tag{4.173}$$

where

$$\nu^2 = \frac{1}{2}\frac{(\kappa c_0)^2}{(\frac{c_1}{c_0})s_0 + q_0}(q_0 + s_0\frac{c_1^3}{c_0^3}), \tag{4.174}$$

$$a_0 = \frac{(\frac{c_1}{c_0})s_0 q_0}{(\frac{c_1}{c_0})s_0 + q_0}\left[\frac{\kappa^2(c_0^2 - c_1^2)}{(\frac{c_1}{c_0})s_0 + q_0}\right]. \tag{4.175}$$

As expected, (4.170) and (4.173) gives a zero solution for $R(t;\kappa)$ when there is no mismatch at the surface $z = 0$, i.e. $c_1 = c_0$, $s_0 = q_0 = 1$.

4.3.4.3 *A point source above a homogeneous dissipative half-space*

For a homogeneous dissipative half-space $z > 0$, one has b = constant, $a = \exp[-bc_1 z/2]$, $c_z = 0$. In this case (4.167) and (4.168) will not give a trivial solution, i.e. $G^{-}(z,t;\kappa) \neq 0$ (cf. (4.168), which gives $G^{-}(z,0) = -\frac{1}{4}bc_1^2 \neq 0$ in this case). This is because the splitting (4.121) does not have a direct physical interpretation in the homogeneous dissipative half-space $z > 0$. Of course, (4.167) and (4.168) can nevertheless be used to solve the wave propagation problem in the dissipative half-space. However, a more direct method of solution, analogous to the one used above for the non-dissipative half-space, is possible also in the present case, since a splitting which has a direct physical interpretation inside the dissipative half-space $z > 0$ is available. Such a splitting is (He 1992a)

$$\hat{p}^{\pm} = (1/2)[\bar{p} \mp \mathcal{K}\partial\bar{p}], \tag{4.176}$$

where

$$\mathcal{K}f(z,t;\kappa) = c_1 J_0(a_1 t)e^{-b_1 t} * f(z,t;\kappa), \qquad z > 0, \tag{4.177}$$

$$\mathcal{K}^{-1}f(z,t;\kappa) = \frac{1}{c_1}\partial_t f + \frac{bc_1}{2}f + \frac{a_1}{c_1}[J_1(a_1 t)e^{-b_1 t}/t] * f(z,t;\kappa), \tag{4.178}$$

with

$$b_1 = \frac{1}{2}bc_1^2, \tag{4.179}$$

$$a_1 = \begin{cases} c_1(\kappa^2 - \frac{1}{4}b^2c_1^2)^{\frac{1}{2}}, & \text{if } \kappa \geq \frac{1}{2}bc, \\ ic_1(\frac{1}{4}b^2c_1^2 - \kappa^2)^{\frac{1}{2}}, & \text{if } \kappa < \frac{1}{2}bc. \end{cases} \tag{4.180}$$

The physical significance of this splitting comes from the fact that it corresponds to a factorization of the whole wave equation (the velocity is a constant c_1), i.e.

$$(\partial_z - \mathcal{K}^{-1})(\partial_z + \mathcal{K}^{-1})u = 0. \tag{4.181}$$

Furthermore, $\hat{p}^{\pm}(0^-,t;\kappa) = \bar{p}^{\pm}(0^-,t;\kappa)$ since the half-space $z < 0$ is non-dissipative.

The Green's functions related to this splitting are defined in complete analogy with (4.155) and (4.156) by

$$\hat{p}^{+}(z,t+\tau;\kappa) = a(z)t_1[\hat{p}^{+}(0^-,t;\kappa) + \mathcal{G}^{+}(z,t;\kappa) * \hat{p}^{+}(0^-,t;\kappa)], \tag{4.182}$$

$$\hat{p}^{-}(z,t+\tau) = a(z)t_1\mathcal{G}^{-}(z,t;\kappa) * \hat{p}^{+}(0^-,t;\kappa). \tag{4.183}$$

In a homogeneous dissipative half-space $z > 0$, one obtains (He 1992a), by differentiating the expressions for the Green's functions and combining them with the dynamic equations for the split fields in the by now well-established procedure,

$$\mathcal{G}_z^- = \frac{2}{c}\mathcal{G}_t^- + bc\mathcal{G}^- + [\frac{a_1}{c}J_1(a_1 t)e^{-b_1 t}/t] * \mathcal{G}^-,$$

$$\mathcal{G}^-(z,0) = 0,$$

which gives the trivial solution

$$\mathcal{G}^-(z,t;\kappa) = 0. \tag{4.184}$$

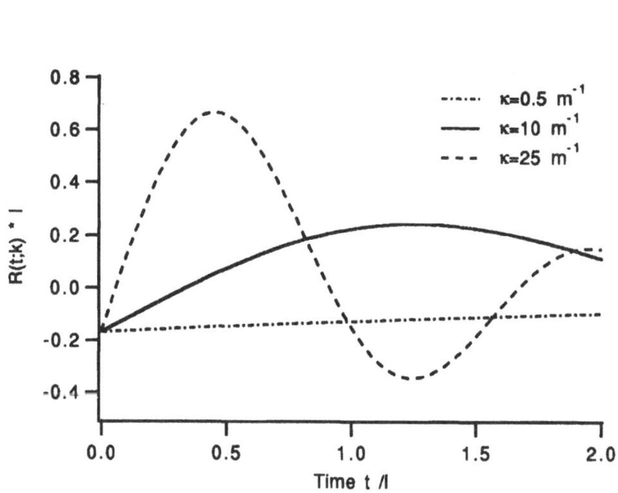

FIG. 4.10. The reflection kernel $R(t;\kappa)$ for a homogeneous dissipative half-space for three different κ values.

Through a derivation similar to the one given previously for the case of a homogeneous non-dissipative half-space, one obtains the following Volterra equation of the second kind (He et al. 1993b)

$$R(t;\kappa) + C_4 * R(t;\kappa) + D_4 = 0, \tag{4.185}$$

where

$$C_4 = -\frac{c_0}{(\frac{c_1}{c_0})s_0 + q_0}\left\{\kappa q_0 J_1(\kappa c_0 t) + s_0\frac{c_1}{c_0^2}[a_1 J_1(a_1 t) + b_1 J_0(a_1 t)]e^{-b_1 t}\right\}, \tag{4.186}$$

$$D_4 = -\frac{2(\frac{c_1}{c_0})s_0 q_0}{(\frac{c_1}{c_0})s_0 + q_0}\left[\frac{c_0}{(\frac{c_1}{c_0})s_0 + q_0}\right] \times\left\{\kappa J_1(\kappa c_0 t) - \frac{1}{c_0}[a_1 J_1(a_1 t) + b_1 J_0(a_1 t)]e^{-b_1 t}\right\}. \tag{4.187}$$

Equation (4.185) gives the solution of the reflection kernel for a point source over a homogeneous dissipative half-space in the Hankel transformed space. As expected, (4.185) reduces to (4.170) when $b = 0$. The reflection kernel for a homogeneous dissipative half-space is presented in Fig. 4.10 for $\kappa = 0.5$, 10, 25 m^{-1}, respectively, by solving the Volterra equation (4.185) of the second kind (the parameters are chosen as $s_0 = 1$, $q_0 = \epsilon(0^-)/\epsilon(0^+) = 0.25$, $c_0 = 3.0 \times 10^8$ $\mathrm{m\,s}^{-1}$, $b = 0.04$ mho/m; the reflection kernel is scaled by $l = 6.66 \times 10^{-10}$ s).

One notes that the incident field $\bar{p}^i(z,t;\kappa)$ due to a point source in the Hankel transformed space can be envisaged as a combination of oblique plane-wave-like constituents with a weighted distribution that is a function of the angle of incidence. The weighted distribution function changes as the κ-value changes. When $\kappa = 0$, it consists only of a normally incident plane-wave-like constituent. As κ gets larger, the distribution function will have larger weight on the oblique constituents.

4.4 Applications of the explicit form of the wave-splitting to some electromagnetic wave propagation problems

The explicit form of the wave-splitting given by (4.121), (4.127), and (4.130) has many applications in wave propagation problems. In this section, the applications to the electromagnetic wave propagation in homogeneous plasmas, dispersive media, and waveguides are described.

4.4.1 *Wave propagation in a homogeneous plasma*

Consider the propagation of a plane wave through the ionosphere. Assume that the plasma is homogeneous (i.e., the density of the free electrons does not depend on the altitude z), the earth's magnetic field is negligible, and the mean free path of the electrons is long enough. Then one has the following equation for the electric field E polarized in the transverse direction (Stratton 1941)

$$\left(\partial_z^2 - \frac{1}{c^2}\partial_t^2 - \kappa^2\right) E(z,t) = 0, \tag{4.188}$$

where $c = 1/\sqrt{\mu_0\epsilon_0}$, $\kappa^2 = \frac{\omega_p^2}{c^2}$, and ω_p is the plasma cutoff frequency given by

$$\omega_p^2 = \frac{ne^2}{m\epsilon_0} \tag{4.189}$$

(n is the density of free electrons, m is the mass of an electron, and e is the charge).

The plasma wave equation (4.188) has the same form as the Hankel transformed wave equation (4.119) in the homogeneous region. Equation (4.188) can thus be factorized as $\left(\partial_z - K^{-1}\right)\left(\partial_z + K^{-1}\right) E = 0$, where the operator K^{-1} is given by the explicit expression (4.130), i.e.

$$K^{-1} f(z,t) = \frac{1}{c}\partial_t f + A(t;\kappa) * f(z,t),$$

where $A(t;\kappa) = \kappa\frac{J_1(c\kappa t)}{t}$. Assume that the transient wave propagates downwards, i.e., in the $+z$ direction. If $E(0,t)$ is known (it is identically zero for

$t < 0$), one can calculate the transient field at any arbitrary position $z \geq 0$ by the following formula

$$E(z, t + z/c) = E(0, t) + G(z, t) * E(0, t), \tag{4.190}$$

where $G(z, t)$ is the regular part of the fundamental solution (corresponding to an impulsive field at $z = 0$). In fact, $G(z, t) = G^+(z, t) + G^-(z, t) = G^+(z, t)$ (note that the up-going component is $G^-(z, t) \equiv 0$ since the medium is homogeneous). $E(z, t)$ satisfies the following down-going wave condition,

$$(\partial_z + K^{-1})E = 0. \tag{4.191}$$

Differentiating (4.190) with respect to z, and using the above condition and the fact that $E(0, t)$ is an arbitrary field, onc obtains

$$\partial_z G = -A(t; \kappa) - A(t; \kappa) * G, \tag{4.192}$$

which has the following solution (through the Laplace transform; note that $G(0, t) = 0$)

$$G(z, t) = -\kappa z \frac{J_1\left(\kappa c \sqrt{t^2 + 2zt/c}\right)}{\sqrt{t^2 + 2zt/c}}. \tag{4.193}$$

Consider the following double exponential pulse (also called the Bell Laboratory waveform)

$$E(0, t) = a_0 \left(e^{-\alpha_1 t} - e^{-\alpha_2 t}\right) \Theta(t), \tag{4.194}$$

where a_0 is a constant, and $\Theta(t)$ is the Heaviside unit step function. Such a double exponential pulse has been commonly used in the study of propagation of ultra-wide-band electromagnetic pulses. From (4.190) (at a fixed position $z = z_0 > 0$) one then obtains

$$\begin{aligned} E(z_0, t + z_0/c) &= a_0\Theta(t) \left\{ e^{-\alpha_1 t} - e^{-\alpha_2 t} + \int_0^t [e^{-\alpha_1 t'} - e^{-\alpha_2 t'}] G(z_0, t - t')\, dt' \right\} \\ &\left(\equiv a_0\Theta(t) \left\{ e^{-\alpha_1 t} - e^{-\alpha_2 t} + \int_0^t f(t, t')\, dt' \right\} \right), \end{aligned} \tag{4.195}$$

or, equivalently (by changing the integration variable $t' \to t - t'$ in the above equation)

$$\begin{aligned} E(z_0, t + z_0/c) &= a_0\Theta(t) \left\{ e^{-\alpha_1 t} - e^{-\alpha_2 t} \right. \\ &\quad \left. + e^{-\alpha_1 t} \int_0^t e^{\alpha_1 t'} G(z_0, t')\, dt' - e^{-\alpha_2 t} \int_0^t e^{\alpha_2 t'} G(z_0, t')\, dt' \right\} \\ &\left(\equiv a_0\Theta(t) \left\{ e^{-\alpha_1 t} - e^{-\alpha_2 t} + e^{-\alpha_1 t} \int_0^t f_1(t')\, dt' - e^{-\alpha_2 t} \int_0^t f_2(t')\, dt' \right\} \right). \end{aligned} \tag{4.196}$$

We point out that in order to compute the transient field at the position $z = z_0$ for a period of time, one should use the expression (4.196), rather than (4.195) (cf.

the discussion in He (1996b)). The CPU time is mainly consumed by the evaluation of the special functions (the values of these functions are computed numerically by series expansion). Assume that the time period of interest is discretized with N gridpoints, i.e., $t = 0, t_1, t_2, ..., t_N$. The calculation of $\int_0^t f(t, t')\, dt'$ in the expression (4.195) for $t = 0, t_1, t_2, ..., t_N$, requires $\sum_{i=1}^{N} i = N(N+1)/2$ times of evaluation of special functions, while the calculation of the integral $\int_0^t f_j(t')\, dt'$, $j = 1$ or 2, in (4.196) for $t = 0, t_1, t_2, ..., t_N$, requires only N times of evaluation of special functions (since $\int_0^{t_i} f_1(t')\, dt' = \int_0^{t_{i-1}} f_1(t')\, dt' + \int_{t_{i-1}}^{t_i} f_1(t')\, dt' \approx \int_0^{t_{i-1}} f_1(t')\, dt' + f_1(t_i)[t_i - t_{i-1}]$, one only needs to add a small additional term to the computed result of such an integral at the previous time t_{i-1} to obtain the result at the next discretized time t_i; one can use a more accurate approximation for $\int_0^{t_i} f_1(t')\, dt'$ to improve the accuracy; there is no similar formula for $\int_0^t f(t, t')\, dt'$). Therefore, the expression (4.196) is approximately $N/4$ times faster to complete the computation than the expression (4.195). The time convolution in the representation (4.190) is in general a t' integral of a function of two variables (t and t'), which is computationally intensive, since the integral has to be performed over many time steps for each time t_i, $1 \leq i \leq N$, in the time period of interest. However, if $E(0, t)$ is a sum of exponentials, one can always rewrite the time convolution in the representation (4.190) as a sum of t' integrals of functions of a single variable t', and improve the computational efficiency by an order of N. In fact, for many other forms of $E(0, t)$ (e.g., $E(0, t) = a_0 e^{-\alpha t} \sin(at)\Theta(t)$ or $a \cos(\omega_0 t)\Theta(t)$), one can use a so-called ODE representation to improve greatly (by an order of N) the computational efficiency of the time convolution (see e.g. Joseph et al. (1991)).

4.4.2 *First precursor in Lorentz dispersive media*

For a plane wave propagating in a homogeneous dispersive medium, the electric field $E(z, t)$ (transversely polarized) satisfies the following equation (cf. Section 3.3.1)

$$\partial_z^2 E - \frac{1}{c^2} \partial_t^2 E - \frac{1}{c^2} \partial_t^2 [\chi * E] = 0, \tag{4.197}$$

where $c = 1/\sqrt{\epsilon \mu_0}$, and $\chi(t)$ is the susceptibility kernel. For the Lorentz medium, one has

$$\chi(t) = \frac{\omega_p^2}{\nu_0} e^{-\nu t/2} \sin \nu_0 t \Theta(t), \tag{4.198}$$

where ω_p, ν are the plasma and collision frequencies of the medium, respectively, and $\nu_0 = \sqrt{\omega_0^2 - \nu^2/4}$ (ω_0 is the resonance frequency of the medium). Assume that the transient wave propagates in the $+z$ direction (starting from the position $z = 0$ where the field is identically zero for $t < 0$). Brillouin has examined the wave propagation of a harmonically oscillating signal (starting at $t = 0$ and $z = 0$) in a dispersive medium (Brillouin 1960; Stratton 1941). During the transition from vanishingly small amplitudes at the wavefront to the relatively large values of the signal, the main body of the signal is preceded by a first precursor which travels with the velocity c. The first precursor arrives with zero amplitude, and then grows slowly both in period and in amplitude. The amplitude then decreases as the period approaches the natural period of the electrons (which is when the second precursor arrives, having travelled with a velocity less than c). The short time behaviour of $\chi(t)$ for the Lorentz medium is

$$\chi(t) \approx \omega_p^2 t \Theta(t), \qquad \text{for small } t. \tag{4.199}$$

Therefore, (4.197) reduces to

$$\partial_z^2 E - \frac{1}{c^2}\partial_t^2 E - \frac{\omega_p^2}{c^2} E = 0, \qquad \text{for small } t, \tag{4.200}$$

which has the same form as the plasma wave equation (4.188). Therefore, one has the following formula for the early arrival of the signal (see (4.190) and (4.193))

$$E(z, t + z/c) = E(0,t) - \frac{\omega_p}{c} z \left[\frac{J_1\left(\omega_p \sqrt{t^2 + 2zt/c}\right)}{\sqrt{t^2 + 2zt/c}} \right] * E(0,t). \tag{4.201}$$

Equation (4.201) gives the same result as (3.133) for the Lorentz medium. For a large distance z, and for $E(0,t) = a\frac{t^m}{m!}\Theta(t)$ $(m > 0)$, the above expression reduces to (Kristensson 1995)

$$E(z, t + z/c) = a \left[\frac{t}{\xi} \right]^{m/2} J_m(2\sqrt{\xi t}), \tag{4.202}$$

where $\xi = z\omega_p^2/(2c)$. The above equation is consistent with the classical result for the first precursor given in Jackson (1975). The expression (4.201) is thus a generalization of the classical result for the first precursor. More general results on precursors obtained by the expansion of the wave propagator (Section 2.4) can be found in Karlsson and Rikte (1997).

4.4.3 *Wave propagation in cylindrical waveguides*

Consider a cylindrical waveguide (i.e., a hollow metallic cylindrical tube with a cross-section that is uniform along the z axis) with an arbitrary cross-section Ω (denote the boundary curve of the cross-section by $\partial\Omega$). The waveguide is filled with a homogeneous isotropic medium with electromagnetic parameters ϵ and μ.

Since the surface $\partial\Omega$ is assumed to be a perfect conductor, one has the following boundary conditions

$$\vec{\mathbf{E}} \times \hat{\mathbf{n}} = 0, \quad \vec{\mathbf{H}} \cdot \hat{\mathbf{n}} = 0, \qquad \mathbf{r} \in \partial\Omega, \tag{4.203}$$

where $\hat{\mathbf{n}}$ is the unit outward normal to $\partial\Omega$. The initial condition

$$\vec{\mathbf{E}} = \vec{\mathbf{H}} = 0, \qquad \text{at } t = 0, \tag{4.204}$$

will be imposed on the field quantities.

Rather than applying the splitting to a general solution of Maxwell's equations, the splitting that will be used will apply to the individual TM and TE modes represented by the E_3 and H_3 components, respectively (the TEM mode will not be considered).

For a homogeneous medium, the components E_3, H_3 satisfy the wave equation

$$\begin{cases} \nabla^2 E_3 - \frac{1}{c^2}\partial_t^2 E_3 = 0, \\ \\ \nabla^2 H_3 - \frac{1}{c^2}\partial_t^2 H_3 = 0, \end{cases} \tag{4.205}$$

where the velocity $c = (\mu\epsilon)^{-1/2}$ is constant. The first two components of Maxwell's equations can be written in the following form (cf. (4.60))

$$\partial_z \begin{bmatrix} \mathbf{E}_T \\ \mathbf{H}_T \end{bmatrix} = \begin{bmatrix} 0 & -\mu I_2 \partial_t \\ I_2 \epsilon \partial_t & 0 \end{bmatrix} \begin{bmatrix} \mathbf{E}_T \\ \mathbf{H}_T \end{bmatrix} + \begin{bmatrix} \nabla_T E_3 \\ \nabla_T H_3 \end{bmatrix}, \tag{4.206}$$

where $\mathbf{E}_T$, $\mathbf{H}_T$ are the tangential electric and magnetic fields, respectively.

It is well-known that the transverse fields can be expressed in terms of the z-components of the fields, E_3 and H_3 (see e.g. Jackson (1975)). Thus, one only needs to concentrate on the calculation of E_3 and H_3. For the TM mode (represented by E_3), $H_3 \equiv 0$ and E_3 satisfies the Dirichlet boundary condition $E_3 = 0$ on $\partial\Omega$. For the TE mode (represented by H_3), $E_3 \equiv 0$ and H_3 satisfies the Neumann boundary condition $\partial_n H_3 = 0$ on $\partial\Omega$ (here ∂_n denotes differentiation in the direction of the outward pointing normal $\hat{\mathbf{n}}$ on the conducting surface $\partial\Omega$). We use $\mathbf{r} = (x, y, z) \equiv (\boldsymbol{\rho}, z)$ and write for the TM and TE modes respectively

$$\begin{cases} E_3 = v(\boldsymbol{\rho})u(z,t), \\ \\ H_3 = 0, \end{cases} \qquad \text{for TM mode,} \tag{4.207}$$

$$\begin{cases} H_3 = w(\boldsymbol{\rho})u(z,t), \\ \\ E_3 = 0, \end{cases} \qquad \text{for TE mode.} \tag{4.208}$$

Substituting the above equations into (4.205) and the boundary conditions, one obtains the following eigenvalue problems,

$$\begin{cases} (\nabla_T^2 + \lambda^2)v(\boldsymbol{\rho}) = 0, \; \boldsymbol{\rho} \in \Omega, \\ \\ v(\boldsymbol{\rho}) = 0, \qquad\qquad \boldsymbol{\rho} \in \partial\Omega, \end{cases} \qquad \text{for TM mode,} \tag{4.209}$$

$$\begin{cases} (\nabla_T^2 + \lambda^2) w(\boldsymbol{\rho}) = 0, \ \boldsymbol{\rho} \in \Omega, \\ \hat{n} \cdot \nabla_T w(\boldsymbol{\rho}) = 0, \qquad \boldsymbol{\rho} \in \partial\Omega, \end{cases} \qquad \text{for TM mode,} \tag{4.210}$$

where $\nabla_T^2 = \partial_x^2 + \partial_y^2$, and λ^2 is the eigenvalue of $-\nabla_T^2$ (note that the eigenvalue $\lambda_0 = 0$, which corresponds to a TEM mode, will be excluded in the analysis).

Let λ_n^2, $n = 1, 2, 3, \ldots$, be the *non-zero* eigenvalues associated with (4.209) or (4.210), and v_n, w_n the corresponding normalized eigenfunctions satisfying the conditions ($\lambda_n \neq 0$)

$$\begin{cases} \displaystyle\int_\Omega v_n(\boldsymbol{\rho}) v_m(\boldsymbol{\rho}) d\boldsymbol{\rho} = \frac{1}{\lambda_n^2} \delta_{nm}, \\ \displaystyle\int_\Omega w_n(\boldsymbol{\rho}) w_m(\boldsymbol{\rho}) d\boldsymbol{\rho} = \frac{1}{\lambda_n^2} \delta_{nm}, \end{cases} \tag{4.211}$$

where δ_{nm} is the Kronnecker delta function. From (4.205), (4.207)–(4.210) one thus obtains the following one-dimensional Klein–Gordon equation (Morse and Feshbach 1953):

$$\left(\partial_z^2 - \frac{1}{c^2} \partial_t^2 - \lambda_n^2 \right) u_n(z, t) = 0. \tag{4.212}$$

From the initial condition $E_3 = H_3 = 0$ at $t = 0$, it follows that $u_n = 0$ at $t = 0$, and from the system (4.206) and initial conditions $\mathbf{E}_T = \mathbf{H}_T = 0$ at $t = 0$ it follows that $\partial_t u_n = 0$ at $t = 0$. Thus, u_n must be solutions of (4.212) satisfying the initial conditions

$$u_n = \partial_t u_n = 0. \tag{4.213}$$

For the TM mode, the first equation of the system (4.206) yields the following

$$\begin{aligned} (\partial_z^2 - \frac{1}{c^2} \partial_t^2) \mathbf{E}_T &= \nabla_T v_n \partial_z u_n(z, t) \\ &= \frac{1}{\lambda_n^2} (\partial_z^2 - \frac{1}{c^2} \partial_t^2)(\nabla_T v_n \partial_z u_n(z, t)), \end{aligned}$$

which gives the solution

$$\mathbf{E}_T = \frac{1}{\lambda_n^2} \partial_z u_n(z, t) \nabla_T v_n(\boldsymbol{\rho}).$$

The second equation of the system (4.206) yields

$$\mathbf{H}_T = -\frac{\epsilon}{\lambda_n^2} \partial_t u_n(z, t) \hat{\mathbf{z}} \times \nabla_T v_n(\boldsymbol{\rho}),$$

using the fact that

$$I_2 \begin{bmatrix} f_1 \\ f_2 \end{bmatrix} = -\hat{\mathbf{z}} \times (f_1\hat{\mathbf{x}} + f_2\hat{\mathbf{y}}).$$

Thus the TM modes are given by

$$\begin{cases} \vec{\mathbf{E}} = \dfrac{1}{\lambda_n^2}\partial_z u_n(z,t)\nabla_T v_n(\rho) + \hat{\mathbf{z}} u_n(z,t) v_n(\rho), \\ \\ \vec{\mathbf{H}} = -\dfrac{\epsilon}{\lambda_n^2}\partial_t u_n(z,t)\hat{\mathbf{z}} \times \nabla_T v_n(\rho). \end{cases} \tag{4.214}$$

In a similar manner it can be shown that the TE modes are given by

$$\begin{cases} \vec{\mathbf{E}} = \dfrac{\mu}{\lambda_n^2}\partial_t u_n(z,t)\hat{\mathbf{z}} \times \nabla_T w_n(\rho), \\ \\ \vec{\mathbf{H}} = \dfrac{1}{\lambda_n^2}\partial_z u_n(z,t)\nabla_T w_n(\rho) + \hat{\mathbf{z}} u_n(z,t) w_n(\rho). \end{cases} \tag{4.215}$$

Because of the orthogonality property and the chosen normalized factor for the set $\{v_n\}$, $n = 1, 2, 3, ..., \infty$, it follows that the vectors $\nabla_T v_n$ in the tangential component $\mathbf{E}_T$ of the TM wave form an orthonormal set over Ω,

$$\int\int_\Omega \nabla_T v_n \cdot \nabla_T v_m d\rho = \delta_{nm},$$

and the vectors $\hat{\mathbf{z}} \times \nabla_T v_n$ in the tangential component $\mathbf{H}_T^n$ form an orthonormal set over Ω also. A similar result holds for the corresponding components of the TE mode.

The Klein–Gordon equation (4.212) has the same form as the Hankel transformed wave equation (4.119) in the homogeneous region. Thus, the explicit form of the wave-splitting given by (4.121), (4.127), and (4.130) can be applied directly to the Klein–Gordon equation (4.212) with initial conditions (4.213). Thus, right-going (along the positive z direction) and left-going (along the negative z direction) TM (or TE) waves can be expressed in terms of $u_n(x_3, 0)$ (see (4.190) and (4.193) and Schulz-DuBois (1970), Kristensson (1995))

$$u_n^\pm(z,t) = u_n(0, t - |z|/c) - c\lambda_n|z| \int_0^{t-|z|/c} \frac{J_1(\lambda\sqrt{c^2(t-s)^2 - z^2}\,)}{\sqrt{c^2(t-s)^2 - z^2}} u_n(0,s)\,ds. \tag{4.216}$$

Employing these results in (4.214), the corresponding TM mode tangential components are obtained by

$$\begin{cases} \mathbf{E}_T^\pm = \mp\dfrac{1}{\lambda_n^2} K^{-1} u_n^\pm \nabla_T v_n + \hat{\mathbf{z}} u_n^\pm v_n, \\ \\ \mathbf{H}_T^\pm = -\dfrac{\epsilon}{\lambda_n^2}\partial_t u_n^\pm \hat{\mathbf{z}} \times \nabla_T v_n, \end{cases} \tag{4.217}$$

where $u_n^{\pm}$ are given by (4.216). A similar result holds for the TE mode tangential components.

In the wave propagation problem for a cylindrical waveguide with a given cross-section, if the field $E_z(\rho, z, t)$ (or $H_z(\rho, z, t)$) at $z = 0$ is known for a specific TM (or TE) mode, one can calculate the z-components of the fields at an arbitrary $z > 0$ for this mode. This can be carried out as follows. Since the cross-section is known, one first solves the eigenvalue problem (4.209) (or (4.210)) to obtain the eigen transverse pattern $v_n(\rho)$ (or $w_n(\rho)$) for a specific eigenvalue λ_n corresponding to this TM (or TE) mode. Then $u_n(0,t) = E_3(\rho, z = 0, t)/v_n(\rho)$ (or $H_3(\rho, z = 0, t)/w_n(\rho)$). From (4.216), one obtains $u_n(z,t)$. The z-components of the fields at an arbitrary $z > 0$ for this mode can thus be calculated by (4.207) (or (4.208)). Once the z-components of the fields are known, one also knows the transverse fields at this position z for this mode (cf. (4.217)). The multi-mode propagation problem can be solved by summing the corresponding solutions over all possible modes in the waveguide. Using this explicit form of wave-splitting, the wave propagation problem for a cylindrical waveguide filled with a medium stratified in the direction of propagation can be solved with the wave-splitting approach (Bernekorn et al. 1996).

Exercises for Chapter 4

Exercise 4.1. Show that for an imbedding geometry, the split moments are connected to each other by the moments of the reflection kernel through relations of the forms given by (4.14)–(4.16).

Exercise 4.2. Show that the moments of the split fields are given by (4.17)–(4.20).

Exercise 4.3. Consider the following anisotropic wave equation

$$\partial_t^2 u - [c_1(z)]^2 \partial_x^2 u - [c_2(z)]^2 \partial_y^2 u - [c(z)]^2 \partial_z^2 u = 0,$$

in a stratified half-space $z > 0$, where in general the three phase velocities $c_1(z)$, $c_2(z)$ and $c(z)$ are different from each other, i.e. waves travel at different velocities in different directions. Use the x and y direction second moments of the field $u(x, y, z, t)$ defined by

$$u_2^{(1)}(z,t) = \int\int_{\mathcal{R}^2} x^2 u(x, y, z, t)\, dx\, dy,$$

$$u_2^{(2)}(z,t) = \int\int_{\mathcal{R}^2} y^2 u(x, y, z, t)\, dx\, dy,$$

respectively, instead of the type of second moment defined in (4.74) (the definition for the zeroth moment remains the same as given by (4.73)).

(1) Show that

$$u_2^{(i)\pm} = (1/2)[u_2^{(i)} \mp c\partial_t^{-1} u_{2z}^{(i)}] \mp \frac{1}{2} c c_i^2 \partial_t^{-3} u_{0z}, \qquad i = 1, 2,$$

$$u_2^{(i)-}(z,t) = R_2^{(i)}(z,t) * u_0^+(z,t) + R_0(z,t) * u_2^{(i)+}(z,t), \quad i = 1, 2,$$

where $R_2^{(1)}$ and $R_2^{(2)}$ are x and y direction second moments of the reflection kernel, respectively.

(2) Derive the imbedding equations for R_0, $R_2^{(1)}$ and $R_2^{(2)}$, and initial conditions that are useful for reconstruction.

Exercise 4.4. Show the formula (4.125) using the relation $\left[K^{-1}\right]^2 = \frac{1}{c^2}\partial_t^2 + \kappa^2$.

Exercise 4.5. Derive the closed-form solution (4.148) for the reflection kernel in the Hankel transformed space for an acoustic half-space with a specifically stratified density as given by (4.147) (with $\alpha_1 \neq 1$; the velocity is assumed to be a constant in the whole space; the upper half-space $z < 0$ is homogeneous with a constant density ρ_0).

Hint: Use the following transformation,

$$\bar{p}(z,t;\kappa) = \varphi(z,t;\kappa)\frac{e^{-\beta z}(1+\alpha)}{e^{-2\beta z}+\alpha}, \quad z > 0,$$

$$\bar{p}(z,t;\kappa) = \varphi(z,t;\kappa), \quad z < 0.$$

Then derive the expression for the reflection through a Laplace transform and a wave-splitting that splits the total wave φ into a down-going wave φ^+ and an up-going wave φ^-.

5

WAVE-SPLITTINGS COMBINED WITH OPTIMIZATION TECHNIQUES

Recent developments and applications of various optimization methods have provided efficient tools for obtaining numerical solutions to various types of inverse problems (see, e.g. Frank and Balanis (1987), Kleinman and van den Berg (1990), Chew and Otto (1992), He and Kabanikhin (1995)). Optimization methods can be grouped into two types, namely, global search methods and gradient search methods. Global search methods include simulated annealing (see, e.g. Garnero et al. (1991)), neural network methods (see, e.g. Lu and Berryman (1990)), and genetic algorithms (see, e.g. Weile and Michielssen (1997)). A global search method is based on a stochastic algorithm, and its convergence can be very slow. A gradient search method is usually based on a deterministic algorithm, and it converges rapidly (though it may converge to a local minimum). In this chapter, wave-splittings are combined with gradient search optimization techniques. The procedure of deriving an explicit expression for the gradient of an objective functional through dual functions is described. The conjugate gradient method is used to reconstruct the line parameters, or spatial source distributions on a non-uniform line. The time domain inverse problem of reconstructing the temporal behaviours of a bi-isotropic slab is also considered, and a matrix formalism is introduced to derive an explicit expression for the gradient. A multi-dimensional acoustic inverse problem is also considered, and numerical reconstruction of a two-dimensional density profile is given. The Newton–Kantorovich procedure with a wave-splitting process is applied to a time domain inverse problem for a quasi-linear acoustic wave equation. Finally, a measurement system is described and the measurement data are used to reconstruct the susceptibility kernel of a polar liquid (1-buthanol).

5.1 Explicit expressions for gradients and parameter reconstruction

In order to apply a gradient search optimization approach to an inverse problem, one introduces a suitable objective functional first, and then computes the gradient of this functional. The gradient of the functional gives the direction of fastest increase for the functional, and thus the opposite direction gives the direction of fastest decrease. Once the gradient of the objective functional has been computed, one can use a conventional steepest descent method or conjugate gradient method to minimize the objective functional in an iterative way. The steepest descent method usually exhibits slower convergence rates than the conjugate gradient method. If one does not have an explicit expression for the gradient, one usually has to choose a complete set of orthonormal basis functions to compute the gradient numerically, which slows down the computation speed (see the end of Section 5.1.4 for a computation time comparison). In this section, dual functions are introduced to derive an explicit expression for the gradient, and the computation becomes faster and simple. Generally speaking, there are two different types of dual functions depending on the way that they are introduced. For the first type, they are introduced through appropriate boundary conditions (see e.g. He and Kabanikhin (1995)), and the other type is introduced using a δ-function type of sources (see e.g. Takenaka et al. (1997)). In this chapter, we concentrate on the first type of dual functions. It is shown that by using wave-splitting one obtains a simple expression for the gradient and a robust algorithm.

5.1.1 *Reconstruction of source distributions on a line*

Consider a non-uniform LCRG transmission line of length l between $x = 0$ and $x = l$. The non-uniform line is assumed to be terminated by two uniform LC semi-infinite lines at the two ends. In Section 3.1.1 we considered the inverse problem of reconstructing the parameters of a transmission line, which has no internal source. In this subsection, we assume that an external electromagnetic field (e.g. from a stroke of lightning) hits the line and one has the following model equations (see e.g. Nucci et al. (1993))

$$\frac{\partial V}{\partial x} + L(x)\frac{\partial I}{\partial t} + R(x)I = f(t)E^i(x), \tag{5.1}$$

$$\frac{\partial I}{\partial x} + C(x)\frac{\partial V}{\partial t} + G(x)V = f(t)H^i(x), \tag{5.2}$$

where $f(t)$ is a dimensionless common time-dependent function of the sources, and $E^i(x)$ and $H^i(x)$ are the spatial source distribution functions induced by a non-uniform electromagnetic field.

The total voltages are decomposed into the right-moving voltage V^+ and left-moving voltage V^-, as defined by (3.3). The dynamic equations for the split voltages are

$$\frac{\partial}{\partial x}\begin{bmatrix} V^+ \\ V^- \end{bmatrix} + \frac{1}{c(x)}\frac{\partial}{\partial t}\begin{bmatrix} V^+ \\ -V^- \end{bmatrix} - \begin{bmatrix} \alpha & \beta \\ \gamma & \vartheta \end{bmatrix}\begin{bmatrix} V^+ \\ V^- \end{bmatrix} = f(t)\begin{bmatrix} E^+(x) \\ E^-(x) \end{bmatrix}, \tag{5.3}$$

where α, β, γ, ϑ are given by (3.9)–(3.12), and

$$E^{\pm}(x) = (1/2)[E^i(x) \pm Z(x)H^i(x)] \tag{5.4}$$

($Z = \sqrt{L/C}$ is the impedance).

The boundary and initial conditions for the split voltages are

$$V^+(0,t) = V^-(l,t) = 0, \tag{5.5}$$

$$V^\pm(x,0) = 0. \tag{5.6}$$

The boundary condition (5.5) is due to the fact that there is no right-moving (left-moving) wave but only a left-moving (right-moving) transmitted wave in the region $x < 0$ ($x > l$). The initial condition (5.6) is due to the assumption that the external electromagnetic field does not reach the non-uniform line until the time $t = 0$. Equations (5.3), (5.5) and (5.6) constitute a set of equations from which $V^+(x,t)$ and $V^-(x,t)$ can be computed using the method of characteristics.

The inverse source problem is to reconstruct the source distribution functions $E^i(x)$ and $H^i(x)$ using the received transient voltages at $x = 0$ and/or $x = l$ (the time-dependent function $f(t)$ and the parameters of the non-uniform line are assumed to be known). We wish to solve the inverse problem by means of a suitable optimization approach. To this end we define an objective functional as follows,

$$J(E^i, H^i) = \int_0^T \left(w_1(t)[V^-(0,t) - V_1(t)]^2 + w_2(t)[V^+(l,t) - V_2(t)]^2\right) dt, \tag{5.7}$$

where $V_1(t)$, $V_2(t)$ are the received voltages at the endpoints $x = 0$ and $x = l$, respectively, which are assumed to be known from measurements, and $w_1(t)$, $w_2(t)$ are two non-negative weight functions. The inverse problem is to find functions $E^i(x)$ and $H^i(x)$, or equivalently $E^+(x)$ and $E^-(x)$, which minimize the objective functional $J(E^i, H^i)$. $E^+(x)$ and $E^-(x)$ are reconstructed through an optimization approach, and $E^i(x)$ and $H^i(x)$ are then obtained from the relation (5.4). Thus, hereafter we denote the objective function (defined by (5.7)) by $J(E^+, E^-)$ instead of $J(E^i, H^i)$. The minimum of $J(E^+, E^-)$ is zero if the inverse problem has a solution.

An explicit expression for the gradient of $J(E^+, E^-)$ can be derived as follows. Let $\tilde{V}^\pm = \tilde{V}^\pm(x, t; E^+ + \delta E^+, E^- + \delta E^-)$ be the solution to the system (5.3) with boundary and initial conditions (5.5) and (5.6) when $E^\pm$ have small increments $\delta E^\pm$. Then $\delta V^\pm \equiv \tilde{V}^\pm - V^\pm$ satisfy the following system of equations

$$\frac{\partial}{\partial x}\begin{bmatrix}\delta V^+ \\ \delta V^-\end{bmatrix} + \frac{1}{c(x)}\frac{\partial}{\partial t}\begin{bmatrix}\delta V^+ \\ -\delta V^-\end{bmatrix} - \begin{bmatrix}\alpha & \beta \\ \gamma & \vartheta\end{bmatrix}\begin{bmatrix}\delta V^+ \\ \delta V^-\end{bmatrix} = f(t)\begin{bmatrix}\delta E^+(x) \\ \delta E^-(x)\end{bmatrix}, \tag{5.8}$$

with the initial and boundary conditions

$$\delta V^\pm(x,0) = 0, \tag{5.9}$$

$$\delta V^+(0,t) = \delta V^-(l,t) = 0. \tag{5.10}$$

The corresponding increment of the functional $J(E^+, E^-)$ can then be written as follows:

$$\begin{aligned}\delta J &\equiv J(E^+ + \delta E^+, E^- + \delta E^-) - J(E^+, E^-)\\ &= 2\int_0^T \{\delta V^-(0,t)w_1(t)[V^-(0,t) - V_1(t)] + \delta V^+(l,t)w_2(t)[V^+(l,t) - V_2(t)]\}dt\\ &\quad + o(\delta E^+, \delta E^-),\end{aligned} \tag{5.11}$$

where

$$\lim_{\|\delta E^\pm\|\to 0} \frac{o(\delta E^+, \delta E^-)}{\|\delta E^\pm\|} = 0$$

($\|\ \|$ denotes the L_2 norm of a function, i.e., $\|\delta E^\pm\| = \{\int_0^l [\delta E^\pm(x)]^2\,dx\}^{1/2}$). If the increment δJ can be rewritten in the inner product form

$$\begin{aligned}\delta J &= \int_0^l (g^+\cdot\delta E^+ + g^-\cdot\delta E^-)\,dx + o(\delta E^+, \delta E^-)\\ &\equiv \langle g^+, \delta E^+\rangle + \langle g^-, \delta E^-\rangle + o(\delta E^+, \delta E^-),\end{aligned} \tag{5.12}$$

where the inner product $<,>$ is defined by $\langle a, b\rangle \equiv \int_0^l a(x)b(x)\,dx$, then $g(E^+, E^-) \equiv (g^+(E^+,E^-), g^-(E^+,E^-))$ (a vector with two elements) is the gradient of the objective functional $J(E^+, E^-)$.

Introduce a pair of dual functions (sometimes also called adjoint or auxiliary functions) $\psi^\pm(x,t)$ that satisfy the following boundary conditions

$$\psi^-(0,t) = w_1(t)[V^-(0,t) - V_1(t)], \tag{5.13}$$

$$\psi^+(l,t) = w_2(t)[V^+(l,t) - V_2(t)]. \tag{5.14}$$

Then one has

$$\delta J = 2\int_0^T \{[(\delta V^-)\psi^-](0,t) + [(\delta V^+)\psi^+](l,t)\}\,dt + o(\delta E^+, \delta E^-). \tag{5.15}$$

Integrating by parts and using (5.10), one obtains

$$\begin{aligned}\delta J = {} & 2\int_0^T\int_0^l \{\,[\partial_x(\delta V^+)\psi^+ + (\delta V^+)\partial_x\psi^+]\\ & -[\partial_x(\delta V^-)\psi^- + (\delta V^-)\partial_x\psi^-]\,\}\,dxdt + o(\delta E^+, \delta E^-).\end{aligned} \tag{5.16}$$

Using the system (5.8) of PDEs for $\delta V^\pm$, one has

$$\begin{aligned}&\int_0^T\int_0^l [\partial_x(\delta V^+)]\psi^+\,dxdt\\ &= \int_0^T\int_0^l [-\frac{1}{c}\partial_t(\delta V^+) + \alpha\delta V^+ + \beta\delta V^- + f(t)\delta E^+]\psi^+\,dxdt\end{aligned}$$

$$= \int_0^T \int_0^l [(\frac{1}{c}\partial_t \psi^+)\delta V^+ + (\alpha\psi^+)\delta V^+ + (\beta\psi^+)\delta V^- + f(t)\psi^+\delta E^+]\, dxdt$$
$$-\frac{1}{c}(\psi^+\delta V^+)(x,T) + o(\delta E^+, \delta E^-),$$

$$\int_0^T \int_0^l [\partial_x(\delta V^-)]\psi^-\, dxdt$$
$$= \int_0^T \int_0^l [\frac{1}{c}\partial_t(\delta V^-) + \gamma\delta V^+ + \vartheta\delta V^- + f(t)\delta E^-]\psi^-\, dxdt$$
$$= \int_0^T \int_0^l [-(\frac{1}{c}\partial_t\psi^-)\delta V^- + (\gamma\psi^-)(\delta V^+) + (\vartheta\psi^-)(\delta V^-) + f(t)\psi^-\delta E^-]\, dxdt$$
$$+\frac{1}{c}(\psi^-\delta V^-)(x,T) + o(\delta E^+, \delta E^-).$$

Therefore, from (5.16) one obtains

$$\delta J = 2\int_0^T \int_0^l \{[(\partial_x + \frac{1}{c}\partial_t)\psi^+ + \alpha\psi^+ - \gamma\psi^-](\delta V^+) - [(\partial_x - \frac{1}{c}\partial_t)\psi^-$$
$$-\beta\psi^+ + \vartheta\psi^-](\delta V^-) + f(t)\psi^+\delta E^+ - f(t)\psi^-\delta E^-\}\, dxdt$$
$$-\frac{2}{c}(\psi^+\delta V^+)(x,T) - \frac{2}{c}(\psi^-\delta V^-)(x,T) + o(\delta E^+, \delta E^-). \tag{5.17}$$

If one chooses $\psi^\pm$ so that they satisfy the following system of PDEs and initial conditions

$$\frac{\partial}{\partial x}\begin{bmatrix}\psi^+\\ \psi^-\end{bmatrix} + \frac{1}{c(x)}\frac{\partial}{\partial t}\begin{bmatrix}\psi^+\\ -\psi^-\end{bmatrix} - \begin{bmatrix}-\alpha & \gamma\\ \beta & -\vartheta\end{bmatrix}\begin{bmatrix}\psi^+\\ \psi^-\end{bmatrix} = 0, \tag{5.18}$$

$$\psi^+(x,T) = \psi^-(x,T) = 0, \tag{5.19}$$

then from (5.17) one has

$$\delta J = 2\int_0^T \int_0^l [f(t)\psi^+\delta E^+ - f(t)\psi^-\delta E^-]\, dxdt, \tag{5.20}$$

i.e.,

$$g^+(E^+, E^-) = 2\int_0^T [f(t)\psi^+]\, dt, \tag{5.21}$$

$$g^-(E^+, E^-) = -2\int_0^T [f(t)\psi^-]\, dt. \tag{5.22}$$

One may prefer to choose the weight functions $w_1(t)$ and $w_2(t)$ as two constants (e.g. $w_1(t) = w_2(t) = 1$). In such a case, in order to avoid any discontinuity in

$\psi^{\pm}$, one should theoretically choose the time duration T to be large enough so that (since $\psi^{-}(0,T) = \psi^{+}(l,T) = 0$, cf. (5.19), (5.13) and (5.14))

$$V^{-}(0,T) - V_1(T) \approx 0, \quad V^{+}(l,T) - V_2(T) \approx 0. \tag{5.23}$$

However, in practice, since the reconstruction algorithm is stable, as has been tested numerically (He 1996a) (a small discontinuity in $\psi^{\pm}$ can be regarded as noise), it is not critical to have these conditions fulfilled very accurately. An alternative way to avoid any discontinuity in $\psi^{\pm}$ is to chooses the weight functions as certain smooth (non-constant) distributions with

$$w_1(T) = w_2(T) = 0. \tag{5.24}$$

When the gradient of the objective functional $J(E^+, E^-)$ has been computed, a standard conjugate gradient method (e.g., the Polack–Ribiere algorithm (Polack 1971)) can be used to minimize $J(E^+, E^-)$ and reconstruct $E^{\pm}(x)$ (the source distribution functions are then obtained through $E^i(x) = E^+(x) + E^-(x)$, $H^i(x) = [E^+(x) - E^-(x)]/Z(x)$). The iterative algorithm is as follows:

Step 0: Set the initial approximation to be zero, i.e., $E^+(x) = E^{+(0)}(x) \equiv 0$, $E^-(x) = E^{-(0)}(x) \equiv 0$.

Step 1: Set $i = 0$. Solve the direct problem for $V^{\pm}$ consisting of (5.3), (5.5) and (5.6), and the direct problem for $\psi^{\pm}$ consisting of (5.18), (5.13), (5.14) and (5.19), with the method of characteristics, and calculate the gradient $g^{\pm(0)} = g^{\pm}(E^{+(0)}(x), E^{-(0)}(x))$ using (5.21) and (5.22). Set the search direction $h^{\pm(0)} = g^{\pm(0)}$.

Step 2: Compute a scalar stepsize $\lambda_i > 0$ (i.e., a line search) such that

$$\begin{aligned} &J(E^{+(i)} - \lambda_i h^{+(i)}, E^{-(i)} - \lambda_i h^{-(i)}) \\ &\quad = \min\{J(E^{+(i)} - \lambda h^{+(i)}, E^{-(i)} - \lambda h^{-(i)}) | \lambda \geq 0\}. \end{aligned} \tag{5.25}$$

Step 3: Improve the reconstruction by setting

$$E^{\pm(i+1)} = E^{\pm(i)} - \lambda_i h^{\pm(i)}. \tag{5.26}$$

Step 4: Compute the new gradient $g^{\pm(i+1)} = g^{\pm}(E^{+(i+1)}, E^{-(i+1)})$ using (5.21) and (5.22) ($V^{\pm}$ and $\psi^{\pm}$ are obtained with the method of characteristics using the improved source distribution functions).

Step 5: If $g^{\pm(i+1)} = 0$, stop; if not, set the new search direction

$$h^{\pm(i+1)} = g^{\pm(i+1)} + p_i h^{\pm(i)}, \tag{5.27}$$

where

$$p_i = \frac{< g^{+(i+1)} - g^{+(i)}, g^{+(i+1)} > + < g^{-(i+1)} - g^{-(i)}, g^{-(i+1)} >}{< g^{+(i)}, g^{+(i)} > + < g^{-(i)}, g^{-(i)} >}, \quad (5.28)$$

and set $i = i + 1$, and go to step 2.

Note that the gradient as well as $\psi^{\pm}$ becomes smaller, as the number of iterations increases. When the reconstructed source distribution functions are identical to the true values, then $\psi^{\pm}(x, t) = 0$ and $g^{\pm}(x) = 0$.

In the numerical computation, all the parameters are scaled to be dimensionless. L, C, R and G are scaled by L_0 (an arbitrary non-zero value with dimension of inductance), C_0 (an arbitrary non-zero value with dimension of capacitance), $R_0 = \frac{1}{l}\sqrt{L_0/C_0}$, and $G_0 = \frac{1}{l}\sqrt{C_0/L_0}$, respectively. x and t are scaled by l and $t_0 = l\sqrt{L_0 C_0}$, respectively. If E^i is scaled by E_0 (an arbitrary non-zero value with dimension of electrical field), V and H^i should be scaled by $V_0 = lE_0$ and $H_0 = E_0\sqrt{C_0/L_0}$, respectively.

The method of characteristics with a forward time-stepping from $t = 0$ to $t = T$ is used to calculate $V^{\pm}(x, t)$ for $x \in [0, l]$ and $t \in [0, T]$ from equations (5.3), (5.5) and (5.6). Similarly, one can compute the dual functions $\psi^{\pm}(x, t)$ for $x \in [0, l]$ and $t \in [0, T]$ from equations (5.18), (5.13), (5.14) and (5.19), using the method of characteristics with a backward time-stepping from $t = T$ to $t = 0$. Note that the method of characteristics is computationally less intensive and more accurate than a finite difference method (see e.g. Lundstedt and He (1997) for a finite difference algorithm) with regard to numerical dispersion errors.

When an explicit expression for the gradient is not available, one may calculate the gradient numerically through small perturbations. In the present reconstruction algorithm with an explicit expression for the gradient, the computation is speeded up greatly (a more detailed computation time comparison is given at the end of Section 5.1.4).

The scaled parameters for the non-uniform line used in the numerical example are shown by the solid lines in Fig. 5.1. The true source distribution functions $E^i(x)$ and $H^i(x)$ are plotted by the solid lines in Fig. 5.2. The dimensionless common time-dependent function of the source is chosen as

$$f(t) = \begin{cases} \sin(2\pi t/t_0), & \text{when } 0 \le t \le \frac{1}{2}t_0, \\ 0, & \text{otherwise.} \end{cases}$$

The received voltages at the two endpoints calculated by the method of characteristics (with 64 points in the x direction and 192 points in the t direction) are shown by the solid lines in Fig. 5.3. To check the direct solver independently, this result has been compared to the corresponding one computed by means of a finite difference method (Lundstedt and He 1997) for the system of equations (5.1) and (5.2) (without using wave-splitting). As Fig. 5.3 shows, the calculated voltages at the endpoints using two different methods are indistinguishable on

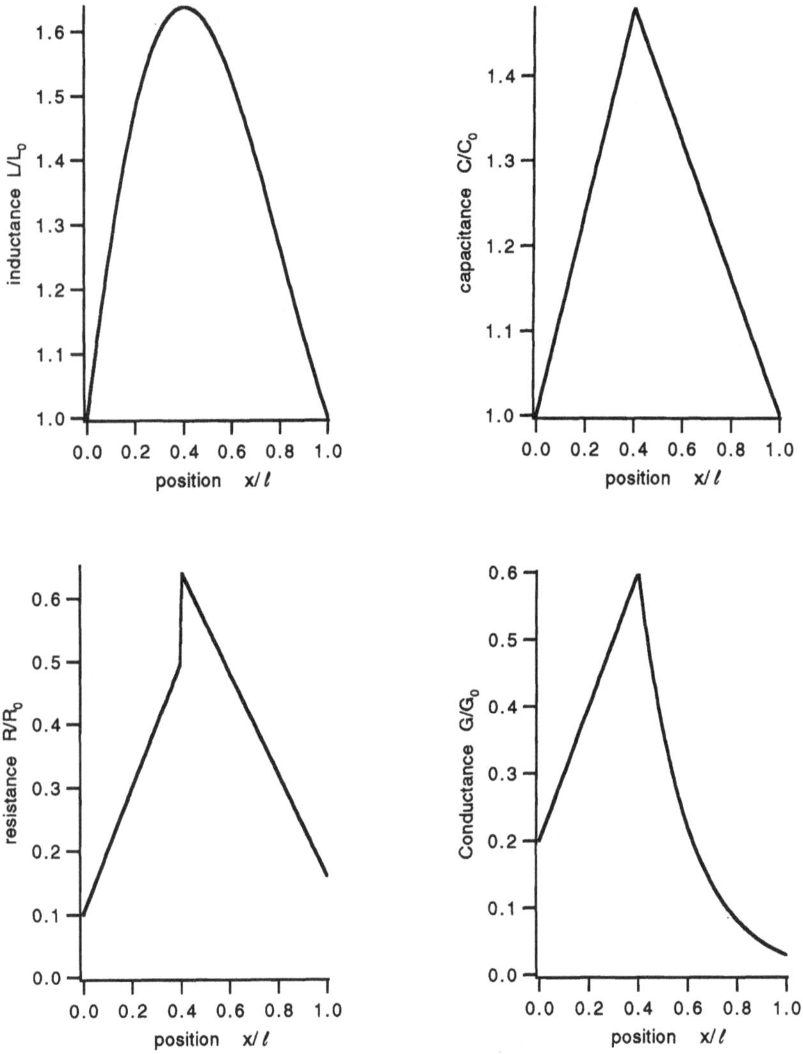

FIG. 5.1. Parameter profiles for the non-uniform transmission line.

the scale of the graph in Fig. 5.3 (for graphical clarity, the computed voltages at the two endpoints from the finite difference method are indicated by circles in Fig. 5.3 only for every fifth point). The present direct solver (based upon the method of characteristics) is certainly faster and more accurate than any finite difference method if the same number of discretization points is used.

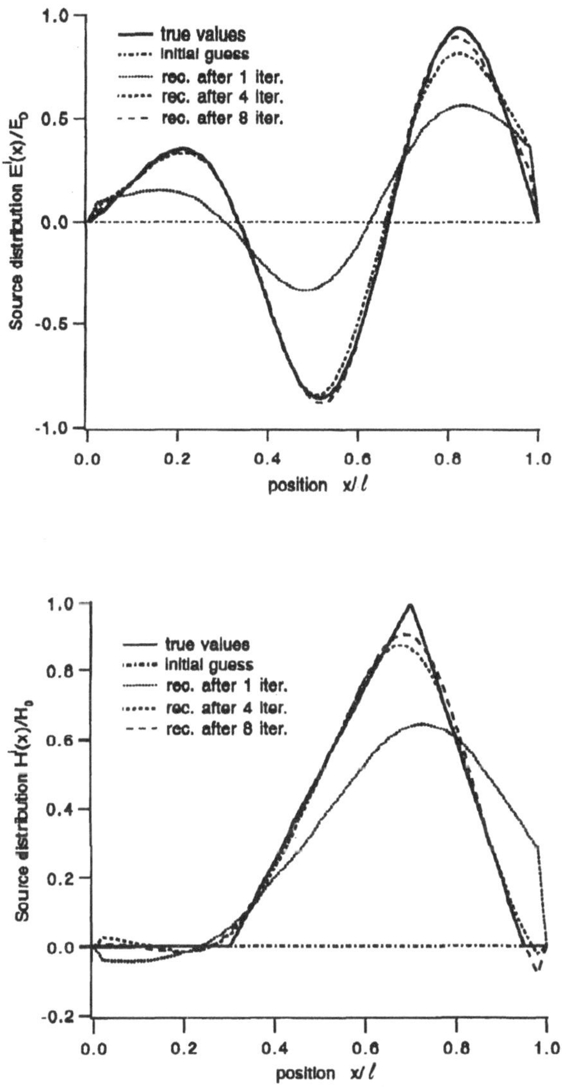

FIG. 5.2. The true source distribution functions (solid lines) and simultaneous reconstructions of $E^i(x)$ and $H^i(x)$ using two-sided data after 0, 1,4 and 8 iterations.

Reconstruction using two-sided measurements. First, consider the case of using two-sided measurements, i.e., setting the weight functions $w_1 = w_2 = 1$. The corresponding simultaneous reconstruction of the source distribution func-

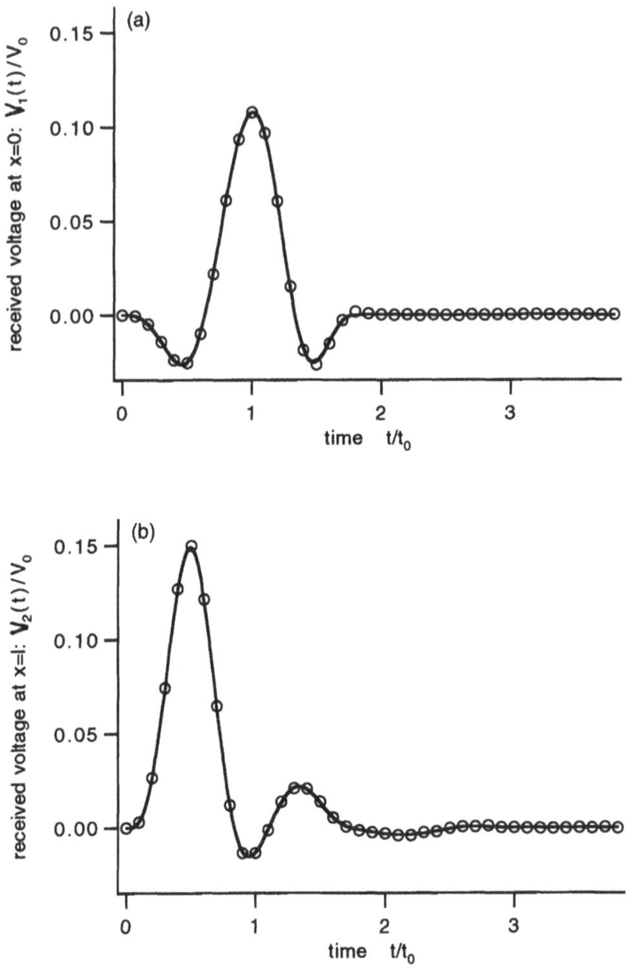

FIG. 5.3. The received voltages at the two endpoints $x = 0$ and $x = l$. The solid lines are calculated with the method of characteristics, and the circles are calculated with the finite difference method.

tions $E^i(x)$ and $H^i(x)$ is given in Fig. 5.2, where the improvement in the reconstruction as the number of iterations increases is illustrated for up to 8 iterations (the initial guess is $E^i(x) = H^i(x) = 0$). Note that any jump or peak can only be reconstructed approximately by a smooth curve, as in the peak in Fig. 5.2(b). As the number of iterations increases, the reconstruction improves further. It appears that the objective functional, defined by (5.7), is convex in a large area centred at the true profile, since we never came across a case in which the reconstruction was trapped in a local minimum. Numerical testing (He 1996a) has shown that the reconstruction algorithm is stable under noise in the measurements, as well as any small perturbation of the line parameters or the common time-dependent function $f(t)$.

Reconstruction using one-sided measurements. For the the case of one-sided measurements, simply set the weight functions to $w_1 = 1$ and $w_2 = 0$ (the measured data is the voltage at the endpoint $x = 0$). Numerical results (He 1996a) show that one can reconstruct one of the source distribution functions accurately using only one-sided measurements, when the other source distribution function is assumed to be known (e.g., zero, cf. Nucci et al. (1993) for a practical case when $H^i = 0$). Numerical results (He 1996a) also show that the simultaneous reconstruction using one-sided measurements is possible when there is an impedance mismatch at the back endpoint $x = l$. However, the reconstruction is less stable under noise in the measurements than the one using two-sided measurements. The role of the impedance mismatch at the back endpoint is that the right-moving signal at $x = l^-$ is reflected by the impedance mismatch, and thus the received signal at $x = 0$ contains this reflected information though it may be attenuated.

5.1.2 *Reconstruction of line parameters*

In this subsection, the inverse problem of reconstructing simultaneously the characteristic impedance, resistance and conductance of a non-uniform transmission line is considered (the wavefront velocity is assumed to be constant; a simultaneous reconstruction of parameters including the wavefront velocity requires certain conditions to be satisfied (Lundstedt and Ström 1996) and will not be considered here). Denote the compact Green's functions for the left-sided incidence as $G^{c\pm}(x,t)$ (the inverse of the attenuation factor in the definition of $G^{c\pm}(x,t)$ is denoted as $a^+(x)$), and the compact Green's functions for the right-sided incidence as $g^{c\pm}(x,t)$ (the inverse of the attenuation factor in the definition of $g^{c\pm}(x,t)$ is denoted as $e^-(x)$). Then $G^{c\pm}(x,t)$ satisfy the system (3.34) of equations (replacing $G_R^{c\pm}$, $a_R(x)$ with $G^{c\pm}$ and $a^+(x)$, respectively), and $g^{c\pm}(x,t)$ satisfy the system (3.39) of equations (replacing $G_L^{c\pm}$, $a_L(x)$ with $g^{c\pm}$ and $a^-(x)$, respectively). The given transient response of the non-uniform line is denoted by a superscript m and a full description of the two-sided transient response is given by

$$\left.\begin{aligned} &G^{c\pm}(0,t) = G^{m\pm}(t) \\ &a^+(0) = a^{m+} \end{aligned}\right\} \quad \text{for the left-sided incidence,}$$

$$\left.\begin{aligned} g^{c\pm}(l,t) &= g^{m\pm}(t) \\ e^{-}(l) &= e^{m-} \end{aligned}\right\} \quad \text{for the right-sided incidence.}$$

Define the following objective functional F which measures the difference between the compact Green's functions (corresponding to the parameters $Z(x)$, $R(x)$, $G(x)$) and the given boundary values,

$$\begin{aligned} J &= \int_0^T \left[\left(G^{c+}(0,t) - G^{m+}(t)\right)^2 + \left(G^{c-}(0,t) - G^{m-}(t)\right)^2 \right] dt \\ &+ \int_0^T \left[\left(g^{c+}(l,t) - g^{m+}(t)\right)^2 + \left(g^{c-}(l,t) - g^{m-}(t)\right)^2 \right] dt \\ &\equiv J^+ + J^-, \end{aligned} \tag{5.29}$$

where $T \geq 2\tau(0,l)$ (hence, according to the compactness of $G^{c\pm}$ and $g^{c\pm}$, the integrands are zero for times larger than T). The objective functionals J^+ and J^- refer to the error in the compact Green's functions for the left- and right-sided incidences, respectively.

After a derivation similar to the one given in Section 5.1.1, one obtains the following explicit expressions for the gradients J^+_α, J^+_β, J^+_γ, and J^+_ϑ (cf. Exercise 5.1)

$$J^+_\alpha(x) = -\int_0^T G^{c+}(x,t)\Psi^+(x,t)dt + \int_0^x \Psi^-(x',0)\gamma(x')a^+(x')dx', \tag{5.30}$$

$$J^+_\beta(x) = -\int_0^T G^{c-}(x,t)\Psi^+(x,t)\,dt, \tag{5.31}$$

$$J^+_\gamma(x) = -\int_0^T G^{c+}(x,t)\Psi^-(x,t)\,dt - a^+(x)\Psi^-(x,0), \tag{5.32}$$

$$J^+_\vartheta(x) = -\int_0^T G^{c-}(x,t)\Psi^-(x,t)\,dt, \tag{5.33}$$

where the dual functions $\Psi^\pm(x,t)$ satisfy the following system of PDEs,

$$\frac{\partial}{\partial x}\begin{bmatrix}\Psi^+(x,t)\\ \Psi^-(x,t)\end{bmatrix} - \frac{2}{c(x)}\frac{\partial}{\partial t}\begin{bmatrix}0\\ \Psi^-(x,t)\end{bmatrix} = \begin{bmatrix}-\alpha(x) & -\gamma(x)\\ -\beta(x) & -\vartheta(x)\end{bmatrix}\begin{bmatrix}\Psi^+(x,t)\\ \Psi^-(x,t)\end{bmatrix}, \tag{5.34}$$

with the boundary conditions

$$\Psi^+(0,t) = 2\left[G^{c+}(0,t) - G^{m+}(t)\right], \tag{5.35}$$

$$\Psi^-(0,t) = 2\left[G^{c-}(0,t) - G^{m-}(t)\right]. \tag{5.36}$$

The dual functions $\Psi^\pm(x,t)$ can be calculated from (5.34)–(5.36).

Similarly, the gradients J_α^-, J_β^-, J_γ^-, and J_ϑ^- are given by

$$J_\alpha^-(x) = \int_0^T g^{c+}(x,t)\psi^+(x,t)\,dt, \tag{5.37}$$

$$J_\beta^-(x) = \int_0^T g^{c-}(x,t)\psi^+(x,t)\,dt + e^-(x)\,\psi^+(x,0), \tag{5.38}$$

$$J_\gamma^-(x) = \int_0^T g^{c+}(x,t)\psi^-(x,t)\,dt, \tag{5.39}$$

$$J_\vartheta^-(x) = \int_0^T g^{c-}(x,t)\psi^-(x,t)\,dt - \int_l^x \psi^+(x',0)\beta(x')e^-(x')\,dx', \tag{5.40}$$

where the dual functions $\psi^\pm(x,t)$ satisfy the following system of PDEs,

$$\frac{\partial}{\partial x}\begin{bmatrix}\psi^+(x,t)\\ \psi^-(x,t)\end{bmatrix} + \frac{2}{c(x)}\frac{\partial}{\partial t}\begin{bmatrix}\psi^+(x,t)\\ 0\end{bmatrix} = \begin{bmatrix}-\alpha(x) & -\gamma(x)\\ -\beta(x) & -\vartheta(x)\end{bmatrix}\begin{bmatrix}\psi^+(x,t)\\ \psi^-(x,t)\end{bmatrix}, \tag{5.41}$$

with the boundary conditions

$$\psi^+(l,t) = 2\left[g^{c+}(l,t) - g^{m+}(t)\right], \tag{5.42}$$

$$\psi^-(l,t) = 2\left[g^{c-}(l,t) - g^{m-}(t)\right]. \tag{5.43}$$

The dual functions $\psi^\pm(x,t)$ can be calculated from (5.41)–(5.43). Thus, the gradients with respect to α, β, γ and ϑ are

$$J_\alpha(x) = J_\alpha^+(x) + J_\alpha^-(x), \qquad J_\beta(x) = J_\beta^+(x) + J_\beta^-(x), \tag{5.44}$$

$$J_\gamma(x) = J_\gamma^+(x) + J_\gamma^-(x), \qquad J_\vartheta(x) = J_\vartheta^+(x) + J_\vartheta^-(x). \tag{5.45}$$

From the dynamic equations for the split voltages, one sees that it is natural to define the functions

$$p_1(x) = Z_x Y, \tag{5.46}$$

$$p_2(x) = GZ - RY, \tag{5.47}$$

$$p_3(x) = GZ + RY. \tag{5.48}$$

These functions represent different physical aspects of the scattering on the line. The function $p_1(x)$ represents the variation of the characteristic impedance, which gives rise to reflections and changes of the signal amplitudes, but does not contribute to the losses. Both the series resistance R and shunt conductance G are dissipative terms which give rise to both scattering and attenuation of waves. The scattering of waves, caused by R and G, is described by the function $p_2(x)$ and the attenuation is described by the function $p_3(x)$. By reconstructing the functions $p_1(x)$, $p_2(x)$ and $p_3(x)$, rather than the parameters $Z(x)$, $R(x)$ and $G(x)$, one can make use of the fact that the characteristic impedance at both ends of the line are known, i.e., $Z(0) = Z_0$ and $Z(l) = Z_l$. Thus, one has

$$\int_0^l p_1(x)\,dx = \int_0^l Z_x Y\,dx = \ln\frac{Z_l}{Z_0}. \tag{5.49}$$

From the knowledge of the total attenuation of the line, one obtains

$$\begin{aligned}\int_0^l p_3(x)\,dx &= \int_0^l (GZ + RY)\,dx \\ &= \ln\left(\frac{(e^{m-})^2 Z_0}{Z_l}\right) = \ln\left(\frac{(a^{m+})^2 Z_l}{Z_0}\right).\end{aligned} \tag{5.50}$$

Using the functions p_1, p_2 and p_3, the functional space in the search for the unknown parameters can be narrowed by the constraints (5.49) and (5.50) (Lundstedt and He 1996). The gradients of F with respect to p_1, p_2 and p_3 are given by

$$J_{p_1}(x) = \frac{1}{2}\left(J_\alpha - J_\beta - J_\gamma + J_\vartheta\right), \tag{5.51}$$

$$J_{p_2}(x) = \frac{1}{2}\left(-J_\beta + J_\gamma\right), \tag{5.52}$$

$$J_{p_3}(x) = \frac{1}{2}\left(-J_\alpha + J_\vartheta\right). \tag{5.53}$$

Using these explicit expressions for the gradients, one can reconstruct the functions p_1, p_2 and p_3 with a conjugate gradient method (see Section 5.1.1). The electrical parameters are then given by

$$Z(x) = Z_0 \exp\left(\int_0^x p_1(x')\,dx'\right) = Z_l \exp\left(\int_l^x p_1(x')\,dx'\right), \tag{5.54}$$

$$R(x) = \frac{1}{2} Z(x)\left[p_3(x) - p_2(x)\right], \tag{5.55}$$

$$G(x) = \frac{1}{2} Y(x)\left[p_3(x) + p_2(x)\right]. \tag{5.56}$$

The transient responses, i.e., $G^{c\pm}(0,t)$, $g^{c\pm}(l,t)$, $a^+(0)$ and $e^-(l)$, of a synthetic non-uniform transmission line are computed first (details of the computational scheme can be found in Section 2.3). The compact Green's functions are

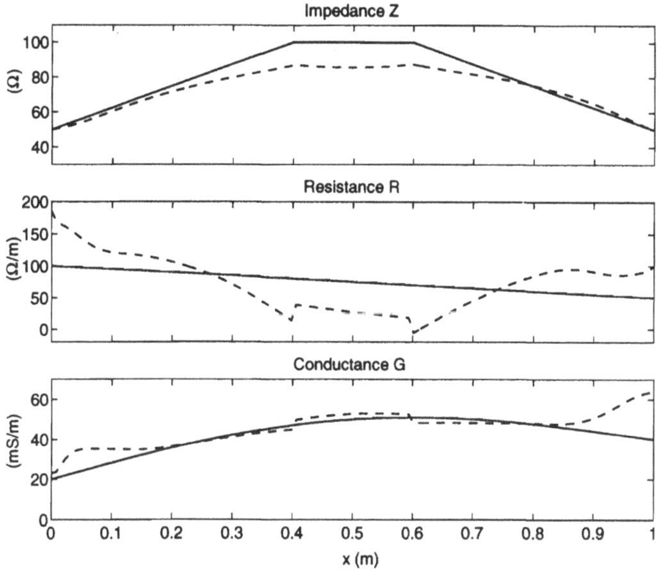

FIG. 5.4. The non-uniqueness of the inverse problem of reconstructing $Z(x)$, $R(x)$ and $G(x)$. Solid lines: true parameters. Dashed lines: reconstructed parameters.

computed with more discretization steps than those used for the parameters, to prevent the objective functional from appearing to be smaller than it actually is (the errors of the compact Green's functions have local minima at each discretization point of the parameters). Also, if the same number of discretization steps is used, after a large number of iterations, when a reasonable reconstruction has been achieved, growth of oscillating errors can sometimes occur. Numerical simulations using the present reconstruction algorithm showed that most of the two-parameter reconstructions are indistinguishable from the true parameters on the scale of the graphs (if the parameters were chosen to satisfy the conditions for a unique reconstruction given in Table 3.1). Numerical simulations also showed that a simultaneous reconstruction of the three parameters $Z(x)$, $R(x)$ and $G(x)$ is in general non-unique (which is consistent with the theoretical result given in Section 3.1.1). Of course, as the numerical calculation of the transient response has a limited accuracy, the numerical test is not a proof of non-uniqueness. We note that the non-uniqueness found in the numerical tests may come from the ill-posedness of the problem. Figures 5.4 and 5.5 show an example of the non-uniqueness of the inverse problem of reconstructing $Z(x)$, $R(x)$ and $G(x)$. In Fig. 5.4, the solid lines are the true parameters, and the dashed lines are the reconstructed parameters (after 100 iterations, the normalized value

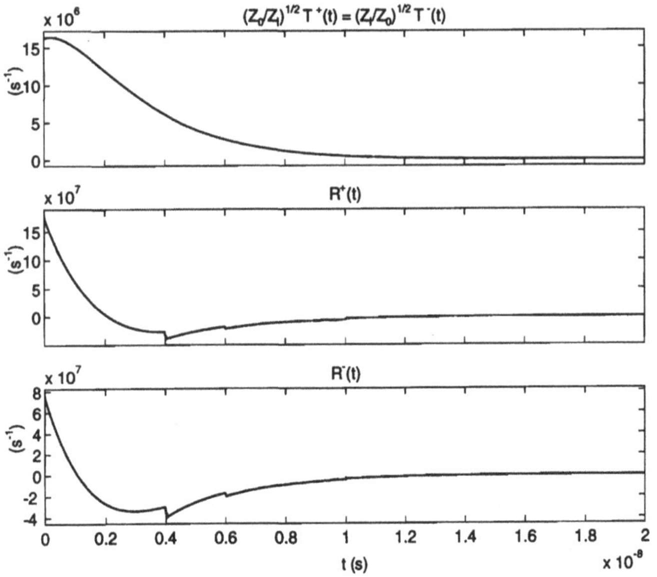

FIG. 5.5. The impulse responses, corresponding to the parameters in Fig. 5.4, in terms of reflection and transmission from both sides. The solid and dashed lines correspond to the true and reconstructed parameters, respectively (they are indistinguishable on the scale of the graphs).

of the objective functional was 1×10^4). For simplicity, the wavefront speed is chosen to be constant with a value of 2×10^8 m/s in the numerical simulation. A comparison between the transient responses of the true and the reconstructed profiles is shown in Fig. 5.5, where the transient responses are represented in the usual way (in terms of the incident signal instead of the transmitted signal). $R^{\pm}(t)$ and $T^{\pm}(t)$ denote the reflection and transmission kernels for the right- and left-sided incidences, respectively. These scattering kernels are computed with the usual Green's function approach. Since the transmission kernels $T^{+}(t)$ and $T^{-}(t)$, for all linear transmission lines, are proportional to each other (this is a generalization of the result referred to in Section 1.6 and it can be derived in an analogous way), they are displayed in a single graph. From Fig. 5.5 one sees that the transient responses for the true and reconstructed profiles in Fig. 5.4 are indistinguishable on the scale of the graphs. Using the present method, one always obtains electrical parameters with the prescribed transient response. Thus, the method is useful, for example, in design problems.

5.1.3 *Reconstruction of the temporal behaviour of a bi-isotropic slab*

The constitutive relations for a bi-isotropic medium are written as follows

$$\vec{\mathbf{D}} = \epsilon(z)\vec{\mathbf{E}} + \epsilon_0 G(t) * \vec{\mathbf{E}} + \frac{1}{c_0} K(t) * \vec{\mathbf{H}}, \tag{5.57}$$

$$\vec{\mathbf{B}} = \mu(z)\vec{\mathbf{H}} + \frac{1}{c_0} L(t) * \vec{\mathbf{E}} + \mu_0 F(t) * \vec{\mathbf{H}}, \tag{5.58}$$

where $c_0 = 1/\sqrt{\epsilon_0\mu_0}$. As shown in Karlsson and Kristensson (1992), (5.57) and (5.58) are the most general time domain constitutive relations for a bi-isotropic, dispersive medium. If the susceptibility kernels $K = L = 0$, the medium is an ordinary isotropic dispersive medium. If $L = -K$, then the medium is reciprocal. K and L describe the cross-coupling between the electric and magnetic field, and thus a medium with non-zero K and/or L is called a bi-isotropic medium. In (5.57) and (5.58), ϵ and μ are the usual static permittivity and permeability, respectively. The parameters ϵ and μ can be depth-dependent and thus the approach is applicable for the situation, e.g., when the permittivity and permeability are mismatched at the endpoints $z = 0$ and $z = l$ (one can use a very steep slope to approximate a discontinuity in the permittivity or permeability in a numerical treatment). Note that all the four susceptibility kernels G, K, L and F have the dimension of $[\text{time}]^{-1}$. One-dimensional time domain inverse problems for bi-isotropic media have been solved by exact wave-splitting methods (without using optimization) based upon the assumption that the boundary values of the scattering kernels or Green's functions are known (see e.g. Kristensson and Rikte (1993), Rikte (in press)). These scattering kernels or Green's functions are obtained from ill-posed deconvolutions. In view of the ill-posedness of the deconvolution that is needed in order to relate to the measured fields, it is of interest to develop more robust inversion methods for the inverse problems in an optimization setting.

Consider a plane wave, normally incident from the left on a bi-isotropic slab. On either side of the slab there is a homogeneous, non-dispersive medium, i.e.,

$$G(t) = K(t) = L(t) = F(t) = 0, \qquad \text{if } z < 0, \text{ or } z > l. \tag{5.59}$$

It is assumed that the incident wave will not reach the position $z = 0$ until time $t = 0$, i.e. $\vec{\mathbf{E}}(z, 0) = \vec{\mathbf{H}}(z, 0) = 0$, for $z > 0$.

The total tangential field is decomposed into the right-moving field $\mathbf{E}^+$ and left-moving field $\mathbf{E}^-$ defined as

$$\mathbf{E}^{\pm}(z,t) = \frac{1}{2}\left[\mathbf{E}(z,t) \pm \sqrt{\mu/\epsilon}\, I_2 \mathbf{H}(z,t)\right], \tag{5.60}$$

where

$$I_2 = \begin{bmatrix} 0 & 1 \\ -1 & 0 \end{bmatrix}, \quad \mathbf{E}^{\pm} \equiv \begin{bmatrix} E_1^{\pm} \\ E_2^{\pm} \end{bmatrix}, \quad \mathbf{E} \equiv \begin{bmatrix} E_1 \\ E_2 \end{bmatrix}, \quad \mathbf{H} \equiv \begin{bmatrix} H_1 \\ H_2 \end{bmatrix}.$$

Note that outside the bi-isotropic slab the E_1^+ and E_1^- (E_2^+ and E_2^-) give the x (y) components of the electric fields of the right- and left-moving waves, respectively. Also note that $\mathbf{E}^+ + \mathbf{E}^- = \mathbf{E}$.

The system of equations for the split fields is (cf. Exercise 5.2)

$$\frac{\partial}{\partial z}\begin{bmatrix}\mathbf{E}^+\\ \mathbf{E}^-\end{bmatrix} = -\frac{1}{c}\frac{\partial}{\partial t}\left\{\begin{bmatrix} I & 0\\ 0 & -I\end{bmatrix}\begin{bmatrix}\mathbf{E}^+\\ \mathbf{E}^-\end{bmatrix}\right\} + \frac{1}{4}\frac{d}{dz}\ln(\frac{\mu}{\epsilon})\begin{bmatrix} I & -I\\ -I & I\end{bmatrix}\begin{bmatrix}\mathbf{E}^+\\ \mathbf{E}^-\end{bmatrix}$$

$$+\frac{1}{2}\frac{\partial}{\partial t}\left\{\begin{bmatrix} M_{11}* & M_{12}*\\ M_{21}* & M_{22}*\end{bmatrix}\begin{bmatrix}\mathbf{E}^+\\ \mathbf{E}^-\end{bmatrix}\right\}, \tag{5.61}$$

where I is the 2×2 identity matrix, $c = 1/\sqrt{\epsilon\mu}$ is the wavefront velocity, and M_{ij}, $i, j = 1, 2$, are 2×2 matrices defined by

$$M_{11} = -(\mu_0\sqrt{\epsilon/\mu}F + \epsilon_0\sqrt{\mu/\epsilon}G)I - (1/c_0)(L - K)I_2, \tag{5.62}$$

$$M_{12} = (\mu_0\sqrt{\epsilon/\mu}F - \epsilon_0\sqrt{\mu/\epsilon}G)I - (1/c_0)(L + K)I_2, \tag{5.63}$$

$$M_{21} = -(\mu_0\sqrt{\epsilon/\mu}F - \epsilon_0\sqrt{\mu/\epsilon}G)I - (1/c_0)(L + K)I_2, \tag{5.64}$$

$$M_{22} = (\mu_0\sqrt{\epsilon/\mu}F + \epsilon_0\sqrt{\mu/\epsilon}G)I - (1/c_0)(L - K)I_2. \tag{5.65}$$

Assume that the electric field of the incident wave is polarized in the x-direction. Then one has the following boundary and initial conditions for the split fields

$$\mathbf{E}^+(0,t) = \begin{bmatrix} E^i(t)\\ 0\end{bmatrix}, \tag{5.66}$$

$$\mathbf{E}^-(l,t) = 0, \tag{5.67}$$

$$\mathbf{E}^\pm(z,0) = 0. \tag{5.68}$$

With the initial and boundary conditions (5.66)–(5.68), $\mathbf{E}^\pm(z,t)$ can be computed from the system (5.61) using the method of characteristics. Note that due to the magneto-electric coupling, the direction of the electric field will be rotated when going through the bi-isotropic slab, and thus in general $E_2^\pm \neq 0$ even though the incident electric field is polarized in the x direction.

Introduce a four-element susceptibility kernel vector $p = (p_1, p_2, p_3, p_4) \equiv (G, K, L, F)$. Define an objective functional as follows,

$$J(p) = \int_0^T \left(\sum_{j=1,2} w_j^r(t)[E_j^-(0,t) - E_j^r(t)]^2 + \sum_{j=1,2} w_j^t(t)[E_j^+(l,t) - E_j^t(t)]^2\right) dt, \tag{5.69}$$

where $E_1^r(t)$ and $E_2^r(t)$ ($E_1^t(t)$ and $E_2^t(t)$) are the x and y components, respectively, of the reflected (transmitted) fields at the endpoints $z = 0$ ($z = l$), which are assumed to be known from measurements, and $w_j^r(t)$, $w_j^t(t)$, $j = 1,2$, are non-negative weight functions. The upper limit T of the integration corresponds to the duration of the measurement. Note that since the incident electric field is polarized in the x direction, $E_2^r(t)$ ($E_2^t(t)$) gives the cross-polarized reflection (transmission) which is due to the non-zero value of K and L.

The explicit expression for the gradients can be derived as follows. Let $\tilde{\mathbf{E}}^{\pm} = \mathbf{E}^{\pm}(z,t;p+\delta p)$ be the solution to the system of equations (5.61), (5.66), (5.67) and (5.68) with the parameter vector $p + \delta p$. Then $\delta\mathbf{E}^{\pm} \equiv \tilde{\mathbf{E}}^{\pm} - \mathbf{E}^{\pm}$ satisfy the following system of equations

$$\frac{\partial}{\partial z}\begin{bmatrix}\delta\mathbf{E}^+\\ \delta\mathbf{E}^-\end{bmatrix} = -\frac{1}{c}\frac{\partial}{\partial t}\{\begin{bmatrix}I & 0\\ 0 & -I\end{bmatrix}\begin{bmatrix}\delta\mathbf{E}^+\\ \delta\mathbf{E}^-\end{bmatrix} + \frac{1}{4}\frac{d}{dz}\ln(\frac{\mu}{\epsilon})\begin{bmatrix}I & -I\\ -I & I\end{bmatrix}\begin{bmatrix}\delta\mathbf{E}^+\\ \delta\mathbf{E}^-\end{bmatrix}$$

$$+\frac{1}{2}\frac{\partial}{\partial t}\{\begin{bmatrix}M_{11}* & M_{12}*\\ M_{21}* & M_{22}*\end{bmatrix}\begin{bmatrix}\delta\mathbf{E}^+\\ \delta\mathbf{E}^-\end{bmatrix}\} + \frac{1}{2}\frac{\partial}{\partial t}\{\begin{bmatrix}\delta M_{11}* & \delta M_{12}*\\ \delta M_{21}* & \delta M_{22}*\end{bmatrix}\begin{bmatrix}\mathbf{E}^+\\ \mathbf{E}^-\end{bmatrix}\} + o(\delta p), \tag{5.70}$$

$$\delta\mathbf{E}^{\pm}(z,0) = 0, \tag{5.71}$$

$$\delta\mathbf{E}^{+}(0,t) = \delta\mathbf{E}^{-}(l,t) = 0, \tag{5.72}$$

where $\delta M_{ij} = M_{ij}(p+\delta p) - M_{ij}(p)$, $i,j = 1,2$.

The corresponding increment of the functional $J(p)$ can then be written as follows,

$$\begin{aligned}\delta J &\equiv J(p+\delta p) - J(p)\\ &= 2\int_0^T \Bigg\{ \sum_{j=1,2} w_j^r(t)\delta E_j^-(0,t)[E_j^-(0,t) - E_j^r(t)]\\ &\quad + \sum_{j=1,2} w_j^t(t)\delta E_j^+(l,t)[E_j^+(l,t) - E_j^t(t)] \Bigg\} dt + o(\delta p).\end{aligned} \tag{5.73}$$

If the increment δJ can be rewritten in an inner product form as

$$\delta J = \sum_{i=1,2,3,4}\int_0^T g_{p_i}\cdot\delta p_i\,dt + o(\delta p) \equiv \sum_{i=1,2,3,4} < g_{p_i}, \delta p_i > +o(\delta p), \tag{5.74}$$

where $<,>$ denotes the inner product, then g_{p_i} is the gradient of the objective functional $J(p)$ with respect to the susceptibility kernel p_i.

Introduce dual functions $\mathbf{U}^{\pm}(x,t) \equiv (U_1^{\pm}, U_2^{\pm})^{\mathsf{T}}$ (the superscript T denotes transposition) that satisfy the following boundary conditions

$$U_j^-(0,t) = -w_j^r(t)[E_j^-(0,t) - E_j^r(t)], \quad j = 1,2, \tag{5.75}$$

$$U_j^+(l,t) = w_j^t(t)[E_j^+(l,t) - E_j^t(t)], \quad j = 1,2 \tag{5.76}$$

(notice the '−' sign at the beginning of the right-hand side of (5.75), which is for the convenience of rewriting δJ in a matrix product form). Integrating by parts, one then obtains from (5.73) that

$$\delta J = 2\int_0^T\int_0^l \Bigg\{ \sum_{j=1,2} [\partial_x(\delta E_j^+)U_j^+ + (\delta E_j^+)\partial_x U_j^+]$$
$$+ \sum_{j=1,2} [\partial_x(\delta E_j^-)U_j^- + (\delta E_j^-)\partial_x U_j^-] \Bigg\}\, dxdt + o(\delta p). \tag{5.77}$$

In the present case of a bi-isotropic medium, the matrix formulation presented below facilitates the derivation of the expression for the gradient. Rewrite the above formula in a matrix product form as follows

$$\delta J = 2\int_0^T\int_0^l \Bigg\{ \left(\frac{\partial}{\partial x}\begin{bmatrix}\mathbf{U}^+\\ \mathbf{U}^-\end{bmatrix}^{\mathsf{T}}\right)\begin{bmatrix}\delta\mathbf{E}^+\\ \delta\mathbf{E}^-\end{bmatrix} + \begin{bmatrix}\mathbf{U}^+\\ \mathbf{U}^-\end{bmatrix}^{\mathsf{T}}\frac{\partial}{\partial x}\begin{bmatrix}\delta\mathbf{E}^+\\ \delta\mathbf{E}^-\end{bmatrix}\Bigg\}\, dzdt$$
$$= 2\int_0^T\int_0^l \Bigg\{ \left(\frac{\partial}{\partial x}\begin{bmatrix}\mathbf{U}^+\\ \mathbf{U}^-\end{bmatrix}^{\mathsf{T}}\right)\begin{bmatrix}\delta\mathbf{E}^+\\ \delta\mathbf{E}^-\end{bmatrix}$$
$$+ \begin{bmatrix}\mathbf{U}^+\\ \mathbf{U}^-\end{bmatrix}^{\mathsf{T}}\left(-\frac{1}{c}\frac{\partial}{\partial t}(\begin{bmatrix}I & 0\\ 0 & -I\end{bmatrix}\begin{bmatrix}\delta\mathbf{E}^+\\ \delta\mathbf{E}^-\end{bmatrix}) + \frac{1}{4}\frac{d}{dz}\ln(\frac{\mu}{\epsilon})\begin{bmatrix}I & -I\\ -I & I\end{bmatrix}\begin{bmatrix}\delta\mathbf{E}^+\\ \delta\mathbf{E}^-\end{bmatrix}\right.$$
$$\left.+\frac{1}{2}\frac{\partial}{\partial t}\{\begin{bmatrix}M_{11}* & M_{12}*\\ M_{21}* & M_{22}*\end{bmatrix}\begin{bmatrix}\delta\mathbf{E}^+\\ \delta\mathbf{E}^-\end{bmatrix}\} + \frac{1}{2}\frac{\partial}{\partial t}\{\begin{bmatrix}\delta M_{11}* & \delta M_{12}*\\ \delta M_{21}* & \delta M_{22}*\end{bmatrix}\begin{bmatrix}\mathbf{E}^+\\ \mathbf{E}^-\end{bmatrix}\}\right)\Bigg\}\, dzdt. \tag{5.78}$$

Introduce the operator $\otimes$, defined as follows

$$f \otimes g = \int_t^T f(\cdot\,, t' - t)g(t')\, dt'. \tag{5.79}$$

Note that in general $f \otimes g \neq g \otimes f$. One can show that

$$\textstyle\int_0^T (F * E)U\, dt = \displaystyle\int_0^T [\int_{t'}^T F(t - t')U(t)\, dt]E(t')\, dt'$$
$$= \int_0^T E[F \otimes U]\, dt, \tag{5.80}$$

$$\int_0^T [\frac{\partial}{\partial t}(F * E)]U\,dt = \int_0^T [F * \frac{\partial}{\partial t}E]U\,dt = \int_0^T (\frac{\partial}{\partial t}E)[F \otimes U]\,dt$$

$$= -\int_0^T E[\frac{\partial}{\partial t}(F \otimes U)]\,dt. \tag{5.81}$$

Taking the conditions for $\mathbf{U}^{\pm}(x,t)$ at $t = T$ to be

$$\mathbf{U}^+(x,T) = \mathbf{U}^-(x,T) = 0, \tag{5.82}$$

and using (5.80) and (5.81), one then obtains the following result from (5.78)

$$\delta J = (\delta J)^T = 2\int_0^T \int_0^l \left\{ \begin{bmatrix} \delta\mathbf{E}^+ \\ \delta\mathbf{E}^- \end{bmatrix}^{\mathsf{T}} \left(\frac{\partial}{\partial x}\begin{bmatrix} \mathbf{U}^+ \\ \mathbf{U}^- \end{bmatrix} + \frac{1}{c}\frac{\partial}{\partial t}(\begin{bmatrix} I & 0 \\ 0 & -I \end{bmatrix}^{\mathsf{T}} \begin{bmatrix} \mathbf{U}^+ \\ \mathbf{U}^- \end{bmatrix}) \right.\right.$$

$$\left. + \frac{1}{4}\frac{d}{dz}\ln(\frac{\mu}{\epsilon}) \begin{bmatrix} I & -I \\ -I & I \end{bmatrix}^{\mathsf{T}} \begin{bmatrix} \mathbf{U}^+ \\ \mathbf{U}^- \end{bmatrix} - \frac{1}{2}\frac{\partial}{\partial t}\{\begin{bmatrix} M_{11}\otimes & M_{12}\otimes \\ M_{21}\otimes & M_{22}\otimes \end{bmatrix}^{\mathsf{T}} \begin{bmatrix} \mathbf{U}^+ \\ \mathbf{U}^- \end{bmatrix}\} \right)$$

$$\left. + \frac{1}{2}\begin{bmatrix} \mathbf{U}^+ \\ \mathbf{U}^- \end{bmatrix}^{\mathsf{T}} \frac{\partial}{\partial t}\{\begin{bmatrix} \delta M_{11}* & \delta M_{12}* \\ \delta M_{21}* & \delta M_{22}* \end{bmatrix} \begin{bmatrix} \mathbf{E}^+ \\ \mathbf{E}^- \end{bmatrix}\} \right\} dzdt.$$

Thus, if the dual functions satisfy the following system of equations

$$\frac{\partial}{\partial x}\begin{bmatrix} \mathbf{U}^+ \\ \mathbf{U}^- \end{bmatrix} = -\frac{1}{c}\frac{\partial}{\partial t}(\begin{bmatrix} I & 0 \\ 0 & -I \end{bmatrix} \begin{bmatrix} \mathbf{U}^+ \\ \mathbf{U}^- \end{bmatrix}) - \frac{1}{4}\frac{d}{dz}\ln(\frac{\mu}{\epsilon}) \begin{bmatrix} I & -I \\ -I & I \end{bmatrix} \begin{bmatrix} \mathbf{U}^+ \\ \mathbf{U}^- \end{bmatrix}$$

$$+ \frac{1}{2}\frac{\partial}{\partial t}\{\begin{bmatrix} M_{11}\otimes & M_{12}\otimes \\ M_{21}\otimes & M_{22}\otimes \end{bmatrix}^{\mathsf{T}} \begin{bmatrix} \mathbf{U}^+ \\ \mathbf{U}^- \end{bmatrix}\}, \tag{5.83}$$

one has

$$\delta J = \int_0^T \int_0^l \left\{ \begin{bmatrix} \mathbf{U}^+ \\ \mathbf{U}^- \end{bmatrix}^{\mathsf{T}} \frac{\partial}{\partial t}\{\begin{bmatrix} \delta M_{11}* & \delta M_{12}* \\ \delta M_{21}* & \delta M_{22}* \end{bmatrix} \begin{bmatrix} \mathbf{E}^+ \\ \mathbf{E}^- \end{bmatrix}\} \right\} dzdt. \tag{5.84}$$

One can rewrite the above equation in the form of (5.74), and obtain the following explicit expression for the gradients (He and Ström 1995c)

$$g_K(t) = \frac{1}{c}\frac{\partial}{\partial t}\int_0^l \{[\mathbf{E}^+ - \mathbf{E}^-]^{\mathsf{T}} \otimes [I_2(\mathbf{U}^+ - \mathbf{U}^-)]\}\,dz, \tag{5.85}$$

$$g_L(t) = -\frac{1}{c}\frac{\partial}{\partial t}\int_0^l \{[\mathbf{E}^+ + \mathbf{E}^-]^{\mathsf{T}} \otimes [I_2(\mathbf{U}^+ + \mathbf{U}^-)]\}\,dz, \tag{5.86}$$

$$g_G(t) = \frac{\partial}{\partial t}\int_0^l \epsilon_0\sqrt{\mu/\epsilon}\{[\mathbf{E}^+ + \mathbf{E}^-]^{\mathsf{T}} \otimes [\mathbf{U}^+ - \mathbf{U}^-]\}\,dz, \tag{5.87}$$

$$g_F(t) = \frac{\partial}{\partial t} \int_0^l \mu_0 \sqrt{\epsilon/\mu} \{[\mathbf{E}^+ - \mathbf{E}^-]^{\mathrm{T}} \otimes [\mathbf{U}^+ + \mathbf{U}^-]\}\, dz. \tag{5.88}$$

Note that the dual functions $\mathbf{U}^{\pm}(z,t)$ can be computed for $z \in [0,l]$ and $t \in [0,T]$ from equations (5.83), (5.75), (5.76) and (5.82), using the method of characteristics with backward time-stepping from $t = T$ to $t = 0$.

In the numerical example, for simplicity we choose $\epsilon_r \equiv \epsilon/\epsilon_0 = 3$ and $\mu \equiv \mu/\mu_0 = 2$ everywhere. All the parameters are scaled to be dimensionless, i.e., $\epsilon_0 = \mu_0 = 1$, and the length of the slab is $l = 1$. The synthetic profiles for the susceptibility kernels of the bi-isotropic slab are chosen as (a Debye–Lorentz model, see e.g. Robin et al. (1994) for the form of these kernels or their Fourier transforms)

$$\begin{cases} K = 0.15\sin(2t)\, \exp(-0.4t), \\ L = [0.05\cos(3t) + 0.1\sin(3t)]\exp(-0.5t), \\ G = 0.1\exp(-0.7t) + 0.45\sin(4t)\, \exp(-0.5t), \\ F = 0.35\sin(3t)\, \exp(-0.3t). \end{cases}$$

These profiles are given by the solid lines in Fig. 5.6. The incident electric field (polarized in the x-direction) is chosen to be

$$E^i(t) = \begin{cases} \sin(\pi t), & \text{when } 0 \le t \le 1, \\ 0, & \text{otherwise.} \end{cases}$$

The corresponding reflected and transmitted electric fields are calculated by the method of characteristics, and are given by the solid lines in Fig. 5.7. The simultaneous reconstruction of all the four susceptibility kernels using these clean scattering data is given by the dotted lines in Fig. 5.6, after 64 iterations in a standard conjugate gradient method. The starting guess for these kernels is set to be identically zero. In this example the weight functions are chosen as $w_1^r = 36$, $w_2^r = 16$, $w_1^t = 1$, $w_2^t = 4$. In order to illustrate how noise influences the reconstruction, 5 percent Gaussian random noise (with respect to the peak value) is added to the co- and cross-polarized scattered fields (with different samples of noise for E_j^r and E_j^t, $j = 1, 2$). The noise-contaminated scattered fields are shown by the dashed lines in Fig. 5.7. The corresponding reconstructions of the susceptibility kernels using these noise-contaminated scattering data are given by the dashed lines in Fig. 5.6.

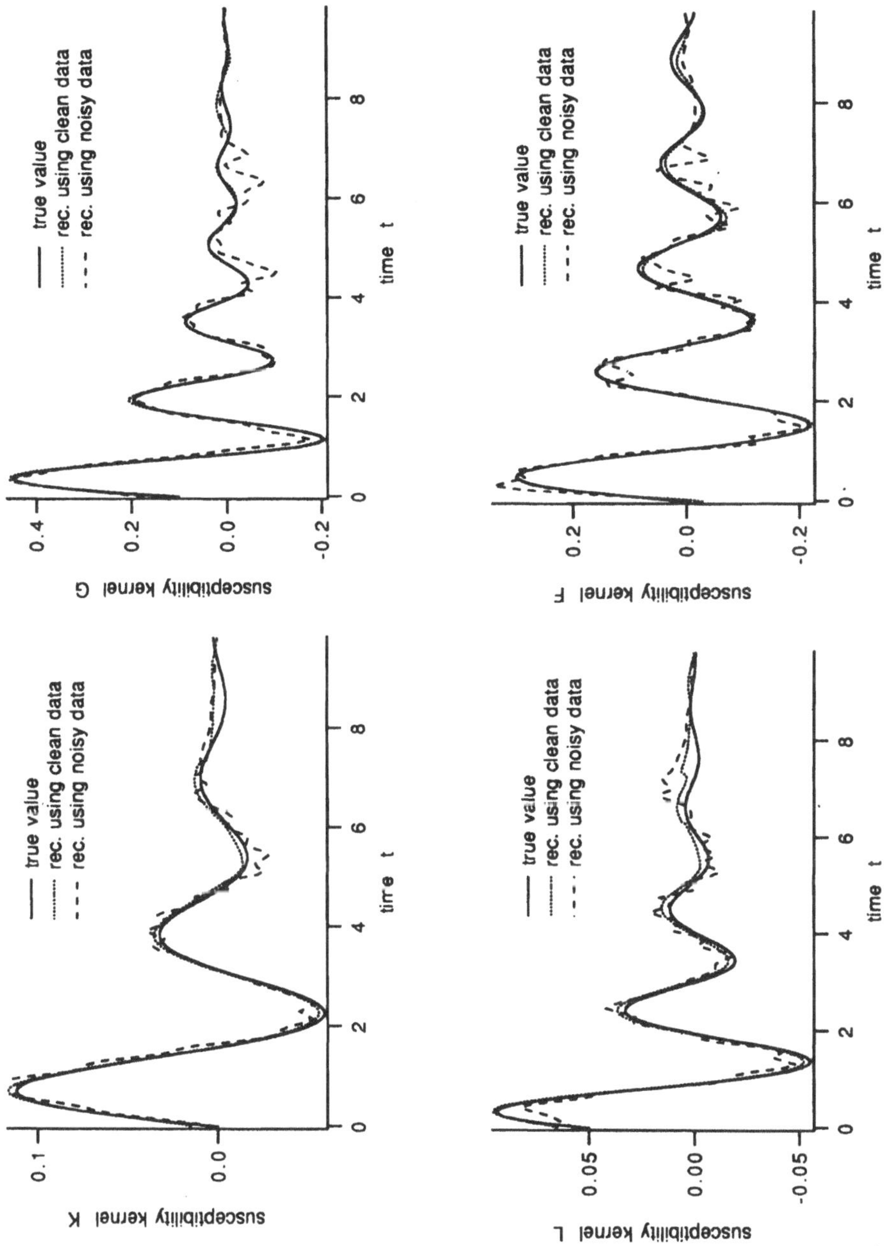

FIG. 5.6. The true susceptibility kernels (solid lines) and the simultaneous reconstructions. The dotted lines are the reconstructions using the clean scattering data. The dashed lines are the reconstructions using the noisy data shown by the dashed lines in Fig. 5.7.

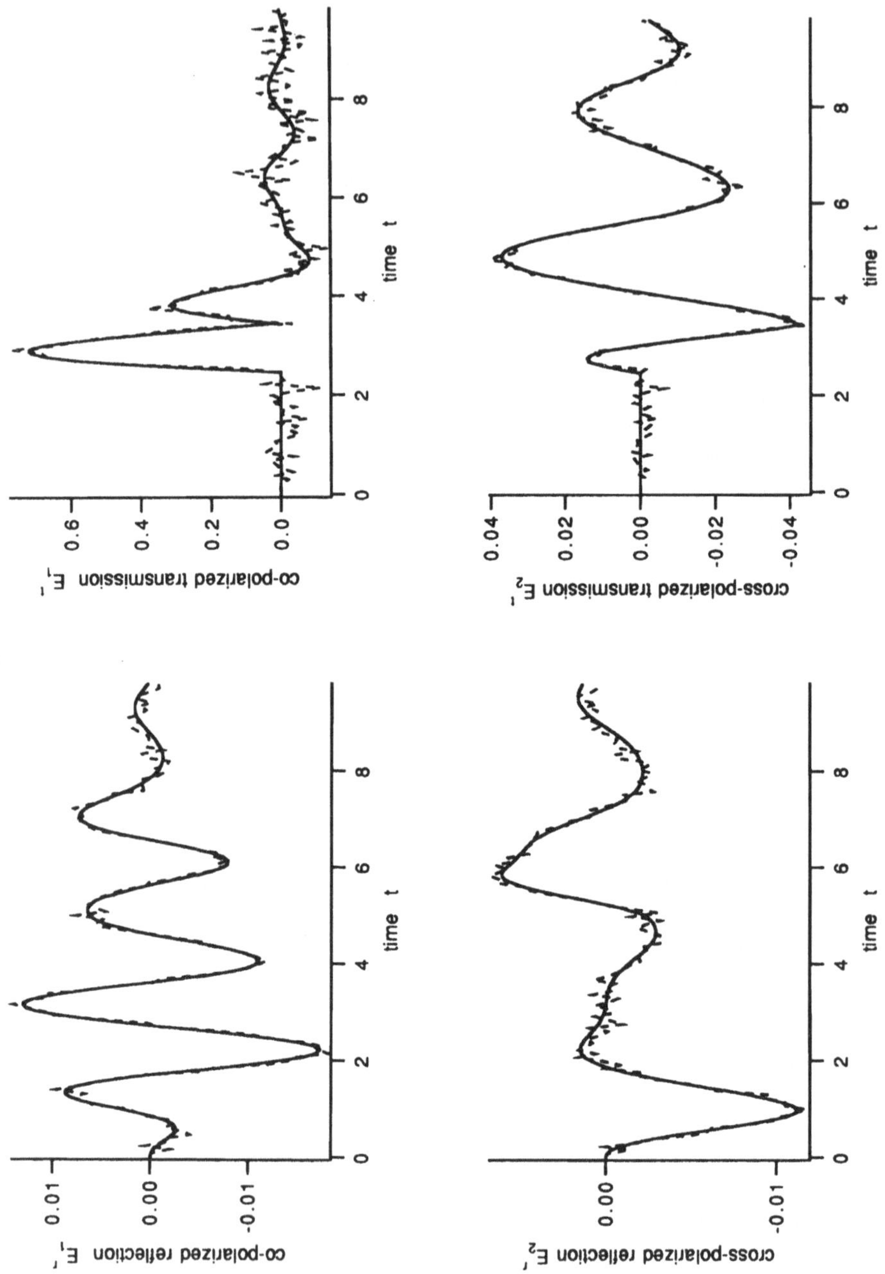

FIG. 5.7. The reflected and transmitted electric fields corresponding to a bi-isotropic slab with susceptibility kernels shown in Fig. 5.6, and $\epsilon_r = 3$, $\mu_r = 2$, $l = 1$.

5.1.4 *Reconstruction of density and/or sound speed in two or three dimensions*

Consider acoustic wave propagation in an inhomogeneous region Ω in two or three dimensions, which is described by the following stress-strain equations for fluids (cf. (4.110), (4.111) and Friedlander (1958))

$$\frac{1}{\rho c^2}\partial_t p + \nabla \cdot \mathbf{v} = 0, \tag{5.89}$$

$$\rho \partial_t \mathbf{v} + \nabla p = 0, \tag{5.90}$$

where p is the pressure, $\mathbf{v}$ is the velocity (the time derivative of the displacement), ρ is the fluid density and c is the sound speed in the medium. Outside the region Ω, the medium is homogeneous with sound speed c_0 and density ρ_0.

It is assumed that the incident wave will not reach the surface Γ of the inhomogeneous domain Ω until the time $t = 0$. This leads to the following initial conditions

$$p(x,0) = 0, \qquad \mathbf{v}(x,0) = 0, \qquad x \in \Omega. \tag{5.91}$$

In the inverse problem, it is assumed that the pressure p and the normal component of the velocity, v_n, are known for $t \in [0,T]$ on the surface Γ, i.e.,

$$p(x,t) = p^{(m)}(x,t), \qquad v_n(x,t) \equiv \mathbf{v} \cdot \hat{\mathbf{n}} = v_n^{(m)}(x,t), \qquad x \in \Gamma, \quad t \in [0,T], \tag{5.92}$$

where $\hat{\mathbf{n}}$ is the unit outward normal vector of the surface Γ, $p^{(m)}(x,t)$ and $v_n^{(m)}(x,t)$ are known on the surface Γ and for the period of time $t \in [0,T]$. The inverse problem is to determine the density $\rho(x)$ and/or the sound speed $c(x)$ in the inhomogeneous region Ω.

Introduce the following split pressures $p^{\pm}$ at the boundary Γ:

$$p^{\pm} = \frac{1}{2}[p \mp \rho_0 c_0 v_n], \qquad x \in \Gamma. \tag{5.93}$$

Note that such a splitting has a physical meaning as incident and scattered pressures only in a homogeneous region (characterized by c_0 and ρ_0) and in a local sense, in the case of an incident wave that is locally a normally incident plane wave. One of the main advantage of using such a wave-splitting is that it gives the best error estimate in the gradient calculation (Gustafsson and He 1998). By using such a splitting, one can obtain a simple, stable, and robust algorithm for the direct solver (Gustafsson and He 1997). From the above definition for the wave-splitting, one obtains

$$p = p^+ + p^-, \tag{5.94}$$

$$v_n = \frac{1}{\rho_0 c_0}[p^- - p^+]. \tag{5.95}$$

Define an objective functional $J(\rho, c)$ as follows,

$$J(\rho,c) \equiv \int_0^T \int_\Gamma \Phi(x,t)\left\{\left[p-p^{(m)}\right]^2 + \rho_0^2 c_0^2\left[v_n - v_n^{(m)}\right]^2\right\} dSdt \quad (5.96)$$

$$= 2\int_0^T \int_\Gamma \Phi\left\{\left[p^+ - p^{(m)+}\right]^2 + \left[p^- - p^{(m)-}\right]^2\right\} dSdt, \quad (5.97)$$

where $\Phi(x,t)$ is a weight function, dS is a surface area element, and

$$p^{(m)\pm} = \frac{1}{2}\left[p^{(m)} \mp \rho_0 c_0 v_n^{(m)}\right], \qquad x \in \Gamma. \quad (5.98)$$

The inverse problem is to find a function $\rho(x)$ and/or $c(x)$ which minimizes the objective functional $J(\rho,c)$. Mathematically we treat $p^+(x,t)$ as a known source on Γ, i.e.,

$$p^+ = p^{(m)+}, \qquad x \in \Gamma, \quad t \in [0,T], \quad (5.99)$$

and $p^-(x,t)$ as the corresponding measurement on Γ. Below we derive an explicit expression for the gradient using the Gauss divergence theorem (Gustafsson and He 1997).

Let $\tilde{p}$, $\tilde{\mathbf{v}}$ be the solution to the system of equations (5.89), (5.90), (5.91) and (5.99) with perturbed parameters $\rho+\delta\rho$, $c+\delta c$. One then obtains the following system of equations for $\delta p \equiv \tilde{p} - p$ and $\delta \mathbf{v} \equiv \tilde{\mathbf{v}} - \mathbf{v}$ (by taking the first-order approximations):

$$\begin{cases} \dfrac{1}{\rho c^2}\partial_t(\delta p) + \nabla\cdot\delta\mathbf{v} = -\left(\dfrac{\delta\rho}{\rho} + \dfrac{2\delta c}{c}\right)\nabla\cdot\mathbf{v}, \\ \rho\partial_t\delta\mathbf{v} + \nabla\delta p = \dfrac{\delta\rho}{\rho}\nabla p, \\ \delta p^+ = 0, \qquad x\in\Gamma, \quad t\in[0,T], \\ \delta p(x,0) = 0, \qquad \delta\mathbf{v}(x,0) = 0, \qquad x\in\Omega. \end{cases} \quad (5.100)$$

The corresponding increment of the objective functional $J(\rho,c)$ can be written as follows,

$$\delta J(\rho,c) \equiv J(\rho+\delta\rho, c+\delta c) - J(\rho,c)$$

$$= 2\int_0^T\int_\Gamma \Phi\Big[(p^+ + \delta p^+ - p^{(m)+})^2 + (p^- + \delta p^- - p^{(m)-})^2$$

$$-(p^+ - p^{(m)+})^2 - (p^- - p^{(m)-})^2\Big]\, dSdt$$

$$= 2\int_0^T\int_\Gamma \Phi \left[\delta p^+(2p^+ - 2p^{(m)+} + \delta p^+) + \delta p^-(2p^- - 2p^{(m)-} + \delta p^-)\right] dSdt$$

$$= 4\int_0^T\int_\Gamma \Phi \left[\delta p^+(p^+ - p^{(m)+}) + \delta p^-(p^- - p^{(m)-})\right] dSdt + o(\delta\rho, \delta c)$$

$$= 4\int_0^T\int_\Gamma \delta p^- \left[\Phi(p^- - p^{(m)-})\right] dSdt + o(\delta\rho, \delta c). \tag{5.101}$$

We use a direct solver for the following forward problem,

$$\begin{cases} \dfrac{1}{\rho c^2}\partial_t p + \nabla\cdot\mathbf{v} = 0, \\ \rho\partial_t\mathbf{v} + \nabla p = 0, \\ p^+ = p^{(m)+}, \qquad x\in\Gamma, \quad t\in[0,T], \\ p(x,0) = 0, \mathbf{v}(x,0) = 0, \end{cases} \tag{5.102}$$

and the following backward dual problem,

$$\begin{cases} \dfrac{1}{\rho c^2}\partial_t q + \nabla\cdot\mathbf{u} = 0, \\ \rho\partial_t\mathbf{u} + \nabla q = 0, \\ q^- = \Phi[p^- - p^{(m)-}], \qquad x\in\Gamma, \quad t\in[0,T], \\ q(x,T) = 0, \qquad \mathbf{u}(x,T) = 0. \end{cases} \tag{5.103}$$

The system (5.102) is solved by a finite difference method with a forward time stepping, and the system (5.103) is solved by a finite difference method with a backward time stepping. In order to avoid any inconsistency between the boundary and initial conditions in the system (5.103), one should choose the weight function such that

$$\Phi(x,T) = 0. \tag{5.104}$$

Using the differential equations in the systems (5.100), (5.102) and (5.103), one obtains

$$\nabla\cdot(\delta p\mathbf{u} + q\delta\mathbf{v}) = [\nabla(\delta p)]\cdot\mathbf{u} + \delta p\nabla\cdot\mathbf{u} + (\nabla q)\cdot\delta\mathbf{v} + q\nabla\cdot(\delta\mathbf{v})$$

$$= -\partial_t\left[\frac{1}{\rho c^2}(\delta p)q + \rho(\delta\mathbf{v})\cdot\mathbf{u}\right] + \frac{\delta\rho}{\rho}[-q\nabla\cdot\mathbf{v} + (\nabla p)\cdot\mathbf{u}] - \frac{2\delta c}{c}q\nabla\cdot\mathbf{v}$$

$$= -\partial_t \left[\frac{1}{\rho c^2}(\delta p)q + \rho(\delta \mathbf{v}) \cdot \mathbf{u} \right] + \frac{\delta \rho}{\rho} \left[-q\nabla \cdot \mathbf{v} + \nabla \cdot (p\mathbf{u}) - p\nabla \cdot \mathbf{u} \right] - \frac{2\delta c}{c} q\nabla \cdot \mathbf{v}$$

$$= -\partial_t \left[\frac{1}{\rho c^2}(\delta p)q + \rho(\delta \mathbf{v}) \cdot \mathbf{u} - \frac{\delta \rho}{\rho^2 c^2}(qp) \right] + \frac{\delta \rho}{\rho} \nabla \cdot (p\mathbf{u}) - \frac{2\delta c}{c} q\nabla \cdot \mathbf{v}.$$

Integrating the above equation over the time period $[0, T]$ and the region Ω, and using the Gauss divergence theorem, one obtains

$$\int_0^T \int_\Gamma (\delta p \mathbf{u} + q\delta \mathbf{v}) \cdot \hat{\mathbf{n}} dS dt = \int_0^T \int_\Omega \frac{\delta \rho}{\rho} \nabla \cdot (p\mathbf{u})\, dx dt - 2\int_0^T \int_\Omega \frac{\delta c}{c} q\nabla \cdot \mathbf{v}\, dx dt. \tag{5.105}$$

Since (cf. (5.101) and noting that $\delta p^+ = 0$ and $q^- = \Phi(p^- - p^{(m)-})$ on Γ)

$$\int_0^T \int_\Gamma (\delta p \mathbf{u} + q\delta \mathbf{v}) \cdot \hat{\mathbf{n}} dS dt = \int_0^T \int_\Gamma (\delta p u_n + q\delta v_n) dS dt$$

$$= -\frac{1}{\rho_0 c_0} \int_0^T \int_\Gamma \left[(\delta p^+ + \delta p^-)(q^+ - q^-) + (\delta p^+ - \delta p^-)(q^+ + q^-) \right] dS dt$$

$$= -\frac{2}{\rho_0 c_0} \int_0^T \int_\Gamma \left[\delta p^+ q^+ - \delta p^- q^- \right] dS dt = \frac{2}{\rho_0 c_0} \int_0^T \int_\Gamma (\delta p^-) q^- dS dt$$

$$= \frac{2}{\rho_0 c_0} \int_0^T \int_\Gamma (\delta p^-) \Phi \left[p^- - p^{(m)-} \right] dS dt = \frac{1}{2\rho_0 c_0} \delta J(\rho, c) + o(\delta \rho, \delta c),$$

it follows from (5.105) that

$$\delta E = < g_\rho, \delta \rho > + < g_c, \delta c > + o(\delta \rho, \delta c), \tag{5.106}$$

where $<,>$ denotes the inner product (with respect to the spatial domain Ω) and

$$g_\rho = \frac{2\rho_0 c_0}{\rho} \int_0^T \nabla \cdot (\mathbf{u} p)\, dt, \qquad x \in \Omega, \tag{5.107}$$

$$g_c = -\frac{4\rho_0 c_0}{c} \int_0^T q\nabla \cdot \mathbf{v}\, dt = \frac{4\rho_0 c_0}{\rho c^3} \int_0^T q\partial_t p\, dt, \qquad x \in \Omega. \tag{5.108}$$

Equations (5.107) and (5.108) give the explicit expressions for the gradients of the objective functional $J(\rho, c)$ with respect to the density $\rho(x)$ and the sound speed $c(x)$, respectively.

Note that the results given in this subsection are valid also for an arbitrary number n of space dimensions, $n \geq 2$. In a similar way one can derive an explicit gradient expression in the electromagnetic case when the permittivity and conductivity are to be reconstructed (Gustafsson and He 1998).

Numerical reconstruction of a two-dimensional density profile is shown in Fig. 5.8 (the sound speed is a constant $c = c_0$). The inhomogeneous region Ω is a square discretized by 72×72 points. The 'measured' data are simulated with a Lax–Wendroff scheme and the direct solver used in the optimization is based on a Leapfrog scheme. Since the 'measured' data are calculated in a way that is different from what is used for the direct solver, we avoid the 'inverse crime' when solving the inverse problem (Colton and Kress 1992). One Gaussian pulse is used as a source illuminating the square from the left side. The scattered fields are measured on all four sides of the square. The contour density plot of the density profile is shown in Fig. 5.8(a), and the corresponding reconstruction after 25 iterations is given in Fig. 5.8(b) (the starting guess is $\rho = \rho_0 = 1$ everywhere). The reconstruction took about 20 minutes on a Sun workstation (Ultra 1). The reconstruction can be improved further if 4 Gaussian pulses at four different sides of the square are used, and the corresponding reconstructions are given in Fig. 5.8(c) (using noisy data with a signal to noise ratio of 73) and Fig. 5.8(d) (using noisy data with a signal to noise ratio of 11). A standard Tikhonov regularization is used by adding a term $\lambda \int_\Omega |\nabla \rho|^2 dx$ in the definition (5.97) of the objective functional (we choose $\lambda = 0.01$ for the reconstructions shown in Fig. 5.8(c) and (d)). The maximum absolute error in the density reconstruction is about 0.29 in Fig. 5.8(b), 0.034 in Fig. 5.8(c), and 0.06 in Fig. 5.8(d). The reconstruction of the the sound speed can be performed in a completely analogous way, and an example is shown in Fig. 5.9(a) (the true profile of the sound speed) and (b) (the corresponding reconstruction). In this reconstruction, noisy data with a signal to noise ratio of 11 are used, and the starting guess is $c = c_0 = 1$ everywhere (the density is constant, $\rho = \rho_0 = 1$). We refer to Gustafsson and He (1997) for more details.

We compare the computation times for the reconstructions obtained with and without the analytic gradients. When an analytic gradient is not available, one can calculate the gradient numerically through the following set of step functions

$$\psi_i(x) = \begin{cases} 1, & x \in V_i, \\ 0, & \text{otherwise}, \end{cases}$$

where V_i, $i = 1, 2, ..., N$, are the descretized volume cells. Assume that $\rho(x) = \sum_i P_i \psi_i(x)$ ($\rho(x)$ is the density to be reconstructed), and the corresponding gradient $G_\rho(x) = \sum_i g_i \psi_i(x)$, then one can calculate the coefficient G_i numerically through a small perturbation as follows,

$$G_i \approx \frac{J(< P_0, P_1, ..., P_i + \varepsilon, ..., P_N >) - J(< P_0, P_1, ..., P_i, ..., P_N >)}{\varepsilon},$$

where ε is a small quantity. In the present reconstruction algorithm, the computation time is consumed mainly by calling the forward direct solver for the system (5.102) and the backward direct solver for the system (5.103) (which

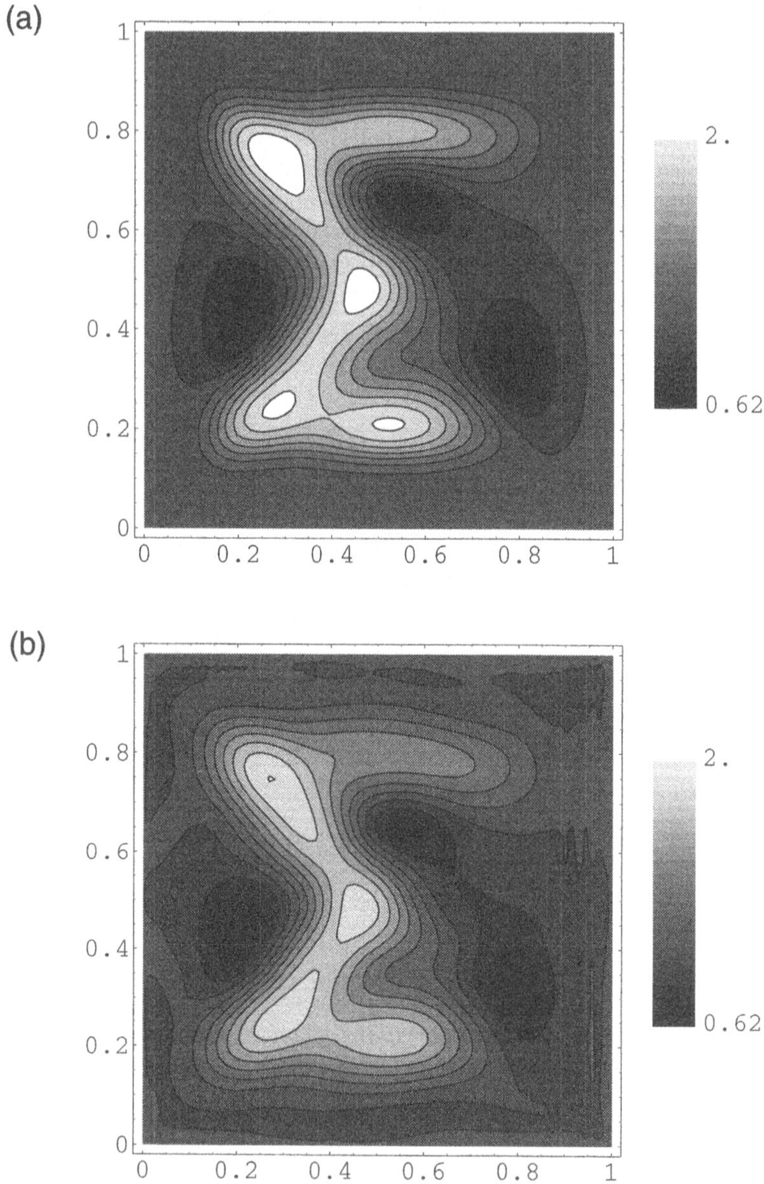
(a)
1
0.8
0.6
0.4
0.2
0
0
0.2
0.4
0.6
0.8
1
2.
0.62
(b)
1
0.8
0.6
0.4
0.2
0
0
0.2
0.4
0.6
0.8
1
2.
0.62

(c)

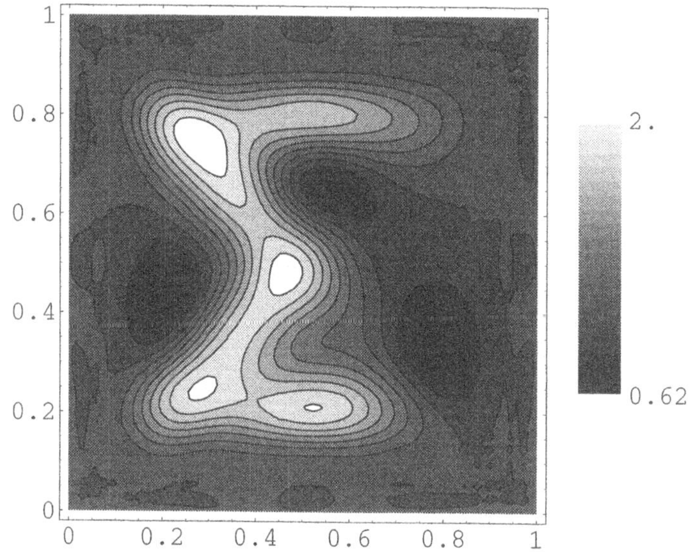

(d)

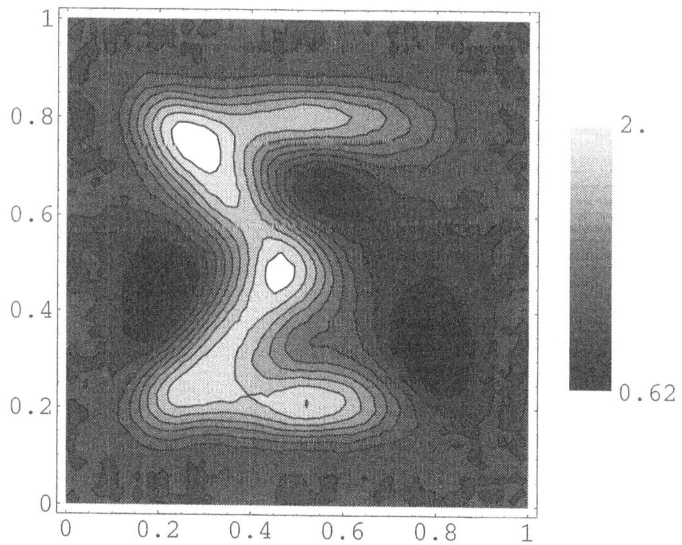

FIG. 5.8. The reconstruction of a two-dimensional density profile (the contour density plot). The starting guess is $\rho = \rho_0 = 1$ everywhere. (a) The true density profile. (b) The reconstructed density profile using 1 Gaussian pulse and clean data. (c) The reconstructed density profile using 4 Gaussian pulses and noisy data with a signal to noise ratio of 73. (d) The reconstructed density using 4 Gaussian pulses and noisy data with a signal to noise ratio of 11.

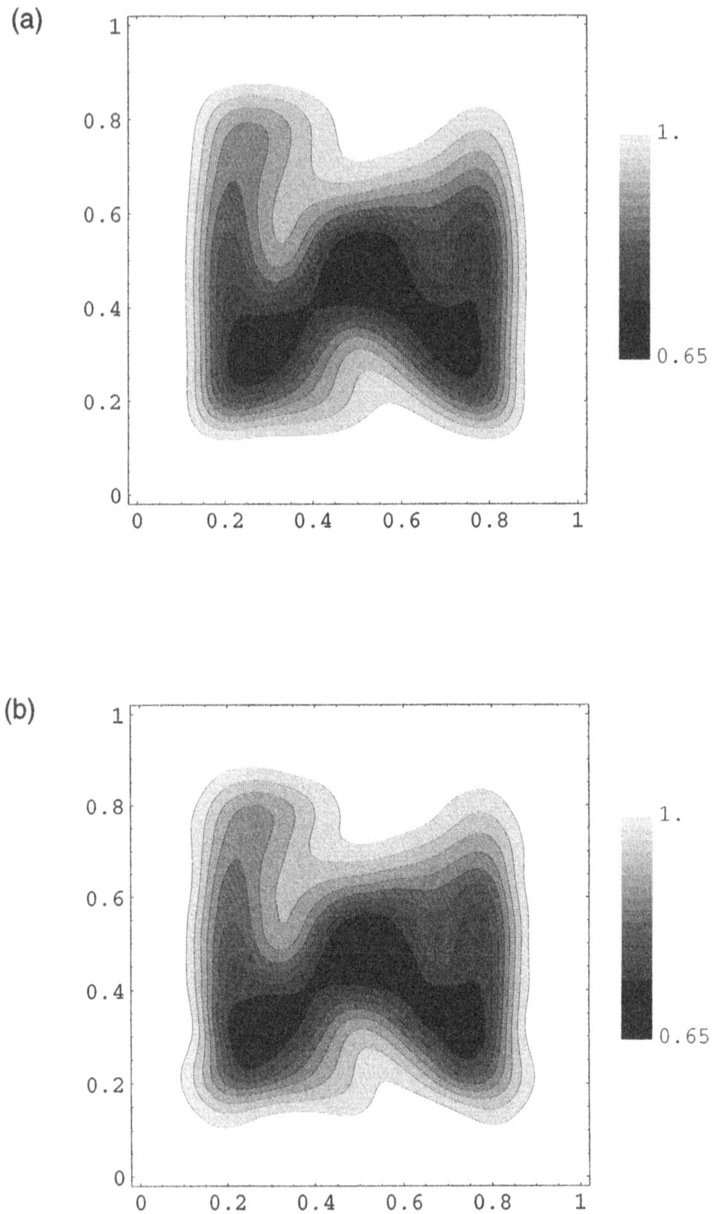

FIG. 5.9. The reconstruction of a two-dimensional sound speed. (a) The true profile. (b) The reconstructed profile.

takes a time that is roughly equal to that for the forward direct solver). To calculate the gradient, one has to call the forward direct solver N times when the numerical perturbation is used, while using the present analytic expression, one only needs to call the forward and backward direct solvers once. Therefore, the ratio of the computation times for the gradient with and without the analytic expression is about $2/N$. In fact, the ratio of the overall computation time for the reconstruction (if the same number of iterations is required) with and without the analytic gradients is approximately $\frac{2+s}{N+1+s}$, where s is the average number of steps in the line search within each iteration ($s \approx 2$ if an optimal line search program is used). In the multi-dimensional case, N is large (if one takes N_x descretization points along each orthogonal space direction, then $N = (N_x)^n$, where $n = 2$, 3 in the two- and three-dimensional cases, respectively). Therefore, the computation time required for the reconstruction can be greatly improved in the multi-dimensional case if an analytic gradient is used.

5.2 Newton–Kantorovich approach for a quasi-linear wave equation

The Newton–Kantorovich method is one of the most efficient methods for obtaining numerical solution to non-linear problems (see e.g. Kantorovich and Akilov (1964), Romanov and Kabanikhin (1994)). To apply the Newton–Kantorovich approach to an inverse problem, one introduces a suitable operator which maps a selected parameter (to be reconstructed) to the scattering data (reflection data or transmission data). If the inverse of the Fréchet derivative (see e.g. Luenberger (1969)) of the mapping operator is computed numerically, one can reconstruct the parameter in an iterative way according to the Newton–Kantorovich scheme. At each iterative step, the inverse of the Fréchet derivative of the mapping operator is calculated by solving an equivalent inverse problem which has been linearized. In this section we use the Newton–Kantorovich method to solve an inverse problem for a quasi-linear wave equation in a wave-splitting setting.

Consider the following quasi-linear acoustic wave equation for the pressure $u(x,t)$,

$$\mathcal{L}u = F(q, u, \partial_x u, \partial_t u, x, t), \tag{5.109}$$

where $F(q, u, \partial_x u, \partial_t u, x, t)$ is a quasi-linear source term (a functional non-linear in u and/or $\partial_x u$ and/or $\partial_t u$), and

$$\mathcal{L}u = \frac{1}{c^2}\partial_t^2 u - \partial_x^2 u + b\partial_t u + (\frac{d}{dx}\ln\rho)\partial_x u, \tag{5.110}$$

and where c is the velocity of the acoustic wave, ρ is the fluid density, and b is the dissipation coefficient. The parameter q in the quasi-linear source term $F(q, u, \partial_x u, x, t)$ denotes the amplitude of the non-linearity. All the parameters c, b, ρ and q are assumed to be depth-dependent, but independent of the field.

The velocity $c(x)$ is piecewisely differentiable. Furthermore, the medium outside the slab $[0, l]$ is assumed to be homogeneous and linear, i.e.

$$F = 0, \quad c = c_0, \quad \rho = \rho_0, \quad b = 0, \qquad x \leq 0 \text{ or } x > l. \tag{5.111}$$

A transient acoustic plane wave is assumed to impinge normally on the stratified slab from the left side $x < 0$. It is also assumed that the field is identically zero in the stratified medium for the time $t \leq 0$, i.e.

$$u|_{t\leq 0} = 0, \quad (\partial_x u)|_{t\leq 0} = 0, \qquad \text{for } x \leq 0. \tag{5.112}$$

The inverse problem is to reconstruct the spatial distribution of the non-linearity, i.e., $q(x)$, from the measured reflection data at $x = 0$, or transmission data at $x = l$, if the parameters $c(x)$, $b(x)$ and $\rho(x)$, as well as the form of $F(q, u, \partial_x u, x, t)$, are known.

The total field $u(x, t)$ is decomposed into the right-moving component $u^+(x, t)$ and left-moving component $u^-(x, t)$ defined as (see Section 1.1)

$$\begin{bmatrix} u^+ \\ u^- \end{bmatrix}(x,t) = \frac{1}{2}\begin{bmatrix} 1 & -c\partial_t^{-1} \\ 1 & c\partial_t^{-1} \end{bmatrix}\begin{bmatrix} u \\ \partial_x u \end{bmatrix}(x,t), \tag{5.113}$$

where ∂_t^{-1} denotes the time integral, i.e., $\partial_t^{-1} g(x,t) = \int_0^t g(x,t')\,dt'$.

The dynamic equations for the split fields are

$$\frac{\partial}{\partial x}\begin{bmatrix} u^+ \\ u^- \end{bmatrix} + \frac{1}{c}\frac{\partial}{\partial t}\begin{bmatrix} u^+ \\ -u^- \end{bmatrix} - \begin{bmatrix} \alpha & \beta \\ \gamma & \vartheta \end{bmatrix}\begin{bmatrix} u^+ \\ u^- \end{bmatrix} = E(q, u^+, u^-, x, t)\begin{bmatrix} 1 \\ -1 \end{bmatrix}, \tag{5.114}$$

where

$$\begin{aligned} &E(q, u^+, u^-, x, t) \\ &\quad = \frac{1}{2}\partial_t^{-1} F(q, u^+ + u^-, \frac{1}{c}\partial_t u^- - \frac{1}{c}\partial_t u^+, \partial_t u^- + \partial_t u^+, x, t), \end{aligned} \tag{5.115}$$

$$\alpha = -\frac{bc}{2} + \frac{1}{2}\frac{d}{dx}\ln\rho + \frac{1}{2}\frac{d}{dx}\ln c, \tag{5.116}$$

$$\beta = -\frac{bc}{2} - \frac{1}{2}\frac{d}{dx}\ln\rho - \frac{1}{2}\frac{d}{dx}\ln c, \tag{5.117}$$

$$\gamma = \frac{bc}{2} - \frac{1}{2}\frac{d}{dx}\ln\rho - \frac{1}{2}\frac{d}{dx}\ln c, \tag{5.118}$$

$$\vartheta = \frac{bc}{2} + \frac{1}{2}\frac{d}{dx}\ln\rho + \frac{1}{2}\frac{d}{dx}\ln c. \tag{5.119}$$

The system (5.114) of quasi-linear first-order hyperbolic differential equations gives the dynamic equations for the split fields in the stratified quasi-linear medium.

The boundary and initial conditions for the split fields are

$$u^+(0,t) = u^i(t), \tag{5.120}$$

$$u^-(l,t) = 0, \tag{5.121}$$

$$u^\pm(x,0) = 0, \qquad x \in [0,l], \tag{5.122}$$

where $u^i(t)$ is the incident field at the surface $x = 0$.

The direct solver. Equations (5.114), (5.120), (5.121) and (5.122) constitute a set of equations from which $u^+(x,t)$ and $u^-(x,t)$ can be computed using the method of characteristics.

The inverse problem. The Newton–Kantorovich method is then applied to the reconstruction of the non-linearity parameter $q(x)$ using scattering data. Define two operators A and B which maps the non-linearity parameter $q(x)$ to the reflected and transmitted data, respectively (cf. the direct solver), i.e.,

$$A(q) = u^-(0,t), \tag{5.123}$$

$$B(q) = u^+(l,t). \tag{5.124}$$

Then according to the Newton–Kantorovich scheme, one has (see e.g. Romanov and Kabanikhin (1994))

$$q^{(n+1)} = q^{(n)} - \left[A'(q^{(n)})\right]^{-1} [A(q^{(n)}) - u^r(t)], \quad n = 0,1,2,..., \tag{5.125}$$

or

$$q^{(n+1)} = q^{(n)} - \left[B'(q^{(n)})\right]^{-1} [B(q^{(n)}) - u^t(t)], \quad n = 0,1,2,..., \tag{5.126}$$

where A', B' are the Fréchet derivatives of the operators A and B, respectively, $u^r(t)$ is the reflected field at $x = 0$, and $u^t(t)$ is the transmitted field at $x = l$. In the inverse problem, it is assumed that either $u^r(t)$ or $u^t(t)$ is known for $t \in [0,T]$ from measurements. The above Newton–Kantorovich scheme has a local quadratic convergence.

$A(q^{(n)})$ and $B(q^{(n)})$ can be computed directly from the direct solver. The remaining problem is to calculate the inverse of the Fréchet derivative, $\left[A'(q^{(n)})\right]^{-1}$ or $\left[B'(q^{(n)})\right]^{-1}$. Let $\tilde{u}^\pm = u^\pm(x,t;q+q_1)$ be the solution to the system of equations (5.114), (5.120), (5.121) and (5.122) with a non-linearity parameter $q + q_1$. Then $v^\pm \equiv \tilde{u}^\pm - u^\pm$ satisfy the following system of equations (to order $o(q_1)$)

$$\frac{\partial}{\partial x}\begin{bmatrix} v^+ \\ v^- \end{bmatrix} + \frac{1}{c}\frac{\partial}{\partial t}\begin{bmatrix} v^+ \\ -v^- \end{bmatrix} - \begin{bmatrix} \alpha & \beta \\ \gamma & \vartheta \end{bmatrix}\begin{bmatrix} v^+ \\ v^- \end{bmatrix} = [E'_q q_1 + E'_{u^+} v^+ + E'_{u^-} v^-]\begin{bmatrix} 1 \\ -1 \end{bmatrix}, \tag{5.127}$$

with the following initial and boundary conditions

$$v^{\pm}(x,0) = 0, \qquad x \in [0,l], \tag{5.128}$$

$$v^+(0,t) = v^-(l,t) = 0. \tag{5.129}$$

E'_q, E'_{u^+}, E'_{u^-} in (5.127) are the Fréchet derivatives of the functional E with respect to q, u^+ and u^-, respectively, i.e., $E(q+q_1, u^+, u^-, x, t) - E(q, u^+, u^-, x, t) = E'_q q_1 + o(q_1)$, etc. The system of equations (5.127)–(5.129) for $v^{\pm}$ can be solved by the method of characteristics, when $q_1(x)$, $q(x)$ and $u^{\pm}(x,t)$ are known (and thus E'_q, E'_{u^+}, and E'_{u^-} are known). From the definition of the Fréchet derivative of a functional, one has

$$A'(q)q_1 = v^-(0,t),$$

$$B'(q)q_1 = v^+(l,t),$$

i.e.,

$$q_1 = [A'(q)]^{-1} v^-(0,t), \tag{5.130}$$

or,

$$q_1 = [B'(q)]^{-1} v^+(l,t). \tag{5.131}$$

Therefore, $[A'(q^{(n)})]^{-1}[A(q^{(n)}) - u^r(t)]$ or $[B'(q^{(n)})]^{-1}[B(q^{(n)}) - u^t(t)]$ can be calculated by solving the linearized inverse problem of determining $q_1(x)$ from (5.127)–(5.129) ($u^{\pm}$ are understood to be the solution of (5.114), (5.120), (5.121) and (5.122) with $q = q^{(n)}$) with the boundary condition

$$v^-(0,t) = A(q^{(n)}) - u^r(t), \tag{5.132}$$

or

$$v^+(l,t) = B(q^{(n)}) - u^t(t). \tag{5.133}$$

The solution of the linearized inverse problem, $q_1(x)$, then gives the following value which is required in the Newton–Kantorovich procedure (cf. (5.125) and (5.126))

$$\left[A'(q^{(n)})\right]^{-1}[A(q^{(n)}) - u^r(t)] = q_1(x), \tag{5.134}$$

or,

$$\left[B'(q^{(n)})\right]^{-1} [B(q^{(n)}) - u^t(t)] = q_1(x). \tag{5.135}$$

The linearized inverse problem for $q_1(x)$, at the nth iterative step of the Newton–Kantorovich procedure, can be solved by minimizing the following objective functional

$$J(q_1) = \int_0^T \Big(w_1(t)\{v^-(0,t) - [A(q^{(n)}) - u^r(t)]\}^2 \\ + w_2(t)\{v^+(l,t) - [B(q^{(n)}) - u^t(t)]\}^2 \Big)\, dt, \tag{5.136}$$

where the weight function $w_2(t) = 0$ when one uses the reflection data, or $w_1(t) = 0$ when one uses the transmission data. The upper limit T of the integration describes the duration of the measurement of the scattering data. The explicit expression for the gradient of the objective functional $J(q_1)$ is found to be (He 1996c)

$$J'(x) = 2\int_0^T [(E'_q)^*(\psi^+ + \psi^-)]\, dt, \tag{5.137}$$

where the dual functions $\psi^{\pm}$ satisfy the following system of differential equations

$$\frac{\partial}{\partial x}\begin{bmatrix}\psi^+\\ \psi^-\end{bmatrix} + \frac{1}{c}\frac{\partial}{\partial t}\begin{bmatrix}\psi^+\\ -\psi^-\end{bmatrix} - \begin{bmatrix}-\alpha & \gamma\\ \beta & -\vartheta\end{bmatrix}\begin{bmatrix}\psi^+\\ \psi^-\end{bmatrix} = [(E'_{u^+})^*\psi^+ + (E'_{u^-})^*\psi^-]\begin{bmatrix}1\\ -1\end{bmatrix}, \tag{5.138}$$

together with the following conditions

$$\psi^{\pm}(x,T) = 0, \tag{5.139}$$

$$\psi^-(0,t) = w_1\{v^-(0,t) - [A(q^{(n)}) - u^r(t)]\}, \tag{5.140}$$

$$\psi^-(l,t) = w_2\{v^+(l,t) - [B(q^{(n)}) - u^t(t)]\}. \tag{5.141}$$

Note that the adjoint operators $(E'_q)^*$ and $(E'_{u^\pm})^*$ in (5.137) and (5.138) are defined as follows,

$$\int_0^l \{(E'_q)^*[g(x,t)]\}f(x)\, dx = \int_0^l \{E'_q[f(x)]\}g(x,t)\, dx,$$

$$\int_0^T \{(E'_{u^\pm})^*[g_2(x,t)]\}g_1(x,t)\, dt = \int_0^T \{E'_{u^\pm}[g_1(x,t)]\}g_2(x,t)\, dt.$$

The dual functions $\psi^{\pm}(x,t)$ can be computed for $x \in [0,l]$ and $t \in [0,T]$ from equations (5.138)–(5.141), using the method of characteristics with backward time-stepping from $t = T$ to $t = 0$. When $u^{\pm}(x,t)$ and $\psi^{\pm}(x,t)$ have been computed, one can use the explicit expression (5.137) to compute the gradient. Once the gradient of the objective functional has been computed, one can use a conventional conjugate gradient method to minimize the objective functional in an iterative way.

A numerical example. The reconstruction algorithm is illustrated with a simple numerical example when the quasi-linear term F has the following quadratic form,

$$F = q(x)[u(x,t)]^2. \tag{5.142}$$

Thus, one has

$$E = \frac{1}{2}q(x)\int_0^t [u^+(x,t') + u^-(x,t')]^2 dt'.$$

We choose $c = c_0 = 1$, $\rho = \rho_0$, $b = 0$, $l = 1$, and $q(x) = 5\sin^2(2\pi x/l) \cdot e^{-2x/l}$ in this example. The incident pulse is chosen as

$$u^i(t) = \begin{cases} \sin(\pi t/1.2), & \text{when } 0 \le t \le 1.2, \\ 0, & \text{otherwise.} \end{cases}$$

The corresponding reflected and transmitted fields are calculated by solving the system of equations (5.114), (5.120), (5.121) and (5.122) with the method of characteristics, and are given by the solid lines in Fig. 5.10. In order to check the accuracy of the direct solver, we first use the calculated reflection data $u^r(t)$ to generate the Neumann data at $x = 0$ through $(\partial_x u)(0,t) = \frac{1}{c}\frac{d}{dt}[u^r(t) - u^i(t)]$. Then we use a finite difference method to solve directly the quasi-linear second-order hyperbolic differential equation (5.109) with this Neumann data given as the boundary condition at the left endpoint (one can obtain the boundary condition at a right endpoint simply by putting the field to be zero at $x = x\max \ge T \cdot \max\{c\}$). The computed field $u(x,t)$ for $t \in [0,T]$ at $x = l$ is given by the dashed line in Fig. 5.10(b), which is consistent with the transmitted field calculated by the direct solver (the solid line in Fig. 5.10(b)). The Newton–Kantorovich procedure is then used to reconstruct the non-linearity parameter $q(x)$. For the quasi-linear term F given by (5.142), one has

$$E'_q[f(x)] = \frac{1}{2}f(x)\int_0^t [u^+(x,t') + u^-(x,t')]^2 dt',$$

$$E'_{u^\pm}[g(x,t)] = q(x)\int_0^t g(x,t')[u^+(x,t') + u^-(x,t')]dt'.$$

The corresponding adjoint operators can be expressed as

$$(E'_q)^*[g(x,t)] = \frac{1}{2}g(x,t)\int_0^t [u^+(x,t') + u^-(x,t')]^2 dt',$$

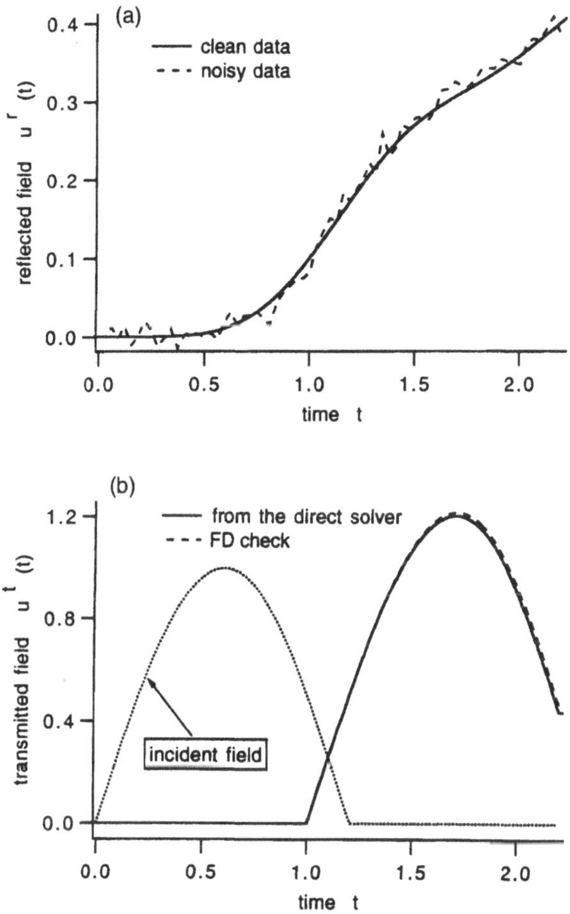

FIG. 5.10. The reflected and transmitted fields from a stratified slab characterized by a quadratic quasi-linear source term $F = q(x)[u(x,t)]^2$, where $q(x)$ is given by the solid line in Fig. 5.11.

$$(E'_{u^\pm})^*[g(x,t)] = q(x) \int_t^T g(x,t')[u^+(x,t') + u^-(x,t')]dt'.$$

At each iterative step of the Newton–Kantorovich procedure, $q_1(x)$ is determined by minimizing the objective functional $J(q_1)$ with a conventional conjugate gradient method (see subsection 8.1.1), with the expression for the gradient given explicitly by (5.137). In this numerical example, the time duration is chosen to be $T = 2.2$. The starting guess for q is $q(x) \equiv 0$. The circles in Fig. 5.11 are

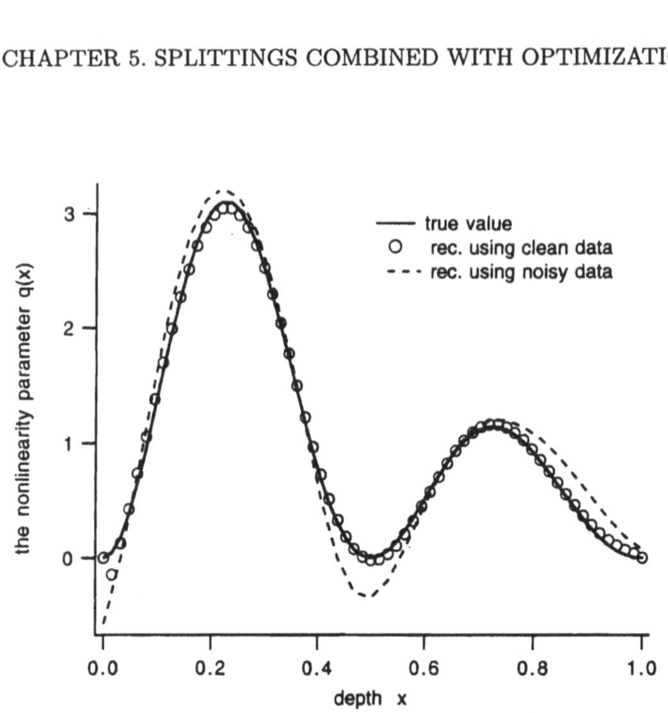

FIG. 5.11. Reconstruction of the spatial distribution of the non-linearity using the reflection data after 10 iterations (the starting guess is $q(x) \equiv 0$).

the reconstruction of the non-linearity parameter $q(x)$ after 10 iterations in the Newton–Kantorovich procedure using the reflected data, i.e., by choosing $w_1 = 1$ and $w_2 = 0$ (the solid line in Fig. 5.11 is the true profile for $q(x)$). Numerical simulation has also shown that the non-linearity parameter $q(x)$ cannot be reconstructed using only the transmitted data (i.e., by choosing $w_1 = 0$ and $w_2 = 1$). In order to test the stability of the algorithm, 3 percent Gaussian random noise is added to the reflected data, as given by the dashed line in Fig. 5.10(a). The corresponding reconstruction is given by the dashed line in Fig. 5.11, which indicates that the reconstruction is stable under noise in the reflection measurement.

5.3 Reconstruction of the susceptibility kernel of 1-buthanol from experimental data

In this section we describe a measurement system, and use the optimization method presented in Section 5.1.3 to reconstruct the susceptibility kernel χ of a polar liquid (1-buthanol) using measured data.

It is well known that polar liquids are strongly dispersive in the microwave regime. The scattering from a normally incident plane wave can be simulated in a coaxial structure. The dispersive polar liquid is introduced into a U-shaped sampleholder (a slotline) which is connected with a coaxial cable at both endpoints. The slotline sampleholder is designed to have matched characteristic impedance

FIG. 5.12. The experimental set-up for measuring the transmission data for a polar liquid. From left to right: (1) impulse generator; (2) slotline sampleholder (in front of the bottle); (3) sampling head HP 54124A; (4) oscilloscope mainframe HP 54120B; (5) desktop computer.

(50 Ohms) at the two ends to the coaxial cable when the sampleholder is empty. The U-shaped sampleholder, which consists of two parallel plates and an elliptic inner conductor, can be obtained as a conformal mapping of a 50 Ohms coaxial line (Wholey and Eldred 1950). The length of the sampleholder is $l = 9.173$ mm. A sampling head HP 54124A (connected with an oscilloscope mainframe HP 54120B) is used to measure the incident, reflected, and transmitted pulses. These pulses are continuously monitored, and the measured transient data are stored under the control of a desktop computer (see Fuks et al. (1994), He et al. (1996) for more detailed description). The experimental set-up is shown by the photograph in Fig. 5.12. The incident short pulse is generated by an impulse generator (Picosecond Pulse Labs Model 3050A), and is measured and plotted in Fig. 5.13(a) (the halfwidth is about 10^{-10} s). The reflected and transmitted waves are measured and plotted in Fig. 5.13(b). The permittivity ϵ of the polar liquid 1-buthanol can be obtained by measuring the time delay between the leading edge of the incident pulse through the air slotline vs. that of the transmitted pulse. The permittivity is determined to be $\epsilon = 3.6\epsilon_0$. Since the measurement of the transmitted field is easier and has higher accuracy compared to the measure-

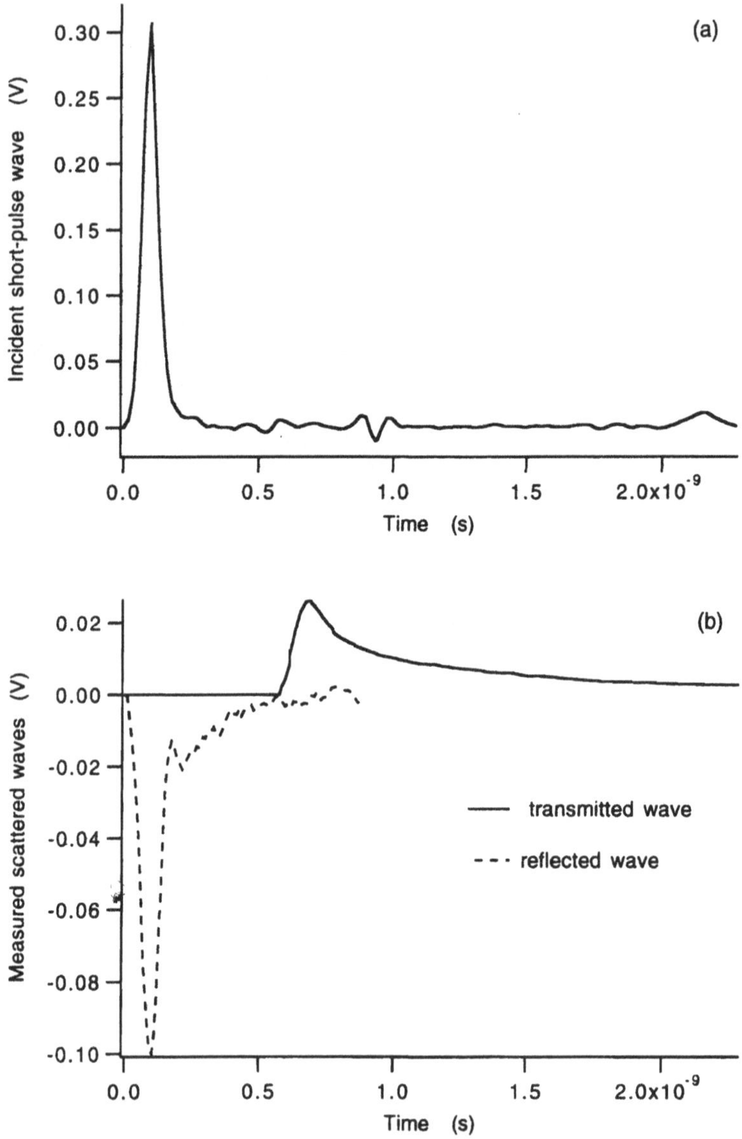

FIG. 5.13. The measured incident, reflected, and transmitted data for a polar liquid, 1-buthanol, in a sampleholder of length $l = 9.173$ mm.

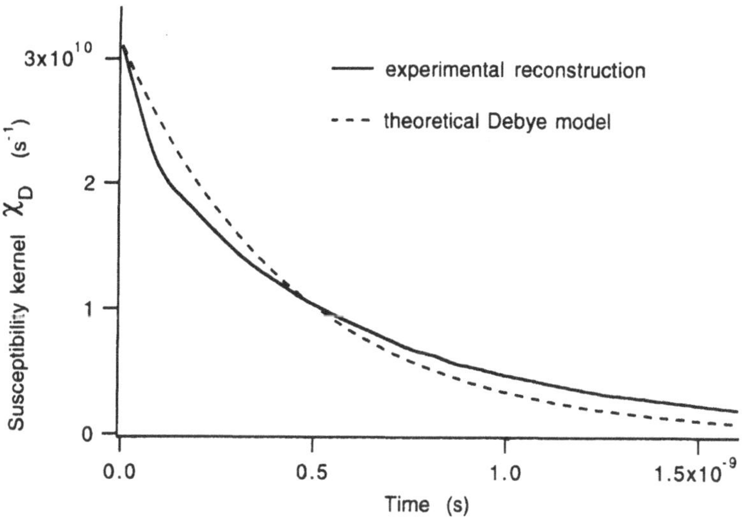

FIG. 5.14. The susceptibility kernel of the polar liquid 1-buthanol reconstructed from the measured transmission data.

ment of the reflected field, we use the measured transmitted data to reconstruct the susceptibility kernel $\chi(t)$. Such an inverse problem is a special case of the problem considered in Section 5.1.3, i.e.,

$$G(t) = \chi(t),$$

$$K(t) = L(t) = F(t) = 0,$$

in (5.57) and (5.58) ($\mu = \mu_0$). Thus, the reconstruction approach and algorithm presented there can be applied directly. The reconstruction is shown by the solid line in Fig. 5.14 (see He et al. (1996) for details). The dashed line in Fig. 5.14 is a susceptibility kernel $\chi_D(t)$ given by the following Debye model (see Section 3.3.1)

$$\chi(t) = ae^{-t/\tau}, \tag{5.143}$$

with $a = 3.1 \times 10^{10}$ s^{-1} and $\tau = 0.45 \times 10^{-9}$ s. From Fig. 5.14, one sees that the dispersion of the polar liquid 1-buthanol can be roughly described by such a Debye model (note that the starting value of $\chi(t)$ has an important contribution to the attenuation of the transmitted field, and thus one should approximate the reconstruction with a Debye model that has the same starting value).

Exercises for Chapter 5

Exercise 5.1. Derive the explicit expressions (5.30)–(5.33) for the gradients in detail.

Exercise 5.2. Derive the dynamic system (5.61) of the split fields for a stratified bi-isotropic medium.

6

TIME-HARMONIC WAVE-SPLITTING APPROACHES

In the previous chapters we have treated time domain wave-splitting approaches and that is the main theme of the present book. However, the time domain results have their frequency domain counterparts, which can be obtained by means of a Fourier transform. One may also develop wave-splitting approaches directly in the frequency domain. The frequency domain treatments are of interest in themselves and they also serve to illustrate additional aspects of the time domain treatments. In view of this, several frequency domain wave-splitting techniques are described in this chapter. Besides direct and inverse scattering problems, some design problems will also be considered in this chapter. A design problem may also be formulated as an inverse problem, in which prescribed scattering data or patterns are used instead of measured scattering data. Relevant design problems are, for example, design of a wide-angled high-pass radome in a given frequency band, a frequency selective anti-reflection coating, etc. There are a few aspects in which a design problem differs from an inverse problem. In an inverse problem, measured (physical) data are used as given data. Usually at least one solution exists if the data are free of noise and the physical model is accurate enough. However, non-uniqueness of the solution can often not be excluded, and in many cases the inverse problem is ill-posed (which means in particular that noise-contaminated data may give an erroneous 'reconstruction'). In a design problem, exact prescribed data are used, and thus the problems associated with noisy data do not appear. Non-uniqueness may still appear, but that means that several alternative designs are possible, and this is not a problem but rather an advantage. However, the existence of a solution may not be guaranteed in a design problem (in many cases one can only reach an optimal 'solution', instead of an exact solution). Another concern in a design problem is the feasibility of the solution obtained, from both a physical and a practical point of view (for instance, the specified data as well as the reconstructed parameters have to obey certain general conditions, which follow from, e.g. passivity and causality).

We concentrate on one-dimensional cases. The procedure of identifying the left- and right-moving modes is first described for a chiral medium. The invariant imbedding approach is applied to calculate the scattered fields, and the time-harmonic Green's function approach is used to calculate the internal fields. In addition, a transmission Green's function approach, which is the frequency domain counterpart of the time domain compact Green's function approach, is described. The direct problem of scattering and propagation in a stratified composite structure consisting of different types of complex (bi-anisotropic) media is solved by a vacuum-splitting approach. A characteristic feature of the vacuum-splitting approach is that it is based on a wave-splitting that is not related to the media which make up the composite structure, but is always defined with respect to the vacuum. The frequency-domain inverse problem for a non-uniform LCRG transmission line is considered using band-limited reflection and/or transmission data. A trace formalism is introduced to derive an explicit gradient in a compact form for the reconstruction/design of a stratified bi-anisotropic slab/coating. Finally, the time-harmonic vacuum-splitting approach is generalized to a special two-dimensional case, namely, scattering from a laterally periodic inhomogeneous structure.

6.1 Time-harmonic wave scattering and propagation in chiral media

In this section, time-harmonic wave-splitting approaches are described for the scattering and propagation of time-harmonic electromagnetic waves in chiral media. Intensive investigations have been carried out recently on wave propagation, radiation and guidance in chiral media in view of their potential usefulness in a variety of applications (cf. e.g. Cory (1995) and references given there). Electromagnetic chirality addresses the effects of handedness in electrodynamics, which involves magnetoelectric coupling in the constitutive relations. Chiral media have been manufactured in several laboratories by placing either left- or right-handed small metal wire helices randomly in an isotropic host material (Cory 1995).

Maxwell's equations for fields with harmonic time dependence $\exp(j\omega t)$ give

$$\nabla \times \vec{\mathbf{E}} = -j\omega\vec{\mathbf{B}}, \qquad \nabla \times \vec{\mathbf{H}} = j\omega\vec{\mathbf{D}}. \tag{6.1}$$

The constitutive relations for the chiral medium are taken to be (Lindell et al. 1994)

$$\vec{\mathbf{D}} = \epsilon\vec{\mathbf{E}} - j\kappa\sqrt{\epsilon_0\mu_0}\,\vec{\mathbf{H}}, \tag{6.2}$$

$$\vec{\mathbf{B}} = \mu\vec{\mathbf{H}} + j\kappa\sqrt{\epsilon_0\mu_0}\,\vec{\mathbf{E}}, \tag{6.3}$$

where ϵ is the permittivity, μ is the permeability, and the parameter κ (dimensionless) gives the degree of chirality. The constants ϵ_0, μ_0 are the permittivity and permeability in vacuum, respectively. From the constitutive relations (6.2) and (6.3), one sees that the medium is reciprocal (Kong 1986). For practical cases, κ is small and satisfies $\kappa^2 < (\epsilon\mu)/(\epsilon_0\mu_0)$. For a lossless medium, all three parameters are real numbers.

In this and the next sections, the following shorthand notation for some 2×2 constant matrices will be used,

$$I = \begin{bmatrix} 1 & 0 \\ 0 & 1 \end{bmatrix}, \qquad I_1 = \begin{bmatrix} 0 & 1 \\ 1 & 0 \end{bmatrix}, \qquad I_2 = \begin{bmatrix} 0 & 1 \\ -1 & 0 \end{bmatrix}. \tag{6.4}$$

We consider plane wave incidence and without loss of generality, the wave vector $\vec{\mathbf{k}}_0$ of the incident wave is assumed to be parallel to the xz plane and makes an angle θ_i with the z axis (and thus the fields are y independent). The x and t dependence of the electromagnetic fields is $e^{j(wt-k_x x)}$, where the propagation constant k_x is given by

$$k_x = k_0 \sin\theta_i, \tag{6.5}$$

and where k_0 is the wave number of the incident wave. Using the third components of Maxwell's equations and the constitutive relations (6.2) and (6.3), one can express the third components of $\vec{\mathbf{E}}$ and $\vec{\mathbf{H}}$ in terms of the field components in the xy plane (denoted by $\mathbf{E}$ and $\mathbf{H}$) as follows

$$\begin{bmatrix} E_3 \\ H_3 \end{bmatrix} = P \begin{bmatrix} \mathbf{E} \\ \mathbf{H} \end{bmatrix}, \tag{6.6}$$

where the 2×4 matrix P is given by

$$P = \frac{1}{\epsilon\mu - \kappa^2 \epsilon_0 \mu_0} \left(\frac{k_x}{\omega}\right) \begin{bmatrix} 0 & j\kappa\sqrt{\epsilon_0\mu_0} & 0 & -\mu \\ 0 & \epsilon & 0 & j\kappa\sqrt{\epsilon_0\mu_0} \end{bmatrix}. \tag{6.7}$$

Adopting the constitutive relations (6.2) and (6.3), the first two components of Maxwell's equations can be rewritten in the following form

$$\partial_z \begin{bmatrix} \mathbf{E} \\ \mathbf{H} \end{bmatrix} = j\omega \begin{bmatrix} 0 & -j\kappa\sqrt{\epsilon_0\mu_0} & 0 & -\mu \\ j\kappa\sqrt{\epsilon_0\mu_0} & 0 & \mu & 0 \\ 0 & \epsilon & 0 & -j\kappa\sqrt{\epsilon_0\mu_0} \\ -\epsilon & 0 & j\kappa\sqrt{\epsilon_0\mu_0} & 0 \end{bmatrix} \begin{bmatrix} \mathbf{E} \\ \mathbf{H} \end{bmatrix}$$

$$-jk_x \begin{bmatrix} 1 & 0 \\ 0 & 0 \\ 0 & 1 \\ 0 & 0 \end{bmatrix} \begin{bmatrix} E_3 \\ H_3 \end{bmatrix}. \tag{6.8}$$

Substituting (6.6) into (6.8) yields

$$\partial_z \begin{bmatrix} \mathbf{E} \\ \mathbf{H} \end{bmatrix} = D \begin{bmatrix} \mathbf{E} \\ \mathbf{H} \end{bmatrix}, \tag{6.9}$$

where

$$D = j\omega \begin{bmatrix} 0 & -j\kappa\sqrt{\epsilon_0\mu_0} & 0 & -\mu \\ j\kappa\sqrt{\epsilon_0\mu_0} & 0 & \mu & 0 \\ 0 & \epsilon & 0 & -j\kappa\sqrt{\epsilon_0\mu_0} \\ -\epsilon & 0 & j\kappa\sqrt{\epsilon_0\mu_0} & 0 \end{bmatrix}$$

$$-j\omega q \begin{bmatrix} 0 & j\kappa\sqrt{\epsilon_0\mu_0} & 0 & -\mu \\ 0 & 0 & 0 & 0 \\ 0 & \epsilon & 0 & j\kappa\sqrt{\epsilon_0\mu_0} \\ 0 & 0 & 0 & 0 \end{bmatrix}, \tag{6.10}$$

with

$$q = \frac{1}{\epsilon\mu - \kappa^2\epsilon_0\mu_0}\left(\frac{k_x^2}{\omega^2}\right). \tag{6.11}$$

Equation (6.9) expresses Maxwell's equations for a chiral medium in terms of the tangential electric and magnetic fields.

6.1.1 *Right- and left-moving modes in a homogeneous chiral medium*

In this subsection the right-moving modes $\mathbf{F}^+$ (propagating in the positive z direction) and left-moving modes $\mathbf{F}^-$ (propagating in the negative z direction) will be identified in a homogeneous chiral medium characterized by the parameters ϵ, μ and κ in the xyz coordinate system introduced above. For a plane wave mode (denoted ϕ) with harmonic time dependence $\exp(jwt)$, the right-moving condition is $\partial_z\phi = -jk_z\phi$, and the left-moving condition is $\partial_z\phi = jk_z\phi$ (k_z is the z component of the wave vector). Let

$$\begin{bmatrix} \mathbf{F}^+ \\ \mathbf{F}^- \end{bmatrix} = T \begin{bmatrix} \mathbf{E} \\ \mathbf{H} \end{bmatrix}, \tag{6.12}$$

where $\mathbf{F}^\pm = \begin{bmatrix} F_1^\pm \\ F_2^\pm \end{bmatrix}$, and the transformation matrix T will be determined later. For a homogeneous medium, the matrix T is independent of z. Therefore, from (6.8) and (6.12) one has

$$\partial_z \begin{bmatrix} \mathbf{F}^+ \\ \mathbf{F}^- \end{bmatrix} = TDT^{-1} \begin{bmatrix} \mathbf{F}^+ \\ \mathbf{F}^- \end{bmatrix},$$

where T^{-1} is the left inverse of T. In order to allow a physical interpretation of $\mathbf{F}^+$ and $\mathbf{F}^-$ as right-moving and left-moving modes, TDT^{-1} should be diagonal. From the usual process of diagonalizing a matrix, it follows that the diagonal elements are the eigenvalues of the matrix D and the matrix T^{-1} consists of the eigenvectors of the matrix D (the eigenvectors should be ordered appropriately so that $\mathbf{F}^+$ and $\mathbf{F}^-$ have the physical interpretation as right-moving and left-moving modes, respectively). Performing the indicated derivation, one can obtain the following diagonalized system (He et al. 1994b)

$$\partial_z \begin{bmatrix} \mathbf{F}^+ \\ \mathbf{F}^- \end{bmatrix} = j\omega \begin{bmatrix} -\lambda_1 & 0 & 0 & 0 \\ 0 & -\lambda_2 & 0 & 0 \\ 0 & 0 & \lambda_1 & 0 \\ 0 & 0 & 0 & \lambda_2 \end{bmatrix} \begin{bmatrix} \mathbf{F}^+ \\ \mathbf{F}^- \end{bmatrix}, \tag{6.13}$$

where

$$\lambda_1 = \left[(\sqrt{\epsilon\mu} + \kappa\sqrt{\epsilon_0\mu_0})^2 - k_x^2/\omega^2\right]^{1/2}, \tag{6.14}$$

$$\lambda_2 = \left[(\sqrt{\epsilon\mu} - \kappa\sqrt{\epsilon_0\mu_0})^2 - k_x^2/\omega^2\right]^{1/2}, \tag{6.15}$$

and where one takes the square roots that give λ_i, $i = 1, 2$, positive real parts. The matrix T is given by

$$T = \frac{1}{2} \begin{bmatrix} 1 & j/\nu_1 & -j\sqrt{\mu/\epsilon} & \sqrt{\mu/\epsilon}/\nu_1 \\ j/\nu_2 & 1 & -\sqrt{\mu/\epsilon}/\nu_2 & j\sqrt{\mu/\epsilon} \\ 1 & -j/\nu_1 & -j\sqrt{\mu/\epsilon} & -\sqrt{\mu/\epsilon}/\nu_1 \\ -j/\nu_2 & 1 & \sqrt{\mu/\epsilon}/\nu_2 & j\sqrt{\mu/\epsilon} \end{bmatrix}, \tag{6.16}$$

and its inverse is

$$T^{-1} = \frac{1}{2} \begin{bmatrix} 1 & -j\nu_2 & 1 & j\nu_2 \\ -j\nu_1 & 1 & j\nu_1 & 1 \\ j\sqrt{\epsilon/\mu} & -\sqrt{\epsilon/\mu}\nu_2 & j\sqrt{\epsilon/\mu} & \sqrt{\epsilon/\mu}\nu_2 \\ \sqrt{\epsilon/\mu}\nu_1 & -j\sqrt{\epsilon/\mu} & -\sqrt{\epsilon/\mu}\nu_1 & -j\sqrt{\epsilon/\mu} \end{bmatrix}, \tag{6.17}$$

where

$$\nu_1 = \frac{\sqrt{\epsilon\mu} + \kappa\sqrt{\epsilon_0\mu_0}}{\lambda_1}, \tag{6.18}$$

$$\nu_2 = \frac{(\sqrt{\epsilon\mu} - \kappa\sqrt{\epsilon_0\mu_0}) - q(\sqrt{\epsilon\mu} + \kappa\sqrt{\epsilon_0\mu_0})}{\lambda_2}. \tag{6.19}$$

The total electromagnetic fields are thus decomposed into the right- and left-moving modes as follows

$$\begin{bmatrix} \mathbf{E} \\ \mathbf{H} \end{bmatrix} = T^{-1} \begin{bmatrix} \mathbf{F}^+ \\ \mathbf{F}^- \end{bmatrix}. \tag{6.20}$$

Note that these eigenmodes are circularly polarized. According to (6.13) the eigenmodes consist of two pairs. The first pair (F_1^+, F_1^-) are the right-moving and left-moving modes with $1/\lambda_1$ as the phase velocity in the z direction. The second pair (F_2^+, F_2^-) are the right-moving and left-moving modes with $1/\lambda_2$ as the phase velocity in the z direction. The two pairs of modes have different phase velocities, which causes a change of polarization of the fields in the chiral medium. In this section, these eigenmodes are used to solve the propagation and scattering problem for chiral media.

6.1.2 *Reflection from a vacuum–chiral interface and Brewster angles*

In this subsection the propagating eigenmodes (derived in the previous subsection) will be used to calculate the reflection at the interface $z = 0$ between a vacuum and a homogeneous chiral half-space.

TE and TM modes in vacuum. Since the medium in the region $z < 0$ is a vacuum, we have $\lambda_1 = \lambda_2 = \sqrt{\varepsilon_0\mu_0}\cos^2\theta_i$, and any linear combination of eigenmodes will still be an eigenmode in this region. Assume that θ_i is the incident angle, and $\phi_i \in [0, 2\pi]$ is the angle between the x axis and the projection of the incident wave vector $\vec{\mathbf{k}}_0$ on the xy plane. To obtain the eigenmodes corresponding to TE and TM modes, we introduce a new coordinate system $x'y'z'$ in which the plane of incidence coincides with the $x'z'$ plane, and we have

$$\begin{bmatrix} x' \\ y' \end{bmatrix} = R_0 \begin{bmatrix} x \\ y \end{bmatrix}, \qquad z' = z, \tag{6.21}$$

where the rotation matrix R_0 is given by

$$R_0 = \begin{bmatrix} \cos\phi_i & \sin\phi_i \\ -\sin\phi_i & \cos\phi_i \end{bmatrix}. \tag{6.22}$$

In the coordinate system (x', y', z'), one easily sees that the tangential electric fields for the split TM and TE modes correspond, respectively, to the following $E_1^{\pm}$ and $E_2^{\pm}$ (cf. (3.61) which gives an identical expression for the split fields in the time domain)

$$\begin{bmatrix} \mathbf{E}^+ \\ \mathbf{E}^- \end{bmatrix} (0^-) = T_0' \begin{bmatrix} \mathbf{E}' \\ \mathbf{H}' \end{bmatrix} (0^-) \tag{6.23}$$

($\mathbf{E}'$, $\mathbf{H}'$ are the tangential electric and magnetic fields in the $x'y'z'$ coordinate system) where

$$T_0' = \frac{1}{2}\begin{bmatrix} I & Z_0 \\ I & -Z_0 \end{bmatrix}, \tag{6.24}$$

and where

$$Z_0 = \sqrt{\frac{\mu_0}{\epsilon_0}} \begin{bmatrix} 0 & \cos\theta_i \\ -\dfrac{1}{\cos\theta_i} & 0 \end{bmatrix}. \tag{6.25}$$

The amplitudes of the tangential fields in different coordinate systems are related to each other through

$$\begin{bmatrix} \mathbf{E}' \\ \mathbf{H}' \end{bmatrix}(0^-) = \begin{bmatrix} R_0 & 0 \\ 0 & R_0 \end{bmatrix} \begin{bmatrix} \mathbf{E} \\ \mathbf{H} \end{bmatrix}(0^-). \tag{6.26}$$

Thus one obtains the following vacuum wave-splitting at $z = 0^-$ (Norgren and He 1994)

$$\begin{bmatrix} \mathbf{E}^+ \\ \mathbf{E}^- \end{bmatrix}(0^-) = T_0 \begin{bmatrix} \mathbf{E} \\ \mathbf{H} \end{bmatrix}(0^-), \tag{6.27}$$

where

$$T_0 = T_0' \begin{bmatrix} R_0 & 0 \\ 0 & R_0 \end{bmatrix} = \frac{1}{2} \begin{bmatrix} R_0 & Z_0 R_0 \\ R_0 & -Z_0 R_0 \end{bmatrix}. \tag{6.28}$$

From the above derivation, one sees that in the vacuum region, the E_1^+, E_1^- modes are right- and left-moving TM modes, respectively, and the E_2^+, E_2^- modes are right- and left-moving TE modes, respectively ($E_i^\pm$, $i = 1, 2$, give the amplitudes of the *tangential* components of the electric field for these modes). In the present case, the plane of incidence is the xz plane, and thus $\phi_i = 0$, i.e., $R_0 = I$.

The reflection coefficient matrix is defined as

$$\mathbf{E}^-(0^-) = \mathbf{r}\mathbf{E}^+(0^-) \equiv \begin{bmatrix} r_{11} & r_{12} \\ r_{21} & r_{22} \end{bmatrix} \mathbf{E}^+(0^-). \tag{6.29}$$

The coefficients of $\mathbf{r}$ relate the amplitudes of the *tangential* components of the reflected and incident electric field to each other for TM and TE modes.

By the continuity of the tangential fields $\mathbf{E}$ and $\mathbf{H}$ across the interface $z = 0$, one has

$$\begin{aligned} \begin{bmatrix} \mathbf{F}^+ \\ \mathbf{F}^- \end{bmatrix}(0^+) &= T \begin{bmatrix} \mathbf{E} \\ \mathbf{H} \end{bmatrix}(0^+) = T \begin{bmatrix} \mathbf{E} \\ \mathbf{H} \end{bmatrix}(0^-) \\ &= TT_0^{-1} \begin{bmatrix} \mathbf{E}^+ \\ \mathbf{E}^- \end{bmatrix}(0^-) = TT_0^{-1} \begin{bmatrix} \mathbf{E}^+(0^-) \\ \mathbf{r}\mathbf{E}^+(0^-) \end{bmatrix} \\ &\equiv \begin{bmatrix} \mathbf{a} & \mathbf{b} \\ \mathbf{m} & \mathbf{n} \end{bmatrix} \begin{bmatrix} \mathbf{E}^+(0^-) \\ \mathbf{r}\mathbf{E}^+(0^-) \end{bmatrix}, \end{aligned} \tag{6.30}$$

where the 2×2 matrices $\mathbf{a}$, $\mathbf{b}$, $\mathbf{m}$ and $\mathbf{n}$ are obtained from the matrix evaluation of TT_0^{-1} (T_0^{-1} is the inverse of T_0).

Since $\mathbf{F}^-(0^+) = 0$ (which is due to the fact that there is no left-moving mode throughout the homogeneous chiral half-space $z > 0$), from (6.30) it follows that

$$\mathbf{n}\mathbf{r} + \mathbf{m} = 0, \tag{6.31}$$

i.e.,

$$\mathbf{r} = -\mathbf{n}^{-1}\mathbf{m} = -\frac{1}{n_{11}n_{22} - n_{12}n_{21}} \begin{bmatrix} n_{22} & -n_{12} \\ -n_{21} & n_{11} \end{bmatrix} \begin{bmatrix} m_{11} & m_{12} \\ m_{21} & m_{22} \end{bmatrix} \tag{6.32}$$

($m_{ij}, n_{ij}, i, j = 1, 2$, are the elements of the 2×2 matrices $\mathbf{m}$ and $\mathbf{n}$, respectively). Equation (6.32) gives the reflection coefficient matrix at a planar interface between the vacuum and a chiral medium for an obliquely incident plane wave.

Brewster angles. The Brewster angle is the incident angle at which there is no reflected power for a certain polarized mode (cf. e.g. Kong (1986)). In the vacuum region $z < 0$, the two pairs of eigenmodes have the same eigenvalues, i.e. $\lambda_1 = \lambda_2$. Therefore, any linear combination of the eigenmodes E_1^+ and E_2^+ (or E_1^- and E_2^-) is still a right-moving (or left-moving) eigenmode. Thus, one can choose propagating eigenmodes in the region $z < 0$ as

$$\begin{bmatrix} \tilde{\mathbf{E}}^+ \\ \tilde{\mathbf{E}}^- \end{bmatrix} = \begin{bmatrix} \Gamma & 0 \\ 0 & \Gamma \end{bmatrix} \begin{bmatrix} \mathbf{E}^+ \\ \mathbf{E}^- \end{bmatrix}, \quad z < 0 \tag{6.33}$$

(Γ is a 2×2 matrix) and define the corresponding reflection coefficient matrix as

$$\tilde{\mathbf{E}}^-(0^-) = \tilde{\mathbf{r}}\tilde{\mathbf{E}}^+(0^-) \equiv \begin{bmatrix} \tilde{r}_{11} & \tilde{r}_{12} \\ \tilde{r}_{21} & \tilde{r}_{22} \end{bmatrix} \tilde{\mathbf{E}}^+(0^-). \tag{6.34}$$

In particular, if Γ is such that $\Gamma\mathbf{r}\Gamma^{-1}$ is diagonal, then the eigenmodes $\tilde{\mathbf{E}}^\pm$ defined by (6.33) are the eigenmodes with *eigenpolarizations*. For the eigenpolarizations, one has (He 1993c)

$$\tilde{r}_{11} = \frac{1}{2}\left[(r_{11} + r_{22}) + \sqrt{(r_{11} + r_{22})^2 - 4(r_{11}r_{22} - r_{12}r_{21})}\right], \tag{6.35}$$

$$\tilde{r}_{22} = \frac{1}{2}\left[(r_{11} + r_{22}) - \sqrt{(r_{11} + r_{22})^2 - 4(r_{11}r_{22} - r_{12}r_{21})}\right], \tag{6.36}$$

$$\tilde{r}_{12} = \tilde{r}_{21} = 0, \tag{6.37}$$

and

$$\Gamma = \begin{bmatrix} 1 & r_{12}/(r_{11} - \tilde{r}_{22}) \\ r_{21}/(r_{22} - \tilde{r}_{11}) & 1 \end{bmatrix}. \tag{6.38}$$

If there is a certain incident angle $\theta_i = \theta_B$ (for a certain fixed φ_i) that satisfies $\tilde{r}_{11} = 0$ or $\tilde{r}_{22} = 0$, the angle θ_B is a Brewster angle for the eigenpolarization. For a chiral medium, there are two Brewster angles (see e.g. Lindell et al. (1994)).

6.1.3 *Reflection and transmission from a homogeneous chiral slab*

Scattering from a homogeneous chiral slab is considered in this subsection . The chiral slab occupies the region $0 < z < l$, and the regions $z < 0$ and $z > l$ are the vacuum. A plane wave is obliquely incident on the slab from the region $z < 0$.

The split fields at $z = 0^+$ and $z = 0^-$ are related to each other by (6.30). In an analogous way one can obtain the following equation at $z = l$

$$\begin{bmatrix} \mathbf{E}^+ \\ \mathbf{E}^- \end{bmatrix} (l^+) = T_0 T^{-1} \begin{bmatrix} \mathbf{F}^+ \\ \mathbf{F}^- \end{bmatrix} (l^-). \tag{6.39}$$

Inside the homogeneous chiral slab, one solves the system (6.13) of ordinary differential equations (ODEs) and obtains

$$\begin{bmatrix} \mathbf{F}^+ \\ \mathbf{F}^- \end{bmatrix} (l^-) = \begin{bmatrix} L^{-1} & 0 \\ 0 & L \end{bmatrix} \begin{bmatrix} \mathbf{F}^+ \\ \mathbf{F}^- \end{bmatrix} (0^+), \tag{6.40}$$

where

$$L = \begin{bmatrix} \exp(j\omega\lambda_1 l) & 0 \\ 0 & \exp(j\omega\lambda_2 l) \end{bmatrix}, \qquad L^{-1} = \begin{bmatrix} \exp(-j\omega\lambda_1 l) & 0 \\ 0 & \exp(-j\omega\lambda_2 l) \end{bmatrix}. \tag{6.41}$$

It thus follows from (6.30), (6.39) and (6.40) that

$$\begin{aligned} \begin{bmatrix} \mathbf{E}^+ \\ \mathbf{E}^- \end{bmatrix} (l^+) &= T_0 T^{-1} \begin{bmatrix} L^{-1} & 0 \\ 0 & L \end{bmatrix} T T_0^{-1} \begin{bmatrix} \mathbf{E}^+ \\ \mathbf{E}^- \end{bmatrix} (0^-) \\ &= \begin{bmatrix} \tilde{\mathbf{a}} & \tilde{\mathbf{b}} \\ \tilde{\mathbf{m}} & \tilde{\mathbf{n}} \end{bmatrix} \begin{bmatrix} \mathbf{E}^+ \\ \mathbf{E}^- \end{bmatrix} (0^-), \end{aligned} \tag{6.42}$$

where the 2×2 matrices $\tilde{\mathbf{a}}$, $\tilde{\mathbf{b}}$ $\tilde{\mathbf{m}}$, and $\tilde{\mathbf{n}}$ are obtained from the matrix evaluation of $T_0 T^{-1} \begin{bmatrix} L^{-1} & 0 \\ 0 & L \end{bmatrix} T T_0^{-1}$. Equation (6.42) relates the split fields at two sides of the slab to each other.

The reflection coefficient matrix for the homogeneous chiral slab is defined by (6.29), and the transmission coefficient matrix is defined as follows

$$\mathbf{E}^+(l^+) = \mathbf{t}\mathbf{E}^+(0^-) \equiv \begin{bmatrix} t_{11} & t_{12} \\ t_{21} & t_{22} \end{bmatrix} \mathbf{E}^+(0^-). \tag{6.43}$$

Since there is no left-moving mode in the region $z > l$, one has $\mathbf{E}^-(l^+) = 0$, which gives (cf. (6.29), (6.42) and (6.43))

$$\begin{bmatrix} \mathbf{t}\mathbf{E}^+(0^-) \\ 0 \end{bmatrix} = \begin{bmatrix} \tilde{\mathbf{a}} & \tilde{\mathbf{b}} \\ \tilde{\mathbf{m}} & \tilde{\mathbf{n}} \end{bmatrix} \begin{bmatrix} \mathbf{E}^+(0^-) \\ \mathbf{r}\mathbf{E}^+(0^-) \end{bmatrix}. \tag{6.44}$$

It thus follows from (6.44) that

$$\mathbf{r} = -\tilde{\mathbf{n}}^{-1}\tilde{\mathbf{m}}, \tag{6.45}$$

and

$$\mathbf{t} = \tilde{\mathbf{a}} + \tilde{\mathbf{b}}\mathbf{r}. \tag{6.46}$$

The above two expressions give the scattering coefficient matrices for a homogeneous chiral slab.

In a similar way, one can derive explicit expressions for the scattering coefficient matrices for a homogeneous bi-anisotropic slab (such explicit expressions can be used in the analysis of, e.g. a reflectionless coating of a homogeneous bi-anisotropic layer on a perfect conductor (Norgren and He 1997)).

In the next three subsections we describe three approaches for the scattering and propagation problem for a stratified chiral slab, namely, the invariant imbedding approach, the Green's function approach, and the transmission Green's function approach.

6.1.4 *Invariant imbedding approach for a stratified chiral slab*

For a plane-stratified chiral slab, the material parameters ϵ, μ and κ are functions of z. In the rest of this section, the parameters $\epsilon(z)$ and $\mu(z)$ are assumed to be piecewise continuous and piecewise continuously differentiable, i.e. they may have multiple discontinuities inside the slab and also be discontinuous at the surfaces $z = 0$ and $z = l$. Note that all the material parameters ϵ, μ and κ may be frequency dependent (however, the frequency dependence of the parameters is suppressed in their arguments for simplicity of notation). To make the description simpler, the case of normal incidence is considered in the rest of this section. The slab occupies the region $0 \leq z \leq l$ and the media on either side of this slab are homogeneous, but may be chiral. The plane wave impinges on the slab from the left side $z < 0$. For normal incidence, $E_3 = H_3 = 0$ and the matrix T becomes (cf. (6.16) and (6.17))

$$T = \frac{1}{2}\begin{bmatrix} 1 & j & -j\sqrt{\mu/\epsilon} & \sqrt{\mu/\epsilon} \\ j & 1 & -\sqrt{\mu/\epsilon} & j\sqrt{\mu/\epsilon} \\ 1 & -j & -j\sqrt{\mu/\epsilon} & -\sqrt{\mu/\epsilon} \\ -j & 1 & \sqrt{\mu/\epsilon} & j\sqrt{\mu/\epsilon} \end{bmatrix}, \tag{6.47}$$

$$T^{-1} = \frac{1}{2}\begin{bmatrix} 1 & -j & 1 & j \\ -j & 1 & j & 1 \\ j\sqrt{\epsilon/\mu} & -\sqrt{\epsilon/\mu} & j\sqrt{\epsilon/\mu} & \sqrt{\epsilon/\mu} \\ \sqrt{\epsilon/\mu} & -j\sqrt{\epsilon/\mu} & -\sqrt{\epsilon/\mu} & -j\sqrt{\epsilon/\mu} \end{bmatrix}.$$

Using (6.12) one can rewrite (6.9) in terms of the modes $\mathbf{F}^{\pm}$ (note that the matrix T is z dependent for a stratified slab)

$$\partial_z \begin{bmatrix} \mathbf{F}^+ \\ \mathbf{F}^- \end{bmatrix} = TDT^{-1} \begin{bmatrix} \mathbf{F}^+ \\ \mathbf{F}^- \end{bmatrix} + (\partial_z T)T^{-1} \begin{bmatrix} \mathbf{F}^+ \\ \mathbf{F}^- \end{bmatrix}$$
$$= \begin{bmatrix} \alpha\ \beta \\ \gamma\ \vartheta \end{bmatrix} \begin{bmatrix} \mathbf{F}^+ \\ \mathbf{F}^- \end{bmatrix}, \tag{6.48}$$

where

$$\begin{cases} \alpha = -j\omega \begin{bmatrix} \lambda_1 & 0 \\ 0 & \lambda_2 \end{bmatrix} + \frac{1}{4}(\frac{\mu_z}{\mu} - \frac{\epsilon_z}{\epsilon})I, \\ \beta = -\frac{j}{4}(\frac{\mu_z}{\mu} - \frac{\epsilon_z}{\epsilon})I_1, \\ \gamma = \frac{j}{4}(\frac{\mu_z}{\mu} - \frac{\epsilon_z}{\epsilon})I_1, \\ \vartheta = j\omega \begin{bmatrix} \lambda_1 & 0 \\ 0 & \lambda_2 \end{bmatrix} + \frac{1}{4}(\frac{\mu_z}{\mu} - \frac{\epsilon_z}{\epsilon})I, \end{cases} \tag{6.49}$$

and where $\lambda_1 = \sqrt{\epsilon\mu} + \kappa\sqrt{\epsilon_0\mu_0}$, $\lambda_2 = \sqrt{\epsilon\mu} - \kappa\sqrt{\epsilon_0\mu_0}$.

The system (6.48) of ODEs is equivalent to Maxwell's equations in terms of the propagating modes $\mathbf{F}^{\pm}$. One notes that $\mathbf{F}^{\pm}$ are the physical right- and left-moving modes in the homogeneous chiral region $z > 0$ or $z < l$ (i.e. outside the inhomogeneous slab).

One can use the invariant imbedding approach (cf. Chapter 1) to calculate the reflection coefficient matrix for the stratified chiral slab. In that approach one considers a subslab between z and l, $0 \leq z \leq l$, and assumes that all the parameters ϵ, μ and κ at the left side of the subslab are temporarily redefined to be constants, continuously terminated to the subslab at the left endpoint z. The associated reflection coefficient matrix $\mathbf{r}(z)$ for this imbedding geometry is defined by

$$\mathbf{F}^-(z) = \mathbf{r}(z)\mathbf{F}^+(z) \equiv \begin{bmatrix} r_{11}(z)\ r_{12}(z) \\ r_{21}(z)\ r_{22}(z) \end{bmatrix} \mathbf{F}^+(z). \tag{6.50}$$

From the above definition one notes that $\mathbf{r}(0^-)$ is the physical reflection coefficient matrix for the slab. The matrix $\mathbf{r}(z)$ does not depend on the incident fields, but only on the parameters of the inhomogeneous slab in the region $[z, l]$.

Substituting (6.50) into (6.48), one obtains the following system of ODEs for the $\mathbf{r}$ matrix

$$\partial_z \mathbf{r} = \gamma + (\vartheta\mathbf{r} - \mathbf{r}\alpha) - \mathbf{r}\beta\mathbf{r}, \tag{6.51}$$

which is a matrix differential equation of Riccati type. In order to solve the nonlinear system (6.51) of ODEs to get the value of the matrix $\mathbf{r}$ at $z = 0$, one needs to provide boundary conditions for $\mathbf{r}$ at $z = l$. Considering the modes at $z = l^{\pm}$

and using the fact that the tangential fields $\mathbf{E}$ and $\mathbf{H}$ are continuous across this surface, one obtains

$$\begin{aligned}\begin{bmatrix}\mathbf{F}^+(l^+)\\ \mathbf{F}^-(l^+)\end{bmatrix} &= T(l^+)\begin{bmatrix}\mathbf{E}(l^+)\\ \mathbf{H}(l^+)\end{bmatrix} = T(l^+)\begin{bmatrix}\mathbf{E}(l^-)\\ \mathbf{H}(l^-)\end{bmatrix}\\ &= T(l^+)T^{-1}(l^-)\begin{bmatrix}\mathbf{F}^+(l^-)\\ \mathbf{F}^-(l^-)\end{bmatrix}\\ &= T(l^+)T^{-1}(l^-)\begin{bmatrix}\mathbf{F}^+(l^-)\\ \mathbf{r}(l^-)\mathbf{F}^+(l^-)\end{bmatrix}.\end{aligned} \tag{6.52}$$

Since there is no left-moving mode but only right-moving modes at the right side of the slab $z > l$, one has

$$\mathbf{F}^-(l^+) = 0. \tag{6.53}$$

Thus, from (6.52) and (6.53) one obtains the following boundary condition for the $\mathbf{r}$ matrix

$$\mathbf{r}(l^-) = j\frac{\sqrt{\epsilon(l^+)/\mu(l^+)} - \sqrt{\epsilon(l^-)/\mu(l^-)}}{\sqrt{\epsilon(l^+)/\mu(l^+)} + \sqrt{\epsilon(l^-)/\mu(l^-)}}I_1\,. \tag{6.54}$$

If both the permittivity and permeability are continuous across the right surface $z = l$, the above boundary condition reduces to $\mathbf{r}(l^-) = 0$.

A discontinuity ('mismatch') of the permittivity or permeability at the position z (inside the slab) will cause a discontinuity in the $\mathbf{r}$ matrix across that z plane. From the continuity of the tangential electric fields $\mathbf{E}(z)$ and magnetic field $\mathbf{H}(z)$, one has

$$\begin{bmatrix}\mathbf{F}^+(z^-)\\ \mathbf{r}(z^-)\mathbf{F}^+(z^-)\end{bmatrix} = T(z^-)T^{-1}(z^+)\begin{bmatrix}\mathbf{F}^+(z^+)\\ \mathbf{r}(z^+)\mathbf{F}^+(z^+)\end{bmatrix} \quad \left(=\begin{bmatrix}\mathbf{F}^+(z^-)\\ \mathbf{F}^-(z^-)\end{bmatrix}\right), \tag{6.55}$$

i.e. (cf. T, T^{-1} are given by (6.47))

$$\begin{aligned}&\mathbf{F}^+(z^-)\\ &= \left[\frac{1}{2}\left(1+\sqrt{\frac{\epsilon(z^+)/\mu(z^+)}{\epsilon(z^-)/\mu(z^-)}}\right)I + \frac{i}{2}\left(1-\sqrt{\frac{\epsilon(z^+)/\mu(z^+)}{\epsilon(z^-)/\mu(z^-)}}\right)I_1\mathbf{r}(z^+)\right]\mathbf{F}^+(z^+),\\ &\mathbf{r}(z^-)\mathbf{F}^+(z^-)\\ &= \left[-\frac{i}{2}\left(1-\sqrt{\frac{\epsilon(z^+)/\mu(z^+)}{\epsilon(z^-)/\mu(z^-)}}\right)I_1 + \frac{1}{2}\left(1+\sqrt{\frac{\epsilon(z^+)/\mu(z^+)}{\epsilon(z^-)/\mu(z^-)}}\right)\mathbf{r}(z^+)\right]\mathbf{F}^+(z^+).\end{aligned}$$

From the above two equations, one obtains the following formula

$$\mathbf{r}(z^-) = \left[-i\left(1-\sqrt{\frac{\epsilon(z^+)/\mu(z^+)}{\epsilon(z^-)/\mu(z^-)}}\right)I_1 + \left(1+\sqrt{\frac{\epsilon(z^+)/\mu(z^+)}{\epsilon(z^-)/\mu(z^-)}}\right)\mathbf{r}(z^+)\right]$$
$$\times\left[\left(1+\sqrt{\frac{\epsilon(z^+)/\mu(z^+)}{\epsilon(z^-)/\mu(z^-)}}\right)I + i\left(1-\sqrt{\frac{\epsilon(z^+)/\mu(z^+)}{\epsilon(z^-)/\mu(z^-)}}\right)I_1\mathbf{r}(z^+)\right]^{-1}$$
$$\equiv \mathcal{M}(z^-, z^+)\mathbf{r}(z^+), \tag{6.56}$$

where the mismatch operator $\mathcal{M}(z^-, z^+)$ maps the $\mathbf{r}$ value at the right side of the mismatch plane z to the $\mathbf{r}$ value at the left side of the mismatch plane (if both the permittivity and permeability are continuous across the surface z, the mismatch operator $\mathcal{M}(z^-, z^+)$ reduces to an identity operator (matrix)). Therefore, one obtains a solution for $\mathbf{r}(0^+)$ by integrating (6.51) along the $-z$ direction starting from $z = l^-$ (cf. (6.54)). In particular, one needs to use the mismatch operator $\mathcal{M}$ at any plane where there is a mismatch of permittivity and/or permeability.

The physical reflection coefficient matrix $\mathbf{r}(0^-)$ is in general not equal to $\mathbf{r}(0^+)$ when the permittivity or permeability has a discontinuity across the surface $z = 0$. According to (6.56), $\mathbf{r}(0^-)$ is given by

$$\mathbf{r}(0^-) = \mathcal{M}(0^-, 0^+)\mathbf{r}(0^+). \tag{6.57}$$

When the reflection coefficient matrix $\mathbf{r}(0^-)$ is known, the reflected modes in the region $z < 0$ are obtained from

$$\mathbf{F}^-(z) = \begin{bmatrix} \{r_{11}(0^-)F_1^+(0^-) + r_{12}(0^-)F_2^+(0^-)\}\exp(i\omega\lambda_1^{(1)}z) \\ \\ \{r_{21}(0^-)F_1^+(0^-) + r_{22}(0^-)F_2^+(0^-)\}\exp(i\omega\lambda_2^{(1)}z) \end{bmatrix}, \qquad z < 0, \tag{6.58}$$

where $\lambda_1^{(1)} = \sqrt{\epsilon_1\mu_1} + \kappa_1\sqrt{\epsilon_0\mu_0}$, $\lambda_2^{(1)} = \sqrt{\epsilon_1\mu_1} - \kappa_1\sqrt{\epsilon_0\mu_0}$ (ϵ_1, μ_1, κ_1 are the permittivity, permeability and chirality in the homogeneous region $z < 0$, respectively). The total fields at the left side of the slab consist of two parts, namely the incident fields (denoted $\mathbf{E}^i$, $\mathbf{H}^i$) and the reflected fields (denoted $\mathbf{E}^r$, $\mathbf{H}^r$),

$$\begin{bmatrix}\mathbf{E}\\ \mathbf{H}\end{bmatrix}(z) = T^{-1}\begin{bmatrix}\mathbf{F}^+\\ \mathbf{F}^-\end{bmatrix}(z)$$
$$= \begin{bmatrix}\mathbf{E}^i\\ \mathbf{H}^i\end{bmatrix}(z) + \begin{bmatrix}\mathbf{E}^r\\ \mathbf{H}^r\end{bmatrix}(z), \qquad z < 0, \tag{6.59}$$

where

$$\begin{bmatrix}\mathbf{E}^i\\ \mathbf{H}^i\end{bmatrix}(z) = T^{-1}\begin{bmatrix}\mathbf{F}^+\\ 0\end{bmatrix}(z), \qquad \begin{bmatrix}\mathbf{E}^r\\ \mathbf{H}^r\end{bmatrix}(z) = T^{-1}\begin{bmatrix}0\\ \mathbf{F}^-\end{bmatrix}(z), \qquad z < 0.$$

The imbedding approach can also be used to derive equations and boundary conditions for the transmission coefficient matrix. This aspect is taken up in connection with the discussion of the vacuum-splitting in Section 6.2 (cf. Eq. (6.93)).

6.1.5 *The Green's function approach and the internal fields*

To calculate the internal fields, two matrices $\mathbf{G}^+(z)$ and $\mathbf{G}^-(z)$ of Green's functions are introduced to express the internal $\mathbf{F}^\pm$ modes in terms of the right-moving incident modes $\mathbf{F}^+(0^-)$ (He 1992b)

$$\begin{bmatrix} \mathbf{F}^+ \\ \mathbf{F}^- \end{bmatrix}(z) = \begin{bmatrix} \mathbf{G}^+(z)\mathbf{F}^+(0^-) \\ \mathbf{G}^-(z)\mathbf{F}^+(0^-) \end{bmatrix}, \qquad 0 < z < l, \tag{6.60}$$

where

$$\mathbf{G}^\pm(z) = \begin{bmatrix} G^\pm_{11}(z) & G^\pm_{12}(z) \\ G^\pm_{21}(z) & G^\pm_{22}(z) \end{bmatrix}.$$

Note that the Green's functions $G^\pm_{ij}(z)$, $i, j = 1, 2$, are defined only for $0 < z < l$ (i.e. inside the slab) and thus $G^\pm_{ij}(0)$ stand for the limit $z \to 0^+$ of $G^\pm_{ij}(z)$, etc. The two matrices $\mathbf{G}^+(z)$ and $\mathbf{G}^-(z)$ of Green's functions do not depend on the incident fields but only on the parameters of the inhomogeneous slab. The physical interpretation of these quantities is as follows: the Green's function $G^+_{ij}(z)$ (or $G^-_{ij}(z)$), $i, j = 1, 2$, is the amplitude of the ith type of $\mathbf{F}^+$ (or $\mathbf{F}^-$) mode at the position z due to a unit amplitude jth type of right-moving incident mode at the surface $z = 0^-$ (with the other type of right-moving incident mode put equal to zero). The ODEs for $\mathbf{G}^\pm(z)$ are obtained as follows. Differentiating (6.60) with respect to z, yields

$$\partial_z \begin{bmatrix} \mathbf{F}^+ \\ \mathbf{F}^- \end{bmatrix}(z) = \begin{bmatrix} \partial_z\mathbf{G}^+(z)\mathbf{F}^+(0^-) \\ \partial_z\mathbf{G}^-(z)\mathbf{F}^+(0^-) \end{bmatrix}.$$

From (6.48) and (6.60) it then follows that

$$\begin{bmatrix} \partial_z\mathbf{G}^+(z)\mathbf{F}^+(0^-) \\ \partial_z\mathbf{G}^-(z)\mathbf{F}^+(0^-) \end{bmatrix} = \begin{bmatrix} \alpha & \beta \\ \gamma & \vartheta \end{bmatrix} \begin{bmatrix} \mathbf{G}^+(z)\mathbf{F}^+(0^-) \\ \mathbf{G}^-(z)\mathbf{F}^+(0^-) \end{bmatrix}.$$

Since the incident modes $\mathbf{F}^+(0^-)$ are arbitrary, one obtains the following ODEs for the Green's functions

$$\partial_z\mathbf{G}^+ = \alpha\mathbf{G}^+ + \beta\mathbf{G}^-, \tag{6.61}$$

$$\partial_z\mathbf{G}^- = \gamma\mathbf{G}^+ + \vartheta\mathbf{G}^-. \tag{6.62}$$

In order to solve the two linear coupled ODE systems (6.61) and (6.62), one needs to provide boundary conditions for $\mathbf{G}^\pm$. After a derivation similar to the one which leads to (6.52), one has

$$\begin{bmatrix} \mathbf{G}^+(0)\mathbf{F}^+(0^-) \\ \mathbf{G}^-(0)\mathbf{F}^+(0^-) \end{bmatrix} = \begin{bmatrix} \mathbf{F}^+(0^+) \\ \mathbf{F}^-(0^+) \end{bmatrix} = T(0^+)T^{-1}(0^-) \begin{bmatrix} \mathbf{F}^+(0^-) \\ \mathbf{r}(0^-)\mathbf{F}^+(0^-) \end{bmatrix}. \tag{6.63}$$

From the above equation with T, T^{-1} given by (6.47), one obtains the following boundary conditions for $\mathbf{G}^{\pm}$ at $z = 0$,

$$\mathbf{G}^+(0) = \frac{1}{2}\left[1 + \sqrt{\frac{\epsilon(0^-)/\mu(0^-)}{\epsilon(0^+)/\mu(0^+)}}\right] I + \frac{j}{2}\left[1 - \sqrt{\frac{\epsilon(0^-)/\mu(0^-)}{\epsilon(0^+)/\mu(0^+)}}\right] I_1\mathbf{r}(0^-), \tag{6.64}$$

$$\mathbf{G}^-(0) = -\frac{j}{2}\left[1 - \sqrt{\frac{\epsilon(0^-)/\mu(0^-)}{\epsilon(0^+)/\mu(0^+)}}\right] I_1 + \frac{1}{2}\left[1 + \sqrt{\frac{\epsilon(0^-)/\mu(0^-)}{\epsilon(0^+)/\mu(0^+)}}\right] \mathbf{r}(0^-). \tag{6.65}$$

If both the permittivity and permeability are continuous across the surface $z = 0$, the above two boundary conditions reduce to $\mathbf{G}^+(0) = I$, $\mathbf{G}^-(0) = \mathbf{r}(0^-)$.

A mismatch of the permittivity or permeability at the position z (inside the slab) will cause a discontinuity in the $\mathbf{G}^{\pm}$ matrices across this z plane. From the continuity of the tangential electric fields $\mathbf{E}(z)$ and magnetic fields $\mathbf{H}(z)$, one has

$$\begin{bmatrix} \mathbf{G}^+(z^+)\mathbf{F}^+(0^-) \\ \mathbf{G}^-(z^+)\mathbf{F}^+(0^-) \end{bmatrix} = T_1(z^+)T_1^{-1}(z^-) \begin{bmatrix} \mathbf{G}^+(z^-)\mathbf{F}^+(0^-) \\ \mathbf{G}^-(z^-)\mathbf{F}^+(0^-) \end{bmatrix} \left(= \begin{bmatrix} \mathbf{F}^+(z^+) \\ \mathbf{F}^-(z^+) \end{bmatrix}\right), \tag{6.66}$$

i.e.

$$\begin{aligned} &\mathbf{G}^+(z^+) \\ &= \frac{1}{2}\left[1 + \sqrt{\frac{\epsilon(z^-)/\mu(z^-)}{\epsilon(z^+)/\mu(z^+)}}\right] \mathbf{G}^+(z^-) + \frac{j}{2}\left[1 - \sqrt{\frac{\epsilon(z^-)/\mu(z^-)}{\epsilon(z^+)/\mu(z^+)}}\right] I_1\mathbf{G}^-(z^-) \\ &\equiv \mathcal{M}^+(z^+, z^-) \begin{bmatrix} \mathbf{G}^+ \\ \mathbf{G}^- \end{bmatrix}(z^-), \end{aligned} \tag{6.67}$$

$$\begin{aligned} &\mathbf{G}^-(z^+) \\ &= -\frac{j}{2}\left[1 - \sqrt{\frac{\epsilon(z^-)/\mu(z^-)}{\epsilon(z^+)/\mu(z^+)}}\right] I_1\mathbf{G}^+(z^-) + \frac{1}{2}\left[1 + \sqrt{\frac{\epsilon(z^-)/\mu(z^-)}{\epsilon(z^+)/\mu(z^+)}}\right] \mathbf{G}^-(z^-) \\ &\equiv \mathcal{M}^-(z^+, z^-) \begin{bmatrix} \mathbf{G}^+ \\ \mathbf{G}^- \end{bmatrix}(z^-), \end{aligned} \tag{6.68}$$

where the mismatch operators $\mathcal{M}^{\pm}(z^+, z^-)$ map the $\mathbf{G}^{\pm}$ values at the left side of the mismatch plane z to the $\mathbf{G}^{\pm}$ values at the right side of this plane.

Therefore, if the physical reflection coefficient matrix $\mathbf{r}(0^-)$ is known, the ODE systems (6.61) and (6.62) for the Green's functions can be solved. The solution for the Green's functions $G_{ij}^{\pm}(z)$, $i, j = 1, 2$, is obtained by integrating (6.61) and (6.62) along the z direction starting from $z = 0$ (cf. (6.64), (6.65)). In particular, one needs to use the mismatch operators $\mathcal{M}^{\pm}$ at any plane where there is a mismatch of permittivity and/or permeability. In a frequency-domain approach, one expects the values of the Green's matrices $\mathbf{G}^{\pm}(z)$ at the position z to depend on the parameter profile of the whole slab. This is indeed the case in the present approach since the physical reflection coefficient matrix $\mathbf{r}(0^-)$ depends on the parameter profile of the whole slab, and $\mathbf{G}^{\pm}(z)$ depend on the values of $\mathbf{r}(0^-)$, as seen from the above-stated algorithm (cf. (6.64), (6.65)). Once the Green's function matrices are determined, the internal $\mathbf{F}^{\pm}$ modes are known through (6.60). The internal fields are then obtained by

$$\begin{bmatrix} \mathbf{E} \\ \mathbf{H} \end{bmatrix}(z) = T^{-1} \begin{bmatrix} \mathbf{F}^+ \\ \mathbf{F}^- \end{bmatrix}(z). \tag{6.69}$$

Since the incident modes $\mathbf{F}^+(0^-)$ are arbitrary, from (6.60) and (6.50) one has the following relation between the Green's function matrices and the $\mathbf{r}$ matrix inside the slab

$$\mathbf{G}^-(z) = \mathbf{r}(z)\mathbf{G}^+(z), \qquad 0 < z < l. \tag{6.70}$$

The transmitted fields. To the right side of the slab, the right-moving (transmitted) modes $\mathbf{F}^+(l^+)$ are linearly related to the right-moving incident modes $\mathbf{F}^+(0^-)$ through a matrix $\mathbf{t}$ of transmission coefficients, i.e.

$$\mathbf{F}^+(l^+) = \mathbf{t}\mathbf{F}^+(0) \equiv \begin{bmatrix} t_{11} & t_{12} \\ t_{21} & t_{22} \end{bmatrix} \mathbf{F}^+(0^-). \tag{6.71}$$

The transmission coefficients t_{ij}, $i, j = 1, 2$, do not depend on the incident fields but only on the parameters of the inhomogeneous slab. They are related to the Green's functions at $z = l$, as shown below. Using the continuity of the tangential fields $\mathbf{E}$ and $\mathbf{H}$ across the the right surface $z = l$, one has

$$\begin{aligned} \begin{bmatrix} \mathbf{F}^+(l^+) \\ \mathbf{F}^-(l^+) \end{bmatrix} &= T(l^+)T^{-1}(l^-) \begin{bmatrix} \mathbf{F}^+(l^-) \\ \mathbf{F}^-(l^-) \end{bmatrix} \\ &= T(l^+)T^{-1}(l^-) \begin{bmatrix} \mathbf{G}^+(l)\mathbf{F}^+(0^-) \\ \mathbf{G}^-(l)\mathbf{F}^+(0^-) \end{bmatrix} \end{aligned} \tag{6.72}$$

(note that $\mathbf{F}^-(l^+) = 0$ in the above equation). Substituting (6.71) into (6.72), one obtains the following formula for the transmission coefficient matrix

$$\mathbf{t} = \frac{1}{2}\left[1 + \sqrt{\frac{\epsilon(l^-)/\mu(l^-)}{\epsilon(l^+)/\mu(l^+)}}\right]\mathbf{G}^+(l) + \frac{j}{2}\left[1 - \sqrt{\frac{\epsilon(l^-)/\mu(l^-)}{\epsilon(l^+)/\mu(l^+)}}\right] I_1\mathbf{G}^-(l). \tag{6.73}$$

If both the permittivity and permeability are continuous across the surface $z = 0$, the above equation reduces to the simple result $\mathbf{t} = \mathbf{G}^+(l)$. When the transmission coefficient matrix $\mathbf{t}$ is known, the transmitted right-moving modes in the region $z > l$ are obtained from

$$\begin{aligned}\mathbf{F}^+(z) &\equiv \begin{bmatrix} F_1^+(z) \\ F_2^+(z) \end{bmatrix} \\ &= \begin{bmatrix} \{t_{11}F_1^+(0^-) + t_{12}F_2^+(0^-)\}\exp[-j\omega\lambda_1^{(2)}(z-l)] \\ \\ \{t_{21}F_1^+(0^-) + t_{22}F_2^+(0^-)\}\exp[-j\omega\lambda_2^{(2)}(z-l)] \end{bmatrix}, \quad z > l,\end{aligned} \tag{6.74}$$

where $\lambda_1^{(2)} = \sqrt{\epsilon_2\mu_2} + \kappa_2\sqrt{\epsilon_0\mu_0}$, $\lambda_2^{(2)} = \sqrt{\epsilon_2\mu_2} - \kappa_2\sqrt{\epsilon_0\mu_0}$ (ϵ_2, μ_2, κ_2 are the permittivity, permeability and chirality in the homogeneous region $z > l$, respectively). The transmitted fields (denoted $\mathbf{E}^t$, $\mathbf{H}^t$) are given by

$$\begin{bmatrix} \mathbf{E}^t \\ \mathbf{H}^t \end{bmatrix}(z) = T^{-1}\begin{bmatrix} \mathbf{F}^+(z) \\ 0 \end{bmatrix}, \qquad z > l. \tag{6.75}$$

If one is not interested in the internal fields, but only the scattered fields, one can also calculate the transmission coefficient matrix by an invariant imbedding approach (cf. (6.93) in the next section).

6.1.6 *The transmission Green's function approach*

In the usual Green's function approach one expresses the internal split fields in terms of the incident field. In such an approach one needs to know in advance the physical reflection coefficient matrix $\mathbf{r}(0)$ (cf. (6.64) and (6.65)), which is obtained by solving the non-linear imbedding equation (6.51) for $\mathbf{r}(z)$. In order to obtain a linear approach which does not require that one first solves a non-linear equation (this becomes a very important aspect in scattering from, e.g. gratings, cf. Section 6.5 below), a modified Green's function approach in which one expresses the internal split fields in terms of the transmitted field can be introduced (such an approach is called the *compact Green's function approach* in the time domain; cf. Section 2.3). Thus, one defines

$$\begin{bmatrix} \mathbf{F}^+ \\ \mathbf{F}^- \end{bmatrix}(z) = \begin{bmatrix} \mathbf{g}^+(z)\mathbf{F}^+(l^-) \\ \mathbf{g}^-(z)\mathbf{F}^+(l^-) \end{bmatrix}, \qquad 0 < z < l, \tag{6.76}$$

where the transmission Green's function matrices $\mathbf{g}^{\pm}(z)$ are 2×2 matrices. Like the usual Green's functions, $\mathbf{g}^{\pm}(z)$ are defined only for $0 < z < l$. If all the parameters are continuous at $z = l$, $\mathbf{F}^+(l^-)$ equals the transmitted component $\mathbf{F}^+(l^+)$ (otherwise, they are related through (6.52)). Obviously, $\mathbf{g}^{\pm}(z)$ satisfy the following system of ODEs

$$\frac{d}{dz}\begin{bmatrix} \mathbf{g}^+ \\ \mathbf{g}^- \end{bmatrix} = \begin{bmatrix} \alpha & \beta \\ \gamma & \vartheta \end{bmatrix}\begin{bmatrix} \mathbf{g}^+ \\ \mathbf{g}^- \end{bmatrix}, \tag{6.77}$$

with the following two boundary conditions at $z = l^-$

$$\mathbf{g}^+(l) = I, \tag{6.78}$$

$$\mathbf{g}^-(l) = \mathbf{r}(l^-), \tag{6.79}$$

and where $\mathbf{r}(l^-)$ is given by the expression (6.54). Therefore, one obtains the solution for $\mathbf{g}(z)$ by integrating the linear ODE system (6.77) along the $-z$ direction starting from $z = l^-$ (cf. (6.78) and (6.79)). In particular, one needs to use the mismatch operators at any z plane where there is a mismatch of permittivity or permeability, viz.

$$\mathbf{g}^+(z^-) = \mathcal{M}^+(z^-, z^+)\begin{bmatrix} \mathbf{g}^+ \\ \mathbf{g}^- \end{bmatrix}(z^+), \qquad \mathbf{g}^-(z^-) = \mathcal{M}^+(z^-, z^+)\begin{bmatrix} \mathbf{g}^+ \\ \mathbf{g}^- \end{bmatrix}(z^+), \tag{6.80}$$

where the mismatch operators $\mathcal{M}^+(z^-, z^+)$ are defined by (6.67) and (6.68) (note that z^- and z^+ should be interchanged in the expressions (6.67) and (6.68) for $\mathcal{M}^{\pm}$ due to the interchange of the two arguments). Note that the matrices $\mathbf{g}^{\pm}(z)$ do not depend on the incident field but only on the parameters of the inhomogeneous slab in the region $[z, l]$. Once $\mathbf{g}^{\pm}(z)$ for $z \in (0, l)$ have been computed, the internal fields are determined as well as the scattered fields (through the continuity of the tangential electric and magnetic fields). For example, when the parameters are continuous at the two end points $z = 0$ and $z = l$, one obtains the physical scattering coefficient matrices through

$$\mathbf{t} = [\mathbf{g}^+(0)]^{-1}, \qquad \mathbf{r} = \mathbf{g}^-(0)[\mathbf{g}^+(0)]^{-1}. \tag{6.81}$$

One can easily find various relations among $\mathbf{g}^{\pm}(z)$, $\mathbf{G}^{\pm}(z)$, and the imbedding scattering coefficient matrices (Norgren 1997).

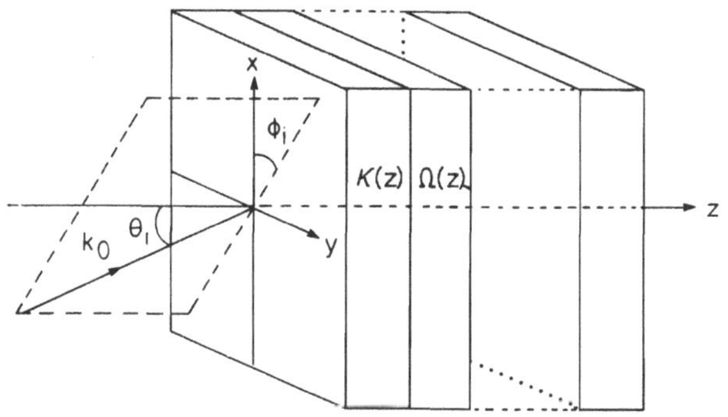

FIG. 6.1. The scattering configuration for a composite structure.

6.2 Vacuum-splitting and scattering from stratified composite structures

6.2.1 *Vacuum-splitting in bi-anisotropic media*

In this section a general scheme is given for the computation of the reflected, transmitted and internal fields due to an electromagnetic plane wave obliquely incident on a composite structure consisting of multiple layers of different types of complex media (the parameters in each layer are functions of z). The complex media that are considered here can be modelled by the following general constitutive relations

$$\vec{\mathbf{D}}(z) = \overline{\overline{\epsilon}}\,(z)\vec{\mathbf{E}}(z) + \overline{\overline{\xi}}\,(z)\vec{\mathbf{H}}(z), \tag{6.82}$$

$$\vec{\mathbf{B}}(z) = \overline{\overline{\mu}}\,(z)\vec{\mathbf{H}}(z) + \overline{\overline{\zeta}}\,(z)\vec{\mathbf{E}}(z), \tag{6.83}$$

where $\overline{\overline{\epsilon}}$ (the permittivity tensor), $\overline{\overline{\mu}}$ (the permeability tensor), $\overline{\overline{\xi}}$ and $\overline{\overline{\zeta}}$ are three-dimensional Cartesian tensors (the above equations are often referred to as the general constitutive relations for bi-anisotropic materials). These parameter tensors will have different forms for different types of complex materials (e.g. chiral materials (Lindell et al. 1994), and Ω materials (Saadoun and Engheta 1992), etc.). Different types of complex materials have different properties and applications, and there is a widespread interest in the design of composite structures with new and improved properties.

Consider an electromagnetic plane wave with harmonic time dependence $\exp(j\omega t)$ obliquely impinging from the left side $z < 0$ on a composite structure located in the region $[0, l]$. The medium outside the composite structure is a vacuum, cf. Fig. 6.1. The composite structure consists of multiple layers of different types of complex media. The parameters in each layer are arbitrary functions of z, which do not need to be continuous or differentiable. These complex materials are in general described by the constitutive relations (6.82) and (6.83).

Assume that the x, y and t dependence of the electromagnetic fields is $\exp[j(wt - k_x x - k_y y)]$, where the propagation constants k_x and k_y are given by

$$k_x = k_0 \sin\theta_i \cos\phi_i, \quad k_y = k_0 \sin\theta_i \sin\phi_i, \tag{6.84}$$

and where $k_0 = \omega/c_0 = \sqrt{\epsilon_0\mu_0}\omega$ is the wave number of the incident field (ϵ_0, μ_0 are the permittivity and permeability of the vacuum, respectively).

From the constitutive relations (6.82) and (6.83) and the third components of Maxwell's equations (6.1), one can always express the third components of $\vec{\mathbf{E}}$ and $\vec{\mathbf{H}}$ in terms of the tangential fields as (cf. the previous section for the case of a chiral medium)

$$\begin{bmatrix} E_3 \\ H_3 \end{bmatrix} = P \begin{bmatrix} \mathbf{E} \\ \mathbf{H} \end{bmatrix}, \tag{6.85}$$

where P is a 2×4 matrix.

Using the above equation, one can rewrite the first two components of Maxwell's equations in terms of the tangential electric and magnetic fields

$$\partial_z \begin{bmatrix} \mathbf{E} \\ \mathbf{H} \end{bmatrix} = \begin{bmatrix} D_1 & D_2 \\ D_3 & D_4 \end{bmatrix} \begin{bmatrix} \mathbf{E} \\ \mathbf{H} \end{bmatrix} \equiv D \begin{bmatrix} \mathbf{E} \\ \mathbf{H} \end{bmatrix}, \tag{6.86}$$

where D_i, $i = 1, 2, 3, 4$, are 2×2 matrices, and D is a 4×4 matrix. Different types of complex materials have different D matrices.

The eigenmodes corresponding to TE and TM modes in vacuum are given by (6.27). As shown in Norgren and He (1995), it is advantageous to calculate the scattered fields from the composite structure by using this vacuum-splitting. Thus, inside the chiral slab one defines

$$\begin{bmatrix} \mathbf{E}^+ \\ \mathbf{E}^- \end{bmatrix}(z) = T_0 \begin{bmatrix} \mathbf{E} \\ \mathbf{H} \end{bmatrix}(z), \tag{6.87}$$

where the matrix T_0 is given by (6.28). With $\mathbf{E}^{\pm}$ as the new basis, one can rewrite Maxwell's equations as follows (cf. (6.86), (6.87); note that T_0 is z independent)

$$\partial_z \begin{bmatrix} \mathbf{E}^+ \\ \mathbf{E}^- \end{bmatrix} = T_0 D T_0^{-1} \begin{bmatrix} \mathbf{E}^+ \\ \mathbf{E}^- \end{bmatrix} \equiv \begin{bmatrix} \alpha & \beta \\ \gamma & \vartheta \end{bmatrix} \begin{bmatrix} \mathbf{E}^+ \\ \mathbf{E}^- \end{bmatrix}, \tag{6.88}$$

where $\boldsymbol{\alpha}$, $\boldsymbol{\beta}$, $\boldsymbol{\gamma}$ and $\boldsymbol{\vartheta}$ are 2×2 matrices. The above system of equations is equivalent to Maxwell's equations in terms of the split fields $\mathbf{E}^{\pm}$. The vacuum-splitting only has physical meaning in the exterior vacuum region, but not inside the composite structure. Inside the composite structure (6.87) is nevertheless a useful change of unknowns for Maxwell's equations from the variables $\mathbf{E}$, $\mathbf{H}$ to $\mathbf{E}^{\pm}$.

Since the tangential fields are continuous across any plane, the split fields $\mathbf{E}^{\pm}$ defined by (6.87) are continuous across any plane even though the parameters may be discontinuous across the plane.

6.2.2 *Reflection, transmission and internal fields*

The reflection and transmission coefficient matrices for the composite structure can be determined by a modified imbedding method. In this modified method one considers an imbedding geometry, i.e. a substructure $[z, l]$ of the original structure $[0, l]$, and assumes that the material on the left side of the substructure is temporarily replaced with a vacuum (in the usual invariant imbedding, the material on the left side of the substructure is temporarily redefined to be constant, *continuously terminated* to the substructure $[z, l]$ at the left endpoint z; see Section 6.1.4). For this imbedding geometry, the reflection coefficient matrix (denoted $\mathbf{r}(z)$) and the transmission coefficient matrix (denoted $\mathbf{t}(z)$) are defined as follows

$$\mathbf{E}^{-}(z) = \mathbf{r}(z)\mathbf{E}^{+}(z) \equiv \begin{bmatrix} r_{11}(z) & r_{12}(z) \\ r_{21}(z) & r_{22}(z) \end{bmatrix} \mathbf{E}^{+}(z), \tag{6.89}$$

$$\mathbf{E}^{+}(l) \equiv \mathbf{t}(z)\mathbf{E}^{+}(z) \equiv \begin{bmatrix} t_{11}(z) & t_{12}(z) \\ t_{21}(z) & t_{22}(z) \end{bmatrix} \mathbf{E}^{+}(z). \tag{6.90}$$

Since the split fields $\mathbf{E}^{\pm}$ are continuous across any plane, the $\mathbf{r}(z)$ and $\mathbf{t}(z)$ matrices defined by the above equations are continuous across any plane even though the parameters may be discontinuous across that plane. From the above definitions one notices that $\mathbf{r}(0)$, $\mathbf{t}(0)$ are the physical reflection and transmission coefficient matrices, respectively, for the whole composite structure.

The physical reflection coefficient matrix $\mathbf{r}(0)$ can be obtained by solving the following matrix Riccati equation (obtained by introducing the definition (6.89) into (6.88)) together with a suitable boundary condition

$$\begin{cases} \partial_z \mathbf{r} = \gamma + (\vartheta \mathbf{r} - \mathbf{r}\alpha) - \mathbf{r}\beta\mathbf{r}, \\ \mathbf{r}(l) = 0. \end{cases} \tag{6.91}$$

The solution for $\mathbf{r}(z)$ is obtained by integrating the matrix Riccati equation along the $-z$ direction starting from $z = l$.

If the composite structure is backed by a perfectly conducting surface at $z = l$, the boundary condition in (6.91) should be replaced by $\mathbf{r}(l) = -I$. If the composite structure is backed by a dielectric half-space (with a permittivity $\bar{\epsilon}$ and a permeability μ_0) at $z = l$, then the boundary condition in (6.91) should be replaced by the following boundary condition

$$\mathbf{r}(l) = \begin{bmatrix} -\dfrac{1-\dfrac{\sqrt{\epsilon_0(\bar{\epsilon}-\epsilon_0\sin^2\theta_i)}}{\bar{\epsilon}\cos\theta_i}}{1+\dfrac{\sqrt{\epsilon_0(\bar{\epsilon}-\epsilon_0\sin^2\theta_i)}}{\bar{\epsilon}\cos\theta_i}} & 0 \\ 0 & \dfrac{1-\dfrac{\sqrt{\bar{\epsilon}-\epsilon_0\sin^2\theta_i}}{\sqrt{\epsilon_0}\cos\theta_i}}{1+\dfrac{\sqrt{\bar{\epsilon}-\epsilon_0\sin^2\theta_i}}{\sqrt{\epsilon_0}\cos\theta_i}} \end{bmatrix}. \tag{6.92}$$

Similarly, one has the following imbedding equation and boundary condition for the transmission coefficient matrix $\mathbf{t}(z)$

$$\begin{cases} \partial_z \mathbf{t}(z) = -\mathbf{t}(z)\boldsymbol{\alpha}(z) - \mathbf{t}(z)\boldsymbol{\beta}(z)\mathbf{r}(z), \\ \mathbf{t}(l) = I. \end{cases} \tag{6.93}$$

Note that the imbedding equation for $\mathbf{t}$ is linear in $\mathbf{t}$ but coupled to the reflection coefficient matrix $\mathbf{r}$. The physical transmission coefficient matrix $\mathbf{t}(0)$ can be obtained by integrating the above imbedding equation along the $-z$ direction starting from $z = l$ (after $\mathbf{r}(z)$ has been computed by solving (6.91)).

Note that $\mathbf{r}(z)$ and $\mathbf{t}(z)$ are the physical scattering matrices for the substructure $[z, l]$ of the composite structure (imbedded in vacuum). The elements of $\mathbf{r}$ and $\mathbf{t}$ relate the amplitudes of the *tangential* components of the scattered and incident electric fields to each other for TM and TE modes. As is customary, the reflection coefficients for TM and TE modes are defined as

$$r_{XY} = \frac{E^r_{TX}}{E^i_{TY}}, \quad t_{XY} = \frac{E^t_{TX}}{E^i_{TY}}, \quad X, Y = E \text{ or } M, \tag{6.94}$$

where E^i_{TE}, E^r_{TE}, E^t_{TE} are the amplitudes of the incident, reflected and transmitted electric fields for the TE mode, respectively, etc. r_{MM}, r_{EM} (t_{MM}, t_{EM}) are the co- and cross-polarized reflection (transmission) coefficients for the substructure $[z, l]$ for TM mode incidence, respectively, and r_{EE}, r_{ME} (t_{EE}, t_{ME}) are the co- and cross-polarized reflection (transmission) coefficients for the substructure $[z, l]$ for TE mode incidence, respectively. It is easy to see that (Norgren and He 1994)

$$r_{MM} = r_{11}, \quad r_{ME} = \frac{r_{12}}{\cos\theta_i}, \quad r_{EM} = r_{21}\cos\theta_i, \quad r_{EE} = r_{22}, \tag{6.95}$$

$$t_{MM} = t_{11}, \quad t_{ME} = \frac{t_{12}}{\cos\theta_i}, \quad t_{EM} = t_{21}\cos\theta_i, \quad t_{EE} = t_{22}. \tag{6.96}$$

Energy conservation requires that

$$|r_{XY}(z)| \le 1, \quad |t_{XY}(z)| \le 1, \qquad X, Y = E, M. \tag{6.97}$$

To calculate the internal fields, one introduces two matrices $\mathbf{G}^+(z)$ and $\mathbf{G}^-(z)$ of Green's functions to express the internal split fields $\mathbf{E}^\pm$ in terms of the right-moving incident field $\mathbf{E}^+(0)$,

$$\begin{bmatrix} \mathbf{E}^+ \\ \mathbf{E}^- \end{bmatrix}(z) = \begin{bmatrix} \mathbf{G}^+(z)\mathbf{E}^+(0) \\ \mathbf{G}^-(z)\mathbf{E}^+(0) \end{bmatrix}, \qquad 0 \le z \le l. \tag{6.98}$$

The solution for $\mathbf{G}^\pm(z)$ is obtained by solving the following system of equations (cf. (6.61) and (6.62))

$$\begin{cases} \partial_z \mathbf{G}^+ = \boldsymbol{\alpha}\mathbf{G}^+ + \boldsymbol{\beta}\mathbf{G}^-, \\ \partial_z \mathbf{G}^- = \boldsymbol{\gamma}\mathbf{G}^+ + \boldsymbol{\vartheta}\mathbf{G}^-, \\ \mathbf{G}^+(0) = I, \\ \mathbf{G}^-(0) = \mathbf{r}(0). \end{cases} \tag{6.99}$$

The solution for the Green's functions $\mathbf{G}^\pm(z)$ is obtained by integrating the ODEs for $\mathbf{G}^\pm(z)$ along the z direction starting from $z = 0$. When the Green's function matrices have been determined, the internal split fields $\mathbf{E}^\pm$ are known through (6.98). The internal field is then obtained from the following equation (cf. (6.87))

$$\begin{bmatrix} \mathbf{E} \\ \mathbf{H} \end{bmatrix}(z) = T_0^{-1} \begin{bmatrix} \mathbf{E}^+ \\ \mathbf{E}^- \end{bmatrix}(z). \tag{6.100}$$

Note that the boundary values of $\mathbf{G}^+$ at $z = l$ give the transmission coefficient matrix for the whole structure, i.e. $\mathbf{G}^+(l) = \mathbf{t}(0)$.

We also note that as an alternative, one can easily apply the transmission Green's function approach (cf. Section 6.1.6) in the vacuum-splitting setting.

A numerical example. As a numerical illustration, the scheme described above is used to calculate the reflection and transmission for a composite structure consisting of a stratified chiral layer and a stratified Ω layer. The D matrix for the chiral medium is given by (6.10). The Ω material is assumed to have the following parameter tensors

$$\overline{\overline{\epsilon}} = \begin{bmatrix} \epsilon_1 & 0 & 0 \\ 0 & \epsilon_2 & 0 \\ 0 & 0 & \epsilon_3 \end{bmatrix}, \qquad \overline{\overline{\mu}} = \begin{bmatrix} \mu_1 & 0 & 0 \\ 0 & \mu_2 & 0 \\ 0 & 0 & \mu_3 \end{bmatrix}, \tag{6.101}$$

$$\overline{\overline{\xi}} = -j\sqrt{\epsilon_0\mu_0} \begin{bmatrix} 0 & 0 & 0 \\ 0 & 0 & \Omega \\ 0 & 0 & 0 \end{bmatrix}, \qquad \overline{\overline{\zeta}} = j\sqrt{\epsilon_0\mu_0} \begin{bmatrix} 0 & 0 & 0 \\ 0 & 0 & 0 \\ 0 & \Omega & 0 \end{bmatrix}. \tag{6.102}$$

The properties of an Ω medium can be envisaged as arising from a distribution of metal half-loops with two extended arms (Saadoun and Engheta 1992). The corresponding D matrix for the Ω material is (Norgren and He 1994)

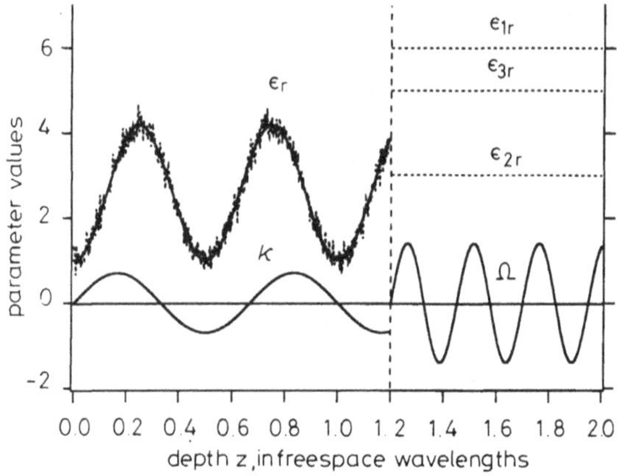

FIG. 6.2. The parameter profiles for a composite structure consisting of a stratified chiral layer and a stratified Ω layer.

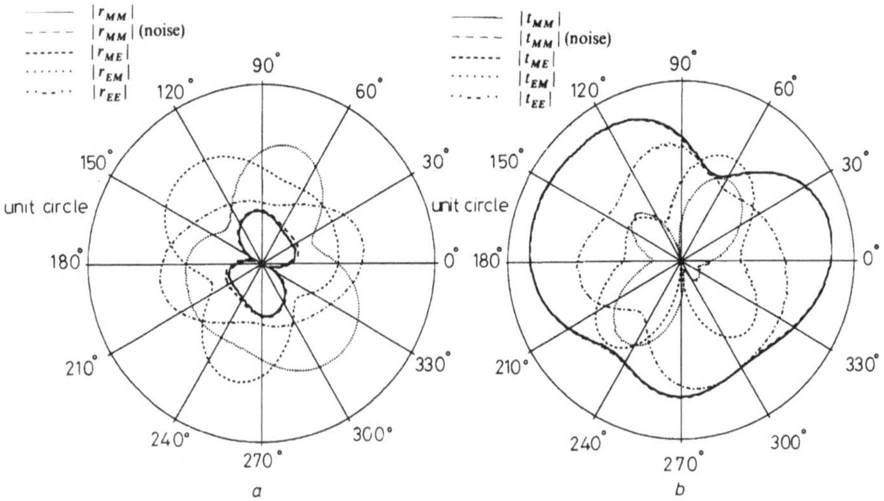

FIG. 6.3. Reflection and transmission coefficients as functions of the incident azimuthal angle ϕ_i for the composite structure given in Fig. 6.2. The incident angle $\theta_i = \pi/4$. (a) co- and cross-polarized reflection coefficients for TE and TM modes; (b) co- and cross-polarized transmission coefficients for TE and TM modes.

$$D = j\omega \begin{bmatrix} 0 & 0 & -b & a_1 \\ 0 & 0 & a_2 & b \\ c_1 & a_3 & 0 & 0 \\ a_4 & c_2 & 0 & 0 \end{bmatrix}, \tag{6.103}$$

where $a_1 = -\mu_2 + \frac{k_x^2}{\epsilon_3\omega^2}$, $a_2 = \mu_1 - \frac{k_y^2}{\epsilon_3\omega^2}$, $a_3 = \epsilon_2 - \frac{k_x^2 + k_0^2\Omega^2}{\mu_3\omega^2}$, $a_4 = -\epsilon_1 + \frac{k_y^2}{\mu_3\omega^2}$, $b = \frac{k_x k_y}{\epsilon_3\omega^2}$, $c_1 = \frac{k_y(k_x + jk_0\Omega)}{\mu_3\omega^2}$ and $c_2 = \frac{k_y(-k_x + jk_0\Omega)}{\mu_3\omega^2}$. The corresponding parameter profiles for this composite structure are given in Fig. 6.2 with $\mu_i(z) = \mu(z) = \mu_0$, $i = 1, 2, 3$. The thickness of the chiral layer is $1.2\lambda_0$, where $\lambda_0 = 2\pi/(\sqrt{\epsilon_0\mu_0}\omega)$ is the free space wavelength. The incident angle is $\theta_i = \pi/4$. Figure 6.3 gives the co- and cross-polarized reflection and transmission as functions of the incident azimuthal angle ϕ_i for a composite structure consisting of a stratified chiral layer and a stratified Ω layer. In this numerical example, noise has been added to the relative permittivity $\epsilon_r(z)$ of the chiral layer to make it non-differentiable (see the dashed line for the 'noisy' permittivity profile in the left portion of Fig. 6.3). The numerical results for $|r_{MM}|$ and $|t_{MM}|$ for the noisy profile are plotted by the dashed lines in Fig. 6.3(a) and (b), respectively (for graphical clarity, the other scattering coefficients for this noisy profile are not given in Fig. 6.3). As one can see, the difference between $|r_{MM}|$ (or $|t_{MM}|$) for the noise-free and noisy profiles respectively is small, which is reasonable physically.

6.2.3 *Discussion and comparison with other approaches*

Compared to approaches using other wave-splittings which come from the diagonalization of the D matrix (see the previous section) or the principal part of the D matrix (He and Hu 1993; He et al. 1994a), the vacuum-splitting approach presented in Subsections 6.2.1 and 6.2.2 has the following two major advantages:

(1) First, the approach using the vacuum-splitting does not require any of the parameters to be differentiable, while the approach using the other splittings will require the permittivity and permeability to be differentiable. One notices that in a realistic case the parameters are more likely to be best described by non-differentiable models.

(2) Secondly, any mismatch in the parameters does not lead to any extra effort in the calculation using the vacuum-splitting, while using the other splittings one has to introduce the mismatch operators, which make the computation cumbersome (see the previous section). One notes that in a composite structure the parameters are in general discontinuous at the interfaces between different

types of complex materials.

Note that one cannot use the vacuum-splitting in a time domain treatment. In the time domain, one has to choose a wave-splitting that factorizes (or diagonalizes) the whole or the principal part of the wave equation (or the D matrix). This is due to the different natures of the elliptic (in the frequency domain) and hyperbolic (in the time domain) partial differential equations. Note also that frequency domain physical wave-splitting (corresponding to the eigenmodes) for bi-anisotropic media are useful for obtaining explicit expressions for scattering coefficient matrices for a homogeneous bi-anisotropic half-space or slab (cf. Section 6.1.2 and Section 6.1.3), and for analysis of surface waves (He 1998).

Unlike most other frequency-domain approaches (e.g., the vector transmission line and circuit theory (Lindell et al. 1994)), the present wave-splitting approach is not limited to piecewise constant stratifications and it can be applied effectively to stratified complex media whose properties vary in a general way as functions of the depth z. Thus the present approach can be expected to be useful in applications where this freedom is of importance.

We also wish to point to the usefulness of bounds of the type obtained for the reflection and transmission coefficients in the wave-splitting formalism (cf. (6.97)). Such bounds are not obtained, e.g. if one uses an imbedding approach based on the impedance matrix but with no splitting. In order to illustrate this, we introduce the impedance matrix $Z(z)$ for the substructure $[z, L]$, i.e.

$$\mathbf{E}(z) = \mathbf{Z}(z)\mathbf{H}(z) \equiv \begin{bmatrix} Z_{11}(z) & Z_{12}(z) \\ Z_{21}(z) & Z_{22}(z) \end{bmatrix} \mathbf{H}(z). \tag{6.104}$$

One then obtains the following imbedding equation and boundary condition for $\mathbf{Z}(z)$ (cf. (6.86))

$$\partial_z \mathbf{Z} = D_2 + (D_1\mathbf{Z} - \mathbf{Z}D_4) - \mathbf{Z}D_3\mathbf{Z}, \tag{6.105}$$

$$\mathbf{Z}(l) = R_0^{-1} Z_0 R_0. \tag{6.106}$$

The matrix Riccati equation (6.105) can be solved by integrating from $z = l$ to $z = 0$. The reflection coefficient matrix for the whole composite structure $[0, l]$ can be obtained from

$$\mathbf{r}(0) = [R_0\mathbf{Z}(0) - Z_0R_0]\,[R_0\mathbf{Z}(0) + Z_0R_0]^{-1}. \tag{6.107}$$

However, unlike the reflection coefficients that are bounded in the electromagnetic case (cf. (6.97)), the impedance coefficients $Z_{ij}(z)$, $i, j = 1, 2$, are not bounded. $Z_{ij}(z)$, $i, j = 1, 2$, can be very large or even singular. In fact, from (6.87), (6.89) and (6.104) one obtains the following relation between the reflection matrix $\mathbf{r}(z)$ and the impedance matrix $\mathbf{Z}(z)$

$$\mathbf{Z} = R_0^{-1}[I - \mathbf{r}]^{-1}(\mathbf{r} + I)Z_0R_0, \quad \text{when } \det[I - \mathbf{r}] \neq 0. \tag{6.108}$$

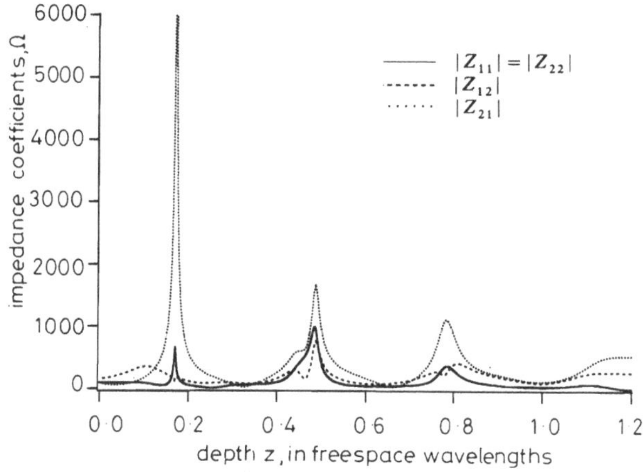

FIG. 6.4. The impedance coefficients for a stratified chiral layer with parameters given by the solid lines in the left portion of Fig. 6.2. The incident angle $\theta_i = \pi/4$. $|Z_{21}|$ becomes very large around the position $z = 0.17$ wavelengths.

$\mathbf{Z}(z)$ becomes singular at the position z when

$$\det[I - \mathbf{r}(z)] = 0. \tag{6.109}$$

Figure 6.4 gives the impedance coefficients for the case of a stratified chiral slab. The corresponding parameters are given in the left portion $z \in [0, 1.2\lambda_0]$ of Fig. 6.2. The incident angle is $\theta_i = \pi/4$ (the impedance matrix $\mathbf{Z}(z)$ is independent of the azimuthal angle ϕ_i in this case). One sees in Fig. 6.4 that the impedance $|Z_{21}|$ becomes very large around the position $z = 0.17\lambda_0$. Since $Z_{ij}(z)$, $i, j = 1, 2$, can be very large or even singular, the approach based on the impedance matrix (without using any wave-splitting) is less suitable for use in numerical computation of the reflection or transmission.

6.2.4 *Fractional linear transformations for the Riccati equation*

The reflection coefficient matrix $\mathbf{r}(z)$ associated with the vacuum-splitting is in fact related to the usual imbedding reflection coefficient matrix (see the previous section) by a certain fractional linear transformation. This will be illustrated in this subsection for a simple case. Consider a plane wave normally incident upon a stratified optical medium that extends from $z = 0$ to $z = l$ (the slab is bounded on each side by a semi-infinite vacuum). Then one has the following Helmholtz equation for the amplitude ϕ,

$$\left[\frac{d^2}{dz^2} + k^2(z)\right]\phi(z) = 0, \tag{6.110}$$

where

$$k(z) = \begin{cases} k_0, & \text{if } z < 0 \quad \text{or} \quad z > l, \\ k_0 n(z), & \text{if } 0 \le z \le l, \end{cases} \tag{6.111}$$

where $n(z)$ is the refractive index of the slab, k_0 is the wave number of the incident wave.

The reflection coefficient for such a stratified slab is usually obtained by solving a Riccati equation for the reflection coefficient. The following conventional Riccati equation for the reflection coefficient can be derived through a variety of techniques

$$\frac{d}{dz}r(z) = \frac{1}{2n(z)}\left[\frac{d}{dz}n(z)\right][1 - r^2(z)] - 2ik_0 n(z) r(z), \tag{6.112}$$

together with the boundary condition

$$r(l^+) = 0. \tag{6.113}$$

Since the above Riccati equation involves $\frac{d}{dz}n(z)$, it requires $n(z)$ to be piecewise continuous and differentiable. The reflection coefficient $r(z)$ has a discontinuity at any position where the refractive index n is discontinuous. Thus the above Riccati equation is not suitable for use when the refractive index $n(z)$ is non-differentiable and/or discontinuous at some positions (one example of a non-differentiable profile is a stratified random medium with a very small correlation length (Ishimaru 1978)).

It is well known that a Riccati equation is formally invariant under fractional linear transformations (Hille 1969). One could therefore hope to be able to obtain a Riccati equation that does not involve $\frac{d}{dz}n(z)$ by a suitable fractional linear transformation. In order to investigate this, consider the following fractional linear transformation

$$\tilde{r}(z) = \frac{r(z) - a(z)}{1 - a(z)r(z)}, \quad a \neq 1, \tag{6.114}$$

i.e.,

$$r(z) = \frac{\tilde{r}(z) + a(z)}{1 + a(z)\tilde{r}(z)}, \tag{6.115}$$

where $a(z)$ will be determined later. Then (6.112) becomes

$$\frac{1 - a^2}{(1 + a\tilde{r})^2}\frac{d}{dz}\tilde{r} = -\frac{1 - \tilde{r}^2}{(1 + a\tilde{r})^2}\left[\frac{d}{dz}a - \frac{1 - a^2}{2n}\frac{d}{dz}n\right] - 2ik_0 n\frac{\tilde{r} + a}{1 + a\tilde{r}}.$$

In order for $\frac{d}{dz}n$ not to appear in the transformed Riccati equation, one should set

$$\frac{d}{dz}a - \frac{1-a^2}{2n}\frac{d}{dz}n = 0,$$

which gives

$$a(z) = \frac{n(z) - a_0}{n(z) + a_0},$$

where a_0 is a constant. In order to retain $\tilde{r}$ as the physical reflection coefficient in the vacuum region outside the slab, one should let

$$\tilde{r}(z_0) = r(z_0), \quad \text{at any position } z = z_0 \text{ if } n(z_0) = 1. \tag{6.116}$$

Thus it immediately follows from the above two equations and (6.114) that $a_0 = 1$, i.e.,

$$a(z) = \frac{n(z) - 1}{n(z) + 1}. \tag{6.117}$$

After such a fractional linear transformation, the Riccati equation (6.112) becomes

$$\frac{d}{dz}\tilde{r} = -\frac{1}{2}ik_0\left[(n^2-1)(1+\tilde{r}^2) + 2(n^2+1)\tilde{r}\right], \tag{6.118}$$

and one has the following boundary condition

$$\tilde{r}(l^-) = \tilde{r}(l^+) = 0. \tag{6.119}$$

Note here that the derivative of n does not appear in the transformed Riccati equation (6.118). The energy conservation law requires that $|r(z)| \leq 1$. As shown in He (1995a), after the fractional linear transformation (6.114) with $a(z)$ defined by (6.117), the amplitude of the transformed reflection coefficient $\tilde{r}$ is bounded by 1, i.e.,

$$|\tilde{r}(z)| \leq 1. \tag{6.120}$$

Comparing with the vacuum-splitting approach described in the previous subsections, one can easily see that $\tilde{r}(z_0)$, $z_0 \in [0, d]$, is the physical reflection coefficient for the subslab $[z_0, d]$ *imbedded in the vacuum.* Note that the conventional reflection coefficient $r(z_0)$ is the physical reflection coefficient for an imbedded subslab $[z_0, d]$ with the refractive index at the left side of the subslab redefined to be constant, *continuously terminated* to the subslab at the left endpoint $z = z_0$. Thus, both $r(z)$ and $\tilde{r}(z)$ have physical meaning.

If the homogeneous region $z > d$ is not a vacuum, one should replace the boundary condition (6.119) with the following one

$$\tilde{r}(l^-) = \tilde{r}(l^+) = -a(l^+), \tag{6.121}$$

which is obtained from (6.114) with $r(l^+) = 0$. Similarly, if the homogeneous region $z < 0$ is not a vacuum, one obtains the physical reflection coefficient $r(0^-)$ as follows (cf. (6.115))

$$r(0^-) = \frac{\tilde{r}(0) + a(0^-)}{1 + a(0^-)\tilde{r}(0)}. \tag{6.122}$$

From the form of the transformed Riccati equation (6.118), one sees that $\tilde{r}(z)$ is a continuous function of z even when $n(z)$ is discontinuous (since $\tilde{r}(z)$ has physical meaning as the physical reflection coefficient for the subslab $[z, l]$ imbedded in a vacuum, the continuity of $\tilde{r}(z)$ can be explained physically by the fact that the physical reflection coefficient changes continuously when adding an additional infinitesimally thin film with an arbitrary finite refractive index to the left surface). Furthermore, n does not need to be differentiable in order to solve the transformed Riccati equation (6.118). Therefore, the transformed Riccati equation should be used when the refractive index $n(z)$ is non-differentiable and/or discontinuous.

A special case: a multi-layered structure. The transformed Riccati equation (6.118) can be used to obtain an explicit solution for a multi-layered structure that occupies the region $[0, l]$ in a straightforward way. Consider a structure consisting of N layers with refractive indices $n_1, n_2, ..., n_N$ (counted from the right; the corresponding thicknesses are $l_1, l_2, ..., l_N$, respectively; the total thickness is $l = l_1 + l_2 + ... + l_N$). Note that the layer with the refractive index n_1 occupies the region $[l - l_1, l]$, and the layer with the refractive index n_q, $q = 2, 3, ..., N$, occupies the region $[l - l_1 - ... - l_q, l - ... - l_{q-1}]$. The homogeneous medium outside the region $[0, l]$ is a vacuum. In a *homogeneous* region with a constant refractive index n, one can verify that the following expression is the solution to the transformed Riccati equation (6.118) with the boundary condition $\tilde{r}(z_0) = \tilde{r}_0$,

$$\tilde{r}(z) = \frac{\dfrac{\tilde{r}_0 + a}{1 + a\tilde{r}_0} e^{-2ink_0(z - z_0)} - a}{1 - a\dfrac{\tilde{r}_0 + a}{1 + a\tilde{r}_0} e^{-2ink_0(z - z_0)}},$$

where $a = \frac{n-1}{n+1}$. Since $\tilde{r}(z)$ is continuous everywhere, it immediately follows from the above solution that the transformed Riccati equation (6.118) for a multi-layered structure has the following solution (cf. (6.119)),

$$\tilde{r}(z) = \begin{cases} \dfrac{a_1(e^{-2in_1k_0(z-l)} - 1)}{1 - a_1^2 e^{-2in_1k_0(z-l)}}, & \text{if} \quad l - l_1 \le z \le l\,, \\[2ex] \dfrac{\dfrac{\tilde{r}_{q-1} + a_q}{1 + a_q\tilde{r}_{q-1}} e^{-2in_qk_0[z-(l-l_1\ldots-l_{q-1})]} - a_q}{1 - a_q \dfrac{\tilde{r}_{q-1} + a_q}{1 + a_q\tilde{r}_{q-1}} e^{-2in_qk_0[z-(l-l_1\ldots-l_{q-1})]}}, & \\[2ex] \quad \text{if} \quad l - l_1 - \ldots - l_q \le z < l - l_1 \ldots - l_{q-1}, & q = 2, 3, \ldots, N, \end{cases} \tag{6.123}$$

where

$$a_q = \frac{n_q - 1}{n_q + 1}, \quad q = 1, 2, 3, \ldots, N, \tag{6.124}$$

and $\tilde{r}_q$ is the physical reflection coefficient for a multi-layered structure consisting of q layers with the refractive indices $n_1, n_2, \ldots, n_q$ (counted from the right; the corresponding thicknesses are $l_1, l_2, \ldots, l_q$, respectively). According to (6.123) one has

$$\tilde{r}_1 \equiv \tilde{r}(l - l_1) = \frac{a_1(e^{2ink_0l_1} - 1)}{1 - a_1^2 e^{2ink_0l_1}}, \tag{6.125}$$

$$\tilde{r}_q \equiv \tilde{r}(l - l_1 - \ldots - l_q)$$

$$= \frac{\dfrac{\tilde{r}_{q-1} + a_q}{1 + a_q\tilde{r}_{q-1}} e^{2in_qk_0l_q} - a_q}{1 - a_q \dfrac{\tilde{r}_{q-1} + a_q}{1 + a_q\tilde{r}_{q-1}} e^{2in_qk_0l_q}}, \quad q = 2, 3, \ldots, N. \tag{6.126}$$

Note that when applying the formula (6.123) to calculate the reflection coefficient $\tilde{r}(z)$, one should evaluate $\tilde{r}(z)$ layer by layer from the right to the left (starting from the layer to the far right, which occupies the region $[l - l_1, l]$).

If one further takes the fractional linear transformation $\hat{r}(z) = \dfrac{1 - \tilde{r}(z)}{1 + \tilde{r}(z)}$, then one obtains the simplest Riccati equation $\frac{d}{dz}\hat{r} = -ik_0(\hat{r}^2 - n^2)$, together with the boundary condition $\hat{r}(l) = 1$. However, one can easily show that $\hat{r}$ is not bounded, and is singular at any position $z = z_0$ if $\tilde{r}(z_0) = -1$ (i.e. $r(z_0) = -1$). Furthermore, $\hat{r}$ does not have any physical meaning. Therefore, it seems that the fractional linear transformation (6.114) with $a(z)$ defined by (6.117) provides the best approach for obtaining the reflection coefficient for a stratified slab when the refractive index $n(z)$ is non-differentiable and/or discontinuous at some positions.

6.3 Reconstruction of transmission line parameters with band-limited scattering data

In this section, the wave-splitting approach to inverse problems using band-limited scattering data is illustrated for a non-uniform LCRG transmission line. Consider a non-uniform transmission line occupying the region $z \in [0, l]$ where the parameters L (the inductance), C (the capacitance), R (the resistance) and G (the shunt conductance) varies with the position z. The telegrapher's equations for the voltage $V(z,\omega)$ and current $I(z,\omega)$ with harmonic time dependence $\exp(j\omega t)$ are (cf. (3.1) and (3.2))

$$\partial_z \begin{bmatrix} V \\ I \end{bmatrix} = \begin{bmatrix} 0 & -R - j\omega L \\ -G - j\omega C & 0 \end{bmatrix} \begin{bmatrix} V \\ I \end{bmatrix} \equiv \mathbf{D} \begin{bmatrix} V \\ I \end{bmatrix}. \tag{6.127}$$

The non-uniform line is excited at $z = 0$ from a uniform (not necessarily lossless) transmission line with a charactaristic impedance Z_0, and is terminated with a load impedance Z_L at $z = l$. Both Z_0 and Z_L may be in general frequency dependent. The inverse problem is to determine one or several of the line parameters in the region $[0, l]$ from the reflection and/or transmission data in a certain frequency band $[\omega_{min}, \omega_{max}]$ (in all the numerical examples in this section, scattering data in a microwave frequency band are used). To enhance the information for reconstruction, one may also use the reflection data at $z = l$ when the line is excited at $z = l$ (right-sided excitation). However, to simplify the notation the formulas are only given for left-sided excitation, since the formalism for right-sided excitation is completely analogous.

6.3.1 *The direct solver*

An optimization approach is used in this section to solve the inverse problem in an iterative way. To apply an optimization approach, one needs a direct solver for the direct problem to calculate the gradient at each iteration. Using the idea of vacuum-splitting described in the previous section, the following wave-splitting is introduced

$$\begin{bmatrix} V^+ \\ V^- \end{bmatrix}(z,\omega) = \frac{1}{2}\begin{bmatrix} 1 & Z_0 \\ 1 & -Z_0 \end{bmatrix}\begin{bmatrix} V \\ I \end{bmatrix}(z,\omega) \equiv \mathbf{T}_0 \begin{bmatrix} V \\ I \end{bmatrix}(z,\omega). \tag{6.128}$$

With this wave-splitting, the split voltages $V^+(z,\omega)$ and $V^-(z,\omega)$ are the right- and left-moving voltages, respectively, on any homogeneous region of a transmission line with a charactaristic impedance Z_0. Thus $V^+(z,\omega)$ and $V^-(z,\omega)$ are the incident and reflected voltages, respectively, in the homogeneous region $z \leq 0$. Using (6.127) and (6.128), one obtains the following system of equations for the split voltages

$$\partial_z \begin{bmatrix} V^+ \\ V^- \end{bmatrix} = \mathbf{T}_0\mathbf{D}\mathbf{T}_0^{-1} \begin{bmatrix} V^+ \\ V^- \end{bmatrix} \equiv \begin{bmatrix} -a & -b \\ b & a \end{bmatrix}\begin{bmatrix} V^+ \\ V^- \end{bmatrix}, \tag{6.129}$$

where

$$a = \frac{1}{2}\left[j\omega\left(CZ_0 + LZ_0^{-1}\right) + \left(GZ_0 + RZ_0^{-1}\right)\right], \tag{6.130}$$

$$b = \frac{1}{2}\left[j\omega\left(CZ_0 - LZ_0^{-1}\right) + \left(GZ_0 - RZ_0^{-1}\right)\right]. \tag{6.131}$$

The reflection and transmission coefficients for the non-uniform transmission line can be determined by the modified imbedding method (cf. Section 6.2.2). In this method one considers an imbedding geometry, i.e. a subline $[z,l]$ of the original line $[0,l]$, and assumes that the subline is temporarily terminated at the left side with a uniform line with a charactaristic impedance Z_0. For this imbedding geometry, the reflection coefficient (denoted $r(z,\omega)$) and the transmission coefficient matrix (denoted $t(z,\omega)$) are defined as follows

$$V^{-}(z,\omega) = r(z,\omega)V^{+}(z,\omega), \tag{6.132}$$

$$V(l,\omega) = t(z,\omega)V^{+}(z,\omega). \tag{6.133}$$

From the above definitions one sees that $r(0,\omega)$, $t(0,\omega)$ are the physical reflection and transmission coefficients, respectively, for the original non-uniform line. From (6.129) and (6.132), one obtains the following imbedding equation for $r(z,\omega)$ (the derivation follows the lines given in Section 6.2.2),

$$\partial_z r = 2ar + b\left(1 + r^2\right), \tag{6.134}$$

together with the boundary condition

$$r(l,\omega) = \frac{Z_L - Z_0}{Z_L + Z_0}. \tag{6.135}$$

By integrating (6.134) in the $-z$ direction (starting from $z = l$), one obtains the reflection coefficient for the original non-uniform line, i.e. $r(0,\omega)$. Similarly, one obtains the following linear imbedding equation for $t(z,\omega)$,

$$\partial_z t = (a + br)\,t, \tag{6.136}$$

together with the following boundary condition

$$t(l,\omega) = \frac{2Z_L}{Z_L + Z_0}. \tag{6.137}$$

From the above two equations, one obtains the following solution for $t(z,\omega)$

$$t(z,\omega) = \frac{2Z_L}{Z_L + Z_0}\exp\left\{-\int_z^l (a + br)\,dz'\right\}. \tag{6.138}$$

6.3.2 *Explicit expression for the gradient*

In the inverse problem, an optimization approach is used to reconstruct various combinations of the line parameters L, C, R and G. Introduce a four-element parameter vector $p = (L, C, R, G)^T$ (the superscript T denotes the transposition). Define an objective functional as follows,

$$J(p) = \sum_{\omega=\omega_{min}}^{\omega_{max}} w_r(\omega)|r(0,\omega) - r_m(\omega)|^2 + w_t(\omega)\,|t(0,\omega) - t_m(\omega)|^2, \tag{6.139}$$

where $r_m(\omega)$, $t_m(\omega)$ are the *measured* reflection and transmission coefficients, respectively, and $w_r(\omega)$, $w_t(\omega)$ are non-negative weight functions describing the weight of using the reflection and transmission data at different frequency points. In the above definition, the summation is performed over discrete frequency points in a certain frequency band $[\omega_{min}, \omega_{max}]$. In order to apply an optimization approach, one needs to compute the gradient of this objective functional.

Let $\tilde{r} = r(z; p + \delta p)$ and $\tilde{t} = t(z; p + \delta p)$ be the solutions to (6.134)–(6.137) with the parameter vector $p + \delta p$. Then $\delta r \equiv \tilde{r} - r$ and $\delta t \equiv \tilde{t} - t$ satisfy the following equations and boundary conditions,

$$\partial_z(\delta r) - 2\,(a + br)\,\delta r = 2r\delta a + \left(1 + r^2\right)\delta b + o\,(\delta p)\,, \tag{6.140}$$

$$\partial_z(\delta t) - (a + br)\,\delta t = (\delta a + r\delta b)\,t + bt\delta r + o\,(\delta p)\,, \tag{6.141}$$

$$\delta r(l) = \delta t(l) = 0, \tag{6.142}$$

where $\delta a = a(p + \delta p) - a(p), \delta b = b(p + \delta p) - b(p)$.

The corresponding increment of the functional $J(p)$ can then be written as follows,

$$\begin{aligned} \delta J\,(p) &\equiv J\,(p + \delta p) - J\,(p) \\ &= 2\mathrm{Re} \sum_{\omega=\omega_{min}}^{\omega_{max}} \{w_r[r\,(0,\omega) - r_m]^*\delta r\,(0,\omega) \\ &\quad + w_t[t\,(0,\omega) - t_m]^*\delta t\,(0,\omega)\} + o\,(\delta p)\,, \end{aligned} \tag{6.143}$$

where the superscript $*$ denotes the complex conjugate, and Re denotes the real part. Introduce a pair of dual functions $U(z,\omega)$ and $W(z,\omega)$ which satisfy the following equations and boundary conditions (Norgren and He 1996b)

$$\partial_z U + 2\,(a + br)\,U = -btW, \tag{6.144}$$

$$U\,(0,\omega) = w_r(\omega)[r\,(0,\omega) - r_m(\omega)]^*, \tag{6.145}$$

$$\partial_z W + (a + br)\,W = 0, \tag{6.146}$$

$$W(0,\omega) = w_t(\omega)[t(0,\omega) - t_m(\omega)]^*. \tag{6.147}$$

From (6.136) and (6.146) it immediately follows that $\frac{d}{dz}(tW) = 0$, i.e., $t(z,\omega)W(z,\omega) = K(\omega)$, where (cf. (6.147))

$$K(\omega) \equiv w_t(\omega)[t(0,\omega) - t_m(\omega)]^* t(0,\omega). \tag{6.148}$$

$U(z,\omega)$ can then be obtained explicitly from (6.144) and (6.145),

$$\begin{aligned} U(z,\omega) &= \exp\left\{-2\int_0^z (a+br)\,dz'\right\} \\ &\times \left[w_r\left(r(0,\omega) - r_m\right)^* - K\int_0^z b\exp\left\{2\int_0^{z'}(a+br)\,dz''\right\}dz'\right]. \end{aligned} \tag{6.149}$$

From (6.140), (6.141), (6.144) and (6.146), one has

$$\partial_z(U\delta r) = \left(2r\delta a + \left(1+r^2\right)\delta b\right)U - btW\delta r + o(\delta p), \tag{6.150}$$

$$\partial_z(W\delta t) = (\delta a + r\delta b)\,tW + btW\delta r + o(\delta p). \tag{6.151}$$

It then follows that (by integrating the sum of (6.150) and (6.151) over $[0,l]$ and using the boundary condition (6.142))

$$\begin{aligned} &U(0,\omega)\,\delta r(0,\omega) + W(0,\omega)\,\delta t(0,\omega) \\ &= -\int_0^l \left\{\left(2r\delta a + \left(1+r^2\right)\delta b\right)U + (\delta a + r\delta b)\,tW\right\}dz + o(\delta p). \end{aligned} \tag{6.152}$$

Therefore, one can write the increment of the functional as

$$\begin{aligned} \delta J(p) &= 2\mathrm{Re}\sum_{\omega=\omega_{min}}^{\omega_{max}} \left\{U(0,\omega)\,\delta r(0,\omega) + W(0,\omega)\,\delta t(0,\omega)\right\} + o(\delta p) \\ &= -2\int_0^l \left[\mathrm{Re}\sum_{\omega=\omega_{min}}^{\omega_{max}} \left\{[2r\delta a + \left(1+r^2\right)\delta b]U + (\delta a + r\delta b)\,K\right\}\right]dz + o(\delta p). \end{aligned} \tag{6.153}$$

If all the line parameters are non-dispersive, one can rewrite (6.153) in an inner product form

$$\begin{aligned} \delta J(p) &= \int_0^l g^T \cdot \delta p dz + o(\delta p) \\ &\equiv \langle g_L, \delta L\rangle + \langle g_C, \delta C\rangle + \langle g_R, \delta R\rangle + \langle g_G, \delta R\rangle + o(\delta p), \end{aligned} \tag{6.154}$$

where $g(z)$ is the gradient of the objective functional. Using (6.130) and (6.131) to express the increments δa and δb in (6.154) in terms of the parameter increments, one can identify the gradients as,

$$g_L(z) = -\mathrm{Re}\sum_{\omega=\omega_{min}}^{\omega_{max}} \frac{j\omega}{Z_0(\omega)}\left\{K(\omega)(1-r(z,\omega)) - (1-r(z,\omega))^2 U(z,\omega)\right\}, \tag{6.155}$$

$$g_C(z) = -\mathrm{Re}\sum_{\omega_{min}}^{\omega_{max}} j\omega Z_0(\omega)\left\{K(\omega)(1+r(z,\omega)) + (1+r(z,\omega))^2 U(z,\omega)\right\}, \tag{6.156}$$

$$g_R(z) = -\mathrm{Re}\sum_{\omega=\omega_{min}}^{\omega_{max}} \frac{1}{Z_0(\omega)}\left\{K(\omega)(1-r(z,\omega)) - (1-r(z,\omega))^2 U(z,\omega)\right\}, \tag{6.157}$$

$$g_G(z) = -\mathrm{Re}\sum_{\omega=\omega_{min}}^{\omega_{max}} Z_0(\omega)\left\{K(\omega)(1+r(z,\omega)) + (1+r(z,\omega))^2 U(z,\omega)\right\}. \tag{6.158}$$

Note that the above gradients (with respect to the line parameters) for the left-sided excitation are only functions of the z variable. Also note that if the scattering data for right-sided excitation, as well as the scattering data for left-sided excitation, are used in an inverse problem, an analogous term should be added in the definition (6.139) of the objective functional. Consequently, in that case the above expressions for the gradients will be modified with additional terms (which are similar to the right-hand side expressions of (6.155)–(6.158)).

6.3.3 *Numerical reconstruction*

With the explicit expression for the gradient of the objective functional available, a standard conjugate gradient method (see Section 5.1.1) is then used to minimize the objective functional. We give some numerical examples below of reconstructions of the line parameters with band-limited scattering data. Using the explicit expression for the gradient, one can also design a non-uniform line which is optimal with respect to different criteria, such as design of filters, design of impedance matching sections, and design of impedance-matched voltage transformers (Norgren and He 1996a).

Numerical example for reconstruction of one parameter from one-sided reflection data. In this numerical example a transmission line with a piecewise constant conductance is considered. The conductance profile is shown

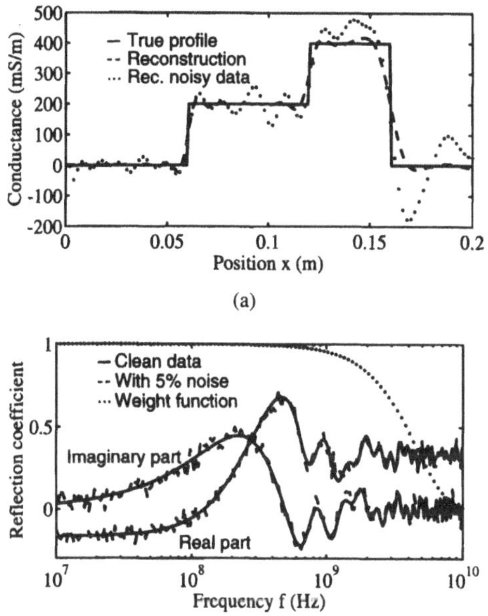

(b)

FIG. 6.5. (a) Reconstruction of the shunt conductance $G(z)$. (b) The reflection coefficient (the solid line) calculated by using (6.159) recursively (the dashed line is the noise-contaminated reflection coefficient; the dotted line is the weight function $w_r(\omega)$).

by the solid line in Fig. 6.5(a) (the length of the line is $l = 0.2$ m). The other three parameters are constants with values of $L = 0.50\,\mu$H/m, $C = 50\,$pF/m and $R = 0$. The non-uniform line is terminated at $z = 0$ with a uniform line with a characteristic impedance $Z_0 = 50\,\Omega$. The load end (at $z = 0.2\,$m) is left open, i.e. $Z_L = \infty$. For a line with piecewise constant parameters, one can easily obtain an explicit solution for the reflection coefficient by using recursively the following formula (Cheng 1993)

$$Z_{in} = Z_{char}\frac{Z_{load} + Z_{char}\tanh(\gamma d)}{Z_{char} + Z_{load}\tanh(\gamma d)}, \tag{6.159}$$

for each subportion of the line, starting at $z = l$. In (6.159), $Z_{char} = \sqrt{\frac{R+j\omega L}{G+j\omega C}}$, $\gamma = \sqrt{(R + j\omega L)(G + j\omega C)}$, d is the length of the subportion, and Z_{in} is the input impedance of the subportion when its load impedance is Z_{load}. When $Z_{in}|_{z=0}$ is calculated, the reflection coefficient can be obtained by $r_m(\omega) = [Z_{in}|_{z=0} - Z_0]/[Z_{in}|_{z=0} + Z_0]$. The calculated reflection coefficient $r_m(\omega)$ in the microwave frequency band $f \in [10\,\text{MHz}, 10\,\text{GHz}]$ is shown in Fig. 6.5(b), and will be used as the measured reflection data for the inverse problem in this example. Since the measured data $r_m(\omega)$ are calculated in a way that is different from the direct solver, the numerical 'marching back effect' is avoided when solving the inverse problem. 201 frequency points are used with a logarithmic spacing in the microwave frequency band $f \in [10\,\mathrm{MHz}, 10\,\mathrm{GHz}]$, and 401 grid-points are used in $z \in [0, l]$. In this example the weight function for the reflection data is chosen to be $w_r(\omega) = 100^{-(\omega/\omega_{max})^2}$, where $\omega_{max} = 2\pi \cdot 10^9\ \mathrm{s}^{-1}$ (see the dotted line in Fig. 6.5(b); $w_t(\omega) \equiv 0$, i.e., no transmission data is used). The dashed line in Fig. 6.5(a) is the reconstruction after 150 iterations (the starting guess is identically zero). To test the stability of the algorithm, Gaussian noise with a standard deviation of $0.05 \cdot \max |r_m(\omega = 2\pi f)|$ is added on both the real and the imaginary parts of $r_m(\omega)$ (see the dashed lines in Fig. 6.5(b)). The dotted line in Fig. 6.5(a) is the corresponding reconstruction using the noisy reflection data. The influence of noise is clearly visible around $z = 0.16$ m where the parameter has a large discontinuity. Nevertheless, one sees from this figure that the reconstruction algorithm is reasonably stable. In the case of reconstructing one parameter, it is found that any of the parameters L, C, R and G can be successfully reconstructed from one-sided reflection data. If two-sided reflection data are used, the rate of convergence can be increased significantly. However, the inclusion of transmission data does not give any improvement of the reconstruction.

Reconstruction of two parameters from two-sided reflection data. Numerical experiments indicate that using only one-sided excitation one cannot obtain a good simultaneous reconstruction of two parameters (the other two are assumed to be known) with the optimization approach. However, if two-sided reflection data are used, one can reconstruct two parameters simultaneously. The reconstruction of a reactive parameter (L or C) together with a dissipative parameter (R or G) appears to be successful with fast convergence. Reconstruction of the two dissipative parameters, R and G, is also successful. In the case of reconstructing the two reactive parameters L and C, the reconstruction is successful (however, with slow convergence) if the non-uniform line is lossy. If $R = G \equiv 0$, unsuccessful reconstruction readily occurs. Here we give a numerical example for a dispersive case. The telegrapher's equations (6.127) are derived with the assumption that the wave in the transmission line is (or is approximately) a TEM-wave, which is true only for the case when the conductivity σ_c of the conductors is much higher than the conductivity σ of the material between the conductors. In such a case, the series resistance R at high frequencies is proportional to the real part of the surface impedance Z_s, which is given by $Z_s = (1+j)\sqrt{\frac{\omega\mu_c}{2\sigma_c}}$ (see e.g. Jackson (1975); μ_c is the permeability of the conductors). The imaginary part of Z_s is due to an *internal* inductance L_i. In view of this physical background, it is interesting to consider a dispersive case for which R in the telegrapher's equations (6.127) is replaced with the following quantity

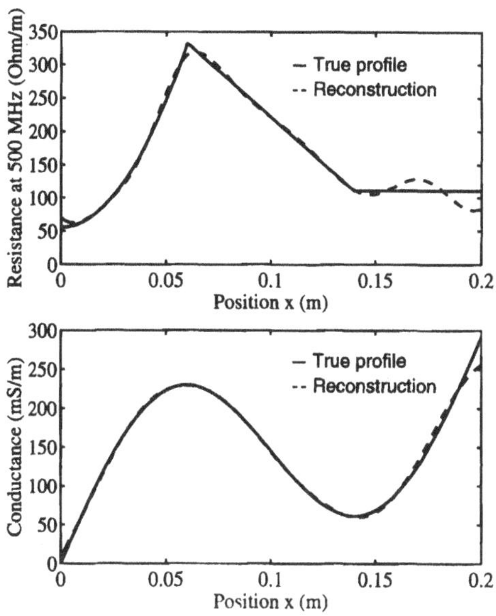

FIG. 6.6. Simultaneous reconstruction of the resistance R (dispersive) and the conductance G.

$$R(z,\omega)+j\omega L_i(z,\omega)\equiv(1+j)\,Q(z)\sqrt{\omega}, \tag{6.160}$$

where $Q(z)$ is proportional to $\sqrt{\mu_c(z)/\sigma_c(z)}$ (Cheng 1993). Consequently, the expression (6.157) for the gradient $g_R(z)$ should be replaced with the following expression for the gradient with respect to the function $Q(z)$,

$$\begin{aligned} g_Q(z) = -\mathrm{Re} \sum_{\omega=\omega_{min}}^{\omega_{max}} &(1+j)\sqrt{\omega}Y_0(\omega) \\ &\cdot\left\{K(\omega)\left[1-r(z,\omega)\right]-\left[1-r(z,\omega)\right]^2 U(z,\omega)\right\}. \end{aligned} \tag{6.161}$$

As a numerical example, the non-uniform dispersive line is always excited from a 50 Ω uniform LC-line and kept open at the other end (i.e., the load impedance is $Z_L=\infty$), for both cases of the left- and right-sided reflection. The length of the non-uniform line is chosen to be $l=0.2$ m. The line parameters are shown by the solid lines in Figs 6.6 and 6.7. The number of gridpoints in z is 101, and 100 frequency points are used over the microwave frequency band $f\in[10\,\mathrm{MHz},1\,\mathrm{GHz}]$ with a linear spacing. The simultaneous reconstruction of R (at the frequency 500 MHz) and G after 100 iterations is shown by the dashed lines in Fig. 6. 6 (the starting guess of the line parameters is $R=G\equiv0$; note that the reconstruction is carried out for the parameter pairs $G(z)$ and $Q(z)$,

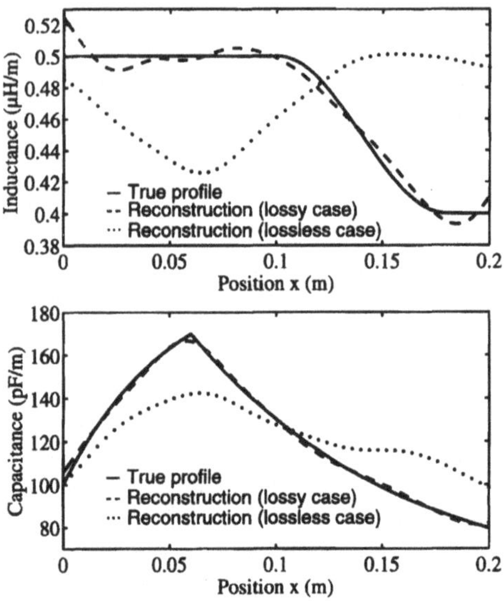

FIG. 6.7. Simultaneous reconstruction of the inductance L and the capacitance C. The true profiles are given by the solid lines. The dashed lines are the reconstructions when R and G are as shown in Fig. 6.6, and the dotted lines are the reconstructions when the line is lossless (i.e. $R = G \equiv 0$).

although the reconstruction result is shown for G and $R \equiv Q\sqrt{\omega}$ in Fig. 6.6). The simultaneous reconstruction of L and C using two-sided reflection data is presented in Fig. 6.7 after 170 iterations (the starting guess of the line parameters is chosen as $L = 0.25$ μH/m and $C = 100$ pF/m). The dashed lines in Fig. 6.7 are the simultaneous reconstructions when the non-uniform line is lossy (with R and G as shown by the solid lines in Fig. 6.6), while the dotted lines are the corresponding reconstructions when the non-uniform line is lossless (i.e., $R = G = 0$). The reconstruction is reasonably good in the lossy case, but fails in the lossless case. It is well known that in the corresponding time domain model, $(L/C)^{1/2}$ is the characteristic impedance and $(LC)^{-1/2}$ gives the velocity of the wavefront. Numerical experiments indicated that in the case of reconstructing L and C when the line is lossless, the failure lies in the reconstruction of the wavefront velocity (the reconstruction of the characteristic impedance is still reliable).

Reconstruction of more than two parameters. For simultaneous reconstruction of three parameters, successful reconstruction using two-sided reflection data can be achieved only when the starting guess of the line parameters is close to their true parameter profiles (otherwise the reconstruction readily fails). Numerical experiments also show that using several different values for the load impedance Z_L does not give any improvement in the reconstruction.

The use of transmission data. The solution to the inverse problem of reconstructing any line parameter using only transmission data is in general not unique. For example, if the load impedance is equal to the impedance of the terminated uniform line, i.e., $Z_L = Z_0$, it can be shown from reciprocity that it is impossible to distinguish a parameter profile $p(z)$ from its centre-image profile $p(l - z)$, when only transmission data are used (i.e., the two profiles generate the same transmission data). Even when the solution to the inverse problem is unique, the reconstruction may not be achieved due to the existence of local minima. In the types of reconstruction (described above) which use only reflection data, we never came across a case in which the reconstruction was trapped by a local minimum (i.e., it appears that the objective functional, defined by (6.139) with $w_t \equiv 0$, is convex in a very large area centred at the true profile). One may like to include the transmission data to make full use of the available information. However, numerical experiments show that including the transmission data will introduce local minima, and thus it is better not to use the transmission data at all in the present optimization approach.

6.4 Trace formalism and explicit gradients

In the previous section, we have shown how to derive an explicit gradient for an objective function related to scattering coefficients obtained by solving, e.g. scalar differential equations of Riccati type. In an inverse scattering problem for a stratified bi-anisotropic slab, the objective function is related to scattering coefficient matrices which are obtained by solving, e.g. a *matrix* differential equation of Riccati type. In order to derive an exact and explicit expressions for the gradients in a simple and compact matrix form, we show in this section how the matrix trace operator can be utilized to achieve this (He et al. 1997). (Note that without using the trace operator one has to rearrange the scattering coefficient matrices into vectors, and the derivation will become more complicated.)

6.4.1 *Riccati equation, trace formalism and explicit gradients*

Consider an electromagnetic plane wave with harmonic time dependence $e^{j\omega t}$ which is obliquely incident on a stratified bi-anisotropic slab (with incident angles (θ, ϕ)). The slab occupies the region $0 \leq z \leq l$. The plane wave impinges on the slab from the left side $z < 0$.

The reflection coefficient for such a stratified slab can be obtained by solving the following equations for the reflection coefficient matrix $\mathbf{r}(z)$ (cf. (6.91))

$$\partial_z \mathbf{r} = \boldsymbol{\gamma} + (\boldsymbol{\vartheta}\,\mathbf{r} - \mathbf{r}\,\boldsymbol{\alpha}) - \mathbf{r}\,\boldsymbol{\beta}\,\mathbf{r}, \tag{6.162}$$

$$\mathbf{r}(l) = \mathbf{r}_l, \tag{6.163}$$

where $\mathbf{r}_l$ is a fixed known matrix (i.e., the reflection coefficient matrix at an interface between two homogeneous media characterized by the parameters at $z = l^-$ and $z = l^+$). It is understood that the elements of any 2×2 matrix in this section are denoted by the same letter as the matrix and have row and column indices as subscripts, i.e.,

$$\mathbf{r} \equiv \begin{bmatrix} r_{11} & r_{12} \\ r_{21} & r_{22} \end{bmatrix}, \qquad \mathbf{A} \equiv \begin{bmatrix} A_{11} & A_{12} \\ A_{21} & A_{22} \end{bmatrix}, \qquad \text{etc.}$$

As shown in the previous sections, the Riccati equation for the reflection coefficient matrix is not unique, rather it depends on what type of splitting that is used. We recommend the Riccati equation associated with the vacuum wave-splitting (cf. Section 6.2), in which spatial derivatives of the constitutive parameters do not appear. The solution for $\mathbf{r}(z)$ is obtained by integrating (6.162) along the $-z$ direction starting from $z = l$. Note that the physical reflection coefficient matrix for the stratified slab is $\mathbf{r}(0)$.

For an inverse or design problem, the parameter vector $p = \{p_1, p_2, ..., p_N\}(z)$ (N is the number of the constitutive parameters to be reconstructed or designed), can be obtained by minimizing a suitable objective functional. Define an objective functional as follows:

$$J(p) = \sum_{\omega,\theta,\phi} \sum_{i,j=1,2} \left| r_{ij}(0; p; \omega, \theta, \phi) - r_{ij}^{(m)}(\omega, \theta, \phi) \right|^2, \tag{6.164}$$

where $r_{ij}^{(m)}(\omega, \theta, \phi)$, $i, j = 1, 2$, are the measured/prescribed reflection coefficients. Here we use the reflection data at a finite set of frequency points and for several different values of the incident angles (θ, ϕ). For, e.g. the design of an anti-reflection bi-anisotropic coating, one simply sets $r_{ij}^{(m)}(\omega, \theta, \phi) \equiv 0$, $i, j = 1, 2$. For simplicity of notation we will suppress the dependence on the frequency, the incident angles and the parameter vector $p(z)$ in the arguments of the scattering coefficients in the rest of this section. In order to apply a minimization approach, one needs to compute the gradient of this objective functional.

We show how the matrix trace operator can be utilized to derive the gradient explicitly in a simple and compact matrix form. We recall that the trace of a $N \times N$ matrix $\mathbf{A} = \{A_{ij}\}$, denoted by $\mathrm{tr}(\mathbf{A})$, is defined by $\mathrm{tr}(\mathbf{A}) = \sum_{i=1}^{N} A_{ii}$. The objective functional defined by (6.164) can be rewritten in the following trace form,

$$J(p) = \sum_{\omega,\theta,\phi} \mathrm{tr} \left\{ [\mathbf{r}(0) - \mathbf{r}^{(m)}][\mathbf{r}(0) - \mathbf{r}^{(m)}]^\dagger \right\}, \tag{6.165}$$

where the superscript $\dagger$ denotes the complex conjugate transpose. Let $\tilde{\mathbf{r}}(z) \equiv \mathbf{r}(z; p + \delta p)$ be the solutions correponding to the parameter profile $p + \delta p$ (δp

is a small change of the profile $p(z)$). Then $\delta\mathbf{r} \equiv \tilde{\mathbf{r}} - \mathbf{r}$ satisfies the following equations:

$$\begin{aligned}\partial_z(\delta\mathbf{r}) &= \delta\boldsymbol{\gamma} + (\delta\boldsymbol{\vartheta})\mathbf{r} + \boldsymbol{\vartheta}(\delta\mathbf{r}) - (\delta\mathbf{r})\boldsymbol{\alpha} - \mathbf{r}(\delta\boldsymbol{\alpha}) \\ &\quad -(\delta\mathbf{r})\boldsymbol{\beta}\mathbf{r} - \mathbf{r}(\delta\boldsymbol{\beta})\mathbf{r} - \mathbf{r}\boldsymbol{\beta}(\delta\mathbf{r}) + o\,(\delta p)\,, \qquad (6.166)\\ (\delta\mathbf{r})(l) &= 0. \qquad (6.167)\end{aligned}$$

The increment in the objective functional can be written as

$$\begin{aligned}\delta J\,(p) &= \sum_{\omega,\theta,\phi} \mathrm{tr}\left\{[(\delta\mathbf{r})(0)][\mathbf{r}(0) - \mathbf{r}^{(m)}]^\dagger + [\mathbf{r}(0) - \mathbf{r}^{(m)}][(\delta\mathbf{r})(0)]^\dagger\right\} + o\,(\delta p) \\ &= 2\mathrm{Re}\ \ \mathrm{tr}\left\{[\mathbf{r}(0) - \mathbf{r}^{(m)}]^\dagger(\delta\mathbf{r})(0)\right\} + o\,(\delta p)\,. \qquad (6.168)\end{aligned}$$

In the above derivation, we have used the following properties of the trace of a matrix:

$$\mathrm{Re}\ [\mathrm{tr}(\mathbf{A})] = \mathrm{Re}\ [\mathrm{tr}(\mathbf{A}^\dagger)] = \frac{1}{2}\,\mathrm{tr}(\mathbf{A} + \mathbf{A}^\dagger), \qquad (6.169)$$

$$\mathrm{tr}(\mathbf{AB}) = \mathrm{tr}(\mathbf{BA}). \qquad (6.170)$$

Here $\mathbf{A}$ and $\mathbf{B}$ are two arbitrary square matrices of the same order.

Introduce a dual matrix function $\mathbf{U}(z)$ which satisfies the following boundary condition:

$$\mathbf{U}(0) = [\mathbf{r}(0) - \mathbf{r}^{(m)}]^\dagger. \qquad (6.171)$$

Thus one has

$$\begin{aligned}\int_0^l \partial_z(\mathbf{U}\delta\mathbf{r})\,dz &= \mathbf{U}(l)(\delta\mathbf{r})(l) - \mathbf{U}(0)(\delta\mathbf{r})(0) \\ &= -\mathbf{U}(0)(\delta\mathbf{r})(0) = -[\mathbf{r}(0) - \mathbf{r}^{(m)}]^\dagger(\delta\mathbf{r})(0),\end{aligned}$$

and

$$\delta J\,(p) = -2\mathrm{Re} \sum_{\omega,\theta,\phi} \int_0^l \mathrm{tr}\,\{\partial_z[\mathbf{U}(\delta\mathbf{r})]\}\ dz + o\,(\delta p)\,. \qquad (6.172)$$

Furthermore, using the differential equation (6.166) and the trace property

$$\mathrm{tr}(\mathbf{ABC}) = \mathrm{tr}(\mathbf{CAB}) \qquad (6.173)$$

(note that $\mathrm{tr}(\mathbf{ABC}) \neq \mathrm{tr}(\mathbf{ACB})$), one obtains

$$\begin{aligned}\operatorname{tr}\{\partial_z[\mathbf{U}(\delta\mathbf{r})]\} &= \operatorname{tr}\{(\partial_z\mathbf{U})(\delta\mathbf{r}) + \mathbf{U}\partial_z(\delta\mathbf{r})\}\\ &= \operatorname{tr}\{(\partial_z\mathbf{U})(\delta\mathbf{r}) + \mathbf{U}[\delta\boldsymbol{\gamma} + (\delta\boldsymbol{\vartheta})\mathbf{r} + \boldsymbol{\vartheta}(\delta\mathbf{r}) - (\delta\mathbf{r})\boldsymbol{\alpha}\\ &\quad -\mathbf{r}(\delta\boldsymbol{\alpha}) - (\delta\mathbf{r})\boldsymbol{\beta}\mathbf{r} - \mathbf{r}(\delta\boldsymbol{\beta})\mathbf{r} - \mathbf{r}\boldsymbol{\beta}(\delta\mathbf{r})]\}\\ &= \operatorname{tr}\{(\partial_z\mathbf{U} + \mathbf{U}\boldsymbol{\vartheta} - \boldsymbol{\alpha}\mathbf{U} - \mathbf{U}\mathbf{r}\boldsymbol{\beta} - \boldsymbol{\beta}\mathbf{r}\mathbf{U})(\delta\mathbf{r})\\ &\quad +\mathbf{U}[\delta\boldsymbol{\gamma} + (\delta\boldsymbol{\vartheta})\mathbf{r} - \mathbf{r}(\delta\boldsymbol{\alpha}) - \mathbf{r}(\delta\boldsymbol{\beta})\mathbf{r}]\}. \end{aligned} \tag{6.174}$$

Thus, if $\mathbf{U}$ satisfies the following differential equation:

$$\partial_z\mathbf{U} = -\mathbf{U}\,\boldsymbol{\vartheta} + \boldsymbol{\alpha}\,\mathbf{U} + \mathbf{U}\,\mathbf{r}\,\boldsymbol{\beta} + \boldsymbol{\beta}\,\mathbf{r}\,\mathbf{U}, \tag{6.175}$$

it follows from (6.172) and (6.174) that

$$\delta J\,(p) = -2\mathrm{Re}\sum_{\omega,\theta,\phi}\int_0^l \operatorname{tr}\{\mathbf{U}[\delta\boldsymbol{\gamma} + (\delta\boldsymbol{\vartheta})\mathbf{r} - \mathbf{r}(\delta\boldsymbol{\alpha}) - \mathbf{r}(\delta\boldsymbol{\beta})\mathbf{r}]\}\,dz + o\,(\delta p)\,. \tag{6.176}$$

The above equation can be rewritten in an inner product form

$$\delta J(p) = \int_0^l \left\{\sum_{n=1}^N \frac{\partial J}{\partial p_n}\,\delta p_n\right\}\,dz + o\,(\delta p)\,, \tag{6.177}$$

where the gradients $\frac{\partial J}{\partial p_n}$ of the objective functional $J(p)$ with respect to the parameters $p_n(z)$, $n = 1, 2, ..., N$, are given by

$$\frac{\partial J}{\partial p_n}(z) = 2\mathrm{Re}\sum_{\omega,\theta,\phi}\operatorname{tr}\left\{\mathbf{U}\left[\mathbf{r}\left(\frac{\partial\boldsymbol{\alpha}}{\partial p_n}\right) + \mathbf{r}\left(\frac{\partial\boldsymbol{\beta}}{\partial p_n}\right)\mathbf{r} - \frac{\partial\boldsymbol{\gamma}}{\partial p_n} - \left(\frac{\partial\boldsymbol{\vartheta}}{\partial p_n}\right)\mathbf{r}\right]\right\}. \tag{6.178}$$

Note that $\mathbf{U}(z)$ is obtained by integrating (6.175) along the z direction starting from $z = 0$ (cf. (6.171)). This trace formalism was introduced in He et al. (1997).

A numerical example. As a numerical example, we consider the reconstruction of the constitutive parameters for a stratified chiral slab. The constitutive relations for the chiral medium are written as

$$\mathbf{D} = \epsilon_0\epsilon_r\mathbf{E} - j\kappa\sqrt{\epsilon_0\mu_0}\,\mathbf{H}, \qquad \mathbf{B} = \mu_0\mu_r\mathbf{H} + j\kappa\sqrt{\epsilon_0\mu_0}\,\mathbf{E},$$

where the parameter κ (dimensionless) gives the degree of chirality (the constants ϵ_0, μ_0 are the permittivity and permeability in vacuum, respectively). Using the vacuum wave-splitting, one has

$$\alpha = \frac{k_0}{2}\{\kappa I_2 + j\epsilon_r \mathbf{Z}_0 I_2 - j\mu_r I_2 \mathbf{Z}_0^{-1} + \kappa \mathbf{Z}_0 I_2 \mathbf{Z}_0^{-1} + \frac{\sin^2\theta}{\epsilon_r\mu_r - \kappa^2}[\kappa I_3 - j\epsilon_r \mathbf{Z}_0 I_3 + j\mu_r I_3 \mathbf{Z}_0^{-1} + \kappa \mathbf{Z}_0 I_3 \mathbf{Z}_0^{-1}]\},$$

$$\beta = \frac{k_0}{2}\{\kappa I_2 + j\epsilon_r \mathbf{Z}_0 I_2 + j\mu_r I_2 \mathbf{Z}_0^{-1} - \kappa \mathbf{Z}_0 I_2 \mathbf{Z}_0^{-1} + \frac{\sin^2\theta}{\epsilon_r\mu_r - \kappa^2}[\kappa I_3 - j\epsilon_r \mathbf{Z}_0 I_3 - j\mu_r I_3 \mathbf{Z}_0^{-1} - \kappa \mathbf{Z}_0 I_3 \mathbf{Z}_0^{-1}]\},$$

$$\gamma = \frac{k_0}{2}\{\kappa I_2 - j\epsilon_r \mathbf{Z}_0 I_2 - j\mu_r I_2 \mathbf{Z}_0^{-1} - \kappa \mathbf{Z}_0 I_2 \mathbf{Z}_0^{-1} + \frac{\sin^2\theta}{\epsilon_r\mu_r - \kappa^2}[\kappa I_3 + j\epsilon_r \mathbf{Z}_0 I_3 + j\mu_r I_3 \mathbf{Z}_0^{-1} - \kappa \mathbf{Z}_0 I_3 \mathbf{Z}_0^{-1}]\},$$

$$\vartheta = \frac{k_0}{2}\{\kappa I_2 - j\epsilon_r \mathbf{Z}_0 I_2 + j\mu_r I_2 \mathbf{Z}_0^{-1} + \kappa \mathbf{Z}_0 I_2 \mathbf{Z}_0^{-1} + \frac{\sin^2\theta}{\epsilon_r\mu_r - \kappa^2}[\kappa I_3 + j\epsilon_r \mathbf{Z}_0 I_3 - j\mu_r I_3 \mathbf{Z}_0^{-1} + \kappa \mathbf{Z}_0 I_3 \mathbf{Z}_0^{-1}]\},$$

and where $k_0 = \omega\sqrt{\epsilon_0\mu_0}$ is the wave number of the incident field, and

$$\mathbf{Z}_0 = \begin{bmatrix} 0 & \cos\theta \\ -\dfrac{1}{\cos\theta} & 0 \end{bmatrix} = -\mathbf{Z}_0^{-1}, \qquad I_3 = \begin{bmatrix} 0 & 1 \\ 0 & 0 \end{bmatrix}.$$

The parameters are reconstructed with a conjugate gradient method and the results are shown in Fig. 6.8. The solid curves are the true parameters, and the circles are the reconstruction using double-sided reflection (backed by a perfectly conducting surface) and one incident angle $\theta = \pi/4$. The small error (in particular at $z = 0$) is mainly due to the limited frequency band and numerical errors. The objective functional $J(p)$ has been reduced by a factor of 10^{-5} for the reconstruction in Fig. 6.8 compared to the objective functional for the initial guess, which is a vacuum. We refer to Norgren (1996, 1997) for numerical results on the design of anti-reflective bi-anisotropic coatings using explicit gradients.

In the next subsection we will use both reflection and transmission data to reconstruct/design several parameters. When both reflection and transmission data are used, it is useful to introduce weight functions which in a flexible way can model the different influences of the reflection and transmission data in the optimization process (transmission data are usually less suitable for use in a minimization approach due to the appearance of local minima; cf. Section 6.3). If one wishes to introduce weight functions in the definition of the objective functional, (6.164) can be modified as

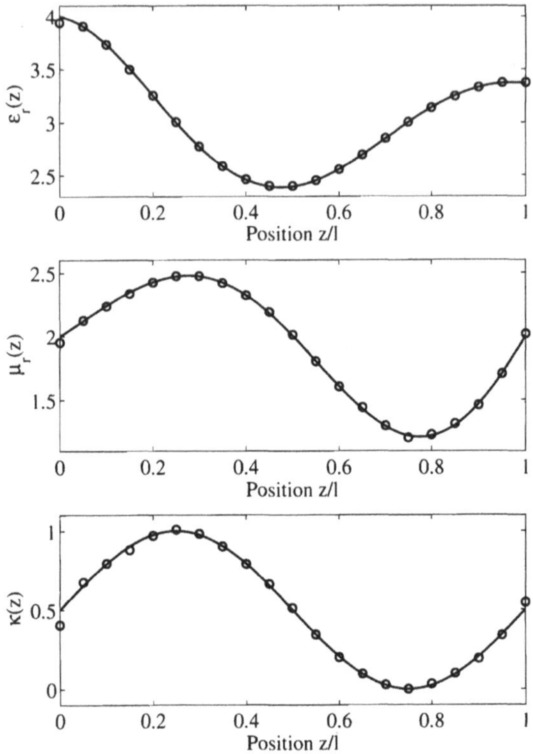

FIG. 6.8. The parameter reconstruction for a stratified chiral slab.

$$\begin{aligned}J(p) &= \sum_{\omega,\theta,\phi} \sum_{i,j=1,2} W^r_{i,j} \left| r_{ij}(0) - r^{(m)}_{ij} \right|^2 \\ &= \sum_{\omega,\theta,\phi} \operatorname{tr} \left\{ \mathbf{W}^r \circ \left[\mathbf{r}(0) - \mathbf{r}^{(m)} \right] \left[\mathbf{r}(0) - \mathbf{r}^{(m)} \right]^+ \right\},\end{aligned}$$

where $W^r_{i,j}$, $i,j = 1,2$, are non-negative weight functions (which can be functions of the frequency and the incident angles), and $\circ$ denotes the dot product of two matrices, i.e.,

$$\begin{bmatrix} A_{11} & A_{12} \\ A_{21} & A_{22} \end{bmatrix} \circ \begin{bmatrix} B_{11} & B_{12} \\ B_{21} & B_{22} \end{bmatrix} = \begin{bmatrix} A_{11}B_{11} & A_{12}B_{12} \\ A_{21}B_{21} & A_{22}B_{22} \end{bmatrix}.$$

The dot product has the following property:

$$(\mathbf{A} \circ \mathbf{B})\mathbf{C} = \mathbf{A} \circ (\mathbf{BC}).$$

Using the above property one can show that the explicit expression for the gradient remains the same, as well as the differential equation for the dual matrix function $\mathbf{U}(z)$. However, the boundary condition for $\mathbf{U}$ becomes (instead of (6.171))

$$\mathbf{U}(0) = \mathbf{W}^r \circ [\mathbf{r}(0) - \mathbf{r}^{(m)}]^+.$$

One can choose a slightly different boundary condition for the dual matrix function to avoid the matrix dot product operator when both the reflection and transmission data are used, as shown in the next two sections.

6.4.2 *Reconstruction/design using both reflection and transmission data*

In this subsection we use both the reflection and transmission data to reconstruct/design several constitutive parameters of a stratified bi-anisotropic slab. The imbedding equation and boundary condition for the transmission coefficient matrix are (cf. (6.93))

$$\partial_z \mathbf{t} = -\mathbf{t}\,(\boldsymbol{\alpha} + \boldsymbol{\beta}\mathbf{r}), \tag{6.179}$$

$$\mathbf{t}(l) = \mathbf{t}_l, \tag{6.180}$$

where $\mathbf{t}_l$ is a fixed known matrix (i.e., the transmission coefficient matrix at an interface between two homogeneous media characterized by the parameters at $z = l^-$ and $z = l^+$). The solution for $\mathbf{t}(z)$ can be obtained by integrating (6.179) along the $-z$ direction starting from $z = l$.

The objective functional is now defined by

$$J(p) = \sum_{\omega,\theta,\phi} \sum_{i,j=1,2} \left\{ W^r_{i,j} \left| r_{ij}(0) - r^{(m)}_{ij} \right|^2 + W^t_{i,j} \left| t_{ij}(0) - t^{(m)}_{ij} \right|^2 \right\}. \tag{6.181}$$

One of the relevant design problems using transmission data is the design of a wide-angled high-pass radome in a given frequency band. The increment in the objective functional due to a small parameter change δp can be written as

$$\begin{aligned} \delta J(p) = {} & 2 \sum_{\omega,\theta,\phi} \sum_{i,j=1,2} \mathrm{Re} \left\{ W^r_{i,j} \left[r_{i,j}(0) - r^{(m)}_{i,j} \right]^* (\delta r_{i,j})(0) \right. \\ & \left. + W^t_{i,j} \left[t_{i,j}(0) - t^{(m)}_{i,j} \right]^* (\delta t_{i,j})(0) \right\} + o(\delta p), \end{aligned} \tag{6.182}$$

where the superscript $*$ denotes the complex conjugate. Choosing the following boundary values for the dual matrix functions $\mathbf{U}(z)$ and $\mathbf{V}(z)$:

$$U_{i,j}(0) = W^r_{i,j} \left[r_{i,j}(0) - r^{(m)}_{i,j} \right]^*, \tag{6.183}$$

$$V_{i,j}(0) = W^t_{i,j}\left[t_{i,j}(0) - t^{(m)}_{i,j}\right]^*, \tag{6.184}$$

one obtains

$$\delta J(p) = 2\mathrm{Re}\sum_{\omega,\theta,\phi} \mathrm{tr}\left\{\mathbf{U}^\top(0)(\delta\mathbf{r})(0) + \mathbf{V}^\top(0)(\delta\mathbf{t})(0)\right\} + o(p), \tag{6.185}$$

where the superscript $\top$ denotes the matrix transpose. The increments $(\delta\mathbf{r})(z)$ and $(\delta\mathbf{t})(z)$ satisfy the following equations:

$$\begin{aligned}
\partial_z(\delta\mathbf{r}) &= (\boldsymbol{\vartheta} - \mathbf{r}\boldsymbol{\beta})(\delta\mathbf{r}) - (\delta\mathbf{r})(\boldsymbol{\alpha} + \boldsymbol{\beta}\mathbf{r}) \\
&\quad + \delta\boldsymbol{\gamma} + (\delta\boldsymbol{\vartheta})\mathbf{r} - \mathbf{r}(\delta\boldsymbol{\alpha}) - \mathbf{r}(\delta\boldsymbol{\beta})\mathbf{r} + o(p), &(6.186)\\
\partial_z(\delta\mathbf{t}) &= -\delta\mathbf{t}(\boldsymbol{\alpha} + \boldsymbol{\beta}\mathbf{r}) - \mathbf{t}\boldsymbol{\beta}(\delta\mathbf{r}) - \mathbf{t}(\delta\boldsymbol{\alpha}) - \mathbf{t}(\delta\boldsymbol{\beta})\mathbf{r} + o(p), &(6.187)\\
\delta\mathbf{r}(l^-) &= \delta\mathbf{t}(l^-) = 0. &(6.188)
\end{aligned}$$

Using (6.185)–(6.188) and the matrix trace properties (6.170) and (6.173), one obtains

$$\begin{aligned}
\delta J(p) &= 2\mathrm{Re}\sum_{\omega,\theta,\phi}\mathrm{tr}\left\{-\int_0^l \partial_z\left[\mathbf{U}^\top(\delta\mathbf{r}) + \mathbf{V}^\top(\delta\mathbf{t})\right]dz\right\} + o(p) \\
&= -2\int_0^l dz\,\mathrm{Re}\sum_{\omega,\theta,\phi}\mathrm{tr}\left\{(\partial_z\mathbf{U}^\top)(\delta\mathbf{r}) + \mathbf{U}^\top\partial_z(\delta\mathbf{r}) + (\partial_z\mathbf{V}^\top)(\delta\mathbf{t})\right.\\
&\quad \left. + \mathbf{V}^\top\partial_z(\delta\mathbf{t})\right\} + o(p) \\
&= -2\int_0^l dz\,\mathrm{Re}\sum_{\omega,\theta,\phi}\mathrm{tr}\left\{(\partial_z\mathbf{U}^\top)(\delta\mathbf{r}) + \mathbf{U}^\top\left[(\boldsymbol{\vartheta} - \mathbf{r}\boldsymbol{\beta})(\delta\mathbf{r}) - (\delta\mathbf{r})(\boldsymbol{\alpha} + \boldsymbol{\beta}\mathbf{r})\right]\right.\\
&\quad + \mathbf{U}^\top\left[\delta\boldsymbol{\gamma} + (\delta\boldsymbol{\vartheta})\mathbf{r} - \mathbf{r}(\delta\boldsymbol{\alpha}) - \mathbf{r}(\delta\boldsymbol{\beta})\mathbf{r}\right] + \left(\partial_z\mathbf{V}^\top\right)(\delta\mathbf{t}) \\
&\quad \left. - \mathbf{V}^\top(\delta\mathbf{t})(\boldsymbol{\alpha} + \boldsymbol{\beta}\mathbf{r}) - \mathbf{V}^\top\mathbf{t}\boldsymbol{\beta}(\delta\mathbf{r}) - \mathbf{V}^\top\mathbf{t}(\delta\boldsymbol{\alpha} + (\delta\boldsymbol{\beta})\mathbf{r})\right\} + o(p) \\
&= -2\int_0^l dz\,\mathrm{Re}\sum_{\omega,\theta,\phi}\mathrm{tr}\left\{\left[\partial_z\mathbf{U}^\top + \mathbf{U}^\top(\boldsymbol{\vartheta} - \mathbf{r}\boldsymbol{\beta}) - (\boldsymbol{\alpha} + \boldsymbol{\beta}\mathbf{r})\mathbf{U}^\top\right.\right.\\
&\quad \left. - \mathbf{V}^\top\mathbf{t}\boldsymbol{\beta}\right](\delta\mathbf{r}) + \left[\partial_z\mathbf{V}^\top - (\boldsymbol{\alpha} + \boldsymbol{\beta}\mathbf{r})\mathbf{V}^\top\right](\delta\mathbf{t}) \\
&\quad \left. - \mathbf{U}^\top\left[\mathbf{r}(\delta\boldsymbol{\alpha}) + \mathbf{r}(\delta\boldsymbol{\beta})\mathbf{r} - \delta\boldsymbol{\gamma} - (\delta\boldsymbol{\vartheta})\mathbf{r}\right] - \mathbf{V}^\top\mathbf{t}\left[\delta\boldsymbol{\alpha} + (\delta\boldsymbol{\beta})\mathbf{r}\right]\right\} + o(p).
\end{aligned}$$

To eliminate $\delta\mathbf{r}$ and $\delta\mathbf{t}$ terms in the above equation, we choose the following differential equations for the dual matrix functions:

$$\begin{aligned}
\partial_z\mathbf{U}^\top &= (\boldsymbol{\alpha} + \boldsymbol{\beta}\mathbf{r})\mathbf{U}^\top - \mathbf{U}^\top(\boldsymbol{\vartheta} - \mathbf{r}\boldsymbol{\beta}) + \mathbf{V}^\top\mathbf{t}\boldsymbol{\beta}, &(6.189)\\
\partial_z\mathbf{V}^\top &= (\boldsymbol{\alpha} + \boldsymbol{\beta}\mathbf{r})\mathbf{V}^\top. &(6.190)
\end{aligned}$$

The increment in the objective functional thus becomes

$$\delta J(p) = 2\int_0^l dz\mathrm{Re}\sum_{\omega,\theta,\phi} \mathrm{tr}\left\{\left(\mathbf{U}^\top\mathbf{r}+\mathbf{V}^\top\mathbf{t}\right)[\delta\alpha+(\delta\beta)\mathbf{r}] \right.$$
$$\left. -\mathbf{U}^\top[\delta\gamma+(\delta\vartheta)\mathbf{r}]\right\}+o(p), \tag{6.191}$$

from which one obtains immediately the following explicit expression for the gradient:

$$\frac{\partial J}{\partial p_n}(z) = 2\mathrm{Re}\sum_{\omega,\theta,\phi} \mathrm{tr}\left\{\left(\mathbf{U}^\top\mathbf{r}+\mathbf{V}^\top\mathbf{t}\right)\left(\frac{\partial\alpha}{\partial p_n}+\frac{\partial\beta}{\partial p_n}\mathbf{r}\right)\right.$$
$$\left. -\mathbf{U}^\top\left(\frac{\partial\gamma}{\partial p_n}+\frac{\partial\vartheta}{\partial p_n}\mathbf{r}\right)\right\}. \tag{6.192}$$

Note that $\mathbf{U}(z)$ and $\mathbf{V}(z)$ in the above expression are computed by integrating (6.189) and (6.190) along the z direction starting from $z = 0$ (cf. (6.183) and (6.184)).

6.4.3 *Reconstruction/design using the transmission Green's function approach*

As an alternative, we derive in this subsection an explicit expression for the gradient in the setting of the transmission Green's functions approach (cf. Section 6.1.6). The main advantage of the transmission Green's function approach is that the measured reflection and transmission information is provided at the same endpoint (i.e., $z = 0$; while in the usual Green's function approach the measured reflection and transmission information is provided at the two different endpoints $z = 0$ and $z = l$), and this can simplify the derivation of the gradient. The differential equation and boundary conditions for the modified Green's functions are (cf. (6.77)–(6.79))

$$\partial_z\begin{bmatrix}\mathbf{g}^+\\ \mathbf{g}^-\end{bmatrix} = \begin{bmatrix}\alpha & \beta\\ \gamma & \vartheta\end{bmatrix}\begin{bmatrix}\mathbf{g}^+\\ \mathbf{g}^-\end{bmatrix} \equiv \mathbf{M}\begin{bmatrix}\mathbf{g}^+\\ \mathbf{g}^-\end{bmatrix}, \tag{6.193}$$
$$\mathbf{g}^+(l) = I, \qquad \mathbf{g}^-(l) = \mathbf{r}_l. \tag{6.194}$$

Note that the differential equations for the Green's functions are linear, and this makes the derivation for the gradient simpler. The inverse/design problem is to minimize the following objective functional

$$J(p) = \sum_{\omega,\theta,\phi}\sum_{i,j=1,2} W^+_{i,j}\left\{\left|g^+_{ij}(0)-g^{(m)+}_{ij}\right|^2 + W^-_{i,j}\left|g^-_{ij}(0)-g^{(m)-}_{ij}\right|^2\right\}, \tag{6.195}$$

where $\mathbf{g}^{(m)+} = [\mathbf{t}^{(m)}]^{-1}\mathbf{t}_l$, $\quad\mathbf{g}^{(m)-} = \mathbf{r}^{(m)}[\mathbf{t}^{(m)}]^{-1}\mathbf{t}_l$ can be treated as mesured or prescribed information.

Choosing the following boundary values for the dual matrix functions $\mathbf{X}^{\pm}$

$$X^{+}_{i,j}(0) = W^{+}_{i,j}[g^{+}_{ij}(0) - g^{(m)+}_{ij}]^{*}, \tag{6.196}$$

$$X^{-}_{i,j}(0) = W^{-}_{i,j}[g^{-}_{ij}(0) - g^{(m)-}_{ij}]^{*}, \tag{6.197}$$

one can write the increment in the objective functional in the following form:

$$\begin{aligned}\delta J(p) &= 2\mathrm{Re}\sum_{\omega,\theta,\phi} \mathrm{tr}\left\{\mathbf{X}^{+^{\top}}(0)\,\delta\mathbf{g}^{+}(0) + \mathbf{X}^{-^{\top}}(0)\,\delta\mathbf{g}^{-}(0)\right\} + o(p) \\ &= 2\mathrm{Re}\sum_{\omega,\theta,\phi} \mathrm{tr}\left\{\mathbf{X}^{\top}(0)\,\delta\mathbf{g}(0)\right\} + o(p),\end{aligned} \tag{6.198}$$

where

$$\mathbf{g} = \begin{bmatrix}\mathbf{g}^{+}\\ \mathbf{g}^{-}\end{bmatrix}, \qquad \mathbf{X} = \begin{bmatrix}\mathbf{X}^{+}\\ \mathbf{X}^{-}\end{bmatrix}.$$

The increment $\delta\mathbf{g}$ satisfies the following equations:

$$\partial_z(\delta\mathbf{g}) = \mathbf{M}(\delta\mathbf{g}) + (\delta\mathbf{M})\mathbf{g} + o(p), \tag{6.199}$$

$$(\delta\mathbf{g})(l) = 0. \tag{6.200}$$

Using (6.198)–(6.200) one obtains

$$\begin{aligned}\delta J(p) &= -2\mathrm{Re}\sum_{\omega,\theta,\phi} \mathrm{tr}\left\{\int_0^l \left[(\partial_z\mathbf{X}^{\top})(\delta\mathbf{g}) + \mathbf{X}^{\top}\partial_z(\delta\mathbf{g})\right] dz\right\} + o(p) \\ &= -2\int_0^l dz\mathrm{Re}\sum_{\omega,\theta,\phi} \mathrm{tr}\left\{\left[\partial_z\mathbf{X}^{\top} + \mathbf{X}^{\top}\mathbf{M}\right](\delta\mathbf{g}) + \mathbf{X}^{\top}(\delta\mathbf{M})\mathbf{g}\right\} + o(p).\end{aligned}$$

To eliminate the $\delta\mathbf{g}$ term in the above equation, one chooses the following differential equation for the dual matrix functions:

$$\partial_z\mathbf{X}^{\top} = -\mathbf{X}^{\top}\mathbf{M}. \tag{6.201}$$

Thus, one has

$$\delta J(p) = -2\int_0^l dz\mathrm{Re}\sum_{\omega,\theta,\phi} \mathrm{tr}\left\{\mathbf{X}^{\top}(\delta\mathbf{M})\mathbf{g}\right\} + o(p), \tag{6.202}$$

from which the following explicit expression for the gradient is obtained:

$$\frac{\partial J}{\partial p_n}(z) = -2\mathrm{Re}\sum_{\omega,\theta,\phi} \mathrm{tr}\left\{\mathbf{X}^{\top}\frac{\partial\mathbf{M}}{\partial p_n}\mathbf{g}\right\}. \tag{6.203}$$

In the above expression $\mathbf{g}(z)$ is computed by integrating (6.193) along the $-z$ direction starting from the boundary condition (6.194) at $z = l$, and $\mathbf{X}(z)$ is computed by integrating (6.201) along the z direction starting from the boundary conditions (6.196) and (6.197) at $z = 0$.

The present trace formalism can be generalized to more complicated systems (involving matrices of larger size), such as those encountered in the design of stratified piezoelectric layers for suppression of sound (Braga et al. 1992).

6.5 Scattering from a laterally periodic inhomogeneous structure

In this last section of this chapter, we generalize the frequency domain wave-splitting method to a special two-dimensional case, namely, scattering from a laterally periodic inhomogeneous structure.

Consider an electromagnetic scattering problem for a laterally periodic inhomogeneous structure situated in $0 \leq z \leq l$. Outside the structure there is a vacuum (with a permittivity ϵ_0 and a permeability μ_0). The permittivity ϵ is assumed to be y-independent and have the following periodicity,

$$\epsilon(x + d, z) = \epsilon(x, z), \qquad 0 \leq z \leq l \tag{6.204}$$

where $d > 0$ is the period of the permittivity in the x direction. Structures with the property (6.204) include gratings, a stack of gratings, and gratings built on a layered substrate, provided all the gratings have the same period. Note that ϵ may be complex due to losses.

A plane wave with the electric field in the y direction (i.e., the TE case) impinges on the laterally periodic inhomogeneous structure from the upper side $z < 0$ with the xz plane as the plane of incidence (the angle of incidence is θ). In most of the literature on gratings, the $e^{-j\omega t}$ time dependence of the fields is used (see e.g. Petit (1980)), and this time dependence is assumed here (note that in all the other sections of this chapter, the $e^{j\omega t}$ time dependence is used). Then the electric field can be written as $\mathbf{E} = E(x, z)\hat{\mathbf{e}}_y$, where $\hat{\mathbf{e}}_y$ is the unit vector along the y direction, and the amplitude $E(x, z)$ satisfies the following Helmholtz equation

$$\nabla^2 E + [k(x, z)]^2 E = 0, \tag{6.205}$$

where

$$[k(x, z)]^2 = \omega^2 \mu_0 \epsilon(x, z). \tag{6.206}$$

The wave vector of the incident field is

$$\mathbf{k}_0 = (k_0 \sin\theta, 0, -k_0 \cos\theta) \equiv (\alpha_0, 0, -\beta_0), \tag{6.207}$$

where $k_0 = \omega\sqrt{\mu_0 \epsilon_0}$ is the wavenumber in vacuum. The periodicity of the geometry implies that $E(x, z)$ is a pseudo-periodic function (i.e., $E(x, z)e^{-j(k_0 \sin\theta)x}$ is a periodic function) in x with period d (Petit 1980). Thus, one has the following Fourier expansion,

$$E(x,z) = \sum_n E_n(z) e^{j(k_0 \sin\theta + n\frac{2\pi}{d})x} = \sum_n E_n(z) e^{j\alpha_n x}, \tag{6.208}$$

where

$$\alpha_n = k_0 \sin\theta + n\frac{2\pi}{d}, \qquad n = 0, \pm 1, \pm 2, \pm 3, \ldots. \tag{6.209}$$

Assume that the periodic function k^2 has the following Fourier expansion,

$$[k(x,z)]^2 = \sum_n K_n(z) e^{jn\frac{2\pi}{d}x}, \tag{6.210}$$

where

$$K_n(z) = \frac{1}{d}\int_0^d [k(x,z)]^2 e^{-jn\frac{2\pi}{d}x}\, dx. \tag{6.211}$$

From (6.205) and (6.208) it follows that

$$\frac{d^2}{dz^2}E_n + [K_0(z) - \alpha_n^2]E_n + \sum_{m\neq n, m=-\infty}^{m=+\infty} K_{n-m}E_m = 0. \tag{6.212}$$

Note that outside the inhomogeneous structure, one has the following decoupled equation for each Fourier component of the electric field,

$$\frac{d^2}{dz^2}E_n + [k_0^2 - \alpha_n^2]E_n = 0, \qquad z \leq 0, \text{ or } z \geq l. \tag{6.213}$$

Note also that (6.212) gives a system of coupled equations involving an infinite number of E_n. In a numerical computation, one truncates the series by setting

$$E_n = 0, \qquad \text{if } |n| \geq N, \tag{6.214}$$

where N is a positive integer. Thus, only a finite number of spatial harmonics are taken into account and the higher order spatial harmonics are neglected. This type of truncation is used in the rest of this section.

Define the following vacuum wave-splitting

$$E_n^{\pm}(z) = \frac{1}{2}\left[E_n \mp \frac{j}{\beta_n}\frac{d}{dz}E_n\right], \tag{6.215}$$

where

$$\beta_n = \begin{cases} \sqrt{k_0^2 - \alpha_n^2}, & \text{if } k_0 \geq |\alpha_n|, \\ j\sqrt{\alpha_n^2 - k_0^2}, & \text{if } k_0 < |\alpha_n|. \end{cases} \tag{6.216}$$

$E_n^{\pm}$ only have physical meaning in the vacuum as down-going and up-going components (assuming $\beta_n \neq 0$; otherwise modifications are required in the treatment), respectively, along the diffracted direction θ_n, which is defined by

$$\sin\theta_n = \sin\theta + n\frac{\lambda_0}{d}, \qquad n = 0, \pm 1, \pm 2, \pm 3, ... \tag{6.217}$$

(λ_0 is the wavelength in vacuum). There is no scattered field at angles other than these discrete diffraction angles. Note that $E_n^+ + E_n^- = E_n$. Note also that E_n^+ and E_n^- have the same $k_x = k_0 \sin\theta + n\frac{2\pi}{d}$, but opposite sign in k_y. In particular, in the vacuum region $z < 0$, $E_n^+ = E^i\,\delta_{n0}$ (E^i is the amplitude of the incident wave and δ_{n0} is Kronnecker's delta function), and E_n^- is the amplitude of the reflected wave at the diffraction angle θ_n. In the vacuum region $z > l$, E_n^+ is the amplitude of the transmitted wave at the diffraction angle θ_n, and the up-going component is zero, i.e., $E_n^-(l) = 0$. For large $|n|$ when $|\alpha_n| = |k_0 \sin\theta + n\frac{2\pi}{d}| > k_0$, $E_n^{\pm}$ represent evanescent waves which propagate along the x direction and which are exponentially damped to zero as $|z|$ increases towards infinity. Inside the inhomogeneous structure (6.215) is also a useful change of basis for (6.212) from the variables E_n, $\frac{d}{dz}E_n$ to $E_n^{\pm}$. A characteristic feature of a vacuum wave-splitting approach is that it is based on a wave-splitting that is not related to the media which make up the inhomogeneous structure, but is always defined with respect to the vacuum.

We rewrite the wave-splitting (6.215) in a matrix form as

$$\begin{bmatrix} \Psi^+ \\ \Psi^- \end{bmatrix} = \frac{1}{2}\begin{bmatrix} I_{2N+1} & -P \\ I_{2N+1} & P \end{bmatrix}\begin{bmatrix} \Psi \\ \frac{d}{dy}\Psi \end{bmatrix} \equiv T_0 \begin{bmatrix} \Psi \\ \frac{d}{dy}\Psi \end{bmatrix}, \tag{6.218}$$

where I_{2N+1} is the identity matrix of rank $2N+1$, and

$$\Psi = \begin{bmatrix} E_{-N} \\ E_{-N+1} \\ \cdots \\ E_{N-1} \\ E_N \end{bmatrix}, \qquad \Psi^{\pm} = \begin{bmatrix} E_{-N}^{\pm} \\ E_{-N+1}^{\pm} \\ \cdots \\ E_{N-1}^{\pm} \\ E_N^{\pm} \end{bmatrix}, \tag{6.219}$$

$$P = j\,\mathrm{diag}\,\{\frac{1}{\beta_{-N}}, \frac{1}{\beta_{-N+1}}, ..., \frac{1}{\beta_0}, ..., \frac{1}{\beta_N}\}. \tag{6.220}$$

Equation (6.212) can be rewritten in the following matrix form

$$\frac{d}{dz}\begin{bmatrix} \Psi \\ \frac{d}{dy}\Psi \end{bmatrix} = \begin{bmatrix} 0 & I_{2N+1} \\ -K+Q & 0 \end{bmatrix}\begin{bmatrix} \Psi \\ \frac{d}{dy}\Psi \end{bmatrix} \equiv W \begin{bmatrix} \Psi \\ \frac{d}{dy}\Psi \end{bmatrix}, \tag{6.221}$$

where

$$K = \begin{bmatrix} K_0 & K_{-1} & \dots & K_{-2N} \\ K_1 & K_0 & \dots & K_{-2N+1} \\ \dots & \dots & \dots & \dots \\ K_{2N} & K_{2N-1} & \dots & K_0 \end{bmatrix}, \tag{6.222}$$

$$Q = \text{diag}\,\{\alpha_{-N}^2, \alpha_{-N+1}^2, ..., \alpha_0^2, ..., \alpha_N^2\}. \tag{6.223}$$

It immediately follows from (6.218) and (6.221) that

$$\frac{d}{dz}\begin{bmatrix} \Psi^+ \\ \Psi^- \end{bmatrix} = T_0 W T_0^{-1} \begin{bmatrix} \Psi^+ \\ \Psi_- \end{bmatrix} \equiv \begin{bmatrix} A & B \\ C & D \end{bmatrix} \begin{bmatrix} \Psi^+ \\ \Psi^- \end{bmatrix}, \tag{6.224}$$

where T_0^{-1} denotes the inverse of T_0, and

$$A = -D = -\frac{1}{2}P^{-1} + \frac{1}{2}P[K - Q], \tag{6.225}$$

$$B = -C = \frac{1}{2}P^{-1} + \frac{1}{2}P[K - Q]. \tag{6.226}$$

One can then apply the imbedding approach to the system (6.224), and obtain the immbedding equation and the boundary condition for the reflection coefficient matrix. Our numerical results have shown that it is very difficult to obtain accurate numerical results by solving the non-linear imbedding equation for the reflection coefficient matrix (which is a matrix of large size), in particular for the case of Maxwell's equations with a general incidence (Forslund and He 1998). Therefore, the imbedding approach is not suitable for use for the present scattering problem. The usual Green's function approach is not suitable for use either, since in this approach one needs to know in advance the physical reflection coefficient matrix at $z = 0$, which is obtained by solving the non-linear imbedding equation. The transmission Green's function approach (cf. Section 6.1.6) is to be preferred for this type of scattering problem when matrices of large size are involved, as is explained below.

Thus, we define the transmission Green's function matrices $g^{\pm}(z)$ which map the transmitted wave back to the internal split waves

$$\Psi^{\pm}(z) = g^{\pm}(z)\Psi^{+}(l). \tag{6.227}$$

Since both E_n and $\frac{d}{dz}E_n$ (due to the continuity of the tangential magnetic fields) are continuous, from the definition (6.215) one sees that the split fields $\Psi^{\pm}$ are continuous across any plane, and thus the matrices $g^{\pm}(z)$ defined by (6.227) are continuous across any plane even though the parameters may be discontinuous across that plane. One can easily show that the matrices $g^{\pm}(z)$ satisfy the following linear system of ODEs (cf. (6.77)–(6.79))

$$\frac{d}{dz}g^{+} = Ag^{+} + Bg^{-}, \tag{6.228}$$

$$\frac{d}{dz}g^- = Cg^+ + Dg^-, \tag{6.229}$$

and have the following boundary conditions

$$g^+(l) = I_{2N+1}, \tag{6.230}$$
$$g^-(l) = 0. \tag{6.231}$$

The numerical solution for the Green's functions $g^{\pm}(z)$ is obtained by integrating (6.228) and (6.229) along the $-z$ direction starting from $z = l$.

Once $g^{\pm}(z)$ have been calculated, one knows the internal fields (note that $\Psi(z) = \Psi^+ + \Psi^-$) as well as the scattered fields outside the inhomogeneous structure. From the definitions for $g^{\pm}$, r and t, one sees that the physical reflection coefficient matrix r and transmission coefficient matrix t can be obtained by

$$r = g^-(0)[g^+(0)]^{-1}, \tag{6.232}$$
$$t = [g^+(0)]^{-1}. \tag{6.233}$$

The present approach can be modified easily to treat the case when the laterally periodic inhomogeneous structure is superimposed on a homogeneous dielectric half-space at $z = l$ (He 1997a).

Numerical results. As a numerical example, we choose the following profile for the permittivity,

$$\epsilon(x, z) = [4.0 + 2.0(1.0 - 0.03j)\cos(2\pi x/d)\sin(\pi z/l)]\epsilon_0, \quad 0 \le z \le l, \tag{6.234}$$

where the period $d = 3\lambda_0$, and the thickness of the inhomogeneous structure $l = 1.2\lambda_0$. The incident angle is $\theta = \pi/6$. Code in Matlab is written to integrate (6.228) and (6.229) along the $-z$ direction starting from $z = l$. To see roughly how large N needs to be to obtain an accurate numerical solution, we first test the code for a homogeneous slab with $\epsilon = 4.0\epsilon_0$ (other parameters are the same as indicated above). For this simple homogeneous case one knows the correct values for the scattering coefficients from an explicit formula (Born and Wolf 1980), i.e., $r_{0,0} = -0.5854 - 0.2181j$, $t_{0,0} = -0.2727 + 0.7317j$. Our code gives $r_{0,0} = -0.5629 - 0.2417j$, $t_{0,0} = -0.3494 + 0.8136j$ when $N = 8$, and $r_{0,0} = -0.5845 - 0.2192j$, $t_{0,0} = -0.2774 + 0.7398j$ when $N = 16$. The error is less than 1% when $N = 16$ in this simple case. Now we apply the code for the inhomogeneous case with the permittivity profile (6.234). The vectors $\mathbf{r}_0 = \{r_{-N,0}, r_{-N+1,0}, \ldots, r_{0,0}, \ldots, r_{N,0}\}^T$ (the superscript T denotes transposition) and $\mathbf{t}_0 = \{t_{-N,0}, t_{-N+1,0}, \ldots, t_{0,0}, \ldots, t_{N,0}\}^T$ have the following values when $N = 16$ (the number of discretization points in z is 512),

$$\mathbf{r}_0 = \begin{bmatrix} \vec{0}_7 \\ -0.0001 - 0.0004j \\ -0.0006 - 0.0033j \\ -0.0028 - 0.0187j \\ -0.0061 - 0.0600j \\ -0.0154 - 0.1375j \\ -0.0761 - 0.1397j \\ -0.1799 + 0.0083j \\ -0.0167 - 0.0184j \\ -0.1270 + 0.1105j \\ -0.1708 + 0.2706j \\ 0.2094 - 0.2344j \\ -0.1582 - 0.1598j \\ -0.1224 - 0.0468j \\ -0.0299 + 0.0206j \\ -0.0045 + 0.0060j \\ -0.0005 + 0.0008j \\ -0.0000 + 0.0001j \\ -0.0000 + 0.0000j \\ -0.0000 + 0.0000j \\ \vec{0}_7 \end{bmatrix}, \qquad \mathbf{t}_0 = \begin{bmatrix} \vec{0}_7 \\ -0.0001 - 0.0001j \\ -0.0009 - 0.0007j \\ -0.0026 - 0.0061j \\ 0.0086 - 0.0339j \\ 0.0351 - 0.0453j \\ 0.0569 + 0.0821j \\ 0.1641 + 0.2580j \\ 0.3041 - 0.3077j \\ -0.4198 - 0.3418j \\ -0.0963 + 0.0546j \\ -0.4373 - 0.1063j \\ -0.4700 - 0.2568j \\ -0.2250 - 0.1406j \\ -0.0790 - 0.0673j \\ -0.0149 - 0.0132j \\ -0.0018 - 0.0015j \\ -0.0002 - 0.0001j \\ -0.0000 - 0.0000j \\ -0.0000 - 0.0000j \\ \vec{0}_7 \end{bmatrix},$$

where $\vec{0}_n$ is a zero vector with n elements. When $N = 32$, $\vec{r}_0$ and $\vec{t}_0$ become

$$\mathbf{r}_0 = \begin{bmatrix} \vec{0}_{23} \\ -0.0001 - 0.0004j \\ -0.0006 - 0.0033j \\ -0.0029 - 0.0185j \\ -0.0063 - 0.0599j \\ -0.0159 - 0.1374j \\ -0.0768 - 0.1392j \\ -0.1803 + 0.0099j \\ -0.0169 - 0.0176j \\ -0.1257 + 0.1107j \\ -0.1694 + 0.2695j \\ 0.2095 - 0.2353j \\ -0.1591 - 0.1604j \\ -0.1226 - 0.0469j \\ -0.0298 + 0.0205j \\ -0.0045 + 0.0060j \\ -0.0005 + 0.0008j \\ -0.0000 + 0.0001j \\ -0.0000 + 0.0000j \\ -0.0000 + 0.0000j \\ \vec{0}_{23} \end{bmatrix}, \qquad \mathbf{t}_0 = \begin{bmatrix} \vec{0}_{23} \\ -0.0001 - 0.0001j \\ -0.0009 - 0.0007j \\ -0.0026 - 0.0061j \\ 0.0084 - 0.0338j \\ 0.0345 - 0.0457j \\ 0.0562 + 0.0799j \\ 0.1625 + 0.2553j \\ 0.3008 - 0.3051j \\ -0.4174 - 0.3359j \\ -0.0932 + 0.0573j \\ -0.4327 - 0.1036j \\ -0.4661 - 0.2543j \\ -0.2234 - 0.1395j \\ -0.0787 - 0.0670j \\ -0.0149 - 0.0132j \\ -0.0018 - 0.0015j \\ -0.0002 - 0.0001j \\ -0.0000 - 0.0000j \\ -0.0000 - 0.0000j \\ \vec{0}_{23} \end{bmatrix},$$

For this example, there are 6 diffraction angles, namely $\theta_{-4} = -\arcsin(5/6)$, $\theta_{-3} = -\pi/6$, $\theta_{-2} = -\arcsin(1/6)$, $\theta_{-1} = \arcsin(1/6)$, $\theta_0 = \pi/6$, and $\theta_1 = \arcsin(5/6)$. The above computed results indicate that if a transverse electric incident wave with a unit amplitude impinges on this laterally periodic inhomogeneous slab at $\theta = \pi/6$, there will be reflected waves with electric field amplitudes $-0.1799+0.0083\jmath$, $-0.0768-0.1392\jmath$, $-0.1803+0.0099\jmath$, $-0.0169-0.0176\jmath$, $-0.1257+0.1107\jmath$, $-0.1694+0.2695\jmath$ and $0.2095-0.2353\jmath$ at the diffraction angles $\theta = -\arcsin(5/6)$, $-\pi/6$, $-\arcsin(1/6)$, $\arcsin(1/6)$, $\pi/6$, and $\arcsin(5/6)$, respectively, and transmitted waves with electric field amplitudes $0.0562+0.0799\jmath$, $0.1625+0.2553\jmath$, $0.3008-0.3051\jmath$, $-0.4174-0.3359\jmath$, $-0.0932+0.0573\jmath$ and $-0.4327-0.1036\jmath$, respectively, at these diffraction angles. The other non-zero elements in $\mathbf{r}_0$ and $\mathbf{t}_0$ give the amplitudes of the evanescent waves (propagating in the x direction with different damping factors $e^{-\sqrt{\alpha_n^2-k_0^2}|z|}$) at the boundary $z = 0$ or $z = l$. Numerical results show that the sum of squares $\sum_m [|r_{m,0}|^2 + |t_{m,0}|^2]$ for all m whose associated wave is non-evanescent does not exceed 1. Note, however, that the sum of squares $\sum_{n=-N}^{N} [|r_{n,0}|^2 + |t_{n,0}|^2]$ may exceed 1 (the energy ratio cannot be represented by such a sum of squares because the evanescent waves are damped in z while the diffracted waves are not). In this example the reflection coefficient $r_{-3,0}$ gives the amplitude of the backward scattering (backward scattering only occurs with a negative value of n and if the incident angle satisfies $\sin\theta = -n\frac{\lambda_0}{2d}$; cf. (6.217) with $\theta_n = -\theta$).

If the direction of the incident plane wave is arbitrary, then in general there is a cross-scattering between TE and TM waves. Consequently, the treatment has to be based directly on Maxwell's equations, as a system of first-order differential equations (see Forslund and He (1997) for electromagnetic scattering from an inhomogeneous grating for arbitrary plane wave incidence using a wave-splitting approach).

Exercises for Chapter 6

Exercise 6.1. Derive the expression (6.103) for the matrix D for a Ω medium characterized by (6.101) and (6.102). Derive the right- and left-going eigenmodes, and use them to obtain an explicit expression for the reflection and transmission coefficients for TE and TM modes from a vacuum–Ω interface and a homogeneous Ω slab.

Exercise 6.2. Consider a TE plane wave obliquely incident (from the left) upon a lossy multi-layered structure that extends from $z = 0$ to $z = l$ (the structure is bounded on each side by a semi-infinite vacuum). Then in each homogeneous layer the amplitude $E(z)$ of the electric field $E(z)e^{j(\omega t - k_0 \sin\theta x)}\hat{\mathbf{e}}_y$ satisfies the following Helmholtz equation:

$$\frac{d^2E}{dz^2} + k_0^2 \frac{(\epsilon - j\sigma/\omega) - \epsilon_0 \sin^2\theta}{\epsilon_0} E = 0, \tag{6.235}$$

where θ is the incidence angle, and ϵ, σ are the permittivity and conductivity of the layer, respectively. Introduce the following vacuum wave-splitting

$$\begin{bmatrix} E^+ \\ E^- \end{bmatrix} = \frac{1}{2} \begin{bmatrix} 1 & \frac{j}{k_0 \cos\theta} \\ 1 & -\frac{j}{k_0\cos\theta} \end{bmatrix} \begin{bmatrix} E \\ \frac{dE}{dz} \end{bmatrix}, \tag{6.236}$$

and derive the following differential equations for the split fields:

$$\frac{d}{dz} \begin{bmatrix} E^+ \\ E^- \end{bmatrix} = \begin{bmatrix} -\alpha & -\beta \\ \beta & \alpha \end{bmatrix} \begin{bmatrix} E^+ \\ E^- \end{bmatrix}, \tag{6.237}$$

where

$$\alpha = \frac{jk_0}{2}\left(\frac{\epsilon - j\sigma/\omega - \epsilon_0\sin^2\theta}{\epsilon_0\cos\theta} + \cos\theta\right), \tag{6.238}$$

$$\beta = \frac{jk_0}{2}\left(\frac{\epsilon - j\sigma/\omega - \epsilon_0\sin^2\theta}{\epsilon_0\cos\theta} - \cos\theta\right). \tag{6.239}$$

Define the transmission Green's function by

$$E^\pm(z) = g^\pm(z)E^+(l), \tag{6.240}$$

and show that inside each homogeneous layer, $g^\pm$ have the following property,

$$\begin{bmatrix} g^+ \\ g^- \end{bmatrix}(z) = P(z_0 - z, \epsilon, \sigma) \begin{bmatrix} g^+ \\ g^- \end{bmatrix}(z_0), \quad z \le z_0, \tag{6.241}$$

where

$$P(z,\epsilon,\sigma) = \frac{1}{a^2 - b^2} \begin{bmatrix} a^2e^{\lambda z} - b^2e^{-\lambda z} & -ab(e^{\lambda z} - e^{-\lambda z}) \\ ab(e^{\lambda z} - e^{-\lambda z}) & a^2e^{-\lambda z} - b^2e^{\lambda z} \end{bmatrix}, \quad z \ge 0, \tag{6.242}$$

and where

$$\lambda = jk_0\sqrt{\frac{\epsilon - j\sigma/\omega - \epsilon_0\sin^2\theta}{\epsilon_0}}, \tag{6.243}$$

$$a(\epsilon,\sigma) = \epsilon_0 \cos 2\theta + \epsilon - j\sigma/\omega + 2\sqrt{\epsilon_0(\epsilon - j\sigma/\omega - \epsilon_0 \sin^2\theta)}\cos\theta, \tag{6.244}$$

$$b(\epsilon,\sigma) = \epsilon_0 - \epsilon + j\sigma/\omega. \tag{6.245}$$

Here one should take the square root in (6.243) with a non-negative real part. For a structure consisting of N layers with permittivities $\epsilon_1, \epsilon_2, ..., \epsilon_N$, and conductivities $\sigma_1, \sigma_2, ..., \sigma_N$ (counted from the left; the corresponding thicknesses are $d_1, d_2, ..., d_N$, respectively; the total thickness is $l = d_1 + d_2 + ... + d_N$), derive the following closed-form formula (He 1997b) for the reflection coefficient r and transmission coefficient t for the whole multi-layered structure

$$r = g^-(0)[g^+(0)]^{-1}, \tag{6.246}$$

$$t = [g^+(0)]^{-1}, \tag{6.247}$$

where $g^\pm(0)$

$$\begin{bmatrix} g^+(0) \\ g^-(0) \end{bmatrix} = P(d_1,\epsilon_1,\sigma_1)P(d_2,\epsilon_2,\sigma_2)...P(d_N,\epsilon_N,\sigma_N)\begin{bmatrix} 1 \\ 0 \end{bmatrix}. \tag{6.248}$$

Exercise 6.3. Consider the same scattering problem as in Exercise 6.2, except now the thickness l of the slab is small compared to the wavelength, and the slab is stratified in an arbitrary fashion (continuous or multi-layered). Introduce the variable

$$\tau = z/l. \tag{6.249}$$

Inside the stratified thin slab, one can expand $E^\pm(z,\omega;l)$ in a power series of the thickness l:

$$E^\pm(z,\omega;l) = E^{(0)\pm}(\tau,\omega) + \frac{l}{\lambda}E^{(1)\pm}(\tau,\omega) + \left(\frac{l}{\lambda}\right)^2 E^{(2)\pm}(\tau,\omega) + \dots. \tag{6.250}$$

Note that $E^{(n)\pm}$, $n = 0, 1, 2, ...$, are independent of l. Show that $E^{(n)\pm}$ satisfy the following differential equations,

$$\frac{d}{d\tau}\begin{bmatrix} E^{(0)+}(\tau,\omega) \\ E^{(0)-}(\tau,\omega) \end{bmatrix} = 0, \tag{6.251}$$

$$\frac{d}{d\tau}\begin{bmatrix} E^{(n)+}(\tau,\omega) \\ E^{(n)-}(\tau,\omega) \end{bmatrix} = \begin{bmatrix} -a(\tau,\omega) & -b(\tau,\omega) \\ b(\tau,\omega) & a(\tau,\omega) \end{bmatrix}\begin{bmatrix} E^{(n-1)+}(\tau,\omega) \\ E^{(n-1)-}(\tau,\omega) \end{bmatrix}, \quad n \geq 1, \tag{6.252}$$

where

$$a(\tau,\omega) = j\pi\left(\frac{\epsilon(\tau) - j\sigma(\tau)/\omega - \epsilon_0\sin^2\theta}{\epsilon_0\cos\theta} + \cos\theta\right), \tag{6.253}$$

$$b(\tau,\omega) = j\pi \left(\frac{\epsilon(\tau) - j\sigma(\tau)/\omega - \epsilon_0 \sin^2\theta}{\epsilon_0 \cos\theta} - \cos\theta \right). \tag{6.254}$$

Here $\sigma(\tau)$ denotes the conductivity at the position $z = \tau l$ (similar notation is used for the other parameters and fields). Using the boundary conditions

$$E^+|_{z=0} = E_0^{in}, \tag{6.255}$$

$$E^-|_{z=l} = 0, \tag{6.256}$$

where E_0^{in} is a constant which gives the amplitude of the incident electric field at the surface $z = 0$, derive the following explicit solution:

$$E^{(n)+}(\tau,\omega) = (-1)^n a_n(\tau,\omega) E_0^{in}, \tag{6.257}$$

$$E^{(n)-}(\tau,\omega) = (-1)^n b_n(\tau,\omega) E_0^{in}, \qquad n = 0, 1, 2, ..., \tag{6.258}$$

where

$$a_0(\tau,\omega) = 1, \qquad b_0(\tau,\omega) = 0, \tag{6.259}$$

$$a_n(\tau,\omega) = \int_0^\tau [a(\tau',\omega)a_{n-1}(\tau',\omega) + b(\tau',\omega)b_{n-1}(\tau',\omega)]\, d\tau', \tag{6.260}$$

$$b_n(\tau,\omega) = \int_\tau^1 [b(\tau',\omega)a_{n-1}(\tau',\omega) + a(\tau',\omega)b_{n-1}(\tau',\omega)]\, d\tau', \quad n = 1, 2, ... \tag{6.261}$$

If the parameter profiles $\epsilon(z/l)$ and $\sigma(z/l)$ are known explicitly (and thus $a(\tau',\omega)$ and $b(\tau',\omega)$ are known explicitly from (6.253) and (6.254)), one can determine $a_n(\tau,\omega)$ and $b_n(\tau,\omega)$, $n = 1, 2, 3, ...$, explicitly by means of the above recursive formulas (6.260) and (6.261). Thus, explicit forms of $E^{(n)\pm}(\tau,\omega)$, $n = 0, 1, 2, 3, ...$, are obtained from (6.257) and (6.258). Show that the explicit asymptotic solution for $E^\pm$ is

$$E^+(z,\omega;l) = \left[1 + \sum_{n=1}^{\infty} (-1)^n a_n(z/l,\omega) \left(\frac{l}{\lambda}\right)^n \right] E_0^{in}, \tag{6.262}$$

$$E^-(z,\omega;l) = \left[\sum_{n=1}^{\infty} (-1)^n b_n(z/l,\omega) \left(\frac{l}{\lambda}\right)^n \right] E_0^{in}. \tag{6.263}$$

The total electric field is then obtained from $E(z,\omega;l) = E^+(z,\omega;l) + E^-(z,\omega;l)$. The above asymptotic solution converges rapidly when

$$\frac{l}{\lambda} << \frac{1}{\max\{|a|,|b|\}}. \tag{6.264}$$

Show that the reflection coefficient r and the transmission coefficient t are (He, personal communication):

$$r = \sum_{n=1}^{\infty} (-1)^n b_n(0,\omega) \left(\frac{l}{\lambda}\right)^n, \tag{6.265}$$

$$t = 1 + \sum_{n=1}^{\infty} (-1)^n a_n(1,\omega) \left(\frac{l}{\lambda}\right)^n . \tag{6.266}$$

Exercise 6.4. In order to use the transmission Green's function approach as a direct solver for the scattering from a non-uniform transmission line, one can define the objective functional as follows (instead of (6.139)),

$$J(p) = \sum_{\omega=\omega_{min}}^{\omega_{max}} w_1(\omega) \left| g^+(0,\omega) - g^{(m)+}(\omega) \right|^2 + w_2(\omega) \left| g^-(0,\omega) - g^{(m)-}(\omega) \right|^2 .$$

Derive an explicit expression for the gradient by introducing appropriate dual functions.

7

THREE-DIMENSIONAL WAVE-SPLITTING FOR THE SCALAR WAVE AND TELEGRAPH EQUATIONS

In view of the results presented in Chapters 1–5, and obtained by application of different versions of the one-dimensional time domain wave-splitting approaches, an obvious question is whether these approaches can be extended to the corresponding three-dimensional problems. In Chapters 7 and 8 we present the current status concerning three-dimensional generalizations. The Green's function approach is emphasized in these chapters. The imbedding approach in the time domain is not given because early investigation (Weston 1990) and subsequent work on the subject indicate that it is not a useful approach to apply in three dimensions.

It is inevitable that the mathematics of the presentation becomes more demanding. An effort is made to bring out, whenever possible and appropriate, analogies with the concepts, methods, and results from the one-dimensional case.

7.1 Planar wave-splitting in $\mathbb{R}^3$ and the associated operators

Recall that in the one-dimensional case, the splitting of the solution u of the wave equation

$$u_{tt} - c^2(x)u_{xx} = 0, \quad x \in \mathbb{R},$$

where the velocity $c(x)$ is a continuously differentiable function of x, is based upon the employment of a local physical splitting. Essentially, one approximates the medium in a piecewise linear manner, where in the infinitesimal interval $x_0 < x < x_0 + \Delta x$, c is given by $c(x) = c(x_0) + (x - x_0)\partial_x c(x_0)$. In this interval, one then employs the local exact physical splitting associated with a medium with $c = c(x_0)$ (constant). In such a medium solutions of the form $u(t - x/c_0)$, $u(t + x/c_0)$, where $c_0 = c(x_0)$, represent waves travelling in the positive and negative x-directions, respectively.

These concepts will be extended to apply to the wave equation in $\mathbb{R}^3$, by approximating the medium where $c = c(x_1, x_2, x_3)$ in a piecewise linear fashion in the variable x_3, by taking $c = c(x_1, x_2, x_3^0) + (x_3 - x_3^0)\partial_{x_3} c(x_1, x_2, x_3^0)$ in the infinitesimal slab $x_3^0 < x_3 < x_3^0 + \Delta x_3$. In each slab an exact physical wave-splitting is employed that is defined in terms of a medium where $c = c(x_1, x_2, x_3^0)$, (velocity depending upon transverse variables only). Thus a key problem will be to define precisely what is meant by a positive-going wave (in the $+x_3$ direction) and a negative-going wave (in the $-x_3$ direction) in a medium where $c = c(x_1, x_2, x_3^0)$. The definition that will be employed for a wave travelling in the $+x_3$ direction, on the plane $x_3 = x_3^0$, will be based upon the physical concept that such waves are produced by sources in the region $x_3 < x_3^0$. Similarly, waves travelling in the $-x_3$ direction on this plane will be produced by sources in the region $x_3 > x_3^0$. Mathematically, the definition of such waves can be defined in terms of a solution of a Dirichlet or Neumann initial-value problem in the source-free part of space.

To illustrate this, consider the one-dimensional wave equation with constant velocity c. The initial value Neumann boundary value problem

$$\left.\begin{array}{ll}\text{(i)} & c^2 u_{xx} - u_{tt} = 0, \quad x > x_0, t > 0, \\ \text{(ii)} & u(x,0) = u_t(x,0) = 0, \quad x > x_0, \\ \text{(iii)} & u_x(x_0, t) = g(t), \quad t > 0,\end{array}\right\} \tag{7.1}$$

where $g \in C^1(\mathbb{R}^+)$ ($C^m(\mathbb{R}^+)$ represents m times continuously differentiable functions on the positive real line), has the solution

$$u(x,t) = -c \int_0^{t+(x_0-x)/c} g(s)\Theta(s)ds, \quad x > x_0$$

where the Heaviside step function $\Theta(t)$ is defined by (2.12). Let $x \to x_0+$, to givc

$$u(x_0,t) + (\mathbf{K}u_x)(t) = 0, \quad x = x_0, t > 0 \tag{7.2}$$

where $\mathbf{K}$ is the integral operator

$$(\mathbf{K}w)(t) = c \int_0^t w(\tau)d\tau. \tag{7.3}$$

Condition (7.2) represents a positive-moving wave at $x = x_0$. A similar condition for a negative-moving wave at $x = x_0$ can be obtained by treating the initial-value boundary-value problem in the half-space $x < x_0$, yielding

$$u(x_0,t) - (\mathbf{K}u_x)(t) = 0. \tag{7.4}$$

An alternative condition for the positive-going wave can be obtained by treating the initial-value Dirichlet problem where condition (iii) in (7.1) is replaced by

$u(x_0, t) = f(t)$, where $f \in C^2(\mathbb{R}^+)$, and $f(0) = f'(0) = 0$. The solution is given by

$$u(x,t) = f(t - (x - x_0)/c)\Theta(t - (x - x_0)/c), \quad x > x_0,$$

yielding

$$\partial_x u(x,t) = -\frac{1}{c} f'(t - (x - x_0)/c)\Theta(t - (x - x_0)/c), \quad x > x_0.$$

Let $x \to x_0+$, to obtain the condition

$$c\partial_x u + \partial_t u = 0, \quad x = x_0,\, t > 0. \tag{7.5}$$

Thus (7.5) is an alternative form for a positive-going wave condition at $x = x_0$, and can be written in the form

$$u_x(x_0, t) + \left(\mathbf{K}^{-1} u\right)(t) = 0, \quad x = x_0 \tag{7.6}$$

where

$$\mathbf{K}^{-1} = \frac{1}{c}\frac{\partial}{\partial t}, \qquad t > 0. \tag{7.7}$$

Note that $\mathbf{K}^{-1}$ is a left-inverse of $\mathbf{K}$ if $u(t)$ is an integrable function on the positive real line, i.e.,

$$\mathbf{K}^{-1}\mathbf{K}u = u(t),$$

and $\mathbf{K}^{-1}$ is a right-inverse if $u(t) \in C^1(R^+)$, and $u(0) = 0$.

The solutions of the wave equation can now be decomposed into positive and negative moving waves. If $u(x,t)$ is a $C^2([a,b] \times [0,\infty))$ solution of the wave equation in a strip $a < x < b$, for $t > 0$

$$u_{xx} - \frac{1}{c^2} u_{tt} = 0,$$

satisfying the initial conditions

$$u(x,0) = u_t(x,0) = 0, \quad a < x < b,$$

then set

$$u^{\pm} = \frac{1}{2}(u \mp \mathbf{K}u_x). \tag{7.8}$$

It immediately follows that

$$u = u^+ + u^-,$$

and it can be easily shown that u^+ and u^- satisfy the positive- and negative-moving wave conditions given by (7.2) and (7.4) at each point x in the interval

$a < x < b$. Thus u^+ and u^- represent a positive and negative moving wave, respectively.

These concepts will now be generalized to apply to the wave equation in $\mathbb{R}^3$ and later to the telegraph equation.

The exact wave-splitting into positive-going and negative-going waves for the solution of the wave equation

$$u_{tt} - c^2\nabla^2 u = 0, \quad c = c(x_1, x_2, \alpha),$$

where the velocity c is a smooth function of the transverse variables x_1, x_2, as well as a parameter α, will be obtained. The velocity will be bounded above, and below by zero, i.e., $0 < \mu < c < \nu$. (These results will be needed later for the wave-splitting of the wave equation where $c = c(x_1, x_2, x_3)$, in which case the parameter α will be replaced by x_3^0.) The solution of initial-value, boundary-value problems will be used to obtain the positive-going and negative-going wave conditions across the plane $x_3 = x_3^0$, for $0 \leq t < T$. Restrictions on the choice of T will be discussed in the subsequent analysis.

For simplification, the following notation will be used to separate out the transverse components of a point x in $\mathbb{R}^3$,

$$x = (x_1, x_2, x_3) = (\mathbf{x}, x_3).$$

The initial-value Neumann boundary-value problem for the half-space $x_3 > x_3^0$ is stated as follows. Find $u(x,t)$ such that

$$\left.\begin{aligned} &u_{tt} - c^2(\mathbf{x}, \alpha)\nabla^2 u = 0, && x_3 > x_3^0, 0 < t < T,\\ &u = u_t = 0, && x_3 > x_3^0, t = 0,\\ &\partial_{x_3} u = u_1(\mathbf{x}, t), && x_3 = x_3^0, 0 < t < T, \end{aligned}\right\} \tag{7.9}$$

where $u_1(\mathbf{x}, t)$ is differentiable in t.

The initial-value Dirichlet boundary-value problem for the half-space $x_3 > x_3^0$, is stated as follows. Find $u(x,t)$ such that

$$\left.\begin{aligned} &u_{tt} - c^2(\mathbf{x}, \alpha)\nabla^2 u = 0, && x_3 > x_3^0,\ 0 < t < T,\\ &u = u_t = 0, && x_3 > x_3^0,\ t = 0,\\ &u = u_0(\mathbf{x}, t), && x_3 = x_3^0,\ 0 < t < T, \end{aligned}\right\} \tag{7.10}$$

where for present purposes it will be assumed that u_0 is twice-differentiable, and that $u_0 = \partial_t u_0 = 0$ at $t = 0$.

As will be seen, these problems as well as the corresponding problems for the half-space $x_3 < x_3^0$ form the basis for the splitting process. Physically the solutions of problems (7.9) and (7.10) can be thought of as arising from sources in the half-space $x_3 < x_3^0$. The solutions of the problems corresponding to (7.9) and (7.10) for the half-space $x_3 < x_3^0$, can be thought of as arising from sources in the half-space $x_3 > x_3^0$.

Solving problem (7.9) for $u(x,t)$, $x_3 > x_3^0$, and then taking the limit as $x_3 \to x_3^0$, one obtains the Dirichlet data $u(\mathbf{x}, x_3^0; t)$. Since these depend linearly upon the Neumann data, one can define the Neumann operator $\mathbf{K}_\alpha$ (which maps Neumann data to Dirichlet data on the same plane) by the following

$$\lim_{x_3 \to x_3^0} u(x,t) = -\mathbf{K}_\alpha u_1. \tag{7.11}$$

This can be placed in the form

$$u + \mathbf{K}_\alpha \partial_3 u = 0, \quad x_3 = x_3^0,\ 0 \le t < T, \tag{7.12}$$

which is a generalization of (7.2) from one to three dimensions.

Similarly, solving the Dirichlet problem (7.10) for $u(x,t)$, $x_3 > x_3^0$, taking the derivative of u with respect to x_3, and taking the limit as $x_3 \to x_3^0$, one can define the operator $\mathbf{K}_\alpha^{-1}$ (which maps Dirichlet data to Neumann data on the same plane) by the relation

$$\lim_{x_3 \to x_3^0} \partial_3 u = -\mathbf{K}_\alpha^{-1} u_0. \tag{7.13}$$

This can be placed in the form

$$\partial_3 u + \mathbf{K}_\alpha^{-1} u = 0, \quad x_3 = x_3^0,\ 0 \le t < T, \tag{7.14}$$

which is a generalization of (7.6) from one to three dimensions.

Because $c = c(\mathbf{x}, \alpha)$ does not depend upon x_3, (7.12) and (7.14) will hold for solutions of the wave equation on any plane $x_3 = x_3^1$, where $x_3^1 \ge x_3^0$ (the sources producing the solution lie below $x_3 = x_3^0$ and hence $x_3 = x_3^1$). Thus (7.12) and (7.14) are interpreted as *down-going wave conditions* (if the x_3-axis is directed downwards) or *positive-going wave conditions.*

If one solves the Neumann or Dirichlet problem corresponding to (7.9) and (7.10) for the half-space $x_3 < x_3^0$ one can obtain the following relations

$$u - \mathbf{K}_\alpha \partial_3 u = 0, \quad x_3 = x_3^0,\ 0 \le t < T, \tag{7.15}$$

$$\partial_3 u - \mathbf{K}_\alpha^{-1} u = 0, \quad x_3 = x_3^0,\ 0 \le t < T. \tag{7.16}$$

In this case, since c is independent of x_3, these equations will hold for solutions of the wave equation on any plane $x_3 = x_3^1 \le x_3^0$ (the sources producing the field lie in the half-space $x_3 > x_3^0$), and (7.15) and (7.16) can be interpreted as *up-going wave conditions* (if the x_3-axis is directed downwards) or more precisely, as *negative-going wave conditions.*

As will be seen, the splitting process depends completely upon the operator $\mathbf{K}_\alpha$ and its properties. *From the standpoint of numerical computation of* $\mathbf{K}_\alpha$ *and* $\mathbf{K}_\alpha^{-1}$ *one could compute the values of* $\mathbf{K}_\alpha u_1$ *and* $\mathbf{K}_\alpha^{-1} u_0$ *by solving systems (7.9) and (7.10) numerically* (Ladyzhevskaya 1985). For further treatment of $\mathbf{K}_\alpha$, the Green's function (fundamental solution) (Courant and Hilbert 1962b) will be employed to get an integral formulation which is useful for obtaining asymptotic results (Weston 1989). But most important, it will yield a useful closed form solution for $\mathbf{K}_\alpha$ for the case $c =$ constant. The fundamental solution $\mathcal{E}^\alpha(\mathbf{x}, \mathbf{y}, x_3 - y_3; t - s)$ is a solution of the system (the notation ∇_x^2 is used to emphasize that the ∇^2 operator here refers to the x coordinate)

$$\left.\begin{aligned} &(\partial_t^2 - c^2(\mathbf{x}, \alpha)\nabla_x^2)\mathcal{E}^\alpha = c^2(\mathbf{x}, \alpha)\delta(x - y)\delta(t - s), \\ &\mathcal{E}^\alpha(\mathbf{x}, \mathbf{y}, x_3 - y_3; t - s) = 0, \quad t - s < 0. \end{aligned}\right\} \tag{7.17}$$

Here the fact has been used that since c is independent of x_3, translational invariance implies that $\mathcal{E}^\alpha$ depends upon the variables x_3 and y_3 in the manner shown. In addition, since the Dirac delta function δ is an even function, $\mathcal{E}^\alpha$ is an even function of $(x_3 - y_3)$, which implies that

$$\frac{\partial \mathcal{E}^\alpha}{\partial x_3} = -\frac{\partial \mathcal{E}^\alpha}{\partial y_3} \quad , \quad \frac{\partial^2 \mathcal{E}^\alpha}{\partial x_3^2} = \frac{\partial^2 \mathcal{E}^\alpha}{\partial y_3^2}.$$

As a function of the variables y and s, the fundamental solution (which behaves like a backward Green's function) satisfies the system

$$\left.\begin{aligned} &(\partial_s^2 - c^2(\mathbf{y}, \alpha)\nabla_y^2)\mathcal{E}^\alpha = c^2(\mathbf{y}, \alpha)\delta(x - y)\delta(s - t), \\ &\mathcal{E}^\alpha(\mathbf{x}, \mathbf{y}, x_3 - y_3, t - s) = 0 \quad , \quad s - t > 0. \end{aligned}\right\} \tag{7.18}$$

More detailed information on the fundamental solution $\mathcal{E}^\alpha$ is given in Weston (1989). In particular, for the time interval $0 < t < T$ (where the restrictions on T are discussed at the end of this paragraph), the fundamental solution of the wave equation with $c = c(\mathbf{x}, \alpha)$ has the form

$$\mathcal{E}^\alpha(\mathbf{x}, \mathbf{y}, x_3 - y_3; t) = \frac{\delta(t - \tau_\alpha(x, y))}{\rho_\alpha(x, y)} + \Theta(t - \tau_\alpha(x, y))A(\mathbf{x}, \mathbf{y}, x_3 - y_3; t).$$

Here $\tau_\alpha(x, y)$, the arrival time of a ray travelling from the point x to point y or vice versa, satisfies the eikonal equation and the asymptotic condition

$$|\nabla_x \tau_\alpha|^2 = \frac{1}{c^2(\mathbf{x}, \alpha)}, \qquad \tau_\alpha(x, y) \overset{x \to y}{\longrightarrow} \frac{1}{c(\mathbf{y}, \alpha)}|x - y|.$$

The amplitude term $\rho_\alpha(x, y)$ satisfies the transport equation and the asymptotic condition

$$\rho_\alpha \nabla_x^2 \tau_\alpha - 2\nabla\tau_\alpha \cdot \nabla\rho_\alpha = 0, \quad \rho_\alpha(x, y) \overset{x \to y}{\longrightarrow} 4\pi|x - y|.$$

The time interval T is chosen so that for arrival times $\tau_\alpha(x, y) < T$, there is only one ray going from the point x to the point y, i.e., there are no caustics.

Additional properties of $\tau_\alpha(x,y)$ are given in Weston (1989) and in particular it is shown that when x and y lie on the same plane $x_3 = y_3$, then the ray going from x to y lies in this plane.

For the special case where $c = c(\alpha)$ (the velocity is constant with respect to position, depending only on the parameter α), then $\tau_\alpha(x,y) = |x-y|/c, \rho_\alpha(x,y) = 4\pi|x-y|$ and the fundamental solution takes the exact simple form

$$\mathcal{E}^\alpha(\mathbf{x},\mathbf{y},x_3-y_3;t) = \frac{\delta(t-|x-y|/c)}{4\pi|x-y|}. \tag{7.19}$$

The solution to the Neumann initial-value problem for the half-space $x_3 > x_3^0$ can be easily obtained using the fundamental solution $\mathcal{E}_N^\alpha$ (Green's function for the Neumann problem) given by

$$\mathcal{E}_N^\alpha = \mathcal{E}^\alpha(\mathbf{x},\mathbf{y},x_3-y_3;t-s) + \mathcal{E}^\alpha(\mathbf{x},\mathbf{y},x_3-2x_3^0+y_3;t-s). \tag{7.20}$$

The fundamental solution $\mathcal{E}_N^\alpha$ satisfies system (7.18) in the half-space $y_3 > x_3^0$, and the boundary condition

$$\frac{\partial \mathcal{E}_N^\alpha}{\partial y_3} = 0, \qquad y_3 = x_3^0. \tag{7.21}$$

To obtain the solution to system (7.9), replace the variables x and t in the first equation of system (7.9) by y and s, then multiply this resulting equation by $\mathcal{E}_N^\alpha$. Subtract from this the equation obtained by multiplying (7.18) (with $\mathcal{E}^\alpha$ replaced by $\mathcal{E}_N^\alpha$) by $u(y,s)$ to obtain

$$\begin{aligned} &u(y,s)\delta(x-y)\delta(t-s) \\ &= -\nabla_y \cdot (u\nabla_y \mathcal{E}_N^\alpha - \mathcal{E}_N^\alpha \nabla_y u) + \frac{1}{c^2(\mathbf{y},\alpha)}\frac{\partial}{\partial s}\left(u\frac{\partial \mathcal{E}_N^\alpha}{\partial s} - \mathcal{E}_N^\alpha \frac{\partial u}{\partial s}\right). \end{aligned}$$

Then integrate with respect to y over the half-space $y_3 > x_3^0$, and over the time interval $0 \le s < \infty$. Use the divergence theorem, the initial conditions for u and $\mathcal{E}_N^\alpha$, and the boundary condition (7.21). Finally noting that $\mathcal{E}_N^\alpha = 2\mathcal{E}^\alpha$ when $y_3 = x_3^0$, one obtains the following solution to the Neumann initial-value problem (7.9),

$$u(x,t) = -2\int_0^t \iint_{\mathbf{R}^2} \mathcal{E}^\alpha(\mathbf{x},\mathbf{y},x_3-x_3^0;t-s)u_1(\mathbf{y},s)d\mathbf{y}ds, \; x_3 > x_3^0. \tag{7.22}$$

Now take the limit $x_3 \to x_3^0$. Since $\mathcal{E}^\alpha$ has a singularity in the form $|x-y|^{-1}$, it tends to behave like a single-layer potential, and the integral given on the right-hand side of (7.22) is continuous across the surface $x_3 = x_3^0$. Thus one obtains

$$u(\mathbf{x},x_3^0;t) = -\int_0^t \iint_{\mathbf{R}^2} k(\mathbf{x},\mathbf{y},\alpha;t-s)u_1(\mathbf{y},s)d\mathbf{y}ds, \tag{7.23}$$

where

$$k(\mathbf{x}, \mathbf{y}, \alpha; t) = 2\mathcal{E}^\alpha(\mathbf{x}, \mathbf{y}, 0; t). \tag{7.24}$$

From (7.11), (7.23), it is seen that the operator $\mathbf{K}_\alpha$ can be defined as the integral operator with kernel $k(\mathbf{x}, \mathbf{y}, \alpha; t)$, i.e.

$$\mathbf{K}_\alpha u = \int_0^t \int_{\mathbb{R}^2} k(\mathbf{x}, \mathbf{y}, \alpha; t - s) u(\mathbf{y}, s) d\mathbf{y} ds. \tag{7.25}$$

For the special case where the velocity c is a constant with respect to position, $c = c(\alpha)$, the operator $\mathbf{K}_\alpha$ has the explicit form (using (7.19) and (7.24))

$$\mathbf{K}_\alpha u[\mathbf{x}, t] = \int\int_{\mathbb{R}^2} \frac{u(\mathbf{y}, t - |\mathbf{x} - \mathbf{y}|/c)}{2\pi |\mathbf{x} - \mathbf{y}|} \Theta(t - |\mathbf{x} - \mathbf{y}|/c) d\mathbf{y}, \tag{7.26}$$

where $c = c(\alpha)$. Replacing the variables of integration by local polar coordinates (r, θ) centred at $\mathbf{x}$, the asymptotic behaviour of $\mathbf{K}_\alpha u$ as $t \to 0$ is easily obtained

$$\mathbf{K}_\alpha u(\mathbf{x}, t) = \frac{1}{2\pi} \int_0^{ct} \int_0^{2\pi} u(\mathbf{x} + r\hat{\theta}, t - r/c) dr d\theta \sim tcu(\mathbf{x}, 0),$$

where $\hat{\theta} = (\cos\theta, \sin\theta)$.

With the fundamental solution, the solution of the Dirichlet problem (7.10) can be used to obtain an integral-differential form for the operator $\mathbf{K}_\alpha^{-1}$. The solution to system (7.10) is given by

$$u(x, t) = \int_0^\infty \int\int_{\mathbb{R}^2} \left\{ \frac{\partial \mathcal{E}_D^\alpha}{\partial y_3} u_0(\mathbf{y}, s) \right\}_{y_3 = x_3^0} d\mathbf{y} ds, \quad x_3 > x_3^0, \tag{7.27}$$

where $\mathcal{E}_D^\alpha$ is the fundamental solution associated with the Dirichlet problem and is given explicitly by

$$\mathcal{E}_D^\alpha(x, y; t - s) = \mathcal{E}^\alpha(\mathbf{x}, \mathbf{y}, x_3 - y_3; t - s) - \mathcal{E}^\alpha(\mathbf{x}, \mathbf{y}, x_3 - 2x_3^0 + y_3; t - s).$$

To simplify notation, the transverse d'Alembertian with respect to variables $\mathbf{x}, t$

$$\square_{\mathbf{x},t}^T = \frac{1}{c^2(\mathbf{x}, \alpha)} \frac{\partial^2}{\partial t^2} - \frac{\partial^2}{\partial x_1^2} - \frac{\partial^2}{\partial x_2^2}, \tag{7.28}$$

is defined. Note that when there is no loss of clarity, the subscripts $\mathbf{x}, t$ will be dropped on $\square^T$.

Using the result

$$\left. \frac{\partial \mathcal{E}_D^\alpha}{\partial y_3} \right|_{y_3 = x_3^0} = -2 \frac{\partial}{\partial x_3} \mathcal{E}^\alpha(\mathbf{x}, \mathbf{y}, x_3 - x_3^0; t - s), \tag{7.29}$$

and the identity

$$\frac{\partial^2 \mathcal{E}^\alpha}{\partial x_3^2} = \Box^T_{x,t}\mathcal{E}^\alpha, \quad x_3 > x_3^0, y_3 = x_3^0,$$

it can be shown that the derivative of expression (7.27) with respect to x_3 is given by

$$\partial_3 u(x,t) = -2\Box^T_{x,t}\int_0^t \int_{\mathbb{R}^2} \mathcal{E}^\alpha(\mathbf{x},\mathbf{y},x_3 - x_3^0; t-s)u_0(y,s)dyds,$$

for $x_3 > x_3^0$. Formally taking the limit $x_3 \to x_3^0$, one obtains

$$\partial_3 u = -\Box^T \mathbf{K}_\alpha u_0,\ x_3 = x_3^0,\ 0 \le t < T \tag{7.30}$$

(for more details, see Weston (1989)). Now on comparing expression (7.30) with (7.13) it is seen that

$$\mathbf{K}_\alpha^{-1} = \Box^T \mathbf{K}_\alpha, \tag{7.31}$$

for a suitable domain of functions. If in particular $w(\mathbf{x},t)$ is twice-continuously differentiable in $\mathbf{x}$ and t, and $w(\mathbf{x},0) = w_t(\mathbf{x},0) = 0$, one obtains the identity

$$w(\mathbf{x},t) - \mathbf{K}_\alpha^2 \Box^T w = 0.$$

This result is rigorously proved (Weston 1992b) for a wider class of second-order hyperbolic differential equations, than just the wave equation. Most importantly, the key identity

$$I = \mathbf{K}_\alpha^2 \Box^T = \Box^T \mathbf{K}_\alpha^2, \tag{7.32}$$

can be deduced for a suitable class of functions. This identity is the cornerstone of the splitting process.

For the case where $c = c(\alpha)$ (constant), (7.26) and (7.31) can be combined to give an expression for $\mathbf{K}_\alpha^{-1}$ suitable for numerical implementation

$$(\mathbf{K}_\alpha^{-1}u)(x,t) = c\partial_t u - \partial_{x_1}^2 L_{11}u + \partial_{x_1}\partial_{x_2}L_{12}u - \partial_{x_2}^2 L_{22}u, \tag{7.33}$$

$$\left.\begin{aligned} L_{ij} &= \frac{c}{2\pi}\int_0^t \int_0^{2\pi} u(x_1 + c(t-s)\cos\theta, x_2 + c(t-s)\sin\theta, x_3; s)b_{ij}(\theta)d\theta ds, \\ b_{11}(\theta) &= \sin^2\theta, \quad b_{12}(\theta) = \sin 2\theta, \quad b_{22}(\theta) = \cos^2\theta. \end{aligned}\right\} \tag{7.34}$$

Similarly, (7.26) gives

$$(\mathbf{K}_\alpha u)(x,t) = \frac{c}{2\pi}\int_0^t \int_0^{2\pi} u(x_1 + c(t-s)\cos\theta, x_2 + c(t-s)\sin\theta, x_3; s)d\theta ds. \tag{7.35}$$

Using the identity (7.32) the exact (physical) splitting process for a medium where $c = c(\mathbf{x},\alpha)$ can now be defined, as stated in the following theorem.

Theorem 7.1 *If $u(x,t)$ is a solution of*

$$\partial_t^2 u - c^2 \nabla^2 u = 0, \qquad x \in \Omega,\ 0 \le t < T,$$

$$u = u_t = 0, \qquad x \in \Omega,\ t = 0,$$

where $c = c(\mathbf{x}, \alpha)$ and Ω is the slab $a < x_3 < b$, then u can be decomposed into a positive-going wave u^+ and a negative-going wave u^- as follows

$$u^{\pm} = \frac{1}{2}[u \mp \mathbf{K}_\alpha \partial_3 u], \tag{7.36}$$

which satisfy

$$(\partial_3 + \mathbf{K}_\alpha^{-1})u^+ = 0, \tag{7.37}$$

$$(\partial_3 - \mathbf{K}_\alpha^{-1})u^- = 0, \tag{7.38}$$

in the region Ω, and

$$u = u^+ + u^-.$$

Proof One just needs to show that u^+ and u^- satisfy (7.37) and (7.38) respectively. It is seen that

$$u^{\pm} \pm \mathbf{K}_\alpha \partial_3 u^{\pm} = \frac{1}{2}[u \mp \mathbf{K}_\alpha \partial_3 u] \pm \frac{1}{2}[\mathbf{K}_\alpha \partial_3 u \mp \mathbf{K}_\alpha \partial_3 \mathbf{K}_\alpha \partial_3 u],$$

and on noting that ∂_3 commutes with $\mathbf{K}_\alpha$, and using the identity (7.32) it follows that

$$u^{\pm} \pm \mathbf{K}_\alpha \partial_3 u^{\pm} = \frac{1}{2}[u - \mathbf{K}_\alpha^2 \partial_3^2 u] = \frac{1}{2}[u - \mathbf{K}_\alpha^2 \Box^T u] = \frac{1}{2}[u - Iu] = 0.$$

□

Next, the operator D_α needs to be introduced which involves the solution of the dynamic equations (7.37), (7.38).

Let $w^+(x,t)$ be a solution of the following system in the half-space $x_3 > x_3^0$

$$\left.\begin{aligned} &(\partial_3 + \mathbf{K}_\alpha^{-1})w^+ = 0, && x_3 > x_3^0,\ 0 < t < T, \\ &w^+ = w_t^+ = 0, && x_3 > x_3^0,\ t = 0, \\ &w^+ = u_0, && x_3 = x_3^0,\ 0 < t < T. \end{aligned}\right\} \tag{7.39}$$

Then define D_α as follows

$$D_\alpha[x_3, x_3^0]u_0 = w^+(x,t), \qquad x_3 > x_3^0. \tag{7.40}$$

Furthermore, let $w^-(x,t)$ be a solution of the system, in the lower half-space $x_3 < x_3^0$

$$\left.\begin{array}{lll} (\partial_3 - \mathbf{K}_\alpha^{-1})w^- = 0, & & x_3 < x_3^0,\ 0 < t < T, \\ w^- = w_t^- = 0, & & x_3 < x_3^0,\ t = 0, \\ w^- = u_0, & & x_3 = x_3^0,\ 0 < t < T. \end{array}\right\} \tag{7.41}$$

Then $D_\alpha[x_3, x_3^0]$ when $x_3 < x_3^0$ is defined as follows

$$D_\alpha[x_3, x_3^0]u_0 = w^-(x,t), \quad x_3 < x_3^0. \tag{7.42}$$

An alternative representation can be obtained. For the case where $x_3 > x_3^0$, operate on the first equation of system (7.39) with $(\partial_3 - \mathbf{K}_\alpha^{-1})$, and employ the identity (7.32) written in the form

$$\mathbf{K}_\alpha^{-2} = \Box^T,$$

to reduce system (7.39) to the form

$$\left.\begin{array}{lll} (\partial_t^2 - c^2(\mathbf{x},\alpha)\nabla^2)w^+ = 0, & & x_3 > x_3^0,\ 0 < t < T, \\ w^+ = w_t^+ = 0, & & x_3 > x_3^0,\ t = 0, \\ w^+ = u_0, & & x_3 = x_3^0,\ 0 < t < T. \end{array}\right\}$$

This implies that $w^+(x,t)$ is the solution of the Dirichlet problem (7.10) for the half-space $x_3 > x_3^0$, and $D_\alpha[x_3, x_3^0]u_0$ is just the mapping of Dirichlet data on the plane $x_3 = x_3^0$ to Dirichlet data on the plane x_3 (where $x_3 > x_3^0$). Using this fact, $D_\alpha[x_3, x_3^0]u_0$ can be computed numerically by a forward (in time) finite-difference formula.

An integral representation can be obtained from (7.27), (7.29) and (7.40)

$$D_\alpha[x_3, x_3^0]u_0 = -2\int_0^t \int_{\mathbb{R}^2} u_0(\mathbf{y}, s)\frac{\partial}{\partial x_3}\mathcal{E}^\alpha(\mathbf{x}, \mathbf{y}, x_3 - x_3^0; t - s)dyds, \tag{7.43}$$

for $x_3 > x_3^0$ (Weston 1992a).

In a similar manner, it can be shown that $w^-(x,t)$ is the solution of the initial-value Dirichlet problem, corresponding to (7.10) for the lower half-space $x_3 < x_3^0$. Again, $D_\alpha[x_3, x_3^0]u_0$ is just the mapping of Dirichlet data on the plane $x_3 = x_3^0$ to a plane $x_3 (x_3 < x_3^0)$. For the case $x_3 < x_3^0$ the corresponding integral representation is given by (7.43) with a difference in sign on the right-hand side.

For the case $c = c(\alpha)$ constant, the integral expression for D_α can be simplified on using (7.19) to give

$$D_\alpha[x_3, x_3^0]u_0$$

$$= \pm \int_{\mathbb{R}^2} \frac{(x_3 - x_3^0)}{2\pi R^3} \left\{ u_0(\mathbf{y}, t - \frac{R}{c}) + \frac{R}{c}\partial_t u_0(\mathbf{y}; t - R/c) \right\} \Theta(t - R/c) d\mathbf{y}, \tag{7.44}$$

where $R^2 = (\mathbf{x} - \mathbf{y})^2 + (x_3 - x_3^0)^2$, and the + and − sign is taken when $x_3 > x_3^0$ and $x_3 < x_3^0$ respectively.

Note that when $x_3 = x_3^0$, D_α is the identity operator, i.e.

$$D_\alpha[x_3^0, x_3^0]u_0 = u_0.$$

This fact can be used to find a solution of the non-homogeneous system (which will be needed later),

$$\left.\begin{aligned} (\partial_3 + \mathbf{K}_\alpha^{-1})w^+ &= v^+, && x_3 > x_3^0,\ 0 < t < T, \\ w^+ = w_t^+ &= 0, && x_3 > x_3^0,\ t = 0, \\ w^+ &= u_0, && x_3 = x_3^0,\ 0 < t < T. \end{aligned}\right\} \tag{7.45}$$

The solution of this system is given by

$$w^+(x, t) = D_\alpha[x_3, x_3^0]u_0 + \int_{x_3^0}^{x_3} D_\alpha[x_3, y_3]v^+ dy_3, \quad x_3 > x_3^0. \tag{7.46}$$

This can easily be verified, by noting that

$$\begin{aligned} &(\partial_3 + \mathbf{K}_\alpha^{-1}) \int_{x_3^0}^{x_3} D_\alpha[x_3, y_3]v^+ dy_3 \\ &\quad = D_\alpha[x_3, x_3]v^+ + \int_{x_3^0}^{x_3} (\partial_{x_3} + \mathbf{K}_\alpha^{-1}) D_\alpha[x_3, y_3]v^+ dy_3 \\ &\quad = v^+(x, t). \end{aligned}$$

Similarly it can be shown that the solution of the system

$$\left.\begin{aligned} (\partial_3 - \mathbf{K}_\alpha^{-1})w^- &= v^-, && x_3 < x_3^0,\ 0 < t < T, \\ w^- = w_t^- &= 0, && x_3 < x_3^0,\ t = 0, \\ w^- &= u_0, && x_3 = x_3^0,\ 0 < t < T, \end{aligned}\right\} \tag{7.47}$$

is given by

$$w^-(x,t) = D_\alpha[x_3, x_3^0]u_0 - \int_{x_3}^{x_3^0} D_\alpha[x_3, y_3]v^- dy_3, \quad x_3 < x_3^0. \tag{7.48}$$

As was pointed out the operator $D_\alpha[x_3, x_3^0]u_0$ can be easily computed using (7.44) for the case c is constant or by solving numerically, by a forward time finite difference scheme, the initial-value boundary-value problem when $x_3 > x_3^0$, or the corresponding lower half-space problem when $x_3 < x_3^0$.

A critical problem needed for the inverse problem is to compute the inverse D_α^{-1}. For the case $x_3^1 < x_3^0$ this problem can be stated as follows:

Solve the equation

$$D_\alpha[x_3^1, x_3^0]u = u_0(\mathbf{x}, x_3^1, t),$$

for $u(\mathbf{x}, x_3^0, t)$ given the data $u_0(\mathbf{x}, x_3^1, t)$ on the surface $x_3 = x_3^1$. As is seen from representation (7.43), this is an integral equation of the first kind. Hence it is ill-posed. Of particular interest to the inverse problem is the solution for a thin infinitesimal slab of thickness h, i.e. with $x_3^0 - x_3^1 = h$, and one would like to have an approximate regularized solution with an error of order h^2. This problem has been studied in detail and numerical results are given for the case where $c = c(\alpha)$ (a constant) in Ren and Weston (1995) and Ren (1995). In particular it was shown that if the data u_0 was sufficiently smooth (at least C^2), then the approximate solution is given by (here $x_3^0 = h$, $x_3^1 = 0$)

$$D_\alpha^{-1}[0, h]u_0 = u(\mathbf{x}, h, t)$$

$$= \frac{1}{2}[u_0(\mathbf{x}, 0; t+h) + u_0(\mathbf{x}, 0; t-h)] + \frac{1}{2}\int_{t-h}^{t+h} \mathbf{K}_\alpha^{-1} u_0[s]ds + O(h^2). \tag{7.49}$$

Since the operator $\mathbf{K}_\alpha^{-1}$ behaves like a differential operator, it needs to be regularized. This is done by taking the convolution of u or $\mathbf{K}_\alpha^{-1}u_0$ in the transverse variables x_1, x_2 with a smooth kernel. One particular choice of kernel is the Fejer kernel (see the numerical results described at the end of Section 7.4)

$$k_d(\mathbf{x}) = \frac{1}{2\pi} \frac{\exp(-\frac{2\beta}{d}\sqrt{|\mathbf{x}|^2 + (2d)^2})}{d\sqrt{|\mathbf{x}|^2 + (2d)^2}}, \tag{7.50}$$

where $|\mathbf{x}|^2 = x_1^2 + x_2^2$, and β is chosen so that

$$2\beta = \exp(-4\beta),$$

which gives $\beta = 0.21315138$. This kernel has the important property that if $u(\mathbf{x}) \in L_1$ and is bounded, then (Cao et al. 1996)

$$\lim_{d \to +0} \int_{R^2} k_d(\mathbf{x} - \mathbf{y})u(\mathbf{y})\, d\mathbf{y} = u(\mathbf{x}). \tag{7.51}$$

The results for D_α^{-1} were extended to the case where c was no longer constant (based upon local analyticity (Ren 1995)) as follows,

$$D_{\alpha}^{-1}[0,h]u_0 = \frac{1}{2}[u_0(\mathbf{x},0;t+\frac{h}{c}) + u_0(\mathbf{x},0;t-\frac{h}{c})]$$

$$+\frac{1}{2}\int_{t-h/c}^{t+h/c} \mathbf{K}_{\alpha}^{-1}u_0(s)ds + O(h^2),$$

where $c = c(x_1, x_2, 0)$, and numerical results were presented.

It was shown (Weston, personal communication) that analyticity in the transverse variables was not needed for the above results to hold, based upon a global estimate. In addition the results where modified for the case of a discontinuous wavefront, i.e., the data u_0 had the behaviour $u_0 \sim (t-\tau)\Theta(t-\tau)U_0 + \ldots$ at the wavefront $t = \tau$.

7.2 The dynamical equations for the split components for the wave and telegraph equations

The wave-splitting concept of positive-going and negative-going waves across the surface $x_3 =$ constant will now be extended to the more general medium $c = c(x_1, x_2, x_3)$ in the manner indicated at the beginning of this section. On each surface $x_3 = x_3^0$, the wave-splitting developed previously for the auxiliary space where $c = c(x_1, x_2, \alpha)$ will be employed with α set equal to x_3^0. Since x_3^0 varies, this gives a one-parameter family of Neumann operators $\mathbf{K}_{\alpha|\alpha=x_3^0}$. For future analysis, to simplify notation, the subscript will be dropped (except for cases that need the subscript for clarification) and thus we define

$$\mathbf{K} = \mathbf{K}_{\alpha|\alpha=x_3^0}. \tag{7.52}$$

The operator $\mathbf{K}$ will retain the properties of the operator $\mathbf{K}_\alpha$. The kernel of $\mathbf{K}$ is given by $k(\mathbf{x}, \mathbf{y}, x_3^0; t)$ and the inverse operator is given by

$$\mathbf{K}^{-1} = \Box^T \mathbf{K}, \tag{7.53}$$

where

$$\Box^T = \frac{1}{c^2(x)}\frac{\partial^2}{\partial t^2} - \frac{\partial^2}{\partial x_1^2} - \frac{\partial^2}{\partial x_2^2}, \tag{7.54}$$

and in particular the commutativity result (cf. (7.32))

$$\Box^T \mathbf{K} = \mathbf{K}\Box^T, \tag{7.55}$$

will hold for functions $w(\mathbf{x}, t) \in C^2(\mathbb{R}^2 \times [0, T])$ such that $w(\mathbf{x}, 0) = w_t(\mathbf{x}, 0) = 0$.

For additional analysis, the derivative of the operator $\mathbf{K}$ with respect to the parameter x_3 will be needed. If $u = u(\mathbf{y}, t)$, then define

$$\left(\frac{\partial}{\partial \alpha}\mathbf{K}_\alpha\right)_{\alpha=x_3} u = \mathbf{K}'u, \tag{7.56}$$

where

$$\mathbf{K}'u = \int_0^\infty \int_{\mathbb{R}^2} \frac{\partial k(\mathbf{x}, \mathbf{y}, \alpha; t-t')}{\partial \alpha} u(\mathbf{y}, t') d\mathbf{y} dt' |_{\alpha=x_3}. \tag{7.57}$$

A rigorous treatment of $\mathbf{K}'$ is given in Weston (1990).

The splitting given by (7.36) for a medium where $c = c(x, \alpha)$ will be employed for $c = c(x)$, by taking $\alpha = x_3^0$ so that on the plane $x_3 = x_3^0$

$$u^\pm = \frac{1}{2}[u \mp K_\alpha \partial_3 u]_{\alpha=x_3^0}, \qquad x_3 = x_3^0.$$

Henceforth, for simplicity replace $K_{x_3^0}$ by K, and the splitting will be written in matrix form

$$\begin{bmatrix} u^+ \\ u^- \end{bmatrix} = T \begin{bmatrix} u \\ \frac{\partial u}{\partial x_3} \end{bmatrix}, \quad T = \frac{1}{2}\begin{bmatrix} \mathbf{I} & -\mathbf{K} \\ \mathbf{I} & \mathbf{K} \end{bmatrix}. \tag{7.58}$$

Note that the inverse of T^{-1} exists and is given by

$$T^{-1} = \begin{bmatrix} \mathbf{I} & \mathbf{I} \\ -\mathbf{K}^{-1} & \mathbf{K}^{-1} \end{bmatrix}. \tag{7.59}$$

The system of dynamical equations that must be satisfied by u^+ and u^- can now be derived using the same procedure as described in Chapter 1, and in Corones et al. (1983a,b).

Let Ω be a slab $a < x_3 < b$, and assume that in Ω, u satisfies the wave equation, expressed in the form

$$\partial_3^2 u = \square^T u, \qquad x \in \Omega,\ 0 < t < T,$$

where $\square^T$ is given by (7.54). The combination of this equation and the identity

$$\partial_3 u = \partial_3 u,$$

yields the following system, expressed in matrix form

$$\partial_3 \begin{bmatrix} u \\ \partial_3 u \end{bmatrix} = \begin{bmatrix} 0 & 1 \\ \square^T & 0 \end{bmatrix} \begin{bmatrix} u \\ \partial_3 u \end{bmatrix}. \tag{7.60}$$

This expression can be transferred into one involving u^+ and u^- by using the inverse transformation

$$\begin{bmatrix} u \\ \partial_3 u \end{bmatrix} = T^{-1} \begin{bmatrix} u^+ \\ u^- \end{bmatrix}.$$

Inserting this expression into (7.60), performing the appropriate differentiation, and premultiplying the resultant system by T, the following is obtained

$$\partial_3 \begin{bmatrix} u^+ \\ u^- \end{bmatrix} = W \begin{bmatrix} u^+ \\ u^- \end{bmatrix}, \tag{7.61}$$

where

$$W = T \begin{bmatrix} 0 & 1 \\ \Box^T & 0 \end{bmatrix} T^{-1} - T\partial_3 T^{-1}.$$

Noting that

$$-T\partial_3 T^{-1} = (\partial_3 T)T^{-1} = \frac{1}{2} \begin{bmatrix} 1 & -1 \\ -1 & 1 \end{bmatrix} \mathbf{K}'\mathbf{K}^{-1},$$

use relation (7.53) for the inverse of $\mathbf{K}^{-1}$ to obtain

$$T \begin{bmatrix} 0 & 1 \\ \Box_T & 0 \end{bmatrix} T^{-1} = \begin{bmatrix} -\mathbf{K}^{-1} & 0 \\ 0 & \mathbf{K}^{-1} \end{bmatrix}.$$

System (7.61) can now be put in the explicit form

$$\partial_3 \begin{bmatrix} u^+ \\ u^- \end{bmatrix} = \begin{bmatrix} -\mathbf{K}^{-1} & 0 \\ 0 & \mathbf{K}^{-1} \end{bmatrix} \begin{bmatrix} u^+ \\ u^- \end{bmatrix} + \frac{1}{2} \begin{bmatrix} 1 & -1 \\ -1 & 1 \end{bmatrix} \mathbf{K}'\mathbf{K}^{-1} \begin{bmatrix} u^+ \\ u^- \end{bmatrix}. \tag{7.62}$$

System (7.62) is the sought-for dynamical system of equations for the wave components u^+ and u^-.

It should be noted that for a medium where c is independent of x_3, the operator $\mathbf{K}'$ vanishes and the system of equations (7.62) uncouples yielding the system

$$\partial_3 \begin{bmatrix} u^+ \\ u^- \end{bmatrix} = \begin{bmatrix} -\mathbf{K}^{-1} & 0 \\ 0 & \mathbf{K}^{-1} \end{bmatrix} \begin{bmatrix} u^+ \\ u^- \end{bmatrix},$$

as expected.

The combination operator $\mathbf{K}'\mathbf{K}^{-1}$ is studied in Weston (1990). For the case where $c = c(\alpha)$ (Weston 1987)

$$\mathbf{K}'_\alpha \mathbf{K}_\alpha^{-1} u = \frac{c'}{c^3} \partial_t^2 \mathbf{K}_\alpha^2 u,$$

where $c'(\alpha) = \partial_\alpha c(\alpha)$, and $\mathbf{K}_\alpha$ is given by (7.26). For the more general case $c = c(\mathbf{x}, \alpha)$ it is shown that $\mathbf{K}'_\alpha \mathbf{K}_\alpha^{-1} u_0$ is equivalent to first solving system (7.10) for $u(x,t)$, then solving the system (with $c' = \partial_\alpha c(\mathbf{x}, \alpha)$)

$$\left.\begin{aligned} &(\partial_t^2 - c^2(\mathbf{x},\alpha)\nabla^2)u' = \tfrac{2c'}{c}\partial_t^2 u, && x_3^0 < x_3,\ 0 < t < T, \\ &u' = u'_t = 0, && x_3^0 < x_3,\ t = 0, \\ &\partial_3 u' = 0, && x_3^0 = x_3,\ 0 < t < T, \end{aligned}\right\} \tag{7.63}$$

for $u'(x,t)$, then taking the limit as $x_3 \to x_3^0$, giving

$$u'(\mathbf{x}, x_3^0, t) = -\mathbf{K}'_\alpha \mathbf{K}_\alpha^{-1} u_0.$$

If necessary, the forward finite-difference formula applied to system (7.10) and (7.63) could be employed to compute the operator combination.

The splitting will now be applied to the telegraph equation in $\mathbb{R}^3$,

$$\partial_t^2 u = \frac{1}{r}\nabla \cdot (p\nabla u) - b\partial_t u + qu, \tag{7.64}$$

where the coefficients p, q, r, b are functions of $x = (x_1, x_2, x_3)$, with p, r being positive and twice-differentiable with bounded derivatives, and where the velocity c is given by

$$p/r = c^2 > 0. \tag{7.65}$$

Rewrite equation (7.64) in the form

$$\partial_3^2 u = (c^{-2}\partial_t^2 - \partial_1^2 - \partial_2^2) - Qu - (\partial_3 \ln p)\partial_3 u, \tag{7.66}$$

where

$$Qu = \sum_{i=1}^{2} (\partial_i \ln p)\partial_i u - \frac{b}{c^2}\partial_t u + \frac{q}{c^2}u. \tag{7.67}$$

Then combine (7.66) and the identity $\partial_3 u = \partial_3 u$ to give the system

$$\partial_3 \begin{bmatrix} u \\ \partial_3 u \end{bmatrix} = \begin{bmatrix} 0 & 1 \\ (\Box^T - Q) & -(\partial_3 \ln p) \end{bmatrix} \begin{bmatrix} u \\ \partial_3 u \end{bmatrix},$$

and follow the same procedure as was used for system (7.60), namely employ relations (7.58) and its inverse to obtain the following dynamical equations for the wave components u^+ and u^-

$$\partial_3 \begin{bmatrix} u^+ \\ u^- \end{bmatrix} = \begin{bmatrix} -\mathbf{K}^{-1} & 0 \\ 0 & \mathbf{K}^{-1} \end{bmatrix} \begin{bmatrix} u^+ \\ u^- \end{bmatrix} + \frac{1}{2}\mathbf{K}'\mathbf{K}^{-1} \begin{bmatrix} 1 & -1 \\ -1 & 1 \end{bmatrix} \begin{bmatrix} u^+ \\ u^- \end{bmatrix}$$
$$+\frac{1}{2}\mathbf{K}Q \begin{bmatrix} 1 & 1 \\ -1 & -1 \end{bmatrix} \begin{bmatrix} u^+ \\ u^- \end{bmatrix}$$
$$-\frac{1}{2}\mathbf{K}(\partial_3 \ln p)\mathbf{K}^{-1} \begin{bmatrix} 1 & -1 \\ -1 & 1 \end{bmatrix} \begin{bmatrix} u^+ \\ u^- \end{bmatrix}. \quad (7.68)$$

For numerical implementation, especially involving the layer stripping approach to inverse scattering, it is best to convert the dynamical system of equations for u^+ and u^- into a system of integral equations. To simplify notation, represent the dynamical systems (7.62) and (7.68) in the general form

$$\left. \begin{aligned} (\partial_3 + \mathbf{K}^{-1})u^+ &= v^+, \\ (\partial_3 - \mathbf{K}^{-1})u^- &= v^-. \end{aligned} \right\} \quad (7.69)$$

The medium will be decomposed into a set of infinitesimal slabs of thickness $\Delta x_3 = h$. In the particular slab $x_3^0 < x_3 < x_3^1$, $x_3^1 = x_3^0 + h$, the operator $\mathbf{K}$ will be approximated by $\mathbf{K}_\alpha$ (where the parameter α satisfies $x_3^0 \le \alpha \le x_3^1$), (7.69) is approximated by

$$\left. \begin{aligned} (\partial_3 + \mathbf{K}_\alpha^{-1})u^+ &= v_\beta^+, \\ (\partial_3 - \mathbf{K}_\alpha^{-1})u^- &= v_\beta^-, \end{aligned} \right\} \quad \begin{aligned} &0 < t < T \\ &x_3^0 < x_3 < x_3^1. \end{aligned} \quad (7.70)$$

Note that the terms v_β^+, v_β^- will involve the operators $\mathbf{K}'_\beta, \mathbf{K}_\beta$ where $x_3^0 \le \beta \le x_3^1$.

Thus the integral formulations (7.46), (7.48) may be employed to give

$$u^+(x,t) = D_\alpha[x_3, x_3^0]u_0^+ + \int_{x_3^0}^{x_3} D_\alpha[x_3, y_3]v_\beta^+ dy_3, \quad (7.71)$$

$$u^-(x,t) = D_\alpha[x_3, x_3^1]u_0^- - \int_{x_3}^{x_3^1} D_\alpha[x_3, y_3]v_\beta^- dy_3, \quad (7.72)$$

for $x_3^0 \le x_3 \le x_3^1$, $0 \le t \le T$, where u_0^+, u_0^- are the values of u^+ on $x_3 = x_3^0$, and u^- on $x_3 = x_3^1$ respectively.

7.3 Inverse problem: the Green's function approach

In the previous section, the orientation of the coordinate system had not been specified. However, for application to the inverse problem in three dimensions, the x_3 direction will be pointed down. The half-space $x_3 \leq 0$ (which is the physical upper half-space) will be taken to be free space (where the velocity is $c = c_0$, a constant). The scattering medium or body will lie in the (physically lower) half-space $x_3 > 0$. The parameters characterizing the scattering medium will be the coefficients p, q, r in the telegraph equation (7.64) where $p/r = c^2 > 0$. In free space $x_3 \leq 0$, $p = 1$, $r = 1/c_0^2$, $b = q = 0$ and elsewhere the coefficients p, q, r, b are assumed to be sufficiently smooth functions of x, and $b \geq 0$.

The incident field will be produced by a point source $y = (0, 0, y_3)$ in free space ($y_3 \leq 0$). In particular, the source term will be given by $\delta(x - y)\Theta(t)$. This implies that the total field is the time integral of the fundamental solution $\mathcal{E}(x, y, t)$ associated with the telegraph equation, i.e.

$$u(x,t) = \int_0^t \mathcal{E}(x, y, s)ds. \tag{7.73}$$

As a preliminary, the fundamental solution (Weston 1992a; Weston and He 1993) and its properties need to be examined next.

With y fixed, $\mathcal{E}(x, y, t)$ is a solution (in the space of distributions) of the system

$$\frac{1}{c^2}(\mathcal{E}_{tt} - \frac{1}{r}\nabla \cdot p\nabla\mathcal{E} + b\mathcal{E}_t - q\mathcal{E}) = \delta(x - y)\delta(t), \tag{7.74}$$

$$\mathcal{E}(x, y, t) = 0,\ t < 0. \tag{7.75}$$

The asymptotic short-time behaviour of $\mathcal{E}(x, y, t)$ is given by

$$\mathcal{E}(x, y, t) = \frac{\delta(t - \tau)}{\rho} + \Theta(t - \tau)a(x, y, t), \tag{7.76}$$

$$a(x, y, t) \sim a_0(x, y) + O(t - \tau),$$

where $O(t)$ stands for terms which behave like t as $t \to 0$. The hypersurfaces $t - \tau = 0$ are the wavefronts. In free space, the arrival-time $\tau(x, y)$ of a ray going from point y to point x is given by $\tau = |x - y|/c_0$, and the amplitude factor ρ is given by $\rho = 4\pi|x - y|$, with $a_0 \equiv 0$. Because the source point y lies in free space, it follows that $\tau(x, y)$ and $\rho(x, y)$ have the following behaviour as $x \to y$

$$\left.\begin{aligned} \tau(x, y) &\sim |x - y|/c_0, \\ \rho(x, y) &\sim 4\pi|x - y|, \end{aligned}\right\} \tag{7.77}$$

so that as $x \to y$

$$-\nabla^2\frac{1}{\rho} \sim -\nabla^2\left(\frac{1}{4\pi|x - y|}\right) = \delta(x - y).$$

Ray coordinates $(\theta_1, \theta_2, \sigma)$ will be employed (Bloom and Kazarinoff 1976; Bleistein 1984) where $\theta_1 = \theta$, $\theta_2 = \phi$ are the polar and azimuthal angle of a local spherical polar coordinate system centred at y (directed in the positive x_3 direction). The ray coordinate σ is given by

$$\sigma = \int_0^s c^{-1} ds, \tag{7.78}$$

where s is the arc-length along the ray from the point y to the point x. Note then that

$$\sigma = \tau(x, y).$$

The metric coefficients h_1, h_2, h_3 associated with the ray coordinates (Weston and He 1993; Bleistein 1984) have the property that

$$h_3 = c, \tag{7.79}$$

$$\left.\begin{array}{l} h_1 \overset{x \to y}{\sim} |x - y| \\ \\ h_2 \overset{x \to y}{\sim} |x - y| \sin \theta_1. \end{array}\right\} \tag{7.80}$$

By restricting the time interval so that no caustics appear, the ray coordinate system $\theta_1(x), \theta_2(x), \sigma(x)$ maps the domain $\{x | 0 < \tau(x, y) < T\}$ into the domain $0 \leq \theta_1 \leq \frac{\pi}{2}$, $0 \leq \theta_2 < 2\pi$, $0 < \sigma < T$ in a $1:1$ fashion (Bloom and Kazarinoff 1976). General forms of the ray equations are given in Bloom and Kazarinoff (1976) and Bleistein (1984). For the particular case where the rays arise from a point more details are given in Weston (1989).

As expected, τ satisfies the eikonal equation

$$|\nabla \tau|^2 = c^{-2}(x), \tag{7.81}$$

and ρ the transport equation

$$2\rho \nabla \tau \cdot \nabla (\frac{1}{\rho}) + [(\frac{1}{p}) \nabla \cdot (p \nabla \tau) + \frac{b}{c^2}] = 0. \tag{7.82}$$

This can be reduced to (Bloom and Kazarinoff 1976)

$$2 \frac{d}{d\sigma} (\frac{1}{\rho}) + (\frac{1}{\rho}) [\frac{d}{d\sigma} \ln (\frac{pJ}{c}) + b] = 0, \tag{7.83}$$

where

$$J = h_1 h_2.$$

The solution of (7.83), combined with initial condition (7.77) along the ray θ_1, θ_2 constant, is given by

$$4\pi(\frac{1}{\rho}) = (\frac{\sin\theta_1 c(\sigma)}{c_0 p(\sigma) J(\sigma)})^{1/2} \exp\left(-\int_0^{\sigma}(\frac{b}{2})d\sigma'\right). \tag{7.84}$$

The remaining term in expression (7.76) satisfies the telegraph equation (7.64) for values of $t > \tau(x, y)$, and the initial condition on the wavefront

$$a(x,y,t) = a_0(x,y), \qquad \text{when } t = \tau(x,y) \text{ and is zero for } t < \tau(x,y).$$

Here $a_0(x, y)$ satisfies the transport equation

$$2\nabla\tau\cdot\nabla a_0 + a_0[(1/p)\nabla\cdot(p\nabla\tau) + b/c^2]$$

$$= q/(\rho c^2) + (1/p)\nabla\cdot[p\nabla(1/\rho)] - \nabla^2(1/4\pi|x-y|),$$

with $a_0(x, y) = 0$ when $x = y$.

One key expression for the asymptotic behaviour of the fundamental solution (7.74) at the wavefront is needed. One has

$$\{\mathbf{K}_\alpha \partial_{x_3}\mathcal{E} + \mathcal{E}\}_{\alpha=x_3=x_3^0} = +\gamma\Theta(t-\tau) + O(t-\tau), \tag{7.85}$$

where

$$\gamma = -(\partial_3\tau)^{-2}\left\{\frac{b}{c^2} + \nabla\ln p\cdot\nabla\tau - (\partial_3 c)/(c^3\partial_3\tau)\right\}(2\rho)^{-1}. \tag{7.86}$$

For its derivation consult Weston (1992a) and Weston and He (1993) (since the x_3-axis points in the opposite direction there, terms involving ∂_3 in the corresponding expressions to (7.85) and (7.86) must be replaced by $-\partial_3$), or for the special case of the wave equation, see Exercise 7.3. The value of γ is used in the inverse problem to determine one of the parameters, $\partial_3 c$, $\partial_3 p$ or b, using knowledge of the other parameters. For the special case of the wave equation, where $p = 1$, $b = 0$, the expression for γ simplifies to

$$\gamma = (\partial_3 c)/[2\rho c^3(\partial_3\tau)^3]. \tag{7.87}$$

The particular inverse problem that will be considered here is the reconstruction of a one-parameter function (velocity, density, etc). Details will be given for the reconstruction of the velocity when the material properties of the medium are characterized by just the wave equation. The modification necessary to reconstruct one of the other parameters of the telegraph equation will be indicated.

The measured data that will be employed will consist of the knowledge of the values of the total or scattered field at all points x on a plane (exterior to the scattering medium) and for finite values of time, i.e., $0 < t < T$. In addition it will be assumed that the incident field is known (at least on the same plane). For the particular problem considered here, the incident field will be produced by a point pulse source (a finite distance away from the scattering medium), and the fields will be quiescent for $t \leq 0$. These conditions imply that the total field will be non-zero only on a compact region of the measurement plane for $0 \leq t \leq T$.

With the coordinate system employed so that the scattering medium lies in the region $x_3 > 0$, and the x_3 axis points down, a point pulse source located at $y^0 = (0, 0, y_3^0)$, $y_3^0 < 0$ of the form

$$\delta(x - y^0)\Theta(t),$$

will be considered. Thus the incident field will be just the time integral of the free space fundamental solution (7.19), i.e.

$$u^i(x,t) = \frac{\Theta(t - |x - y^0|/c_0)}{4\pi|x - y^0|}. \tag{7.88}$$

This will produce a wavefront discontinuity of lower order, just a finite discontinuity in the field. The total field $u = u^i + u^r$ where u^r is the reflected or scattered field, is the time integral of the Green's function defined by (7.74) and (7.75)

$$u(x,t) = g(x,t) = \int_0^t \mathcal{E}(x, y^0, s)ds. \tag{7.89}$$

For the time interval $0 < t < T$, the total field will be zero for points on the plane $x_3 = 0$, that are outside the disk $\mathbf{x}^2 + (x_3 - y_3^0)^2 = T^2 c_0^2$.

For the region $x_3 > 0$, the dynamical equations for the wave components g^+, g^- are given by

$$(\partial_3 + \mathbf{K}^{-1})g^+ = v^+,$$

$$(\partial_3 - \mathbf{K}^{-1})g^- = v^-,$$

where v^+, v^- are given by the right-hand side of equations (7.62) with $u^\pm$ replaced by $g^\pm$

$$v^\pm = \pm\frac{1}{2}\mathbf{K}'\mathbf{K}^{-1}(g^+ - g^-). \tag{7.90}$$

On the surface $x_3 = 0$ (being in free space) $c = c_0, \partial_3 c = 0$. Since $\mathbf{K}' = 0$ on $x_3 = 0$, it follows that g^+ and g^- are uncoupled at $x_3 = 0$, and by definition g^+ is just the incident wave u^i, and g^- will be the reflected wave u^r. This gives

$$g^+(x,t) = u^i(x,t) = \frac{\Theta(t - |x - y^0|/c_0)}{4\pi|x - y^0|}, \; 0 < t < T, \; x_3 = 0, \tag{7.91}$$

where $y^0 = (0, 0, y_3^0)$ and

$$g^-(x,t) = u^r(x,t), \; 0 < t < T, \; x_3 = 0. \tag{7.92}$$

From the relation ($g = g^+ + g^-$)

$$g^- = \frac{1}{2}(g + \mathbf{K}\partial_3 g),$$

together with the expressions (7.85) and (7.87), it follows that g^- has the following asymptotic behaviour in the vicinity of the wave $t = \tau$,

$$g^- = (t-\tau)\Theta(t-\tau)g_0^- + \ldots, \tag{7.93}$$

where

$$g_0^- = \partial_3 c/[4\rho c^3(\partial_3\tau)^3]. \tag{7.94}$$

To the leading term g^+ behaves like g, hence from (7.76) and (7.89), g^+ has the asymptotic behaviour

$$g^+ \sim \frac{1}{\rho}\Theta(t-\tau) + \ldots. \tag{7.95}$$

Recall that in the layer stripping approach to the inverse problem, the region $x_3 > 0$ will be decomposed into a set of infinitesimal slabs of thickness $\Delta x_3 = h$, the nth slab given by $x_3^{n-1} < x_3 < x_3^n$ where $x_3^n = nh$, and a step-by-step procedure is employed to determine the required material property (parameters) over a portion of each slab.

To be specific, the layer stripping approach for the wave equation applied to the nth slab will be stated as follows: given $\tau, \rho, \partial_3\tau, g^+, g^-, c, \partial_3 c$ over a portion Ω_{n-1} of the top surface $x_3 = x_3^{n-1}$ for a time interval $0 < t < T_{n-1}$, to determine the same quantities over a portion Ω_n of the surface $x_3 = x_3^n$ for a time interval $0 < t < T_n$.

As is seen from (7.91) and (7.92) these quantities are known on the initial surface $x_3 = x_3^0 = 0$, for $0 < t < T_0$, $T_0 = T$. ($\partial_3\tau$ is determined from the eikonal equation using the known values of τ and c).

Using the value of $c(\mathbf{x}, x_3^n)$ given by

$$c(\mathbf{x}, x_3^n) = c(\mathbf{x}, x_3^{n-1}) + h\partial_3 c(\mathbf{x}, x_3^{n-1}),$$

and the values of τ, $\partial_3\tau$ on the portion of the surface $x_3 = x_3^{n-1}$ where $\partial_3\tau > 0$, one can use the ray equations (Weston 1992a) to obtain τ, $\partial_3\tau$ on $x_3 = x_3^n$, or else use the approximation $\tau(\mathbf{x}, x_3^n) = \tau(\mathbf{x}, x_3^{n-1}) + h\partial_3\tau(\mathbf{x}, x_3^{n-1})$ where $\partial_3\tau$ is given by

$$\partial_3\tau(x) = [c^{-2}(x) - |\nabla^T\tau(x)|^2]^{1/2}.$$

The calculation of $\nabla^T\tau$ may produce some numerical instability. The evaluation of ρ on $x_3 = x_3^n$ can be obtained from the numerical calculation of the transport equation for $(1/\rho)$ by integrating along the rays or else obtained as the leading coefficient g_0^+ of g^+ after solving for g^+ at $x_3 = x_3^n$ (which is discussed next).

The integral formulation of the dynamical equations for $g^{\pm}$ as given by (7.71) and (7.72) with $u^{\pm}$ replaced by $g^{\pm}$, and $v^{\pm}$ by (7.90) will be employed to obtain the values of $g^{\pm}$ on $x_3 = x_3^n$ from values on $x_3 = x_3^{n-1}$. Using an infinitesimal slab approximation (with the operator D in the integral term replaced by the identity operator) the dynamical equation for points on $x_3 = x_3^n$ away from the wavefront is thus given by

$$g^{+}(\mathbf{x}, x_3^n, t) = D_\alpha[x_3^n, x_3^{n-1}]g^{+} + hv_\beta^{+}(\mathbf{x}, x_3^{n-1}, t), \quad t > \tau(x_3^n), \qquad (7.96)$$

$$\begin{aligned} g^{-}(\mathbf{x}, x_3^{n-1}, t) = {} & D_\alpha[x_3^{n-1}, x_3^n]g^{-} \\ & -hv_\beta^{-}(\mathbf{x}, x_3^{n-1}, t), \quad t > \tau(x_3^{n-1}) + 2h\partial_3\tau(x_3^{n-1}), \qquad (7.97) \end{aligned}$$

where

$$v_\beta^{\pm} = \pm\frac{1}{2}\mathbf{K}'_\beta\mathbf{K}_\beta^{-1}(g^{+} - g^{-}),$$

and $\alpha = x_3^n$, $\beta = x_3^{n-1}$. This means that $v_\beta^{\pm}$ are computed using the values of c, $\partial_3 c$ involved with the operators at $x_3 = x_3^{n-1}$, and the values of g^{+}, g^{-} on $x_3 = x_3^{n-1}$.

For points in the neighbourhood of the wave-front, the integral term in the transport equation (7.96) has to be handled differently. Using the fact that $(g^{+} - g^{-})$ behaves like $\Theta(t-\tau)[1/\rho + O(t-\tau)]$ in the neighbourhood of the wave-front, and using the asymptotic result for the operator $\mathbf{K}'\mathbf{K}^{-1}$ given in Weston and He (1993) (modified with a minus sign to take account of the direction of the wavefront $(\partial_3\tau)$ with respect to the x_3-axis), one obtains the asymptotic behaviour

$$v^{\pm} \sim \frac{1}{2}\mathbf{K}'\mathbf{K}^{-1}(g^{+} - g^{-}) \sim \pm V_0(x)\Theta(t - \tau(x)) + \dots, \qquad (7.98)$$

where

$$V_0(x) = \frac{\partial_3 c}{2\rho(\partial_3\tau)^2 c^3}. \qquad (7.99)$$

The integral term that appears in the transport equation (7.96) for g^{+} is given by

$$\int_{x_3^{n-1}}^{x_3^n} D_\alpha[x_3^n, y_3]v^{+}(y_3)dy_3 \sim V_0(\mathbf{x}, x_3^{n-1}) \int_{x_3^{n-1}}^{x_3^n} D_\alpha[x_3^n, y_3]\Theta(t - \tau(\mathbf{x}, y_3))dy_3.$$

Since the operator $D_\alpha[x_3^n, y_3]$ maps the wavefront $t - \tau(\mathbf{x}, y_3)$ from points on the plane $x_3 = y_3$ to points on the plane $x_3 = x_3^n$ where $x_3^n \geq y_3$, which is in the direction that the wave is travelling, it follows that $D_\alpha[x_3^n, y_3]\Theta(t-\tau) \cong \Theta(t - \tau(\mathbf{x}, x_3^n))$, where $D_\alpha[x_3^n, y_3]\Theta(t-\tau)$ is defined by (7.43) with $u_0(\mathbf{y}, s)$ replaced by $\Theta(s-\tau(\mathbf{y},y_3))$ and x_3, x_3^0 replaced by x_3^n, y_3 respectively. Thus the integral term on the right-hand side of the transport equation (7.96) for g^+ reduces to

$$hV_0(\mathbf{x},x_3^{n-1})\Theta(t-\tau(\mathbf{x},x_3^n)+\dots,$$

in the vicinity of the wave-front. At the wave-front the non-homogeneous term in (7.96) behaves like

$$hV_0(\mathbf{x},x_3^{n-1})\Theta(t-\tau(\mathbf{x},x_3^{n-1}),$$

so the two terms agree apart from the shift in the wavefront. For times $t > \tau(\mathbf{x},x_3^n)$ the terms should differ by order of h^2.

Using (7.96) to compute $g^+(\mathbf{x},x_3^n,t)$ from values of $g^+(\mathbf{x},x_3^{n-1},t)$ is a well-posed problem. The major calculations will be involved with the operator D_α (which involves solving an initial-value boundary value problem for a half-space).

The difficult part arises in the determination of $g^-(x,x_3^n)$ from (7.97). Before considering this problem, a more accurate expression for the non-homogeneous term on the right-hand side of (7.97) needs to be obtained for points in the neighbourhood of the wave-front. Recall that this non-homogeneous term is the infinitesimal thin slab ($h << 1$) approximation to the integral term

$$-\int_{x_3^{n-1}}^{x_3^n} D_\alpha[x_3^{n-1},y_3]v^-(y_3)dy_3,$$

given in (7.72) as applied to the region $x_3^{n-1} \le y_3 \le x_3^n$. Using the asymptotic form (7.95) for v^- in the neighbourhood of the wavefront, this integral term can be approximated by

$$V_0(\mathbf{x},x_3^{n-1})\int_{x_3^{n-1}}^{x_3^n} D_\alpha[x_3^{n-1},y_3]\Theta(t-\tau)dy_3.$$

Since the operator D_α here maps Dirichlet data on the plane $x_3 = y_3$ to Dirichlet data on the plane $x_3 = x_3^{n-1} < y_3$ (in the negative x_3 direction), and the physical wavefront $t-\tau(x)$ is travelling in the opposite direction, the mapping produces a reflected wave. In the geometric optics approximation one can show that

$$\begin{aligned} D_\alpha[x_3^{n-1},y_3]\Theta(t-\tau(y)) &\sim \Theta(t-\tau(y)+(x_3^{n-1}-y_3)\partial_3\tau) \\ &\sim \Theta(t-\tau(x_3^{n-1})-2(y_3-x_3^{n-1})\partial_3\tau(x_3^{n-1})). \end{aligned}$$

Thus the above integral reduces to

$$\sim \frac{V_0(\mathbf{x},x_3^{n-1})}{2\partial_3\tau(x_3^{n-1})}(t-\tau(x_3^{n-1})),\ \tau(x_3^{n-1}) \le t \le \tau^-(x_3^{n-1}) \tag{7.100}$$

and it is equal to zero for $t < \tau(x_3^{n-1})$, where

$$\tau^-(x_3^{n-1}) \sim \tau(x_3^{n-1}) + 2h\partial_3\tau(x_3^{n-1}).$$

The problem of determining g^- can now be considered. Rewrite (7.97) in the form

$$\begin{aligned} D_\alpha[x_3^{n-1}, x_3^n]g^- &= g^-(\mathbf{x}, x_3^{n-1}, t) + hv_\beta^-(\mathbf{x}, x_3^{n-1}, t), \\ &\quad \tau^-(x_3^{n-1}) < t, \end{aligned} \tag{7.101}$$

$$\begin{aligned} &= g^-(\mathbf{x}, x_3^{n-1}, t) - \frac{V_0(x)}{2\partial_3\tau}[t - \tau(x)]\bigg|_{x_3 = x_3^{n-1}}, \\ &\quad \tau(x_3^{n-1}) \le t \le \tau^-(x_3^{n-1}) \end{aligned} \tag{7.102}$$

using the expression (7.100) for the non-homogeneous term that was just developed for the vicinity of the wavefront.

Employing the integral formulation of the operator D_α, it is seen that (7.101) is an integral equation of the first kind (indicating the nature of the ill-posedness of the inverse problem). Furthermore, since g^- must have the asymptotic behaviour at the wavefront $t = \tau$

$$g^- \sim (t - \tau)\Theta(t - \tau)g_0^- + \ldots,$$

it follows that an additional constraint is imposed upon the unknown quantity g^- at $x_3 = x_3^n$, namely that

$$g^- \sim (t - \tau(x_3^n))\Theta(t - \tau(x_3^n))g_0^-, \qquad x_3 = x_3^n.$$

But $D_\alpha[x_3, x_3^n]g^-$ is a wave moving in the $-x_3$ direction so that it must have the asymptotic form $(t - \tau^-)\Theta(t - \tau^-)g_0^-$ for $x_3 < x_3^n$ where $\partial_3\tau^- < 0$, τ^- is a reflected wavefront.

This means that

$$D_\alpha[x_3^{n-1}, x_3^n]g^- = 0, \qquad \text{for } t < \tau^-(x_3^{n-1}).$$

Thus a necessary condition for a solution of (7.101), (7.102) is that the right-hand side of (7.102) must vanish for $\tau(x_3^n) \le t \le \tau^-(x_3^{n-1})$, indicating (using relation (7.99)) that

$$g^-(\mathbf{x}, x_3^{n-1}, t) = \frac{\partial_3 c}{4(\partial_3\tau)^3 c^3 \rho}\bigg|_{x_3^{n-1}} (t - \tau(x_3^{n-1})),$$

for $\tau(x_3^{n-1}) \le t \le \tau^-(x_3^{n-1})$. But this is equivalent to the initial condition (7.93), (7.94) at the wave-front. Thus it is seen that the initial condition at $t = \tau$ must be imposed on g_0^- to solve the inverse problem (for a rigorous development of this point for the telegraph equation when $c = 1$, see Chang and Weston (1997)). Imposing this initial condition one just uses equation (7.101) to solve for g^-.

The results at the end of section (7.1) for the D_α^{-1} operator are used to solve (7.101) for g^-. Extracting the value of g_0^- from the asymptotic behaviour of g^-

$$g^- = (t-\tau)\Theta(t-\tau)g_0^- + \dots, \; x_3 = x_3^n$$

and using the asymptotic expression for g_0^- given by (7.94), it is seen that $\partial_3 c$ can be determined from g_0^- on $x_3 = x_3^n$ when c, ρ, and $\partial_3\tau$ are known there.

Some numerical results on the approximate solution of the dynamical equations are given in Cao et al. (1996) for the case when the value of the velocity c is constant (equal to unity) and the density $\rho = 1/p$ is taken constant. For such a case, the dynamical system (7.68) reduces to the form

$$\begin{aligned}(\partial_z + \mathbf{K}^{-1})u^+ &= f^+,\\ (\partial_z - \mathbf{K}^{-1})u^- &= f^-,\end{aligned}$$

where $f^+ = -f^- = -\frac{1}{2}K[b\partial_t](u^+ + u^-)$. In this case $\mathbf{K}^{-1}$ is given by the expressions (7.33) and (7.34) with $c = 1$. The approximate solution to the dynamical equations for a thin layer of thickness h is given by

$$\begin{aligned} & u^+(\mathbf{x}, x_3 + h; t) \\ & = \frac{1}{2\pi}\int\int_{A(x,y,t)} \frac{h}{R^3}[u^+(\mathbf{x}', x_3; t-R) + R\partial_t u^+(\mathbf{x}', x_3; t-R)]\, dx_1' dx_2' \\ & \qquad + hf^+(\mathbf{x}, x_3; t) + O(h^2), \end{aligned} \tag{7.103}$$

$$u^-(\mathbf{x}, x_3 + h; t) = u^-(\mathbf{x}, x_3; t+h) - h[\mathbf{L}u^- - f^-](\mathbf{x}, x_3; t) + O(h^2), \tag{7.104}$$

where $R = \sqrt{(x_1 - x'_1)^2 + (x_2 - x'_2)^2 + h^2}$, and $A(x_1, x_2, t)$ is a disk area with centre (x_1, x_2) and radius ct at the x_3 plane. The first expression is just the approximation to the operator $D[x_3 + h, x_3]$ for the case when $c = 1$, and the second is just the simple approximation (valid for smooth data) given by a variant of (7.49). The results are given in Fig. 7.1 for an inhomogeneous half-space with the dissipation profile $b(x_1, x_2, x_3) = 100\sin^2(\pi x_3)\exp[-(x_1^2 + x_2^2/4 + x_3^2/9)]$ (for $x_3 \geq 0$). The incident field (produced by a point source) is chosen as

$$u^i = \frac{f(t - |\mathbf{r} - \mathbf{r}_0|)}{4\pi|\mathbf{r} - \mathbf{r}_0|},$$

where $\mathbf{r}_0 = (0, 0, -0.5)$ and

$$f(t) = \begin{cases} \sin^2(\pi t/t_0), & \text{when } 0 \leq t \leq t_0, \\ 0, & \text{otherwise,} \end{cases}$$

with $t_0 = 0.4$.

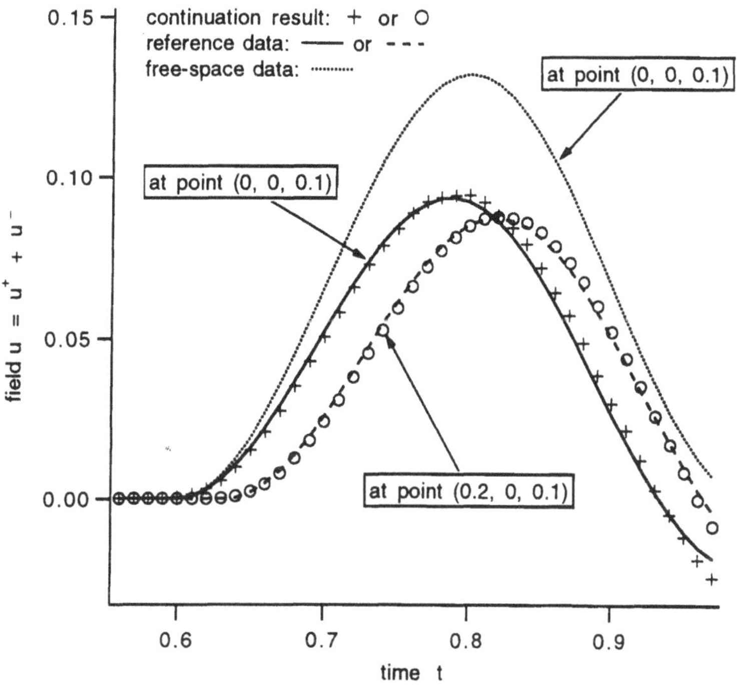

FIG. 7.1. Data continuation using the wave-splitting formulas (7.103) and (7.104) in an inhomogeneous half-space with a synthetic dissipation coefficient profile $b(x, y, z) = 100\sin^2(\pi x_3)\exp[-(x_1^2 + x_2^2/4 + x_3^2/9)]$, $x_3 \geq 0$. The solid and dashed lines are the reference data at points $(0, 0, 0.1)$ and $(0.2, 0, 0.1)$, respectively, calculated with a finite-difference time domain method. The crosses and circles are the corresponding continuation data using the wave-splitting formulas with $\Delta x_1 = \Delta x_2 = \Delta x_3 = 2\Delta t = 0.02$.

7.4 Inverse problem: the continuation method

Instead of using wave-splitting and the associated dynamical equations, the method that will be considered in this section will involve the continuation of the field quantities. In general terms, let $u(x, t)$ be the solution of the telegraph equation in $\mathbb{R}^3$, produced by a source $f(x, t)$ located in a compact domain Ω_0 in $\mathbb{R}^3$, and such that $f = 0$, for $t < 0$,

$$\left.\begin{array}{ll} u_{tt} - \frac{1}{r}\nabla \cdot (p\nabla u) + bu_t - qu = f(x, t), & x \in \mathbb{R}^3,\ 0 < t < T, \\ u = u_t = 0, & x \in \mathbb{R}^3,\ t = 0. \end{array}\right\} \qquad (7.105)$$

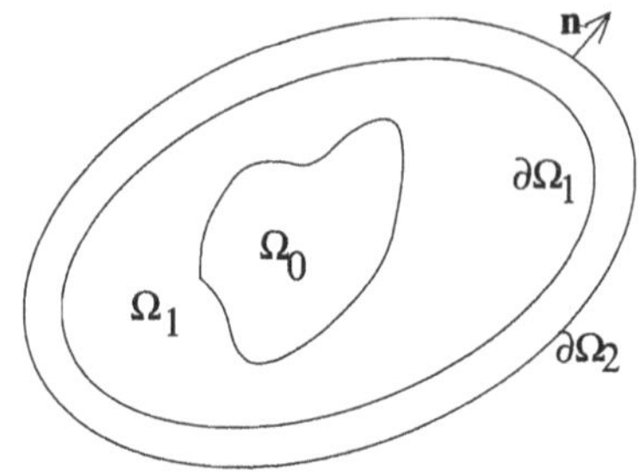

FIG. 7.2. Continuation problem.

Assume that in a region Ω_1 (containing Ω_0) the field quantity $u(x,t)$ is known and in particular the values of $u(x,t)$ and $\partial_n u(x,t) = \hat{n} \cdot \nabla u$ ($\hat{n}$ is the unit outward normal to $\partial\Omega_1$) are known on the surface $\partial\Omega_1$. Most importantly, u and $\partial_n u$ are not independent. If one is known, the other can be obtained by solving an initial-value boundary-value problem. Also assume that in a larger region Ω_2 containing Ω_1 the coefficients p, q, r, and b of the equation are known. Then the continuation will involve the determination of u and $\partial_n u$ on the surface $\partial\Omega_2$ from knowledge of u and/or $\partial_n u$ on the surface $\partial\Omega_1$. See Fig. 7.2.

Let $\mathcal{E}_N(x,y,t-s)$ be the fundamental solution associated with the telegraph equation (x, t fixed)

$$(\partial_s^2 - \frac{1}{r}\nabla_y \cdot (p\nabla_y) - b\partial_s - q)\mathcal{E}_N = \frac{1}{r}\delta(x-y)\delta(t-s),\ y \in \Omega_2,$$

$$\mathcal{E}_N = 0, \quad t - s > 0, \qquad\qquad y \in \Omega_2,$$

and the boundary condition on the surface $\partial\Omega_2$

$$\partial_n \mathcal{E}_N = 0, \quad x \in \partial\Omega_2.$$

Employing Green's theorem for the region Ω_2, and $0 < t < T$, it can be shown that the solution of system (7.105) for $x \in \Omega_2$ is given by

$$\begin{aligned} u(x,t) = & \int_0^t \int_{\Omega_2} \mathcal{E}_N(x,y,t-s) f(y,s) r(y) dy ds \\ & - \int_0^t \int_{\partial\Omega_2} p(y) \mathcal{E}_N(x,y,t-s) \partial_n u(y,s) d\sigma_y ds, \end{aligned} \qquad (7.106)$$

where $d\sigma_y$ is an element of surface area on $\partial\Omega_2$.

If $u(x,t)$ is known for values of $x \in \partial\Omega_1$ and $0 < t < T$, then on letting $x \in \partial\Omega_1$ in the above equation, it follows that one obtains an integral equation of the first kind for the unknown quantity $\partial_n u$ on $x \in \partial\Omega_2$. The resulting system is ill-posed and has to be regularized in the sense of Tikhonov (Tikhonov and Arsenine 1977). If the two surfaces $\partial\Omega_1$ and $\partial\Omega_2$ are sufficiently smooth, and an infinitesimal distance apart, then one can get an approximate solution in much the same manner as was done for the D^{-1} operator at the end of Section 7.1. Once one has a good regularized approximation to $\partial_n u$ on $\partial\Omega_2$, one can obtain the value of u on $\partial\Omega_2$, by letting x in (7.106) approach a point on the surface $\partial\Omega_2$, and then computing $u(x,t)$. Alternatively $u(x,t)$ can be found by solving the initial-value, boundary-value problem in Ω_2,

$$\begin{cases} u_{tt} - \frac{1}{r}\nabla \cdot (p\nabla u) + bu_t - qu = f(x,t),\ x \in \Omega_2,\ 0 < t < T, \\ \\ u = u_t = 0,\ t = 0,\ x \in \Omega_2, \\ \\ \partial_n u \text{ is given on } \partial\Omega_2,\ 0 < t < T. \end{cases} \tag{7.107}$$

This particular problem is well-posed. Hence the only ill-posedness in the continuation process comes from the determination of $\partial_n u$ on $\partial\Omega_2$.

To be specific, the same geometry and same incident field will be used here as was used in Section 7.3. Thus the domains Ω will be half-spaces $x_3 <$ constant. The source term will be a point source with a Heaviside step function time dependence. The inverse problem will consist of applying the continuation problem to determine the values of u and $\partial_3 u$ on a surface $x_3 =$ constant from knowledge of similar values on an adjacent plane surface which is an infinitesimal distance h apart, together with the asymptotic wave-front condition (7.93).

However, instead of using the integral equation approach to determine $\partial_3 u$ and the differential equation approach as indicated by system (7.107) to determine u, a much simpler approach will be used in the continuation procedure. This approach will be based upon local analysis. It greatly reduces the complexity of the computations, but at a cost of much greater instability. The local continuation requires much stronger conditions on the data, namely analyticity in the transverse variables x_1, x_2, except in the vicinity of a wavefront, where asymptotic analysis may be used.

The derivation of the continuation formulas will not be given here, just the results. In the continuation formulas let the value of $u, \partial_3 u$ at $x_3 = 0$ be given as follows

$$u = u_0(x_1, x_2; t),\ \partial_3 u = u_1(x_1, x_2; t),\ x_3 = 0. \tag{7.108}$$

For the wave equation

$$(\partial_t^2 - c^2(x_1, x_2, x_3)\nabla^2)u = 0,$$

the continuation formula for a distance $x_3 = h$ are given by (Ren 1995)

$$u(x_1, x_2, h; t) = \frac{1}{2}[u_0(x_1, x_2; t + \xi) + u_0(x_1, x_2; t - \xi)]$$
$$+\frac{1}{2}\int_{t-\xi}^{t+\xi} u_1(x_1, x_2; s)ds + O(h^2), \tag{7.109}$$

$$\partial_3 u(x_1, x_2, h; t) = \frac{1}{2}\{\, \partial_t[u_0(x_1, x_2; t + \xi) - u_0(x_1, x_2, t - \xi)]$$
$$+u_1(x_1, x_2; t + \xi) + u_1(x_1, x_2; t - \xi)$$
$$-2\xi c^2(x_1, x_2, 0)(\partial^2_{x_1} + \partial^2_{x_2})u_0(x_1, x_2, t)\,\}\eta + O(h^2), \tag{7.110}$$

where

$$\xi = \left[\frac{h}{c} - \frac{h^2}{2c^2}\frac{\partial c}{\partial x_3}\right]_{x_3=0}, \tag{7.111}$$

$$\eta = \left[\frac{1}{c} - \frac{h}{c^2}\frac{\partial c}{\partial x_3}\right]_{x_3=0}. \tag{7.112}$$

For the telegraph equation (with constant unit velocity)

$$\left[\partial_t^2 u - \frac{1}{p}\nabla \cdot (p\ \nabla u) + b\partial_t u\right] = 0,$$

the continuation formulas are given by (Cao et al. 1996)

$$u(x_1, x_2, h; t) = \frac{1}{2}\{\, u_0(x_1, x_2; t + h) + u_0(x_1, x_2; t - h)$$
$$+\int_{t-h}^{t+h} u_1(x_1, x_2; s)ds\,\} + O(h^2), \tag{7.113}$$

$$\partial_3 u(x_1, x_2, h; t) = \frac{1}{2}\{\, \partial_t[u_0(x_1, x_2; t + h) - u_0(x_1, x_2; t - h)]$$
$$+u_1(x_1, x_2, t + h) + u_1(x_1, x_2, t - h)\,\}$$
$$-h(\partial_1^2 + \partial_2^2)u_0(x_1, x_2; t) - h[\partial_1 u_0 \partial \ln p + \partial_2 u_0 \partial_2 \ln p]_{x_3=0}$$
$$-u_1 \partial_3 \ln p|_{x_3=0} + hb\partial_t u_0 + O(h^2). \tag{7.114}$$

Formulas (7.113) and (7.114) require local analyticity in the transverse variables x_1, x_2, and so won't hold in the vicinity of a discontinuity. For a wavefront discontinuity the appropriate results can be obtained using the asymptotic behaviour behind a wavefront. For the telegraph equation (with $c = 1$), the wavefront is given by $t - R = 0$, where R is the distance from the field point $P(x_1, x_2, x_3)$ to the source point $P_0(0, 0, y_3^o)$. The field $u(x;t)$ in the neighbourhood of the wavefront is given by (Friedlander 1958)

$$u(x,t) = \Theta(t-R)[v_0(x) + (t-R)v_1(x) + \ldots], \tag{7.115}$$

where v_0, v_1 satisfy the transport equations

$$2\nabla R \cdot \nabla v_0 + \frac{1}{p}\nabla \cdot (p\nabla R)v_0 + bv_0 = 0,$$

$$2\nabla R \cdot \nabla v_1 + \frac{1}{p}\nabla \cdot (p\nabla R)v_1 + bv_1 = \frac{1}{p}\nabla \cdot (p\nabla v_0).$$

Since the rays are straight lines from the source P_0, it follows that the ray passing through the point $P(x_1, x_2, x_3)$ travels through the point $P'(x_1', x_2', x_3')$ on the adjacent surface

$$x_3' = x_3 + h, \tag{7.116}$$

where

$$x_1' = (1 + h/\ell)x_1,\ x_2' = (1 + h/\ell)x_2,\ \ell = x_3 - y_3^o. \tag{7.117}$$

If R' is the distance of the point P' from the source P_0, then

$$R' = (1 + h/\ell)R. \tag{7.118}$$

Let Δt be the time interval after the wavefront has passed the point P,

$$\Delta t = t - R.$$

It can then be shown on integrating the transport equations along the ray from the point P to the point P', that the value of $u(x';t')$ at the point P' at time $t' = \Delta t + R'$ is related to the value of $u(x;t)$ at the point P at time $t = \Delta t + R$, by the following (Cao et al. 1998)

$$\left.\begin{aligned} u(x', \Delta t + R') &= Au(x, \Delta t + R)\{1 + O(h\Delta t)\}, \\ A &= \sqrt{\tfrac{p(x)}{p(x')}}\ \tfrac{\ell}{(\ell+h)}\ \exp\left\{-\tfrac{1}{2}\int_R^{R'} b(s)ds\right\}. \end{aligned}\right\} \tag{7.119}$$

Furthermore, using the transport equations and the eikonal equation it can be shown with a great deal of work that (Cao et al. 1998) with $t = \Delta t + R$, $t' = \Delta t + R'$,

$$\frac{\partial R'}{\partial x_3'}\frac{\partial u(x',t')}{\partial x_3'} = \frac{A\ell}{(\ell+h)}\frac{\partial R}{\partial x_3}\frac{\partial u(x,t)}{\partial x_3} - \frac{Ah}{(\ell+h)}[1-|\nabla^T R|^2]u_t(x,t)$$

$$-\left[(\nabla^{T'}R'\cdot\nabla^{T'}A)+\frac{A}{2}B(x')-\frac{A\ell B(x)}{2(\ell+h)}\right]u(x,t)$$

$$+\Delta t\Theta(\Delta t)\left[C(x')-\frac{A\ell}{\ell+h}C(x)\right], \qquad (7.120)$$

where

$$B(x) = b(x) + \frac{1}{p(x)}\nabla\cdot(p(x)\nabla R),$$

$$C(x) = \frac{1}{2}\frac{1}{p(x)}\nabla\cdot(p(x)\nabla v_0(x)),$$

with $B(x')$ and $C(x')$ defined similarly.

Numerical results for data continuation for a wave with discontinuous wavefront are shown in Figs 7.3(a) and (b). The incident field is given by (7.88) with $y^0 = (0,0,-0.5)$. The same inhomogeneous dissipative half-space as used in Fig. 7.1 is chosen. The stepsizes for discretization are $\Delta x_1 = \Delta x_2 = \Delta x_3 = 2\Delta t = h = 0.01$. We migrate the surface data (obtained from a time domain finite difference method) to the data at the plane $x_3 = 0.04$ (the fourth layer), and the data continuation results at the point $(0.2, 0, 0.04)$ are shown in Figs 7.3(a) and (b). The first two crosses in Figs 7.3(a) and (b) are obtained by using the wavefront formulas (7.119) and (7.120). The crosses after the fourth cross are obtained by using the local continuation formulas (7.113) and (7.114) (and the regularization (7.51) with $d = 0.005$). The third and fourth crosses in Figs 7.3(a) and (b) are obtained by taking the average of the wavefront formulas and the local continuation formulas. The solid curves are the corresponding reference data obtained by the time domain finite difference method.

If one has a successful scheme for data continuation, one can reconstruct the inhomogeneous parameters layer-by-layer by employing (7.85) and (7.86). Here we first test the condition (7.85) at the wavefront. The inhomogeneous half-space has the following dissipation profile

$$b(x_1,x_2,x_3) = 100\sin^2(\pi x_3)\exp\left[-10\left(x_1^2+\frac{x_2^2}{4}+\frac{x_3^2}{9}\right)\right], \qquad x_3 \geq 0.$$

The data $\mathcal{E}$ and $\partial_{x_3}\mathcal{E}$ at the plane $x_3 = 0.2$ are calculated with a time domain finite difference method (the incident field is given by (7.88) with $y^0 =$

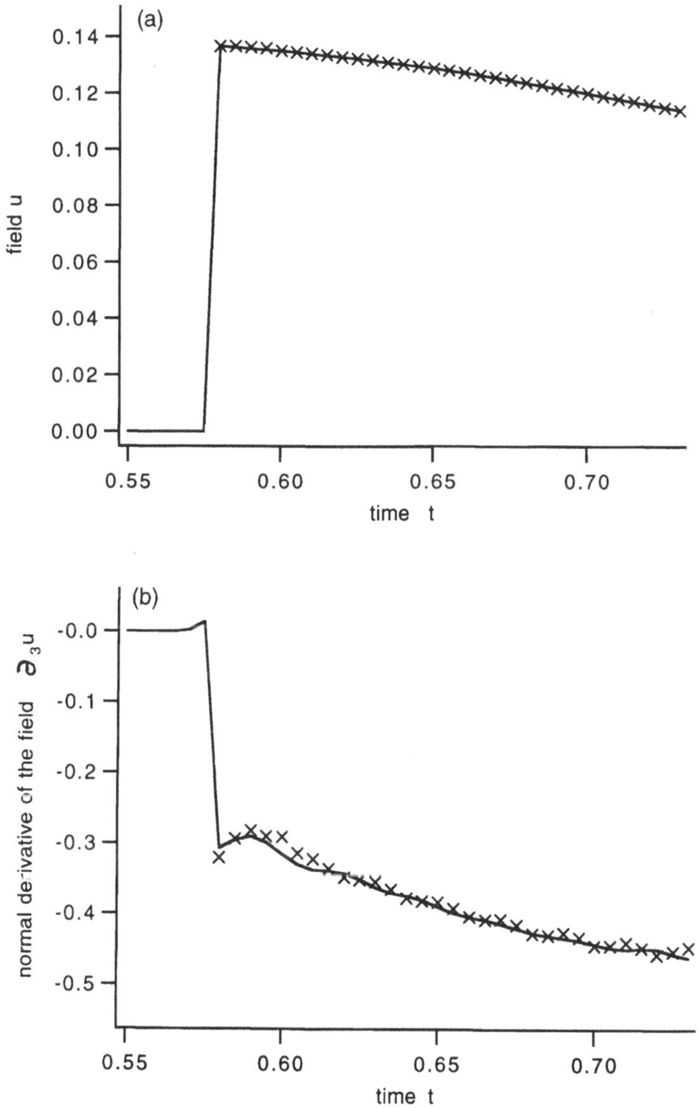

FIG. 7.3. Data continuation for wave with a discontinuous wave front. The crosses are the result of data continuation at the point $(0.2, 0, 0.04)$. The first two crosses are obtained by using the wave front formulas (7.119) and (7.120). The crosses after the fourth cross are obtained by using the local continuation formulas (7.113) and (7.114). The third and fourth crosses are obtained by taking the average of the wave front formulas and the local continuation formulas. The solid curves are the corresponding reference data obtained by the time domain finite difference method.

(a)

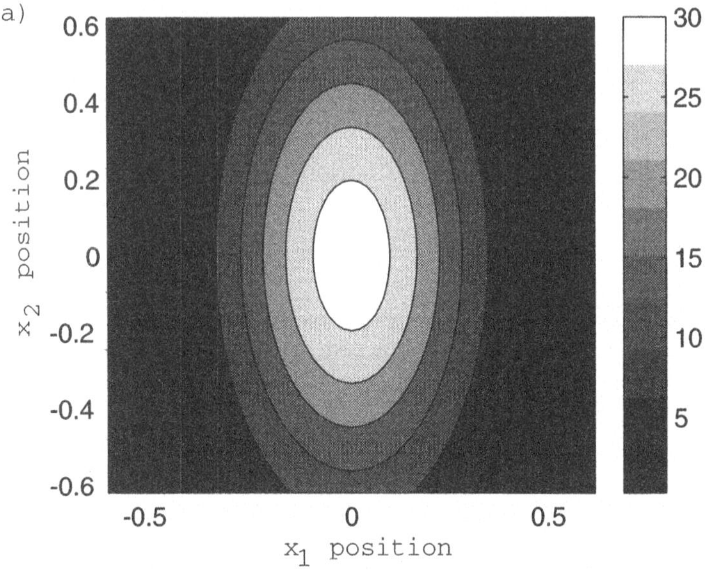
0.6
0.4
0.2
0
-0.2
-0.4
-0.6
x2 position
-0.5
0
0.5
x1 position
30
25
20
15
10
5

(b)

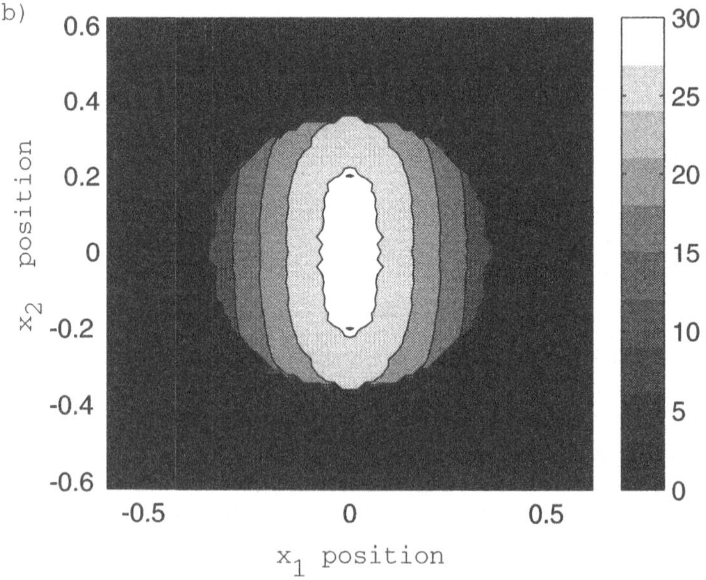
0.6
0.4
0.2
0
-0.2
-0.4
-0.6
x2 position
-0.5
0
0.5
x1 position
30
25
20
15
10
5
0

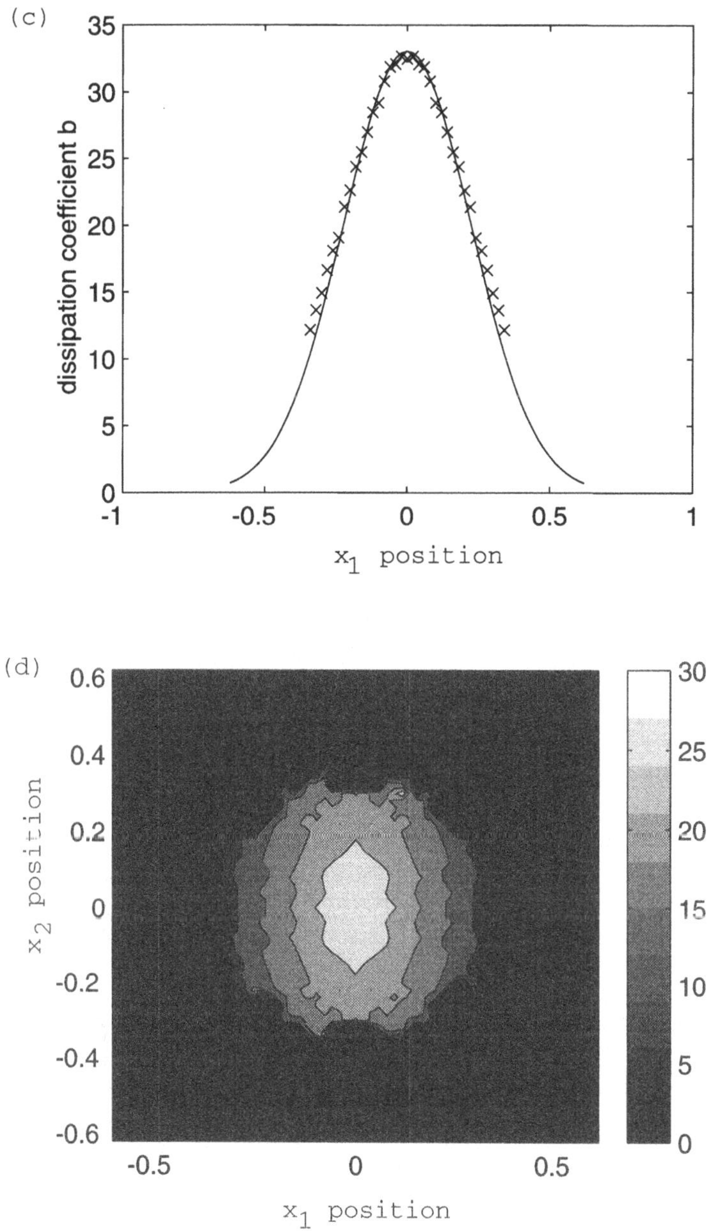

FIG. 7.4. Reconstruction of the dissipation coefficient. (a) The true profile at the plane $x_3 = 0.2$. (b) The dissipation coefficient at the plane $x_3 = 0.2$ determined from the relation (7.86). (c) The dissipation coefficient determined from the relation (7.86) on the x_1 axis of the plane $x_3 = 0.2$. (d) The dissipation coefficient reconstructed at the plane $x_3 = 0.2$ using the data at the upper layer through the continuation formulas.

$(0, 0, -0.5))$. γ at that plane is then obtained from the values of $\{\mathbf{K}_\alpha \partial_{x_3} \mathcal{E} + \mathcal{E}\}$ at the wavefront. The contour density plot of the dissipation coefficient $b(x)$ at that plane determined from the relation (7.86) is shown in Fig. 7.4(b) (the true profile is shown in Fig. 7.4(a)). Figure 7.4(c) shows the dissipation coefficient on the x_1 axis of the plane $x_3 = 0.2$ (the solid line is the true value and the crosses are obtained from the relation (7.86)). Finally we replace the finite difference data with the continuation data $\mathcal{E}$ and $\partial_{x_3}\mathcal{E}$ at the plane $x_3 = 0.2$, which are obtained from the data at the upper layer through the continuation formulas (7.113), (7.114), (7.119), and (7.120). The corresponding reconstruction of the dissipation coefficient $b(x)$ at the plane $x_3 = 0.2$ is shown in Fig. 7.4(d). We refer to Cao et al. (1998) for more details.

7.5 Non-planar wave-splitting for the wave equation

The technique of planar wave-splitting (into positive and negative going waves across a plane) has been generalized to non-planar wave-splitting (into out-going and in-coming waves across a closed surface) in a homogeneous medium (Weston 1988a).

Let S be a smooth orientable surface dividing $\mathbb{R}^3$ into two open regions Ω^+ and Ω^-, with a unit normal $\hat{\mathbf{n}}$ pointing into the region Ω^+ and out of the region Ω^-. The surface S may be an open surface (e.g. a parabolic sheet) or a closed surface enclosing the interior domain Ω^- (which would be bounded); the normal $\hat{\mathbf{n}}$ is then the exterior normal.

Here the decomposition of $u(x, t)$ (a solution of the wave equation in an open neighbourhood of the surface S) into wave components u^+ and u^- will be obtained, where u^+ designates a wave moving in the direction of the prescribed unit normal $\hat{\mathbf{n}}$ (from Ω^- to Ω^+) and u^-, a wave moving in the opposite direction respectively. For a closed surface S, u^+ and u^- represent out-going and in-coming waves, respectively. From a physical standpoint, u^+ can be thought of as a solution of the wave equation generated by sources in the region Ω^-, and u^- by sources in the region Ω^+. The same mathematical approach that was used in Section 7.1 for the planar case could be used. Let $u(x, t)$ be the solution to the exterior Neumann problem

$$\left.\begin{aligned} &\frac{1}{c^2}u_{tt} - \nabla^2 u = 0, \qquad x \in \Omega^+, \qquad 0 < t < T, \\ &u = u_t = 0, \qquad x \in \Omega^+, \qquad t = 0, \\ &\partial_n u = v(x, t), \qquad x \in S, \qquad 0 < t < T. \end{aligned}\right\} \tag{7.121}$$

Then taking the limit as $x \in \Omega^+ \to x \in S$, one obtains a relation of the type

$$\lim_{x \to S} u(x, t) = -\mathbf{K}v. \tag{7.122}$$

This will result in, as in Section 7.1, the positive-going condition

$$u + \mathbf{K}\partial_n u = 0, \qquad x \in S, \qquad 0 < t < T, \tag{7.123}$$

where $\mathbf{K}$ is the Neumann operator.

However, this approach is more difficult for the non-planar surface S than the planar case in Section 7.1. First the operator $\mathbf{K}$ for the case c = constant will not have a simple form like (7.26), but will be an integral operator with the kernel being the fundamental solution satisfying the Neumann boundary condition on the surface S. Secondly and most importantly, the operator $\mathbf{K}$ will not commute with the normal derivative ∂_n (as was the case for the planar surface in Section 7.1). The behaviour of $\partial_n \mathbf{K}$ is needed to demonstrate that the sought for component u^+ indeed satisfies wave condition (7.123).

So instead of using the Neumann operator, a different approach will be used. Employ the time-dependent form of Kirchhoff's formula for a solution to the exterior problem

$$\Delta u - \frac{1}{c^2}\frac{\partial^2 u}{\partial t^2} = 0, \qquad x \in \Omega^+,\ t > 0,$$

$$u(x,0) = u_t(x,0) = 0, \qquad x \in \Omega^+,$$

which is given by

$$u(x,t) = -\frac{1}{4\pi}\int_S \left\{ \frac{1}{r} u_n(y, t - r/c) \right.$$

$$\left. + \frac{1}{r^2}\frac{\partial r}{\partial n_y}[u(y, t - r/c) + \frac{r}{c} u_t(y, t - r/c)] \right\} d\sigma_y, \quad x \in \Omega^+$$

where $r = |x - y|$, the subscript t indicates the partial derivative with respect to time, and $u_n(y,t) = \dfrac{\partial}{\partial n(y)} u(y,t)$. The positive-going wave condition is obtained by considering Kirchhoff's formula for the exterior initial value problem and letting $x \in \Omega^+$, approach S. Since Kirchhoff's formula involves a double layer potential, there is a jump discontinuity on the surface (Günter 1967), leading to the positive-going wave condition

$$(I + M)u + Ku_n = 0, \qquad x \in S, \tag{7.124}$$

where I is the identity operator, and the operators K and M (for any finite $T > 0$) are defined by

$$Ku(x,t) = \frac{1}{2\pi}\int_S \frac{1}{r}u(y,t-r/c)\Theta(t-r/c)d\sigma_y, \tag{7.125}$$

$$Mu(x,t) = \frac{1}{2\pi}\int_S \frac{1}{r^2}\frac{\partial r}{\partial n_y}[u(y,t-r/c) + \frac{r}{c}u_t(y,t-r/c)]\Theta(t-r/c)d\sigma_y. \tag{7.126}$$

The negative-going wave condition is obtained by considering Kirchhoff's formula for the interior Neumann problem in the region Ω^-. The negative-going wave condition is given by

$$(I-M)u - Ku_n = 0, \qquad x \in S. \tag{7.127}$$

Let u be a solution of

$$\frac{1}{c^2}u_{tt} - \nabla^2 u = 0, \qquad 0 < t < T, \qquad x \in \Omega_\epsilon$$

$$u = u_t = 0, \qquad t = 0, \qquad x \in \Omega_\epsilon$$

where Ω_ϵ is an open region in $\mathbb{R}^3$ which contains the surface S. Then the solution u will be split into two components u^+ and u^- as follows

$$\begin{bmatrix} u^+ \\ u^- \end{bmatrix} = \frac{1}{2}\begin{bmatrix} I-M & -K \\ I+M & K \end{bmatrix}\begin{bmatrix} u \\ u_n \end{bmatrix}, \tag{7.128}$$

so that

$$u = u^+ + u^-. \tag{7.129}$$

In order to establish the fact that u^+ and u^- are indeed positive-going and negative-going wave components, it needs to be shown that u^+ and u^- satisfy respectively (7.124) and (7.127). This is much more difficult to do for the curved surface S than for the plane surface as was done in Section 7.1. An outline of the procedure will be indicated (consult Weston (1988a) for details). In order to define the normal derivative of the various operators, one has to set up an orthogonal coordinate system (ξ_1, ξ_2, ξ_3) with metric coefficients h_1, h_2, h_3 so that the surface is given by $\xi_1 = \xi_1^0$. The normal derivative is then given by

$$\frac{\partial}{\partial n} = \frac{1}{h_1}\frac{\partial}{\partial \xi_1}. \tag{7.130}$$

Let y (the variable of integration in the integral representations (7.125) and (7.126)) be given by (ξ_1', ξ_2', ξ_3') and the point x by (ξ_1, ξ_2, ξ_3). On the surface S, $\xi_1' = \xi_1 = \xi_0$. The integral operators can be expressed in terms of integrals like

$$\int_0^t \int\int_\Sigma g(\xi_1, \xi_2, \xi_3; \xi_1', \xi_2', \xi_3'; t-s) u(\xi_1', \xi_2', \xi_3', s) d\xi_2' d\xi_3' ds,$$

where Σ is the domain in the $\xi_2'\xi_3'$ plane characterizing the surface S. By operating on the operators (expressed in this form) with the normal derivative one can establish the values of $\partial_n K = K_n$, and $\partial_n M = M_n$. Using this idea a number of identities were proven in Weston (1988a). Combining these results it can be shown that

$$(I - M^2)u - MKu_n = K\partial_n[Mu + Ku_n] - \frac{1}{h_1}K^2\left(h_1'\left[\nabla^2 u - \frac{1}{c^2}u_{tt}\right]\right), \tag{7.131}$$

where $h_1' = h_1(\xi')$.

This can be used to show that $u^+ = \frac{1}{2}[(1-M)u - Ku_n]$ satisfies condition (7.124). Inserting this expression for u^+ into the left hand side of (7.124) one obtains (after multiplying by 2) and using the identity (7.131)

$$(1+M)[(I-M)u - Ku_n] + K\partial_n[(I-M)u - Ku_n]$$

$$= (I - M^2)u - MKu_n - K\partial_n[Mu + Ku_n]$$

$$= -\frac{1}{h_1}K^2\left(h_1'[\nabla^2 u - \frac{1}{c^2}u_{tt}]\right),$$

which will be zero since u is a solution of the wave equation. Thus u^+ satisfies the positive-going wave condition. A similar result may be shown for u^-.

The splitting given by (7.128) has been employed in connection with the inverse problem for the case where the velocity c is a function of ρ (the radial variable of cylindrical polar coordinates). The dynamical equations corresponding to the splitting into out-going and in-coming waves across a circular cylinder were employed (Kreider 1989b).

Alternatively, the condition for the positive-going wave condition (7.124) can be used as an exact absorbing boundary condition. If one has a solution of the wave equation (c constant) produced by sources in a region Ω^- bounded by a smooth surface S, then one can impose the positive-going wave condition on the surface. This indicates that there will be no reflected wave (negative-going wave) produced by the surface.

7.6 Present and future research

There are a number of problems both theoretical and numerical with regard to the application of three-dimensional wave-splitting to inverse problems, that should be addressed.

One problem that needs attention is the application of the planar wave-splitting to the inverse problem in the region where the primary wavefront (produced by the point pulse source) forms a caustic. It should be pointed out that the Neumann operator **K**, and most importantly, the identity (7.32) which allows one to identify the operator **K** as the square root of the operator $(\frac{1}{c^2}\partial_t^2 - \nabla^2)^{-1}$ and thus define the splitting, depend solely upon the existence of the solutions to the initial-value boundary value problem and the existence of the fundamental solution (Weston 1992b). The restraint that limits the use of the planar wave-splitting analysis to the inverse problem in the region of a caustic of the primary wave is due to the asymptotic analysis used to derive the expression for γ (see (7.86) and (7.87)) which is employed to determine the unknown parameter. Thus additional asymptotic analysis in the caustic region needs to be done to explore the usefulness of the wave-splitting in the inverse (and direct) problem for the caustic region of the primary wave.

Another problem in the use of wave-splitting in the inverse problem is the backward continuation of the fields through various approximations to the operator D^{-1} such as expression (7.49). This expression is not valid in the vicinity of a wavefront with a jump, and the appropriate modification for the telegraph equation in the vicinity of a wavefront with a jump when $c = 1$ is given by (7.119). This wavefront analysis needs to be extended to the general case of non-constant velocity.

An alternative approach that is under investigation is to replace the planar wave-splitting with a wave-front splitting. By wave-front splitting, we mean that we define positive and negative going waves across surfaces which are the wavefronts of the primary waves at different times. Their are several advantages in using such an approach. It places emphasis on a grid system that tracks the wavefront, which is important for numerical work. Also, the expressions for the backward continuation (the inverse D operator) should be simpler and more stable.

Exercises for Chapter 7

Exercise 7.1. Let $\mathcal{E}(x, y, t)$ be the fundamental solution associated with the wave equation (7.74) with $b = q = 0$, $p = 1$, $c^2 r = 1$, for a point source at $y = (0, 0, y_3^0)$, $y_3^0 < 0$ in free space. Using the asymptotic expression (7.76) find the short time asymptotic behaviour (just two terms) for $\dfrac{\partial \mathcal{E}}{\partial x_3}$ on $x_3 = x_3^0 > 0$. From the eikonal equation for τ and transport equation for $1/\rho$, show that

$$\frac{\partial \tau}{\partial x_3}\frac{\partial^2 \tau}{\partial x_3^2} = -\nabla^T \tau \cdot \nabla^T \frac{\partial \tau}{\partial x_3} - \frac{1}{c^3}\frac{\partial c}{\partial x_3}, \quad x_3 = x_3^0$$

$$\frac{\partial \tau}{\partial x_3}\frac{\partial (1/\rho)}{\partial x_3} = -\nabla^T \tau \cdot \nabla^T (1/\rho) - \frac{1}{2\rho}\nabla^2 \tau, \quad x_3 = x_3^0$$

where $\nabla^T \equiv \left(\dfrac{\partial}{\partial x_1}, \dfrac{\partial}{\partial x_2}, 0\right)$ is the transverse gradient.

Exercise 7.2. To find $\mathbf{K}\dfrac{\partial \mathcal{E}}{\partial x_3}$ at $x_3 = x_3^0$, or what is equivalent using the $\mathbf{K}_\alpha$ operator notation $\mathbf{K}_{x_3^0}\dfrac{\partial \mathcal{E}}{\partial x_3}|_{x_3 = x_3^0}$, one has to solve the initial-value, boundary value problem for $w(x, t)$,

$$w_{tt} - c^2(\mathbf{x}, x_3^0)\nabla^2 w = 0, \; x_3^0 < x_3, \quad 0 < t < T,$$

$$w = w_t = 0 \qquad\qquad , \; x_3^0 < x_3, \quad t = 0,$$

$$\frac{\partial w}{\partial x_3} = \frac{\partial \mathcal{E}}{\partial x_3}, \qquad x_3 = x_3^0, \quad 0 < t < T,$$

where $\dfrac{\partial \mathcal{E}}{\partial x_3}$ is given in Exercise 7.1. Then from the definition of the operator $\mathbf{K}$, it follows that $\mathbf{K}\dfrac{\partial \mathcal{E}}{\partial x_3} = -w$, $x_3 = x_3^0$.

(a) Assume an asymptotic solution for $w(x, t)$ of the form $w(x, t) = w_0(x)\delta(t - \tilde{\tau}(x)) + w_1(x)\Theta(t - \tilde{\tau}(x)) + \dots$, $x_3^0 < x_3$, $0 < t < T$, and obtain the eikonal equation for $\tilde{\tau}(x)$ and the transport equation for $w_0(x)$. From these show that

$$\frac{\partial \tilde{\tau}}{\partial x_3}\frac{\partial^2 \tilde{\tau}}{\partial x_3^2} = -\nabla^T \tilde{\tau} \cdot \nabla^T \frac{\partial \tilde{\tau}}{\partial x_3}$$

$$\frac{\partial \tilde{\tau}}{\partial x_3}\frac{\partial w_0}{\partial x_3} = -\nabla^T \tilde{\tau} \cdot \nabla^T w_0 - \frac{1}{2} w_0 \nabla^2 \tilde{\tau}.$$

(b) Match the asymptotic form of $\frac{\partial w}{\partial x_3}$ and $\frac{\partial \mathcal{E}}{\partial x_3}$ given in Exercise 7.1 on the portion of the plane $x_3 = x_3^0$ where $|\nabla\tau| < \frac{1}{c}$, $0 < \tau \le T$, to show that on $x_3 = x_3^0$

$$\tilde{\tau} = \tau, \quad \frac{\partial\tilde{\tau}}{\partial x_3} = \frac{\partial \tau}{\partial x_3}$$

$$w_0 = \frac{1}{\rho}, \quad \frac{\partial \tau}{\partial x_3}(w_1 - a_0) = \frac{\partial w_0}{\partial x_3} - \frac{\partial(1/\rho)}{\partial x_3}.$$

Hence from part (a) and Exercise 7.1 show that

$$\left(\frac{\partial\tau}{\partial x_3}\right)^3 (w_1 - a_0) = -\frac{\partial c}{\partial x_3}[2\rho c^3]^{-1}.$$

(c) Hence deduce from parts (a) and (b) that

$$-\mathbf{K}\frac{\partial \mathcal{E}}{\partial x_3}|_{x_3=x_3^0} = w(\mathbf{x}, x_3^0, t) \sim \frac{1}{\rho}\delta(t-\tau) + [a_0 - \gamma]\Theta(t-\tau), \ldots$$

where $\gamma = \frac{\partial c}{\partial x_3}\left[2\rho c^3\left(\frac{\partial\tau}{\partial x_3}\right)^3\right]^{-1}$.

Exercise 7.3. Use the result of Exercise 7.2 to deduce that the asymptotic behaviour for the negative-going wave component of $\mathcal{E}$ is given by

$$\mathcal{E}^- = \frac{1}{2}\left\{\mathcal{E} + \mathbf{K}\frac{\partial\mathcal{E}}{\partial x_3}\right\}_{x=x_3^0} \sim \frac{\gamma}{2}\Theta(t-\tau) + \ldots$$

8

WAVE-SPLITTING OF MAXWELL'S EQUATIONS IN $\mathbb{R}^3$ AND APPLICATIONS

The concepts of planar wave-splitting developed in Chapter 7 for solutions of the wave and telegraph equation in a source free region is extended to Maxwell's equations

$$\left.\begin{array}{ll} \nabla \times \mathbf{E} = -\partial_t \mathbf{B}, & \nabla \times \mathbf{H} = \partial_t \mathbf{D}, \\ \nabla \cdot \mathbf{D} = 0, & \nabla \cdot \mathbf{B} = 0. \end{array}\right\} \tag{8.1}$$

For the case of a dissipative (or dispersive) medium the splitting is defined in terms of the principal part of the system, where the corresponding isotropic constitutive relations have the form

$$\mathbf{D} = \epsilon(x)\mathbf{E}, \quad \mathbf{B} = \mu(x)\mathbf{H}. \tag{8.2}$$

From the stand-point of physics, the concept of a wave travelling across a plane $x_3 = x_3^0$ in the positive x_3 direction can be thought of having been produced by sources in the region $x_3 < x_3^0$. (A corresponding idea holds for the negative-going wave component.) Since the fictitious sources producing such a positive-going wave are unknown, one can represent the effect of the sources in the region $x_3 < x_3^0$, by specifying the tangential electric field (denoted by $\mathbf{E}_T$) on the surface $x_3 = x_3^0$. Then solving Maxwell's equations in the half-space $x_3 > x_3^0$, for $0 < t < T$, one can obtain the value of the tangential component of $\mathbf{H}$ (denoted by $\mathbf{H}_T$) on the surface $x_3 = x_3^0$. Since this corresponds to the mapping of the tangential field $\mathbf{E}_T$ to the tangential field $\mathbf{H}_T$ on $x_3 = x_3^0$, it can be characterized by a linear boundary condition on the plane $x_3 = x_3^0$. This boundary condition gives the positive-going wave condition. Alternatively one can get another positive-going wave condition by solving for tangential $\mathbf{E}_T$ with tangential $\mathbf{H}_T$ given. The negative wave conditions are developed in a similar manner by solving the corresponding problems in the half-space $x_3 < x_3^0$. These positive and negative-going wave conditions are developed in Section 8.1 for the case where $\epsilon = \epsilon(x_1, x_2, \alpha)$, $\mu = \mu(x_1, x_2, \alpha)$, α being a parameter. In Section 8.2 an appropriate splitting of $\mathbf{E}_T$ into positive and negative-going components $\mathbf{E}_T^+$ and $\mathbf{E}_T^-$, respectively, is obtained. In such a medium where ϵ, μ are independent of x_3, the splitting is an exact mathematical and physical splitting.

In Sections 8.3 and 8.4 the splitting is extended to a medium where $\epsilon = \epsilon(x)$, $\mu = \mu(x)$ and to a dispersive medium. For these cases the splitting is strictly mathematical. The splitting is applied to the inverse problem in Section 8.5. Finally in Section 8.6, non-planar wave-splitting is developed for Maxwell's equations in free space, and the results are applied to obtain a non-planar absorbing boundary condition.

8.1 Positive- and negative-going wave conditions in an isotropic medium with transverse dependence

Here, the positive and negative-going wave conditions on $x_3 = 0$ will be developed for solutions of Maxwell's equations, where the constitutive equations are given by

$$D_i = \epsilon(x_1, x_2, \alpha)E_i, \; B_i = \mu(x_1, x_2, \alpha)H_i, \quad i = 1, 2, 3, \tag{8.3}$$

with ϵ, μ being sufficiently smooth functions of x_1 and x_2, and α is a parameter. To obtain the positive and negative-going wave conditions the initial-value boundary-value problems in the corresponding half-spaces $x_3 > 0$ and $x_3 < 0$ respectively will be employed. Hence, consider the following mixed problem.

Problem P_1: Solve for $\mathbf{E}(x,t)$, $\mathbf{H}(x,t)$ for $x_3 \geq 0, \; 0 < t < T$, assuming that

(i) $\nabla \times \mathbf{E} = -\mu(x_1, x_2, \alpha)\partial_t \mathbf{H}, \; \nabla \times \mathbf{H} = \epsilon(x_1, x_2, \alpha)\partial_t \mathbf{E}, \; x_3 > 0, \; 0 < t < T$,

(ii) $\mathbf{E}(x, 0) = \mathbf{H}(x, 0) = 0, \quad x_3 > 0$,

(iii) $E_i(x_1, x_2, 0, t) = E_i^0(x_1, x_2, t), \quad i = 1, 2, \; x_3 = 0, \; 0 < t < T$,

where the given data E_i^0, $i = 1, 2$, have compact support in space, and are sufficiently smooth for the existence of a solution.

An important result of the condition that the boundary data has compact support (say in a disk $|x_1 - x_1^0|^2 + |y_1 - y_1^0|^2 < p^2$) and a finite time interval T is employed, is that the problem reduces to a mixed problem in a bounded region Ω, where Ω can be taken to be the solid hemisphere in $x_3 > 0$, with centre $(x_1^0, x_2^0, 0)$ and radius $p + c_M T$ where c_M is the maximum velocity. In this case, the tangential electric fields on the hemispherical portion of the bounding surface would be taken to be zero.

Once a solution of Problem P_1 is found, $\lim_{x_3 \to 0+} H_i(x,t)$ for $i = 1, 2$, is taken, and a relation of the following form is obtained

$$\mathbf{H}_T + \frac{1}{\mu} M_\alpha \mathbf{E}_T = 0, \qquad x_3 = 0, \; 0 < t < T, \tag{8.4}$$

where $\mathbf{E}_T = (E_1, E_2)$ and $\mathbf{H}_T = (H_1, H_2)$. This is the sought for positive-going wave condition on the surface $x_3 = 0$. Since ϵ, μ are independent of x_3 the operator M_α is independent of x_3, so that condition (8.4) holds for any surface $x_3 = x_3^0$ (constant) ≥ 0 (when the data is given on $x_3 = x_3^0$, or the sources producing the wave lie in the region $x_3 < x_3^0$). (8.4) defines the operator M_α (its precise form will be obtained later in this section).

By treating the same problem for the half-space $x_3 < 0$, the negative-going wave condition is obtained. It takes the form

$$\mathbf{H}_T - \frac{1}{\mu} M_\alpha \mathbf{E}_T = 0, \qquad x_3 = 0, \ 0 < t < T. \tag{8.5}$$

Alternative forms of the positive and negative-going wave conditions, will be found, so as to show the existence of and to get the form of the *inverse* operator to M_α. For this, consider the following problem.

Problem P_2: Solve for $\mathbf{E}(x,t)$, $\mathbf{H}(x,t)$, for $x_3 > 0$, $0 < t < T$, assuming that

(i) $\nabla \times \mathbf{E} = -\mu(x_1, x_2, \alpha)\partial_t \mathbf{H}$, $\nabla \times \mathbf{H} = \epsilon(x_1, x_2, \alpha)\partial_t \mathbf{E}$, $x_3 > 0$, $0 < t < T$,

(ii) $\mathbf{E}(x, 0) = \mathbf{H}(x, 0) = 0$, $x_3 > 0$,

(iii) $H_i(x_1, x_2, 0, t) = H_i^0(x_1, x_2, t)$, $i = 1, 2$, $x_3 = 0$, $0 < t < T$,

where the boundary data H_i^0, $i = 1, 2$ have compact support and are sufficiently smooth for the existence of a solution.

Once the solution is found, $\lim_{x_3 \to 0+} E_i(x,t)$ is taken, and a condition of the following form is then obtained

$$\mathbf{E}_T - \frac{1}{\epsilon} N_\alpha \mathbf{H}_T = 0, \ x_3 = 0, \ 0 < t < T, \tag{8.6}$$

which defines the operator N_α (its precise form will be obtained later).

This is an alternative form of the positive-going wave condition. Comparing relations (8.4) and (8.6) it can be seen that for a suitable class of functions, $-\frac{1}{\epsilon} N_\alpha$ is the inverse of $\frac{1}{\mu} M_\alpha$. One would then expect

$$-\frac{1}{\epsilon} N_\alpha \left(\frac{1}{\mu} M_\alpha \right) \mathbf{u} = -\frac{1}{\mu} M_\alpha \frac{1}{\epsilon} N_\alpha \mathbf{u} = I\mathbf{u}, \tag{8.7}$$

where I is the identity operator. The precise domain of the tangential vector function $\mathbf{u}$ for which this is valid remains to be investigated.

The corresponding negative-going wave condition obtained by treating Problem P_2 in the half-space $x_3 < 0$ is given by

$$\mathbf{E}_T + \frac{1}{\epsilon} N_\alpha \mathbf{H}_T = 0. \tag{8.8}$$

In summary, the conditions for a wave travelling in the positive x_3 direction is given by (8.4) and (8.6), and the conditions for a wave travelling in the negative x_3 direction is given by (8.5) and (8.8). The operator M_α maps the transverse electric field into the transverse magnetic field on the same surface, whereas N_α is the inverse operation. Both problems P_1 and P_2 could be solved numerically, to obtain the action of the operators M_α and N_α. The next step is to get an explicit form of the operators in terms of the fundamental solutions of Maxwell's equations (i.e., 'dyadic Green's functions' – solutions corresponding to the pulsed electric and magnetic dipoles; see Tai (1971) for properties of the dyadic Green's functions in the frequency domain case). From this the various properties of the operator can be deduced. But most important, a useful explicit form of M_α and N_α for the free space case where μ and ϵ are constant will be obtained.

Let $\mathcal{E}_j(x,y,t_0-t)$, $\mathcal{H}_j(x,y,t_0-t)$ be the fundamental solution corresponding to a pulsed electric dipole at y, satisfying the system

$$\left.\begin{aligned} &\nabla\times\mathcal{E}_j = -\mu(x_1,x_2,\alpha)\partial_t\mathcal{H}_j,\\ &\nabla\times\mathcal{H}_j = \epsilon(x_1,x_2,\alpha)\partial_t\mathcal{E}_j + \hat{\mathbf{i}}_j\delta(x-y)\delta(t-t_0),\\ &\text{where}\quad \mathcal{E}_j = \mathcal{H}_j = 0,\ \text{for } t-t_0>0, \end{aligned}\right\} \tag{8.9}$$

where $\hat{\mathbf{i}}_j$ is the unit vector in the direction of the x_j-axis. Denote the components of $\mathcal{E}_j$ by $\mathcal{E}_{ij}$, i.e.

$$\mathcal{E}_j = (\mathcal{E}_{1j},\mathcal{E}_{2j},\mathcal{E}_{3j}),$$

with similar notation for $\mathcal{H}_j$ (note that $(\mathcal{E}_j,\mathcal{H}_j)$ vanish for $t>t_0$).

In addition to the above, let the pair $\widetilde{\mathcal{E}}(x,y,t_0-t)$, $\widetilde{\mathcal{H}}_j(x,y,t_0-t)$ be the free space fundamental solutions corresponding to a pulsed magnetic dipole at y, satisfying the system

$$\left.\begin{aligned} &\nabla\times\widetilde{\mathcal{E}}_j = -\mu(x_1,x_2,\alpha)\partial_t\widetilde{\mathcal{H}}_j + \hat{\mathbf{i}}_j\delta(x-y)\delta(t-t_0),\\ &\nabla\times\widetilde{\mathcal{H}}_j = \epsilon(x_1,x_2,\alpha)\partial_t\widetilde{\mathcal{E}}_j,\\ &\text{where}\quad \widetilde{\mathcal{E}}_j = \widetilde{\mathcal{H}}_j = 0,\ \text{for } t-t_0>0. \end{aligned}\right\} \tag{8.10}$$

Since ϵ and μ are independent of x_3, both pairs of functions $(\mathcal{E}_j,\ \mathcal{H}_j)$ and $(\widetilde{\mathcal{E}}_j,\widetilde{\mathcal{H}}_j)$ will depend upon the x_3 and y_3 variables in the linear combination (x_3-y_3) (this is due to the invariance under translation in the x_3-direction). It will then be convenient to decompose a point into its transverse components (x_1,x_2) and longitudinal component x_3 by using the notation (as used in Chapter 7)

$$x = (\mathbf{x},x_3),\ \ \mathbf{x} = (x_1,x_2).$$

Since $\delta(x_3 - y_3)$ has even parity with respect to $(x_3 - y_3)$ one can deduce the corresponding parity for the components of $\mathcal{E}$, $\mathcal{H}$, etc., from (8.9) and (8.10). It can be shown that the following components of interest have the following parity: $\mathcal{E}_{ij}$, $\widetilde{\mathcal{H}}_{ij}$, $i, j = 1, 2$, have even parity with respect to $(x_3 - y_3)$, $\mathcal{H}_{ij}, \widetilde{\mathcal{E}}_{ij}, i, j = 1, 2$, have odd parity with respect to $(x_3 - y_3)$. The above parity information can be used to generalize the method of images to a medium where ϵ and μ depend upon the variables x_1 and x_2, so as to obtain a set of fundamental solutions to Maxwell's equation that satisfy the boundary condition that either the tangential electric or the tangential magnetic fields vanish on $x_3 = 0$.

For $j = 1, 2$, and $y_3 > 0$ let

$$\mathcal{E}_j^\dagger(\mathbf{x}, \mathbf{y}, x_3, y_3; t_0 - t) = \mathcal{E}_j(\mathbf{x}, \mathbf{y}, x_3 - y_3; t_0 - t) + \mathcal{E}_j(\mathbf{x}, \mathbf{y}, x_3 + y_3; t_0 - t),$$

with a similar definition for $\mathcal{H}_j^\dagger$, and

$$\widetilde{\mathcal{E}}_j^\dagger(\mathbf{x}, \mathbf{y}, x_3, y_3; t_0 - t) = \widetilde{\mathcal{E}}_j(\mathbf{x}, \mathbf{y}, x_3 - y_3; t_0 - t) + \widetilde{\mathcal{E}}_j(\mathbf{x}, \mathbf{y}, x_3 + y_3; t_0 - t),$$

with a similar definition for $\widetilde{\mathcal{H}_j^\dagger}$.

It then follows from the parity relations given above that for $y_3 > 0$, $i, j = 1, 2$,

$$\mathcal{H}_{ij}^\dagger(\mathbf{x}, \mathbf{y}, 0, y_3; t_0 - t) = \widetilde{\mathcal{E}}_{ij}^\dagger(\mathbf{x}, \mathbf{y}, 0, y_3; t_0 - t) = 0, \tag{8.11}$$

$$\mathcal{E}_{ij}^\dagger(\mathbf{x}, \mathbf{y}, 0, y_3; t_0 - t) = 2\mathcal{E}_{ij}(\mathbf{x}, \mathbf{y}, |y_3|; t_0 - t), \tag{8.12}$$

$$\widetilde{\mathcal{H}}_{ij}^\dagger(\mathbf{x}, \mathbf{y}, 0, y_3; t_0 - t) = 2\widetilde{\mathcal{H}}_{ij}(\mathbf{x}, \mathbf{y}, |y_3|; t_0 - t). \tag{8.13}$$

These results will now be used to get the positive-going wave condition.

Let $(\mathbf{E}, \mathbf{H})$ be a solution of Problem P_2, and use the identity

$$-\nabla \cdot (\mathbf{E} \times \mathcal{H}_j^\dagger + \mathcal{E}_j^\dagger \times \mathbf{H})$$

$$= \partial_t(\epsilon \mathbf{E} \cdot \mathcal{E}_j^\dagger + \mu \mathbf{H} \cdot \mathcal{H}_j^\dagger) + (\mathbf{E} \cdot \hat{\mathbf{i}}_j)\delta(x - y)\delta(t - t_0), \ x_3 > 0$$

by integrating it over the half-space $x_3 > 0$, and with respect to t from 0 to T (where $t_0 \leq T$). Use the divergence theorem and boundary conditions (8.11), (8.12) to obtain

$$E_j(y, t_0) = 2\int_0^{t_0} \int_{\mathbb{R}^2} \hat{\mathbf{i}}_3 \cdot \mathcal{E}_j \times \mathbf{H}|_{x_3=0} d\mathbf{x}dt, \ y_3 > 0, \ j = 1, 2.$$

In a similar manner, using the solution $(\mathbf{E}, \mathbf{H})$ of problem P_1, and the fundamental solution $(\widetilde{\mathcal{E}}^\dagger, \widetilde{\mathcal{H}}^\dagger)$ it can be shown that

$$H_j(y, t_0) = -2\int_0^{t_0}\int_{\mathbb{R}^2} \hat{\mathbf{i}}_3 \cdot \mathbf{E} \times \widetilde{\mathcal{H}}_j|_{x_3=0} d\mathbf{x}dt, \ y_3 > 0, \ j = 1, 2.$$

Taking the limit as $y_3 \to 0+$, the two following forms for the positive-going wave condition can be obtained (on replacing the pair t_0, t by t, s)

$$E_j(\mathbf{y}, 0, t) = 2\int_0^t\int_{\mathbb{R}^2} \{-\mathcal{E}_{2j}(\mathbf{x}, \mathbf{y}, 0, t-s)H_1(\mathbf{x}, 0, s) + \mathcal{E}_{1j}(\mathbf{x}, \mathbf{y}, 0, t-s)H_2(\mathbf{x}, 0, s)\}d\mathbf{x}ds, \tag{8.14}$$

$$H_j(\mathbf{y}, 0, t) = 2\int_0^t\int_{\mathbb{R}^2} \{-\widetilde{\mathcal{H}}_{2j}(\mathbf{x}, \mathbf{y}, 0, t-s)E_1(\mathbf{x}, 0, s) + \widetilde{\mathcal{H}}_{1j}(\mathbf{x}, \mathbf{y}, 0, t-s)E_2(\mathbf{x}, 0, s)\}d\mathbf{x}ds, \tag{8.15}$$

where $j = 1, 2$. Note that one has to be careful in taking the limit as $y_3 \to 0$ (Günter 1967). It will be assumed that ϵ, μ are sufficiently smooth so that the behaviour of the solution in the region of the singularity of $\mathbf{x} = \mathbf{y}$ can be given by the case where ϵ, μ are constant. This special case will be examined next but before doing so the elements of the matrix operators M_α and N_α will be identified.

Identifying (8.15) with (8.4) we see that the operator M_α is given by

$$M_\alpha \mathbf{u} = \int_0^t\int_{\mathbb{R}^2} \begin{bmatrix} m_{11} & m_{12} \\ m_{21} & m_{22} \end{bmatrix} \begin{bmatrix} u_1(\mathbf{x}, s) \\ u_2(\mathbf{x}, s) \end{bmatrix} d\mathbf{x}ds, \tag{8.16}$$

where

$$\left.\begin{aligned} m_{i1} &= 2\mu(\mathbf{y}, \alpha)\widetilde{\mathcal{H}}_{2i}(\mathbf{x}, \mathbf{y}, 0, t-s), \\ m_{i2} &= -2\mu(\mathbf{y}, \alpha)\widetilde{\mathcal{H}}_{1i}(\mathbf{x}, \mathbf{y}, 0, t-s), \end{aligned}\right\} \tag{8.17}$$

and identifying (8.14) with (8.6) we see that

$$N_\alpha \mathbf{u} = \int_0^t\int_{\mathbb{R}^2} \begin{bmatrix} n_{11} & n_{12} \\ n_{21} & n_{22} \end{bmatrix} \begin{bmatrix} u_1(\mathbf{x}, s) \\ u_2(\mathbf{x}, s) \end{bmatrix} d\mathbf{x}ds, \tag{8.18}$$

where

$$\left.\begin{aligned} n_{i1} &= -2\epsilon(\mathbf{y}, \alpha)\mathcal{E}_{2i}(\mathbf{x}, \mathbf{y}, 0, t-s), \\ n_{i2} &= 2\epsilon(\mathbf{y}, \alpha)\mathcal{E}_{1i}(\mathbf{x}, \mathbf{y}, 0, t-s). \end{aligned}\right\} \tag{8.19}$$

The negative-going wave condition can be obtained in a similar manner as the positive-going wave condition, by treating the mixed problems corresponding to problems P_1 and P_2 for the half-space $x_3 < 0$. The resulting condition corresponding to (8.14) and (8.15) is similar except for a difference of sign on the right-hand side of the equation. Thus the negative-going wave conditions are given by (8.5) and (8.8).

For the case where ϵ and μ are constant with respect to position, i.e. $\epsilon = \epsilon(\alpha)$, the fundamental solution $\mathcal{E}_j(x, y, t-s)$, $\mathcal{H}_j(x, y, t-s)$ is given by (with y, t fixed) (Marx and Maystre 1982)

$$\mathcal{H}_j = \nabla \times (\psi \hat{\mathbf{i}}_j), \tag{8.20}$$

$$\epsilon(\alpha)\partial_s \mathcal{E}_j = \nabla \partial_j \psi - c^{-2}(\alpha)\partial_s^2 \psi \hat{\mathbf{i}}_j, \tag{8.21}$$

where

$$\psi = \frac{1}{4\pi r}\delta(t - s - r/c), \quad r = |x - y|, \tag{8.22}$$

is the fundamental solution of the wave equation with velocity $c = c(\alpha)$, i.e.

$$\left.\begin{aligned} &\frac{1}{c^2}\psi_{ss} - \nabla^2\psi = \delta(x - y)\delta(t - s), \\ &\psi = 0, \ t - s < 0. \end{aligned}\right\} \tag{8.23}$$

The limit

$$\begin{aligned} &\lim_{y_3 \to 0+} 2\int_0^t \int_{\mathbb{R}^2} \frac{\delta(t - s - r/c)}{4\pi r}|_{x_3=0} u(\mathbf{x}, s)d\mathbf{x}ds \\ &= \int_{\mathbb{R}^2} \frac{u(\mathbf{x}, t - |\mathbf{x} - \mathbf{y}|/c)}{2\pi|\mathbf{x} - \mathbf{y}|}\, d\mathbf{x} = (\mathbf{K}_\alpha u)(\mathbf{y}, t), \end{aligned} \tag{8.24}$$

where u is a scalar function, defines the operator $\mathbf{K}_\alpha$ which is used in the wave-splitting for the scalar wave equation as given in Section 7.1.

We note that if $u(\mathbf{x}, t)$ is twice differentiable with compact support, the following limit is obtained

$$\begin{aligned} &\lim_{y_3 \to 0+} 2\int_0^t \int_{\mathbb{R}^2} \frac{\partial^2 \psi}{\partial x_i \partial x_j}|_{x_3=0} u(\mathbf{x}, s)d\mathbf{x}ds \\ &= \lim_{y_3 \to 0+} 2\int_0^t \int_{\mathbb{R}^2} \psi|_{x_3=0}\partial_i\partial_j u(\mathbf{x}, s)d\mathbf{x}ds = \mathbf{K}_\alpha \partial_i \partial_j u. \end{aligned} \tag{8.25}$$

If $u(x, t)$ is Hölder continuous it can be shown that with $r = |\mathbf{x} - \mathbf{y}|$,

$$\partial_i \mathbf{K}_\alpha u = \mathbf{K}_\alpha \partial_i u = \frac{1}{2\pi}\int_{\mathbb{R}^2} \frac{1}{r}\,(\partial_{x_i} r)\left[\frac{1}{c}\, u_t(\mathbf{x}, t - r/c) + \frac{1}{r}\, u(\mathbf{x}, t - r/c)\right] d\mathbf{x},$$

where the principal value of the second term on the right-hand side is taken.

On noting that $\partial_s E_j = -\partial_t E_j$, (8.14) can be expressed in the form

$$E_1(\mathbf{y},0,t) = \frac{1}{\epsilon}\partial_t^{-1}\mathbf{K}_\alpha\left\{\partial_1\partial_2 H_1 + \left(\frac{1}{c^2}\partial_s^2 - \partial_1^2\right)H_2\right\}, \tag{8.26}$$

$$E_2(\mathbf{y},0,t) = \frac{1}{\epsilon}\partial_t^{-1}\mathbf{K}_\alpha\left\{\left(-\frac{1}{c^2}\partial_s^2 + \partial_2^2\right)H_1 - \partial_1\partial_2 H_2\right\}, \tag{8.27}$$

and the operator N_α is thus given explicitly by

$$(N_\alpha \mathbf{u})(\mathbf{y},t) = \partial_t^{-1}\mathbf{K}_\alpha\begin{bmatrix}\partial_{x_1}\partial_{x_2} & (\frac{1}{c^2}\partial_t^2 - \partial_{x_1}^2)\\ (\partial_{x_2}^2 - \frac{1}{c^2}\partial_t^2) & -\partial_{x_1}\partial_{x_2}\end{bmatrix}\begin{bmatrix}u_1\\ u_2\end{bmatrix}, \tag{8.28}$$

where $\partial_t^{-1} = \int_0^t dt$, and $c = c(\alpha)$. Since $-\mu\partial_s\widetilde{\mathcal{H}}(x,y,t-s)$ is given by the right hand side of (8.21), it is seen that $-\mu\widetilde{\mathcal{H}}_{ij}(\mathbf{x},\mathbf{y},0;t)$ has the same form $\epsilon\mathcal{E}_{ij}(\mathbf{x},\mathbf{y},0,t)$, hence from (8.17) and (8.19) it follows that

$$M_\alpha = N_\alpha, \text{ when } \epsilon = \epsilon(\alpha),\ \mu = \mu(\alpha). \tag{8.29}$$

Equations (8.28) and (8.29) constitute the sought-for explicit expressions for the operators M_α, N_α when ϵ, μ are constant.

8.2 Exact splitting in a medium where $\epsilon = \epsilon(x_1, x_2, \alpha)$, $\mu = \mu(x_1, x_2, \alpha)$

Using the positive and negative-going wave conditions developed in the previous section, the decomposition of solutions of Maxwell's equations into positive and negative-going waves will be obtained.

Let $(\mathbf{E}, \mathbf{H})$ be solutions of Maxwell's equations with compact support in the slab $a < x_3 < b$, for $0 < t < T$, i.e.

$$\nabla\times\mathbf{E} = -\mu(x_1,x_2,\alpha)\partial_t\mathbf{H}, \quad \nabla\times\mathbf{H} = \epsilon(x_1,x_2,\alpha)\partial_t\mathbf{E},$$

with initial conditions

$$\mathbf{E}(x,0) = \mathbf{H}(x,0) = 0.$$

Since the positive-going and negative-going wave conditions involve the tangential components $\mathbf{E}_T$, $\mathbf{H}_T$, Maxwell's equations will be re-expressed in terms of these components.

As a first step, E_3 and H_3 will be expressed in terms of the tangential components, as follows. The third component of Maxwell's equations gives

$$H_3 = -\frac{1}{\mu}\partial_t^{-1}[\partial_1 E_2 - \partial_2 E_1], \quad E_3 = \frac{1}{\epsilon}\partial_t^{-1}[\partial_1 H_2 - \partial_2 H_1],$$

which yields

$$\begin{bmatrix} \partial_1 H_3 \\ \partial_2 H_3 \end{bmatrix} = -\partial_t^{-1} J_\mu \begin{bmatrix} E_1 \\ E_2 \end{bmatrix}, \quad \begin{bmatrix} \partial_1 E_3 \\ \partial_2 E_3 \end{bmatrix} = \partial_t^{-1} J_\epsilon \begin{bmatrix} H_1 \\ H_2 \end{bmatrix}, \tag{8.30}$$

where

$$J_p = \begin{bmatrix} -\partial_1(\frac{1}{p}\partial_2) & \partial_1(\frac{1}{p}\partial_1) \\ -\partial_2(\frac{1}{p}\partial_2) & \partial_2(\frac{1}{p}\partial_1) \end{bmatrix}, \quad p = \epsilon, \mu. \tag{8.31}$$

The first two components of Maxwell's equation can be written in the form

$$\partial_3 \begin{bmatrix} \mathbf{E}_T \\ \mathbf{H}_T \end{bmatrix} = \begin{bmatrix} 0 & \mu\partial_t P \\ \epsilon\partial_t P & 0 \end{bmatrix} \begin{bmatrix} \mathbf{E}_T \\ \mathbf{H}_T \end{bmatrix} + \begin{bmatrix} \nabla^T E_3 \\ \nabla^T H_3 \end{bmatrix}, \tag{8.32}$$

where the terms $\nabla^T E_3$, $\nabla^T H_3$ are given by (8.30), and P is the matrix

$$P = \begin{bmatrix} 0 & 1 \\ -1 & 0 \end{bmatrix}. \tag{8.33}$$

Insert expression (8.30) into (8.32) to yield

$$\partial_3 \begin{bmatrix} \mathbf{E}_T \\ \mathbf{H}_T \end{bmatrix} = \begin{bmatrix} 0 & (-\mu\partial_t P + \partial_t^{-1} J_\epsilon) \\ (\epsilon\partial_t P - \partial_t^{-1} J_\mu) & 0 \end{bmatrix} \begin{bmatrix} \mathbf{E}_T \\ \mathbf{H}_T \end{bmatrix}. \tag{8.34}$$

Equation (8.34) is the sought for dynamical equation for the tangential components $\mathbf{E}_T, \mathbf{H}_T$. They can be used with the positive and negative-going wave components that will be defined next.

Recall that the positive-going wave condition is given by

$$\mathbf{E}_T - \frac{1}{\epsilon} N_\alpha \mathbf{H}_T = 0, \quad \text{or} \quad \mathbf{H}_T + \frac{1}{\mu} M_\alpha \mathbf{E}_T = 0,$$

and the negative-going wave condition by

$$\mathbf{E}_T + \frac{1}{\epsilon} N_\alpha \mathbf{H}_T = 0, \quad \text{or} \quad \mathbf{H}_T - \frac{1}{\mu} M_\alpha \mathbf{E}_T = 0.$$

Decompose $\mathbf{E}_T$ into components $\mathbf{E}_T^+$ and $\mathbf{E}_T^-$ as follows

$$\mathbf{E}_T = \frac{1}{2}(\mathbf{E}_T + \frac{1}{\epsilon} N_\alpha \mathbf{H}_T) + \frac{1}{2}(\mathbf{E}_T - \frac{1}{\epsilon} N_\alpha \mathbf{H}_T) = \mathbf{E}_T^+ + \mathbf{E}_T^-.$$

This splitting will be expressed in matrix form

$$\begin{bmatrix} \mathbf{E}_T^+ \\ \mathbf{E}_T^- \end{bmatrix} = T_\alpha \begin{bmatrix} \mathbf{E}_T \\ \mathbf{H}_T \end{bmatrix} = \frac{1}{2} \begin{bmatrix} I & \frac{1}{\epsilon} N_\alpha \\ I & -\frac{1}{\epsilon} N_\alpha \end{bmatrix} \begin{bmatrix} \mathbf{E}_T \\ \mathbf{H}_T \end{bmatrix}, \tag{8.35}$$

and since $\frac{1}{\mu} M_\alpha \frac{1}{\epsilon} N_\alpha = -I$, the inverse transformation is given by

$$\begin{bmatrix}\mathbf{E}_T\\ \mathbf{H}_T\end{bmatrix} = T_\alpha^{-1}\begin{bmatrix}\mathbf{E}_T^+\\ \mathbf{E}_T^-\end{bmatrix} = \begin{bmatrix} I & I\\ -\frac{1}{\mu}M_\alpha & \frac{1}{\mu}M_\alpha\end{bmatrix}\begin{bmatrix}\mathbf{E}_T^+\\ \mathbf{E}_T^-\end{bmatrix}. \tag{8.36}$$

In a similar manner, define $\mathbf{H}_T^+$ and $\mathbf{H}_T^-$ as follows

$$\begin{bmatrix}\mathbf{H}_T^+\\ \mathbf{H}_T^-\end{bmatrix} = \frac{1}{2}\begin{bmatrix}-\frac{1}{\mu}M_\alpha & I\\ \frac{1}{\mu}M_\alpha & I\end{bmatrix}\begin{bmatrix}\mathbf{E}_T\\ \mathbf{H}_T\end{bmatrix}. \tag{8.37}$$

It then follows on inserting (8.36) into (8.37) that one obtains

$$\begin{bmatrix}\mathbf{H}_T^+\\ \mathbf{H}_T^-\end{bmatrix} = \begin{bmatrix}-\frac{1}{\mu}M_\alpha & 0\\ 0 & \frac{1}{\mu}M_\alpha\end{bmatrix}\begin{bmatrix}\mathbf{E}_T^+\\ \mathbf{E}_T^-\end{bmatrix}.$$

This indicates that the pair $(\mathbf{E}_T^+, \mathbf{H}_T^+)$ satisfy the positive-going wave condition (8.4), and the pair $(\mathbf{E}_T^-, \mathbf{H}_T^-)$ the negative-going wave condition (8.5). Thus $\mathbf{E}_T^+$ or $\mathbf{H}_T^+$ and $\mathbf{E}_T^-$ or $\mathbf{H}_T^-$ are respectively the positive-going wave and the negative-going wave components. One obtains a set of dynamical equations for the components $\mathbf{E}_T^+, \mathbf{E}_T^-$ by inserting expression (8.36) into system (8.34), differentiating, and then premultiplying the resulting expression by the matrix T_α (defined by (8.35)) to give

$$\partial_3\begin{bmatrix}\mathbf{E}_T^+\\ \mathbf{E}_T^-\end{bmatrix} = W\begin{bmatrix}\mathbf{E}_T^+\\ \mathbf{E}_T^-\end{bmatrix},$$

where

$$W = T_\alpha\begin{bmatrix}0 & (-\mu\partial_t P + \partial_t^{-1}J_\epsilon)\\ (\epsilon\partial_t P - \partial_t^{-1}J_\mu) & 0\end{bmatrix}T_\alpha^{-1} = \frac{1}{2}\begin{bmatrix}A^+ & A^-\\ -A^- & -A^+\end{bmatrix},$$

and

$$A^\pm = \pm(\mu\partial_t P - \partial_t^{-1}J_\epsilon)\frac{1}{\mu}M_\alpha + \frac{1}{\epsilon}N_\alpha(\epsilon\partial_t P - \partial_t^{-1}J_\mu).$$

It is shown in Appendix C of Weston (1993) that

$$\frac{1}{\epsilon}N_\alpha(\epsilon\partial_t P - \partial_t^{-1}J_\mu) = (\mu\partial_t P - \partial_t^{-1}J_\epsilon)\frac{1}{\mu}M_\alpha = -\Lambda_\alpha,$$

where

$$(\Lambda_\alpha\mathbf{u})(\mathbf{x},t) = \lim_{x_3\to 0+} 2\int_0^t\int_{\mathbb{R}}\begin{bmatrix}\partial_{y_3}\mathcal{H}_{21} & -\partial_{y_3}\mathcal{H}_{11}\\ \partial_{y_3}\mathcal{H}_{22} & -\partial_{y_3}\mathcal{H}_{12}\end{bmatrix}_{y_3=0}\begin{bmatrix}u_1(\mathbf{y},t)\\ u_2(\mathbf{y},t)\end{bmatrix} d\mathbf{y}\, ds, \tag{8.38}$$

and $\mathcal{H}_{ij} = \mathcal{H}_{ij}(\mathbf{y},\mathbf{x},y_3 - x_3; t-s)$. Hence it follows that

$$W = \begin{bmatrix}-\Lambda_\alpha(\mathbf{x},t) & 0\\ 0 & \Lambda_\alpha(\mathbf{x},t)\end{bmatrix}.$$

Thus one arrives at an uncoupled system for the wave components, of the form

$$\partial_3 \begin{bmatrix} \mathbf{E}_T^+ \\ \mathbf{E}_T^- \end{bmatrix} = \begin{bmatrix} -\Lambda_\alpha(\mathbf{x},t) & 0 \\ 0 & \Lambda_\alpha(\mathbf{x},t) \end{bmatrix} \begin{bmatrix} \mathbf{E}_T^+ \\ \mathbf{E}_T^- \end{bmatrix} . \tag{8.39}$$

For the case where $\mu = \mu(\alpha)$, $\epsilon = \epsilon(\alpha)$,

$$(\Lambda_\alpha \mathbf{u})(\mathbf{x},t) = \lim_{x_3 \to 0+} 2 \int_0^t \int_{\mathbb{R}} (\partial_{y_3}^2 \psi)\,|_{y_3=0} \begin{bmatrix} u_1(\mathbf{y},s) \\ u_2(\mathbf{y},s) \end{bmatrix} d\mathbf{y}\, ds,$$

where ψ is a solution of (8.23). The expression in the integral yields

$$2 \int_0^t \int_{\mathbb{R}} (\Box^T \psi)\, \mathbf{u}\, d\mathbf{y}\, ds = 2 \int_0^t \int_{\mathbb{R}} \psi \Box^T \mathbf{u}\, d\mathbf{y}\, ds,$$

and on taking the limit as $x_3 \to 0+$, one obtains for the vector function $\mathbf{u}$

$$(\Lambda_\alpha \mathbf{u})(\mathbf{x},t) = \mathbf{K}_\alpha \Box^T \mathbf{u} = \mathbf{K}_\alpha^{-1} \mathbf{u},$$

where $\mathbf{K}_\alpha$ is defined by (8.24). Expression (8.39) thus simplifies to the form

$$\partial_3 \begin{bmatrix} \mathbf{E}_T^+ \\ \mathbf{E}_T^- \end{bmatrix} = \begin{bmatrix} -\mathbf{K}_\alpha^{-1} I & 0 \\ 0 & \mathbf{K}_\alpha^{-1} I \end{bmatrix} \begin{bmatrix} \mathbf{E}_T^+ \\ \mathbf{E}_T^- \end{bmatrix} . \tag{8.40}$$

8.3 Splitting in a medium where $\epsilon = \epsilon(x_1, x_2, x_3)$, $\mu = \mu(x_1, x_2, x_3)$

Here the case where the physical geometry is described by the constitutive equations

$$\mathbf{D} = \epsilon(x_1, x_2, x_3)\mathbf{E}, \quad \mathbf{B} = \mu(x_1, x_2, x_3)\mathbf{H}$$

with ϵ, μ sufficiently smooth functions, will be considered.

The splitting designed for the case of transverse dependence upon x_1, and x_2 and dependence upon the parameter α ($\epsilon = \epsilon(x_1, x_2, \alpha)$, etc.) will be used in a fashion similar to the one used for the scalar wave equation in Chapter 7 to obtain the generalization of the differential system (8.39) for positive- and negative-going waves.

The positive- and negative-going wave decomposition is defined as follows. On each plane $x_3 = x_3^0$, the positive- and negative-going wave component will be defined by (8.35) with the parameter α taken to be x_3^0,

$$\begin{bmatrix} \mathbf{E}_T^+ \\ \mathbf{E}_T^- \end{bmatrix} = T_{x_3^0} \begin{bmatrix} \mathbf{E}_T \\ \mathbf{H}_T \end{bmatrix} . \tag{8.41}$$

This means that the parameter α in the definition of the operators N_α and M_α is replaced by x_3^0. On the plane $x_3 = x_3^0$, Maxwell's equations are expressed in the form

$$\partial_3 \begin{bmatrix} \mathbf{E}_T \\ \mathbf{H}_T \end{bmatrix} = \begin{bmatrix} 0 & (-\mu\partial_t P + \partial_t^{-1} J_\epsilon) \\ (\epsilon\partial_t P - \partial_t^{-1} J_\mu) & 0 \end{bmatrix} \begin{bmatrix} \mathbf{E}_T \\ \mathbf{H}_T \end{bmatrix}, \; x = x_3^0, \quad (8.42)$$

as follows from (8.34), with the parameter α being replaced by x_3^0 in the terms involving ϵ and μ, i.e. $\epsilon = \epsilon(x_1, x_2, x_3^0)$, etc. Then as in the previous section, the splitting given by (8.41) is inserted into (8.42), the appropriate terms differentiated with respect to x_3^0 and finally the result premultiplied by $T_{x_3^0}^{-1}$. The results will be the same as in the previous section, except that there is an additional term due to the derivative of $T_{x_3^0}^{-1}$ (essentially the derivative with respect to the parameter). This term appearing on the right hand side of the equation is given by

$$-T_{x_3^0} \frac{\partial}{\partial x_{3^0}} T_{x_3^0}^{-1} = \left(\frac{\partial T_{x_3^0}}{\partial x_{3^0}} \right) T_{x_3^0}^{-1} = \frac{1}{2} \begin{bmatrix} -\Gamma_{x_3^0} & \Gamma_{x_3^0} \\ \Gamma_{x_3^0} & -\Gamma_{x_3^0} \end{bmatrix},$$

where the matrix operator $\Gamma_{x_3^0}$ is given by (Weston 1993)

$$\Gamma_{x_3^0} = \left[\frac{\partial}{\partial \alpha} (\epsilon^{-1}(x_1, x_2, \alpha) N_\alpha) \right] \mu^{-1}(x_1, x_2, \alpha) M_\alpha, \; \alpha = x_3^0. \quad (8.43)$$

At this point (as with the scalar case) the superscript 0 on x_3^0 will be dropped, and for convenience the following will be defined

$$M = M_{x_3^0}, \quad N = N_{x_3^0}, \quad \Gamma = \Gamma_{x_3^0}, \quad \Lambda = \Lambda_{x_3^0}, \quad K = K_{x_3^0}.$$

The resulting dynamic system thus becomes

$$\partial_3 \begin{bmatrix} \mathbf{E}_T^+ \\ \mathbf{E}_T^- \end{bmatrix} = \begin{bmatrix} -\Lambda & 0 \\ 0 & \Lambda \end{bmatrix} \begin{bmatrix} \mathbf{E}_T^+ \\ \mathbf{E}_T^- \end{bmatrix} + \frac{1}{2} \begin{bmatrix} -\Gamma & \Gamma \\ \Gamma & -\Gamma \end{bmatrix} \begin{bmatrix} \mathbf{E}_T^+ \\ \mathbf{E}_T^- \end{bmatrix}, \quad (8.44)$$

where the operator Λ is given in (8.38).

For the special case where the physical medium is plane-stratified, i.e. $\epsilon = \epsilon(x_3)$, $\mu = \mu(x_3)$, it can be shown that (cf. Exercise 8.2)

$$\Lambda = \mathbf{K}^{-1} I, \quad (8.45)$$

$$\Gamma = \left(\frac{\epsilon'}{\epsilon} + \frac{2c'}{c} - \mathbf{K}'\mathbf{K}^{-1} \right) I + \frac{2c'}{c} \mathbf{K}^2 \begin{bmatrix} \partial_1^2 & \partial_1\partial_2 \\ \partial_1\partial_2 & \partial_2^2 \end{bmatrix}, \quad (8.46)$$

where I is the identity matrix, ϵ', c' are the derivatives of ϵ, c with respect to x_3, and

$$\mathbf{K}'u[\mathbf{x}, t] = \frac{c'(x_3)}{2\pi c^2(x_3)} \int_{\mathbb{R}^2} u_t(\mathbf{y}, t - |\mathbf{x} - \mathbf{y}|/c) dy. \quad (8.47)$$

8.4 Wave-splitting in a medium with dispersion

For the physical medium, the constitutive equations

$$\mathbf{B} = \mu(x_1, x_2, x_3)\mathbf{H},$$

$$\mathbf{D} = \epsilon(x_1, x_2, x_3)\mathbf{E} + G(x_1, x_2, x_3, t) * \mathbf{E}, \tag{8.48}$$

will be assumed, where $*$ denotes convolution in the time variable.

$$G(x, 0) = \sigma(x) \tag{8.49}$$

is the conductivity.

The procedure here is initially the same as in the previous section. The splitting is defined in terms of the principal part (with $G \equiv 0$), and on each plane $x_3 = x_3^0$, the splitting given in Section 8.2 is employed with the parameter $\alpha = x_3^0$. However, instead of deriving the dynamical equations for $\mathbf{E}_T^{\pm}$ similar to system (8.44), the system corresponding to (8.34) is obtained. As a first step, the following component of Maxwell's equation

$$\epsilon\partial_t E_3 + G * \partial_t E_3 = (\partial_1 H_2 - \partial_2 H_1) \tag{8.50}$$

needs to be inverted for $\partial_t E_3$. Note that $E_3 = 0$ for $t \leq 0$. Set

$$\partial_t E_3 = \left(\frac{1}{\epsilon} - \eta *\right)(\partial_1 H_2 - \partial_2 H_1), \tag{8.51}$$

and on inserting this into (8.50) one can show that the unknown quantity $\eta(x, t)$ must be a solution of the Volterra integral equation

$$\epsilon\eta - \epsilon^{-1}G + G * \eta = 0. \tag{8.52}$$

Hence, it follows from Maxwell's equations with constitutive relations given by (8.48) that

$$\begin{bmatrix} \partial_1 E_3 \\ \partial_2 E_3 \end{bmatrix} = [\partial_t^{-1} J_\epsilon - \partial_t^{-1} J_{\frac{1}{\eta}} *] \begin{bmatrix} H_1 \\ H_2 \end{bmatrix}, \tag{8.53}$$

where $J_{\frac{1}{\eta}}$ is given by (8.31) with p replaced by $\frac{1}{\eta}$. This produces an extra term in the right-hand side of the second expression of (8.30). The first equation of (8.30) remains the same. The dispersive term G produces an extra term in the following system obtained from Maxwell's equations

$$\partial_3 \mathbf{H}_T = \partial_3 \begin{bmatrix} H_1 \\ H_2 \end{bmatrix} = (\epsilon\partial_t P - \partial_t^{-1} J_\mu)\mathbf{E}_T + G_t P * \mathbf{E}_T. \tag{8.54}$$

Thus (8.32) has to be modified. On including these additional terms due to the dispersion function G, system (8.34) becomes

$$\partial_3 \begin{bmatrix} \mathbf{E}_T \\ \mathbf{H}_T \end{bmatrix} = \begin{bmatrix} 0 & (-\mu\partial_t P + \partial_t^{-1} J_\epsilon - \partial_t^{-1} J_{\frac{1}{\eta}} *) \\ (\epsilon\partial_t P - \partial_t^{-1} J_\mu + G_t P*) & 0 \end{bmatrix} \begin{bmatrix} \mathbf{E}_T \\ \mathbf{H}_T \end{bmatrix}. \tag{8.55}$$

These equations are then combined with the splitting

$$\begin{bmatrix} \mathbf{E}_T^+ \\ \mathbf{E}_T^- \end{bmatrix} = \frac{1}{2} \begin{bmatrix} I & \frac{1}{\epsilon} N_\alpha \\ I & -\frac{1}{\epsilon} N_\alpha \end{bmatrix} \begin{bmatrix} \mathbf{E}_T \\ \mathbf{H}_T \end{bmatrix}, \quad \alpha = x_3. \tag{8.56}$$

For the special case where $G(x,t)$ is independent of t, i.e.,

$$G(x,t) = \sigma(x), \tag{8.57}$$

there is some simplification of system (8.55). First equation (8.52) takes the form

$$\epsilon\eta - \frac{1}{\epsilon}\sigma(x) + \sigma(x) \int_0^t \eta dt' = 0,$$

which can be solved yielding

$$\eta(x,t) = \frac{\sigma(x)}{\epsilon^2(x)} \exp\left(-\frac{\sigma(x)}{\epsilon(x)} t\right). \tag{8.58}$$

The term G_t in system (8.55) vanishes.

For further application of this wave-splitting, see He and Weston (1997) where it is applied to a bi-isotropic dispersive medium.

8.5 Determination of the permittivity and conductivity in $\mathbb{R}^3$ using the wave-splitting

Here the wave-splitting of Maxwell's equations developed in the previous sections is applied to an inverse problem of determining the permittivity and conductivity in three dimensions (He and Weston 1995).

Consider the following scattering problem for the three-dimensional Maxwell's equations in a cartesian coordinate system $x \equiv (x_1, x_2, x_3)$

$$\nabla \times \mathbf{E} = -\mu_0 \partial_t \mathbf{H}, \tag{8.59}$$

$$\nabla \times \mathbf{H} = \epsilon(x)\partial_t \mathbf{E} + \sigma(x)\mathbf{E} + \hat{\mathbf{1}}\delta(x)\Theta(t), \tag{8.60}$$

or

$$\nabla \times \mathbf{E} = -\mu_0 \partial_t \mathbf{H} + \hat{\mathbf{1}}\delta(x)\Theta(t), \tag{8.61}$$

$$\nabla \times \mathbf{H} = \epsilon(x)\partial_t \mathbf{E} + \sigma(x)\mathbf{E}, \tag{8.62}$$

where $\hat{\mathbf{1}}$ is an arbitrary unit vector associated with the orientation of the electric (or magnetic) dipole and $\Theta(t)$ is the Heaviside step function defined by (2.12).

The reason why $\Theta(t)$ is chosen instead of a $\delta(t)$ type excitation is to avoid the leading singular δ' term in the solution (the singular δ term can be easily removed). The point source is located at $(0,0,0)$. With the x_3 axis pointed down, it will be assumed that the scattering medium lies below (in the physical sense) the plane $x_3 = +h$ ($h > 0$ is a small quantity) and that the region $x_3 \le h$ (upper half region in the physical sense) is free space, i.e.:

$$\epsilon = \epsilon_0, \ \sigma = 0, \ x_3 \le +h,$$

where ϵ_0 is a constant. The permittivity $\epsilon(x)$ and the conductivity $\sigma(x)$ are functions of x in the lower half-space $x_3 > h$ (the permeability is assumed to be a constant μ_0 in the whole space). The analysis becomes simpler if the permittivity $\epsilon(x)$ is continuous everywhere and continuously differentiable in the inhomogeneous half-space $x_3 > h$, and these assumptions are adopted here. The reflected field produced by the point dipole (exterior to the scattering medium) is measured on the surface $x_3 = h$.

It will be assumed that the fields are identically zero in the half-space containing the scattering medium, $x_3 > 0$ for the time $t \le 0$, i.e. the initial conditions are

$$\mathbf{E}(x,0) = 0, \quad \mathbf{H}(x,0) = 0, \qquad \text{for } x_3 > 0.$$

The wave-splitting given in Section 8.2 will be employed in the scattering medium $x_3 > 0$. Thus Maxwell's equations will be replaced by the system (8.55) developed in Section 8.4 for the tangential fields $\mathbf{E}_T$ and $\mathbf{H}_T$

$$\partial_3 \mathbf{E}_T + \mu_0 P \partial_t \mathbf{H}_T - \partial_t^{-1}(J_\epsilon - J_{\frac{1}{\eta}*})\mathbf{H}_T = 0, \quad x_3 > 0, \tag{8.63}$$

$$\partial_3 \mathbf{H}_T - \epsilon P \partial_t \mathbf{E}_T - \sigma P \mathbf{E}_T + \partial_t^{-1} J_{\mu_0} \mathbf{E}_T = 0, \quad x_3 > 0, \tag{8.64}$$

where the matrix operator J_p is given by (8.31) and the matrix P by (8.33), and $\eta(x,t)$ is given by (8.58).

The total tangential field $\mathbf{E}_T$ can be decomposed into a positive-going component $\mathbf{E}_T^+$ and negative-going component $\mathbf{E}_T^-$ by the following

$$\mathbf{E}_T = \mathbf{E}_T^+ + \mathbf{E}_T^-,$$

where

$$\mathbf{E}_T^{\pm} = \frac{1}{2}(\mathbf{E}_T \pm \frac{1}{\epsilon} N_{x_3} \mathbf{H}_T), \tag{8.65}$$

$$\mathbf{E}_T^{\pm} = \begin{bmatrix} \mathbf{E}_1^{\pm} \\ \mathbf{E}_2^{\pm} \end{bmatrix}.$$

The operator N_{x_3} in (8.65) is defined as follows (Problem P_2 in Section 8.1). Consider the following mixed problem in a half-space $x_3 > x_3^0$, $0 < t < T$,

(i) $\nabla \times \mathbf{E} = -\mu_0 \partial_t \mathbf{H}, \quad \nabla \times \mathbf{H} = \epsilon(\mathbf{x}, x_3^0)\partial_t \mathbf{E}, \; x_3 > x_3^0, \quad 0 < t < T,$

(ii) $\mathbf{E}(x, 0) = \mathbf{H}(x, 0) = 0, \quad x_3 > x_3^0,$

(iii) $H_i(\mathbf{x}, x_3^0, t) = h_i(\mathbf{x}, t), \quad i = 1, 2, \; 0 < t < T.$

Then the solution at the limiting plane $x_3 = x_3^0$ will give

$$\mathbf{E}_T = +\frac{1}{\epsilon} N_{x_3^0} \begin{bmatrix} h_1 \\ h_2 \end{bmatrix}, \quad x_3 = x_3^0,$$

i.e.

$$N_{x_3^0} \begin{bmatrix} h_1 \\ h_2 \end{bmatrix} = +\epsilon \mathbf{E}_T, \quad x_3 = x_3^0. \tag{8.66}$$

Note that in the above mixed problem, the medium is not the original physical medium but a medium with only transverse inhomogeneity and with $\sigma = 0$. Therefore, $\mathbf{E}^+$ and $\mathbf{E}^-$ only have physical meaning as positive and negative-going waves in the homogeneous region $+h \geq x_3 > 0$, but not inside the inhomogeneous medium $x_3 > +h$. Note that the input reflection data in the inverse problem are measured on the surface $x_3 = +h$ where $\mathbf{E}^+$ and $\mathbf{E}^-$ have physical meaning as incident and reflected waves, respectively. This wave-splitting will be used to find the condition for the reconstruction, but before doing so the asymptotic short-time behaviour of the fields produced by the point pulse source needs to be developed.

For the point source described by (8.60) or (8.61), the electromagnetic fields will have the following structure

$$\begin{aligned} \mathbf{E}(x, t) = {} & \mathbf{E}^{(0)}(x)\delta(t - \tau) + \mathbf{E}^{(1)}(x)\Theta(t - \tau) \\ & + \mathbf{E}^{(2)}(x) \cdot (t - \tau)\Theta(t - \tau) + \ldots, \; x_3 > 0, \end{aligned} \tag{8.67}$$

$$\begin{aligned} \mathbf{H}(x, t) = {} & \mathbf{H}^{(0)}(x)\delta(t - \tau) + \mathbf{H}^{(1)}(x)\Theta(t - \tau) \\ & + \mathbf{H}^{(2)}(x) \cdot (t - \tau)\Theta(t - \tau) + \ldots, \; x_3 > 0. \end{aligned} \tag{8.68}$$

Since the scattering medium lies in the region $x_3 > +h$ and the input reflection data are measured on the surface $x_3 = +h$, one need only to consider the fields in the region $x_3 > 0$. All equations in the remaining part of this chapter only hold in the region $x_3 > 0$ and this condition will be suppressed in most of the equations hereafter. Substituting (8.67) and (8.68) into (8.59) and (8.60) in the source region $x_3 > 0$ and matching the coefficients of the $\delta'(t-\tau)$, $\delta(t-\tau)$ terms, respectively, yields (see Kline and Kay (1965))

$$\nabla\tau \times \mathbf{E}^{(0)} = \mu_0 \mathbf{H}^{(0)}, \quad \nabla\tau \times \mathbf{H}^{(0)} = -\epsilon(x)\mathbf{E}^{(0)}, \tag{8.69}$$

$$\nabla \times \mathbf{E}^{(0)} - \nabla\tau \times \mathbf{E}^{(1)} = -\mu_0 \mathbf{H}^{(1)}, \tag{8.70}$$

$$\nabla \times \mathbf{H}^{(0)} - \nabla\tau \times \mathbf{H}^{(1)} = \epsilon(x)\mathbf{E}^{(1)} + \sigma(x)\mathbf{E}^{(0)}. \tag{8.71}$$

Since the scattering medium is source-free and $\mu =$ constant, one has

$$\nabla \cdot \mathbf{D} = \nabla \cdot (\epsilon \mathbf{E}) = 0, \quad \nabla \cdot \mathbf{H} = 0, \; x_3 > 0. \tag{8.72}$$

Substituting (8.67) and (8.68) into (8.72), and matching the coefficient of each singular term, yields

$$\mathbf{E}^{(0)} \cdot \nabla\tau = 0, \tag{8.73}$$

$$\mathbf{H}^{(0)} \cdot \nabla\tau = 0, \tag{8.74}$$

$$\mathbf{E}^{(1)} \cdot \nabla\tau = \frac{1}{\epsilon}\nabla \cdot (\epsilon \mathbf{E}^{(0)}), \tag{8.75}$$

$$\mathbf{H}^{(1)} \cdot \nabla\tau = \nabla \cdot (\mathbf{H}^{(0)}) \tag{8.76}$$

(note that $\nabla\tau$ is in the direction of the ray). From (8.69), it follows that

$$\nabla\tau \times (\nabla\tau \times \mathbf{E}^{(0)}) = \mu_0 \nabla\tau \times \mathbf{H}^{(0)} = -\mu_0 \epsilon \mathbf{E}^{(0)},$$

i.e.

$$(\nabla\tau \cdot \mathbf{E}^{(0)})\nabla\tau - (\nabla\tau \cdot \nabla\tau)\mathbf{E}^{(0)} = -\mu_0 \epsilon \mathbf{E}^{(0)},$$

which gives (cf. (8.73)) the eikonal equation (as expected)

$$|\nabla\tau|^2 = \mu_0 \epsilon.$$

Operating with '$\nabla\tau \times$' on both sides of (8.70), yields

$$\nabla\tau \times (\nabla \times \mathbf{E}^{(0)}) - \nabla\tau \times (\nabla\tau \times \mathbf{E}^{(1)}) = -\mu_0 \nabla\tau \times \mathbf{H}^{(1)}. \tag{8.77}$$

Since (cf. (8.75) and (8.71))

$$\nabla\tau \times (\nabla\tau \times \mathbf{E}^{(1)}) = (\nabla\tau \cdot \mathbf{E}^{(1)})\nabla\tau - (\nabla\tau \cdot \nabla\tau)\mathbf{E}^{(1)}$$

$$= \left[\frac{1}{\epsilon}\nabla \cdot (\epsilon \mathbf{E}^{(0)})\right] \nabla\tau - \mu_0 \epsilon \mathbf{E}^{(1)},$$

$$-\mu_0 \nabla\tau \times \mathbf{H}^{(1)} = -\mu_0 \nabla \times \mathbf{H}^{(0)} + \mu_0 \epsilon \mathbf{E}^{(1)} + \mu_0 \sigma \mathbf{E}^{(0)},$$

it follows from (8.77) that (cf. (8.69))

$$\nabla\tau \times (\nabla \times \mathbf{E}^{(0)}) - \left[\frac{1}{\epsilon}\nabla \cdot (\epsilon \mathbf{E}^{(0)})\right] \nabla\tau$$

$$= -\nabla \times (\nabla\tau \times \mathbf{E}^{(0)}) + \mu_0 \sigma \mathbf{E}^{(0)}. \tag{8.78}$$

Using the identities

$$\begin{aligned} &\nabla\tau \times (\nabla \times \mathbf{E}^{(0)}) \\ &\quad = -\mathbf{E}^{(0)} \times (\nabla \times \nabla\tau) + \nabla(\nabla\tau \cdot \mathbf{E}^{(0)}) - (\nabla\tau \cdot \nabla)\mathbf{E}^{(0)} - (\mathbf{E}^{(0)} \cdot \nabla)\nabla\tau \\ &\quad = -(\nabla\tau \cdot \nabla)\mathbf{E}^{(0)} - (\mathbf{E}^{(0)} \cdot \nabla)\nabla\tau, \end{aligned}$$

$$\left[\frac{1}{\epsilon}\nabla \cdot (\epsilon \mathbf{E}^{(0)})\right] \nabla\tau = (\nabla \cdot \mathbf{E}^{(0)})\nabla\tau + \frac{\mathbf{E}^{(0)} \cdot \nabla\epsilon}{\epsilon}\nabla\tau,$$

$$\nabla \times (\nabla\tau \times \mathbf{E}^{(0)}) = (\nabla \cdot \mathbf{E}^{(0)})\nabla\tau - \mathbf{E}^{(0)}(\nabla \cdot \nabla\tau) + (\mathbf{E}^{(0)} \cdot \nabla)\nabla\tau - (\nabla\tau \cdot \nabla)\mathbf{E}^{(0)},$$

one obtains the following equation from (8.78)

$$(\nabla\tau \cdot \nabla)\mathbf{E}^{(0)} = -\frac{1}{2}(\nabla^2\tau + \mu_0\sigma)\mathbf{E}^{(0)} - \frac{\mathbf{E}^{(0)} \cdot \nabla\epsilon}{2\epsilon}\nabla\tau. \tag{8.79}$$

Equation (8.79) is the transport equation for $\mathbf{E}^{(0)}$. In an analogous way one can derive the following transport equation for $\mathbf{H}^{(0)}$

$$(\nabla\tau \cdot \nabla)\mathbf{H}^{(0)} = -\frac{1}{2}(\nabla^2\tau - \frac{\nabla\tau \cdot \nabla\epsilon}{\epsilon} + \mu_0\sigma)\mathbf{H}^{(0)} - \frac{\mathbf{H}^{(0)} \cdot \nabla\epsilon}{2\epsilon}\nabla\tau. \tag{8.80}$$

The transport equations (8.79) and (8.80) will be employed in the reconstruction process. In addition to these, the leading asymptotic behaviour of the negative-going wave $\mathbf{E}_T^-$ in the inhomogeneous medium will be required. With the details given in the appendix to this chapter, it is shown that

$$\mathbf{E}_T^-(x, t = \tau) = \frac{1}{2}\gamma(x) \equiv \frac{1}{2}\begin{bmatrix} \gamma_1(x) \\ \gamma_2(x) \end{bmatrix}, \tag{8.81}$$

where

$$\gamma = \begin{bmatrix} w_1 \\ w_2 \end{bmatrix} + \frac{\sigma}{\epsilon^2}(\partial_3\tau)^{-1} J \mathbf{H}_T^{(0)}, \tag{8.82}$$

$$w_j = -\frac{1}{2}(\partial_3\tau)^{-2}\left\{\left[\frac{\mu_0}{2}(\partial_3\tau)^{-1}E_j^{(0)} + \frac{1}{\epsilon}\partial_j\tau E_3^{(0)}\right]\partial_3\epsilon + \mu_0\sigma E_j^{(0)}\right\}, \ j = 1, 2, \tag{8.83}$$

$$J = \begin{bmatrix} -(\partial_1\tau)(\partial_2\tau) & (\partial_1\tau)^2 \\ -(\partial_2\tau)^2 & (\partial_2\tau)(\partial_1\tau) \end{bmatrix}. \tag{8.84}$$

Since γ_1 and γ_2 are related to the parameters $\partial_3\epsilon$ and σ in (8.82) (cf. (8.83)), one can use the initial condition (8.81) to reconstruct $\partial_3\epsilon$ and σ simultaneously (the permittivity ϵ is obtained by an integration in x_3).

Next a numerical scheme for a simultaneous reconstruction of the permittivity and conductivity in the inhomogeneous half-space using the tangential reflected fields on the surface $x_3 = h$ produced by an impulse dipole at the position $(0,0,0)$ is described.

Let (cf. (8.67) and (8.68))

$$\mathbf{E}(x,t) = \mathbf{E}^{(0)}(x)\delta(t-\tau) + \mathbf{E}^{(r)}(x,t)\Theta(t-\tau), \quad x_3 > 0, \tag{8.85}$$

$$\mathbf{H}(x,t) = \mathbf{H}^{(0)}(x)\delta(t-\tau) + \mathbf{H}^{(r)}(x,t)\Theta(t-\tau), \quad x_3 > 0, \tag{8.86}$$

where $\mathbf{E}^{(r)}$ and $\mathbf{H}^{(r)}$ denote the regular parts of the fields. The reconstruction will be achieved by a layer-stripping approach. The permittivity and the conductivity can be reconstructed by the following steps.

(1) Assume that at the nth stage, the fields $\mathbf{E}^{(0)}, \mathbf{H}^{(0)}, \mathbf{E}^{(r)}, \mathbf{H}^{(r)}$ and the parameters $\epsilon, \partial_3\epsilon, \sigma$ are known at the layer $x_3 = x_3^n$. Note that this assumption is true at $x_3 = h$ (the surface of the inhomogeneous scattering medium).

(2) Use the transport equations (8.79) and (8.80) for $\mathbf{E}^{(0)}$ and $\mathbf{H}^{(0)}$, respectively, to calculate $\mathbf{E}^{(0)}$ and $\mathbf{H}^{(0)}$ at the next layer $x_3 = x_3^{n+1}$, where $x_3^{n+1} = x_3^n + \Delta x_3$, $\Delta x_3 > 0$. This can be done by the following formulas

$$\mathbf{E}^{(0)}(\mathbf{x}, x_3^{n+1}) = \mathbf{E}^{(0)}(\mathbf{x}, x_3^n) + (\partial_3\mathbf{E}^{(0)})(\mathbf{x}, x_3^n)\cdot\Delta x_3 + O((\Delta x_3)^2), \tag{8.87}$$

$$\mathbf{H}^{(0)}(\mathbf{x}, x_3^{n+1}) = \mathbf{H}^{(0)}(\mathbf{x}, x_3^n) + (\partial_3\mathbf{H}^{(0)})(\mathbf{x}, x_3^n)\cdot\Delta x_3 + O((\Delta x_3)^2), \tag{8.88}$$

where $\partial_3\mathbf{E}^{(0)}$ and $\partial_3\mathbf{H}^{(0)}$ at $x_3 = x_3^n$ are evaluated by the transport equations (8.79) and (8.80) (note that the transport equations can be rewritten in the form of $\partial_3\mathbf{E}^{(0)} = \ldots$, etc.).

(3) Determine $\epsilon(\mathbf{x}, x_3^{n+1})$ using the formula

$$\epsilon(\mathbf{x}, x_3^{n+1}) = \epsilon(\mathbf{x}, x_3^n) + (\partial_3\epsilon)(\mathbf{x}, x_3^n)\cdot\Delta x_3 + O((\Delta x_3)^2). \tag{8.89}$$

(4) Calculate $\mathbf{E}^{(r)}$ and $\mathbf{H}^{(r)}$ at the layer $x_3 = x_3^{n+1}$. One can use a finite-difference method to solve the following problem in the region $x_3^n \le x_3 \le x_3^{n+1}$.

(i) $\nabla\times\mathbf{E}^{(r)} = -\mu_0\partial_t\mathbf{H}^{(r)}, \quad \nabla\times\mathbf{H}^{(r)} = \epsilon(\mathbf{x}, x_3^n)\partial_t\mathbf{E}^{(r)}, \quad x_3^n \le x_3 \le x_3^{n+1}$,

(ii) $\mathbf{E}^{(r)}|_{x_3=x_3^n} = \mathbf{E}^{(r)}(\mathbf{x}, x_3^n, t), \ \mathbf{H}^{(r)}|_{x_3=x_3^n} = \mathbf{H}^{(r)}(\mathbf{x}, x_3^n, t), \ x_3^n \le x_3 \le x_3^{n+1}$,

(iii) $\mathbf{E}^{(r)}(x,t) = \mathbf{H}^{(r)}(x,t) = 0, \ t < \tau, \quad x_3^n \le x_3 \le x_3^{n+1}$.

To solve this problem numerically, one may need to use a regularization technique.

(5) Obtain $\mathbf{E}^{(1)}$ and $\mathbf{H}^{(1)}$ at the layer $x_3 = x_3^{n+1}$ using the formulas (cf. (8.67) and (8.68))

$$\mathbf{E}^{(1)}(\mathbf{x}, x_3^{n+1}) = \mathbf{E}^{(r)}(\mathbf{x}, x_3^{n+1}, \tau), \ \mathbf{H}^{(1)}(\mathbf{x}, x_3^{n+1}) = \mathbf{H}^{(r)}(\mathbf{x}, x_3^{n+1}, \tau). \quad (8.90)$$

(6) Calculate $\frac{1}{\epsilon}[N_{x_3}\mathbf{H}_T]^{(1)}$ at the layer $x_3 = x_3^{n+1}$ by solving the following mixed problem in the half-space $x_3 \geq x_3^{n+1}$ (cf. the definition of the operator N_{x_3} and note that $[N_{x_3}\mathbf{H}_T] = [N_{x_3}\mathbf{H}_T]^{(0)}\delta(t-\tau) + [N_{x_3}\mathbf{H}_T]^{(1)}\Theta(t-\tau) + \ldots$)

(i) $\nabla \times \overline{\mathbf{E}} = -\mu_0 \partial_t \overline{\mathbf{H}}, \quad \nabla \times \overline{\mathbf{H}} = \epsilon(\mathbf{x}, x_3^{n+1})\partial_t \overline{\mathbf{E}}, \quad x_3 > x_3^{n+1},$

(ii) $\overline{\mathbf{E}}(x, 0) = \overline{\mathbf{H}}(x, 0) = 0, \quad x_3 > x_3^{n+1},$

(iii) $\overline{\mathbf{H}}_T(\mathbf{x}, x_3^{n+1}, t) = \mathbf{H}_T^{(0)}(\mathbf{x}, x_3^{n+1})\delta(t-\tau) + \mathbf{H}_T^{(1)}(\mathbf{x}, x_3^{n+1})\Theta(t-\tau).$

The solution at the limiting plane $x_3 = x_3^{n+1}$ will give the desired result

$$\frac{1}{\epsilon}[N_{x_3}\mathbf{H}_T]^{(1)} = +\overline{\mathbf{E}}_T^{(1)}, \quad x_3 = x_3^{n+1}. \quad (8.91)$$

(7) Calculate γ at the layer $x_3 = x_3^{n+1}$ (cf. (8.131) and (8.144) of the appendix)

$$\gamma(\mathbf{x}, x_3^{n+1}) = \mathbf{E}_T^{(1)}(\mathbf{x}, x_3^{n+1}, \tau) - \frac{1}{\epsilon}[N_{x_3}\mathbf{H}_T]^{(1)}(\mathbf{x}, x_3^{n+1}, \tau). \quad (8.92)$$

(8) Determine $\partial_3\epsilon$ and σ at the layer $x_3 = x_3^{n+1}$ from the values of γ_1 and γ_2 (calculated in step (7)) using the condition (8.82). Note that the condition (8.82) can be rewritten in the following form

$$\gamma_1(\mathbf{x}, x_3^{n+1}) = a_1 \cdot (\partial_3\epsilon)(\mathbf{x}, x_3^{n+1}) + b_1 \cdot \sigma(\mathbf{x}, x_3^{n+1}),$$

$$\gamma_2(\mathbf{x}, x_3^{n+1}) = a_2 \cdot (\partial_3\epsilon)(\mathbf{x}, x_3^{n+1}) + b_2 \cdot \sigma(\mathbf{x}, x_3^{n+1}),$$

which give a unique solution for $\partial_3\epsilon$ and σ when $a_1 b_2 - a_2 b_1 \neq 0$.

(9) Repeat the sequence of steps (1)–(8) to move one step deeper into the inhomogeneous medium.

Therefore, in the electromagnetic case one can reconstruct the permittivity and conductivity simultaneously in the inhomogeneous half-space using the tangential reflection fields produced by a dipole with a fixed position and orientation.

Note that if one uses a source $\hat{\mathbf{1}} \cdot t\Theta(t)\delta(x)$ instead of a source $\hat{\mathbf{1}} \cdot \Theta(t)\delta(x)$, then the fields will have the following structure

$$\mathbf{E}(x,t) = \mathbf{E}^{(0)}(x)\Theta(t-\tau) + \mathbf{E}^{(1)}(x) \cdot (t-\tau)\Theta(t-\tau) + \ldots, \quad \text{etc.}$$

Thus there is no singularity in the total field, and one does not need to subtract $\mathbf{E}^{(0)}$ in the computation in this case. However, one has to make a time differentiation to get a useful condition for the reconstruction, viz.

$$(\partial_t \mathbf{E}_T^-)(x,\tau) = \frac{1}{2}\gamma(x),$$

where γ is the same as the one given by (8.82) with $\mathbf{E}^{(0)} = \mathbf{E}(x,\tau)$, etc.

8.6 Non-planar wave-splitting and absorbing boundary conditions

In this section, the conditions for out-going and incoming waves on a non-planar surface in free space (ϵ, μ constant) are derived. From this the wave-splitting of Maxwell's equations on a non-planar surface is obtained.

The condition for an out-going wave is then used to derive an absorbing boundary condition, which gives perfect absorption on a non-planar surface for any type of interior transient sources. Absorbing boundary conditions are of growing importance in numerical computation where they are used to terminate the computational domain in a finite element (Silvester and Ferrari 1990) or finite difference (Taflove 1995) solution of the partial differential equations (see e.g. Engquist and Majda (1977), Givoli (1992)). For Maxwell's equations in a Cartesian coordinate system, one usually considers each Cartesian component of the vector electromagnetic fields, which satisfies the wave equation. Consequently, the absorbing boundary conditions for the wave equation (Cao and He 1996) can be applied directly in the electromagnetic computation if the absorbing boundary is planar. However, in many cases it is preferable to use an absorbing boundary which is a curved surface like a spherical surface to avoid spurious reflections from the corners and edges of the introduced boundary. If the absorbing boundary is non-planar, one has to look for absorbing boundary conditions directly applicable to Maxwell's equations since a component of the electromagnetic fields in a curvilinear coordinate system does not satisfy the wave equation.

The derivation of the out-going or incoming wave condition across a non-planar surface is initially similar to the development of the positive and negative-going wave conditions developed in the beginning of this chapter. However, because of the geometry, there are some differences.

Let Ω^- be an open connected set in R^3, bounded by a smooth C^2 surface S. Let the complement of the closure of Ω^- be Ω^+. Thus if Ω^- is a bounded region, Ω^- and Ω^+ are respectively interior and exterior regions to the surface S. Let $\hat{\mathbf{n}}$ be a unit normal of S directed from Ω^- to Ω^+. Consider the following initial-value, boundary-value problems for Maxwell's equations in a homogeneous space (the permittivity ϵ and permeability μ are constant):

Problem P1′: Solve for $\mathbf{E}(x,t)$, $\mathbf{H}(x,t)$

$$\left.\begin{array}{ll}\text{(i)} & \nabla\times\mathbf{E} = -\mu\partial_t\mathbf{H}, \quad \nabla\times\mathbf{H} = \epsilon\partial_t\mathbf{E}, \quad x\in\Omega^+,\ 0<t<T,\\[2ex] \text{(ii)} & \mathbf{E}(x,0) = \mathbf{H}(x,0) = 0, \quad x\in\Omega^+,\\[2ex] \text{(iii)} & \mathbf{E}_T(x,t) = \mathbf{E}_T^0(x,t), \quad x\in S, \quad 0<t<T,\end{array}\right\} \tag{8.93}$$

where the given tangential data $\mathbf{E}_T^0$ has compact support if S is an unbounded surface and is sufficiently smooth for the existence of a solution.

Problem P2′: Solve for $\mathbf{E}(x,t)$, $\mathbf{H}(x,t)$

$$\left.\begin{array}{ll}\text{(i)} & \nabla\times\mathbf{E} = -\mu\partial_t\mathbf{H}, \quad \nabla\times\mathbf{H} = \epsilon\partial_t\mathbf{E}, \quad x\in\Omega^+,\ 0<t<T,\\[2ex] \text{(ii)} & \mathbf{E}(x,0) = \mathbf{H}(x,0) = 0, \quad x\in\Omega^+,\\[2ex] \text{(iii)} & \mathbf{H}_T(x,t) = \mathbf{H}_T^0(x,t), \quad x\in S, \quad 0<t<T,\end{array}\right\} \tag{8.94}$$

where the given tangential data $\mathbf{H}_T^0$ has compact support if S is an unbounded surface and is sufficiently smooth for the existence of a solution.

Let the vectors $\mathcal{E}_j(x,y,t-s), \mathcal{H}_j(x,y,t-s)$ be the free-space fundamental solution satisfying (ϵ, μ constant)

$$\left.\begin{array}{l}\nabla_y\times\mathcal{E}_j = -\mu\partial_s\mathcal{H}_j,\\[2ex] \nabla_y\times\mathcal{H}_j = \epsilon\partial_s\mathcal{E}_j + \hat{\mathbf{i}}_j\delta(x-y)\delta(t-s),\\[2ex] \mathcal{E}_j = \mathcal{H}_j = 0, \quad \text{for } s-t>0.\end{array}\right\} \tag{8.95}$$

where $\hat{\mathbf{i}}_j$, $j=1,2,3$, is the unit vector in the direction of the x_j axis.

Let the vectors $\widetilde{\mathcal{E}}_j(x,y,t-s)$, $\widetilde{\mathcal{H}}_j(x,y,t-s)$ be the free-space fundamental solution satisfying (ϵ, μ constant)

$$\left.\begin{array}{l}\nabla_y\times\widetilde{\mathcal{E}}_j = -\mu\partial_s\widetilde{\mathcal{H}}_j + \hat{\mathbf{i}}_j\delta(x-y)\delta(t-s),\\[2ex] \nabla_y\times\widetilde{\mathcal{H}}_j = \epsilon\partial_s\widetilde{\mathcal{E}}_j,\\[2ex] \widetilde{\mathcal{E}}_j = \widetilde{\mathcal{H}}_j = 0, \quad \text{for } s-t>0.\end{array}\right\} \tag{8.96}$$

Let $\mathcal{E}_j = \{\mathcal{E}_{1j}, \mathcal{E}_{2j}, \mathcal{E}_{3j}\}$, etc. Let $(\mathbf{E}, \mathbf{H})$ be a solution of Problem P_1', with x, t replaced by y, s, respectively. Then we can show that for $x \in \Omega^+$,

$$-\nabla_y \cdot [\mathbf{E}(y,s) \times \mathcal{H}_j + \mathcal{E}_j \times \mathbf{H}(y,s)]$$

$$= \partial_s(\epsilon \mathbf{E} \cdot \mathcal{E}_j + \mu \mathbf{H} \cdot \mathcal{H}_j) + (\mathbf{E} \cdot \hat{\mathbf{i}}_j)\delta(x-y)\delta(t-s).$$

Integrating with respect to y over Ω^+, and with respect to s from $-\infty$ to T (where $t \le T$), and using the divergence theorem and the fact the field at $|x| = +\infty$ is zero, yields

$$E_j(x,t) = \int_0^t \int_S \{\hat{\mathbf{n}} \times \mathbf{E}) \cdot \mathcal{H}_j - (\hat{\mathbf{n}} \times \mathbf{H}) \cdot \mathcal{E}_j\} dS_y ds, \quad x \in \Omega^+, \quad j = 1,2,3, \tag{8.97}$$

where dS_y is an element of surface area of S. Similarly, using the solution $(\mathbf{E}, \mathbf{H})$ of Problem P_2', and the fundamental solution $(\widetilde{\mathcal{E}}_j, \widetilde{\mathcal{H}}_j)$ it can be shown that

$$H_j(x,t) = -\int_0^t \int_S \{\hat{\mathbf{n}} \times \mathbf{E}) \cdot \widetilde{\mathcal{H}}_j - (\hat{\mathbf{n}} \times \mathbf{H}) \cdot \widetilde{\mathcal{E}}_j\} dS_y ds, \quad x \in \Omega^+, \quad j = 1,2,3. \tag{8.98}$$

The fundamental solutions have the following explicit formulas in the homogeneous space (see (8.20) and (8.21))

$$\mathcal{H}_j(x,y,t-s) = \widetilde{\mathcal{E}}_j(x,y,t-s) = (\nabla_y \psi) \times \hat{\mathbf{i}}_j, \tag{8.99}$$

$$\mathcal{E}_j(x,y,t-s) = -\frac{\mu}{\epsilon}\widetilde{\mathcal{H}}_j(x,y,t-s)$$

$$= \frac{1}{\epsilon}\left\{\int_{-\infty}^{s} \nabla_y(\hat{\mathbf{i}}_j \cdot \nabla_y \psi|_{s=s'})ds' - \frac{1}{c^2}(\partial_s \psi)\hat{\mathbf{i}}_j\right\}, \tag{8.100}$$

where $c = 1/\sqrt{\mu\epsilon}$ and ψ (the fundamental solution of the wave equation) is given by (8.22).

Therefore, (8.97) and (8.98) can be rewritten as

$$\hat{\mathbf{a}} \cdot \mathbf{E}(x,t) = I_1(\mathbf{E}, \hat{\mathbf{a}}; x, t) - \tfrac{1}{\epsilon} I_2(\mathbf{H}, \hat{\mathbf{a}}; x, t), \tag{8.101}$$

$$\hat{\mathbf{a}} \cdot \mathbf{H}(x,t) = I_1(\mathbf{H}, \hat{\mathbf{a}}; x, t) - \tfrac{1}{\mu} I_2(\mathbf{E}, \hat{\mathbf{a}}; x, t), \tag{8.102}$$

respectively, where $\hat{\mathbf{a}}$ is an arbitrary tangential unit vector and

$$I_1(\mathbf{F}, \hat{\mathbf{a}}; x, t) = \int_0^\infty \int_S (\hat{\mathbf{n}} \times \mathbf{F}) \cdot [(\nabla_y \psi) \times \hat{\mathbf{a}}] dS_y ds, \tag{8.103}$$

$$\begin{aligned} I_2(\mathbf{F}, \hat{\mathbf{a}}; x, t) &= \int_0^\infty \int_S (\hat{\mathbf{n}} \times \mathbf{F}) \cdot [\nabla_y (\hat{\mathbf{a}} \cdot \nabla_y \int_{-\infty}^s \psi|_{s=s'} ds') - \frac{1}{c^2} (\partial_s \psi) \hat{\mathbf{a}}] dS_y ds \\ &= -\int_0^\infty \int_S \left\{ \int_0^s (\hat{\mathbf{n}} \times \mathbf{F}) ds' \cdot \nabla_y (\hat{\mathbf{a}} \cdot \nabla_y \psi) + \frac{1}{c^2} (\partial_s \psi) \hat{\mathbf{a}} \cdot (\hat{\mathbf{n}} \times \mathbf{F}) \right\} dS_y ds, \end{aligned} \tag{8.104}$$

using the fact that $\mathbf{F} = 0$ when $t < 0$.

Using the relation

$$\nabla_y \psi = -\frac{(y-x)}{4\pi r^3} \{\delta(t - s - r/c) + \frac{r}{c} \delta'(t - s - r/c)\},$$

and the vector relation

$$(\hat{\mathbf{n}} \times \mathbf{F}) \cdot [(\nabla \psi) \times \hat{\mathbf{a}}] = (\hat{\mathbf{a}} \cdot \mathbf{F})(\hat{\mathbf{n}} \cdot \nabla \psi) - (\hat{\mathbf{a}} \cdot \hat{\mathbf{n}})(\mathbf{F} \cdot \nabla \psi),$$

the first integral can be put into the following form (after integrating over the variable s),

$$\begin{aligned} &I_1(\mathbf{F}, \hat{\mathbf{a}}; x, t) \\ &= -\int_S [\hat{\mathbf{a}} \cdot \mathbf{F}(y, t - r/c)] \frac{\hat{\mathbf{n}} \cdot (y - x)}{4\pi r^3} dS_y + \int_S (\hat{\mathbf{a}} \cdot \hat{\mathbf{n}}) \frac{(y - x) \cdot \mathbf{F}(y, t - r/c)}{4\pi r^3} dS_y \\ &+ \int_S \left\{ -[\hat{\mathbf{a}} \cdot \mathbf{F}_t(y, t - r/c)] \frac{\hat{\mathbf{n}} \cdot (y - x)}{4\pi r^2 c} + (\hat{\mathbf{a}} \cdot \hat{\mathbf{n}}) \frac{(y - x) \cdot \mathbf{F}_t(y, t - r/c)}{4\pi r^2 c} \right\} dS_y. \end{aligned} \tag{8.105}$$

For the second integral $I_2(\mathbf{F}, \hat{\mathbf{a}}; x, t)$, the following formula (using Stoke's theorem)

$$\begin{aligned} &\int_0^s \int_S (\hat{\mathbf{n}} \times \hat{\mathbf{F}}) \cdot \nabla \Psi dS_y ds' \\ &= -\int_S \hat{\mathbf{n}} \cdot \nabla \times \int_0^s (\hat{\mathbf{F}} \psi) ds' dS_y + \int_0^s \int_S \Psi (\hat{\mathbf{n}} \cdot \nabla \times \hat{\mathbf{F}}) ds' dS_y \\ &= \int_0^s \int_S \Psi (\hat{\mathbf{n}} \cdot \nabla \times \hat{\mathbf{F}}) dS_y ds', \end{aligned} \tag{8.106}$$

is applied to the first term with $\Psi = \hat{\mathbf{a}} \cdot (\nabla_y \psi)$ and $\hat{\mathbf{F}} = \int_0^s \mathbf{F} ds'$. Then integrating with respect to the variable s and using the properties of the 'delta' function, the second integral takes the form

$$I_2(\mathbf{F}, \hat{\mathbf{a}}; x, t) = \int_S \frac{\hat{\mathbf{a}} \cdot (y - x)}{4\pi r^3} \hat{\mathbf{n}} \cdot \mathbf{G}(y, t - r/c) dS_y$$
$$+ \int_S \frac{\hat{\mathbf{a}} \cdot (y - x)}{4\pi c r^2} \hat{\mathbf{n}} \cdot \mathbf{G}_t(y, t - r/c) dS_y + \int_S \hat{\mathbf{a}} \cdot [\hat{\mathbf{n}} \times \mathbf{F}_t(y, t - r/c)] \frac{1}{4\pi c^2 r} dS_y, \tag{8.107}$$

where

$$\mathbf{G}(y, s) = \int_0^s \nabla \times \mathbf{F}(y, s') ds' = \begin{cases} -\mu \mathbf{H}(y, s), & \text{if } \mathbf{F} = \mathbf{E}, \\ \epsilon \mathbf{E}(y, s), & \text{if } \mathbf{F} = \mathbf{H}, \end{cases} \tag{8.108}$$

and where Maxwell's equations are used.

Note that it is the integration by parts in (8.106) that makes the order of singular-like terms lower.

Let x_0 be a fixed point on S and $\hat{\mathbf{a}}$ be a fixed unit vector which is tangential to S at x_0. The limit $x \to x_0$ can now be taken. Let $\mathbf{F}(x, t)$, $x \in S$, $0 \le t \le T$, be a C^2 electromagnetic vector field on S, where $\mathbf{F} = \mathbf{F}_t = 0$ at $t = 0$. If S is a surface of infinite extent (e.g., a paraboloid of revolution), then $\mathbf{F}(x, t)$ should have compact support on S. Because of the smoothness condition on $\mathbf{F}$ (so that $\mathbf{G}$ will be C^1) and the surface S, there is no problem in the integral for I_2. Most of the terms in I_2 will be of the single layer potential type (integrable singularity of order $|x - y|^{-1}$) (Vladimirov 1971). However, since $\hat{\mathbf{a}}$ is tangential to the surface S at x_0, the first term in the expression for I_2 is of principal value type. Using local polar coordinates (ρ, θ) at the point x_0 on the surface S, and integrating on a disk of radius ε on the tangential plane to x_0, one uses the fact that with $x = x_0 + |x - y| \hat{\mathbf{n}}$,

$$\lim_{x \to x_0} \int_0^{\varepsilon} \int_0^{2\pi} \frac{\hat{\mathbf{a}} \cdot (y - x)}{4\pi |x - y|^3} \hat{\mathbf{n}}(x_0) \cdot \mathbf{G}(x_0, t) \rho d\rho d\phi \to 0 \text{ as } \varepsilon \to 0.$$

Thus, one obtains

$$\lim_{x \to x_0} I_2(\mathbf{F}, \hat{\mathbf{a}}; x, t) = P.V. \int_S \frac{\hat{\mathbf{a}} \cdot (y - x)}{4\pi r^3} \hat{\mathbf{n}} \cdot \mathbf{G}(y, t - r/c) dS_y$$
$$+ \int_S \left\{ \frac{\hat{\mathbf{a}} \cdot (y - x)}{4c\pi r^2} \hat{\mathbf{n}} \cdot \mathbf{G}_t(y, t - r/c) + \hat{\mathbf{a}} \cdot [\hat{\mathbf{n}} \times \mathbf{F}_t(y, t - r/c)] \frac{1}{4\pi c^2 r} \right\} dS_y$$
$$\equiv I_2(\mathbf{F}, \hat{\mathbf{a}}; x_0, t), \tag{8.109}$$

where $\mathbf{G}$ is given by (8.108).

In taking the limit $x \to x_0$ for I_1, the only term that possesses a problem is the first integral which is of a double-layer potential-type integral. Since F is a C^2 function of y and t vanishing for $t < 0$, it follows that (Colton and Kress 1992)

$$\lim_{x \to x_0} \int_S \hat{\mathbf{a}} \cdot \mathbf{F}(y, t - r/c) \frac{\hat{\mathbf{n}} \cdot (y - x)}{4\pi r^3} dS_y$$

$$= \mp \frac{1}{2} \hat{\mathbf{a}} \cdot \mathbf{F}(x_0, t) + P.V. \int_S \hat{\mathbf{a}} \cdot \mathbf{F}(y, t - r_0/c) \frac{\hat{\mathbf{n}} \cdot (y - x_0)}{4\pi r_0^3} dS_y, \quad (8.110)$$

where $r_0 = |x_0 - y|$, and the upper sign is taken when $x \in \Omega_+$ and the lower sign when $x \in \Omega_-$. The resulting limits for $I_1(x)$ can now be stated as

$$\lim_{x \in \Omega_+ \to x_0 \in S} I_1(\mathbf{F}, \hat{\mathbf{a}}; x, t) = \frac{1}{2} \hat{\mathbf{a}} \cdot \mathbf{F}(x_0, t) + I_1(\mathbf{F}, \hat{\mathbf{a}}; x_0, t), \quad (8.111)$$

$$\lim_{x \in \Omega_- \to x_0 \in S} I_1(\mathbf{F}, \hat{\mathbf{a}}; x, t) = -\frac{1}{2} \hat{\mathbf{a}} \cdot \mathbf{F}(x_0, t) + I_1(\mathbf{F}, \hat{\mathbf{a}}; x_0, t), \quad (8.112)$$

where $I_1(\mathbf{F}, \hat{\mathbf{a}}; x_0, t)$ is the principal value of the integral given by (8.105).

Thus one obtains the following out-going wave condition at $x_0 \in S$:

$$\hat{\mathbf{a}} \cdot \mathbf{E}(x_0, t) - 2I_1(\mathbf{E}, \hat{\mathbf{a}}; x_0, t) + \frac{2}{\epsilon} I_2(\mathbf{H}, \hat{\mathbf{a}}; x_0, t) = 0, \quad (8.113)$$

or, equivalently,

$$\hat{\mathbf{a}} \cdot \mathbf{H}(x_0, t) - 2I_1(\mathbf{H}, \hat{\mathbf{a}}; x_0, t) - \frac{2}{\mu} I_2(\mathbf{E}, \hat{\mathbf{a}}; x_0, t) = 0, \quad (8.114)$$

where $I_2(\mathbf{E}, \hat{\mathbf{a}}; x_0, t)$ is given by the limit indicated by (8.109) and $I_1(\mathbf{E}, \hat{\mathbf{a}}; x_0, t)$ is the principal value integral (8.105).

Similarly, by treating problems P_1 and P_2 for the region Ω^-, one can show that the in-coming wave condition is given by

$$\hat{\mathbf{a}} \cdot \mathbf{E}(x_0, t) + 2I_1(\mathbf{E}, \hat{\mathbf{a}}; x_0, t) - \frac{2}{\epsilon} I_2(\mathbf{H}, \hat{\mathbf{a}}; x_0, t) = 0, \quad (8.115)$$

or, equivalently,

$$\hat{\mathbf{a}} \cdot \mathbf{H}(x_0, t) + 2I_1(\mathbf{H}, \hat{\mathbf{a}}; x_0, t) + \frac{2}{\mu} I_2(\mathbf{E}, \hat{\mathbf{a}}; x_0, t) = 0. \quad (8.116)$$

Therefore, the total tangential field in the tangential direction $\hat{\mathbf{a}}$ can be decomposed into an out-going component E_a^+ and an in-coming component E_a^-, i.e., $E_a \equiv \hat{\mathbf{a}} \cdot \mathbf{E}(x_0, t) = E_a^+ + E_a^-$, where

$$E_a^{\pm} = \frac{1}{2} [\hat{\mathbf{a}} \cdot \mathbf{E}(x_0, t) \pm 2I_1(\mathbf{E}, \hat{\mathbf{a}}; x_0, t) \mp \frac{2}{\epsilon} I_2(\mathbf{H}, \hat{\mathbf{a}}; x_0, t)]. \quad (8.117)$$

Technically, one should verify that if $\mathbf{E}, \mathbf{H}$ are solutions of Maxwell's equations in a region Ω_+ containing the surface S, then the linear combination of their tangential components given by E_a^+ and the corresponding relation for H_a^+ should satisfy the out-going relations (8.113) and (8.114). This would require very extensive analysis, where S, a member of a one-parameter family of smooth surfaces, could be described as a coordinate surface of an orthogonal curvilinear coordinate system. One would need to extract various relationships for the operators I_1 and I_2 (see Chapter 7 for the scalar case of wave-splitting on a curved surface). However, for application to the absorbing boundary condition, as considered in this chapter, this extensive analysis is not needed.

The above results will be used to obtain the appropriate absorbing boundary conditions that can be used for the problem where the medium is non-homogeneous in a compact domain Ω_0, and where the exterior region to Ω_0 is free space, and there are no sources in the external region. (For the case when there are sources exterior to Ω_0, the fields produced by these sources can be subtracted out so that one works only with the scattered field component in the exterior region).

Let (ρ, θ, ϕ) be spherical polar coordinates, and assume that the spherical surface $\rho = \rho_0$ encloses the non-homogeneous and source region Ω_0. The concept of out-going waves in the exterior region can be employed to obtain the absorbing boundary condition.

Essentially if the fields are known on the surface $\rho = \rho_0$, conditions (8.101) and (8.102) can be employed to obtain the tangential fields for points on the surface $\rho = \rho_0 + h$, where h is the grid size in the radial variable. Thus the absorbing boundary conditions are given by

$$\hat{\mathbf{a}} \cdot \mathbf{E}(x,t) = I_1(\mathbf{E}, \hat{\mathbf{a}}; x, t) - \tfrac{1}{\epsilon} I_2(\mathbf{H}, \hat{\mathbf{a}}; x, t), \tag{8.118}$$

$$\hat{\mathbf{a}} \cdot \mathbf{H}(x,t) = I_1(\mathbf{H}, \hat{\mathbf{a}}; x, t) + \frac{1}{\mu} I_2(\mathbf{E}, \hat{\mathbf{a}}; x, t), \tag{8.119}$$

where I_1 and I_2 are given by (8.103) and (8.104), respectively, with x being a point on the absorbing (outer surface) a distance h from the surface S.

In the special case of a spherical surface, (8.118) and (8.119) become

$$\begin{aligned}
\hat{\mathbf{a}} \cdot \mathbf{E}(\rho_0 + h, \theta_0, \phi_0; t) &= -\int_S [\hat{\mathbf{a}} \cdot \mathbf{E}(y, t - r/c)] \frac{\hat{\mathbf{n}} \cdot (y - x)}{4\pi r^3} dS_y \\
&+ \int_S (\hat{\mathbf{a}} \cdot \hat{\mathbf{n}}) \frac{(y - x) \cdot \mathbf{E}(y, t - r/c)}{4\pi r^3} dS_y \\
&+ \int_S \left\{ -[\hat{\mathbf{a}} \cdot \mathbf{E}_t(y, t - r/c)] \frac{\hat{\mathbf{n}} \cdot (y - x)}{4\pi c r^2} + (\hat{\mathbf{a}} \cdot \hat{\mathbf{n}}) \frac{(y - x) \cdot \mathbf{E}_t(y, t - r/c)}{4\pi c r^2} \right\} dS_y \\
&- \int_S \frac{\hat{\mathbf{a}} \cdot (y - x)}{4\pi r^3} [\hat{\mathbf{n}} \cdot \mathbf{E}(y, t - r/c) + \frac{r}{c} \hat{\mathbf{n}} \cdot \mathbf{E}_t(y, t - r/c)] dS_y \\
&- \frac{1}{\epsilon} \int_S \hat{\mathbf{a}} \cdot [\hat{\mathbf{n}} \times \mathbf{H}_t(y, t - r/c)] \frac{1}{4\pi c^2 r} dS_y,
\end{aligned} \tag{8.120}$$

$$
\begin{aligned}
&\hat{\mathbf{a}} \cdot \mathbf{H}(\rho_0 + h, \theta_0, \phi_0; t) = -\int_S [\hat{\mathbf{a}} \cdot \mathbf{H}(y, t - r/c)] \frac{\hat{\mathbf{n}} \cdot (y - x)}{4\pi r^3} dS_y \\
&+ \int_S (\hat{\mathbf{a}} \cdot \hat{\mathbf{n}}) \frac{(y - x) \cdot \mathbf{H}(y, t - r/c)}{4\pi r^3} dS_y \\
&+ \int_S \left\{ -[\hat{\mathbf{a}} \cdot \mathbf{H}_t(y, t - r/c)] \frac{\hat{\mathbf{n}} \cdot (y - x)}{4\pi c r^2} + (\hat{\mathbf{a}} \cdot \hat{\mathbf{n}}) \frac{(y - x) \cdot \mathbf{H}_t(y, t - r/c)}{4\pi c r^2} \right\} dS_y \\
&- \int_S \frac{\hat{\mathbf{a}} \cdot (y - x)}{4\pi r^3} [\hat{\mathbf{n}} \cdot \mathbf{H}(y, t - r/c) + \frac{r}{c} \hat{\mathbf{n}} \cdot \mathbf{H}_t(y, t - r/c)] dS_y \\
&+ \frac{1}{\mu} \int_S \hat{\mathbf{a}} \cdot [\hat{\mathbf{n}} \times \mathbf{E}_t(y, t - r/c)] \frac{1}{4\pi c^2 r} dS_y, \qquad (8.121)
\end{aligned}
$$

where S is a spherical surface with radius ρ_0.

As a simple numerical example, we consider the following vertical electric dipole at the origin in free space,

$$
\begin{cases}
\nabla \times \mathbf{E} = -\mu_0 \partial_t \mathbf{H}, \\
\nabla \times \mathbf{H} = \epsilon_0 \partial_t \mathbf{E} + \hat{\mathbf{i}}_3 \, \delta(x) f(t), \\
\mathbf{E} = \mathbf{H} = 0, \quad t < 0,
\end{cases} \qquad (8.122)
$$

where ϵ_0 and μ_0 are the permittivity and permeability in vacuum, respectively (i.e., $\epsilon_0 = 8.854 \times 10^{-12}$ F/m, $\mu_0 = 1.25644 \times 10^{-6}$ H/m), and $f(t) = \sin(\pi t/\tau_0)$ Am, with $\tau_0 = 7.5^{-9}$ s (here all values are in SI units). The corresponding tangential fields can be obtained explicitly by convolving $f(t)$ with the fundamental solutions. To test the absorbing boundary conditions (8.120) and (8.121), choose $\rho_0 = 2m, \ h = 0.2m$. The values for the fields on the surface $\rho = \rho_0$ are evaluated with the explicit exact formula. Then the absorbing boundary conditions (8.120) and (8.121) are used to calculate the tangential fields on the surface $\rho = \rho_0 + h$, and the calculated values at a point with $\theta_0 = \pi/4$ are shown by the dashed curves in Figs 8.1(a) and (b). The explicit exact values for the non-zero tangential components E_θ and H_ϕ (the other tangential components are identically zero in this example) at this point are plotted by the solid curves in Figs 8.1(a) and (b), respectively. In the numerical evaluation for the surface integrals in (8.120) and (8.121), the surface is descretized with the gridsizes $\Delta\theta = \pi/48$, and $\Delta\phi = \pi/36$. The stepsize in time is $\Delta t = 10^{-10}$ s.

A local approximation to the absorbing boundary conditions (8.120) and (8.121) which is useful for small values of time t is derived in He and Weston (1996). This approximation is obtained by representing the surface locally as a portion of a paraboloid of revolution with radius of curvature A at the apex. Applying this result to a sphere of radius $A = \rho_0$, this approximation is given by

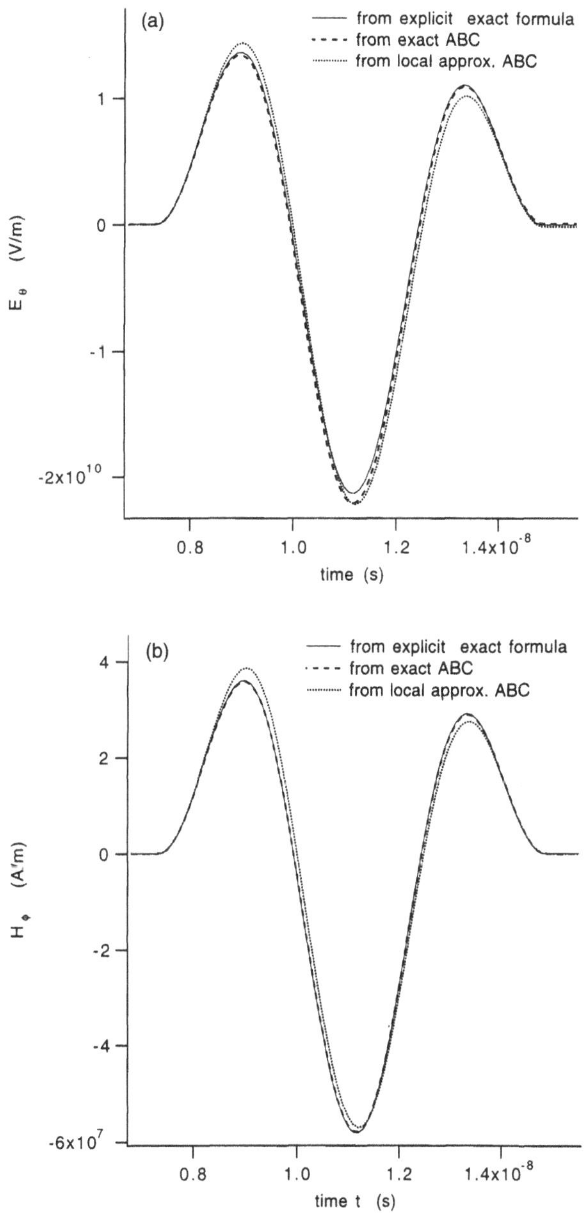

FIG. 8.1. The tangential fields E_θ and H_ϕ at a point with $r = r_0 + h = 2.2$ m and $\theta_0 = \pi/4$. The solid curves show the exact values. The dashed curves show the corresponding fields calculated with the absorbing boundary conditions (8.118) and (8.119). The dotted curves are obtained through the locally approximated absorbing boundary conditions (8.123) and (8.124).

$$\hat{\mathbf{a}}\cdot\mathbf{E}(\rho_0+h,\theta,\phi,t)=\frac{1}{2(1+h/A)}[\hat{\mathbf{a}}\cdot\mathbf{E}(\rho_0,\theta,\phi,t-h/c)]$$

$$+\sqrt{\mu/\epsilon}\frac{1}{2(1+h/A)}\hat{\mathbf{a}}\cdot[\mathbf{H}(\rho_0,\theta,\phi,t-h/c)\times\hat{\mathbf{n}}]$$

$$-\frac{c}{2(1+h/A)^2}\int_0^{t-h/c}\left\{\frac{\hat{\mathbf{a}}\cdot\mathbf{E}(\rho_0,\theta,\phi,s)}{A}+\hat{\mathbf{a}}\cdot\nabla[\hat{\mathbf{n}}\cdot\mathbf{E}(\rho_0,\theta,\phi,s)]\right\}ds, \tag{8.123}$$

$$\hat{\mathbf{a}}\cdot\mathbf{H}(\rho_0+h,\theta,\phi,t)=\frac{1}{2(1+h/A)}[\hat{\mathbf{a}}\cdot\mathbf{H}(\rho_0,\theta,\phi,t-h/c)]$$

$$-\sqrt{\epsilon/\mu}\frac{1}{2(1+h/A)}\hat{\mathbf{a}}\cdot[\mathbf{E}(\rho_0,\theta,\phi,t-h/c)\times\hat{\mathbf{n}}]$$

$$-\frac{c}{2(1+h/A)^2}\int_0^{t-h/c}\left\{\frac{\hat{\mathbf{a}}\cdot\mathbf{H}(\rho_0,\theta,\phi,s)}{A}+\hat{\mathbf{a}}\cdot\nabla[\hat{\mathbf{n}}\cdot\mathbf{H}(\rho_0,\theta,\phi,s)]\right\}ds. \tag{8.124}$$

As a numerical test, the tangential fields on the surface $\rho=\rho_0+h$ are recalculated for the previous numerical example, using these locally approximated absorbing boundary conditions (instead of the exact absorbing boundary conditions (8.120) and (8.121)). The calculated values are shown by the dotted curves in Figs 8.1(a) and (b). As expected, the calculated tangential fields using the locally approximated absorbing boundary conditions are not as good as the ones using the exact absorbing boundary conditions, but are still quite good for this example.

Appendix: Asymptotic behaviour of $\mathbf{E}_T^-$

In this appendix we will show that the negative-going wave $\mathbf{E}_T^-$ inside the inhomogeneous medium has the following property

$$\mathbf{E}_T^-(x,t=\tau)=\frac{1}{2}\gamma(x)\equiv\frac{1}{2}\begin{bmatrix}\gamma_1(x)\\\gamma_2(x)\end{bmatrix}, \tag{8.125}$$

where γ is related to the parameters at the point x, as will be determined below in this section. This condition plays a key role in the reconstruction.

Using the definition of the operator N_{x_3}, one considers the following mixed problem in the half-space $x_3>x_3^0>0$ (cf. (8.63) and (8.64))

(i)
$$\partial_3\overline{\mathbf{E}}_T+\mu_0P\partial_t\overline{\mathbf{H}}_T-\partial_t^{-1}J_\epsilon\overline{\mathbf{H}}_T=0,\quad x_3>x_3^0, \tag{8.126}$$

$$\partial_3\overline{\mathbf{H}}_T-\epsilon P\partial_t\overline{\mathbf{E}}_T+\partial_t^{-1}J_{\mu_0}\overline{\mathbf{E}}_T=0,\quad x_3>x_3^0, \tag{8.127}$$

(ii)
$$\overline{\mathbf{E}}_T(x,0) = \overline{\mathbf{H}}_T(x,0) = 0, \quad x_3 > x_3^0, \tag{8.128}$$

(iii)
$$\overline{\mathbf{H}}_T(\mathbf{x}, x_3^0, t) = \mathbf{H}_T(\mathbf{x}, x_3^0, t) \equiv \mathbf{H}_T^{(0)}\delta(t-\tau) + \mathbf{H}_T^{(1)}\Theta(t-\tau) + \ldots, \; x_3 = x_3^0. \tag{8.129}$$

Then the solution at the limiting plane $x_3 = x_3^0$ will give

$$\overline{\mathbf{E}}_T(\mathbf{x}, x_3^0, t) = +\frac{1}{\epsilon} N_{x_3^0} \mathbf{H}_T, \quad x_3 = x_3^0. \tag{8.130}$$

Note that in the above mixed problem, the medium is not the original physical medium but a medium with only transverse inhomogeneity, i.e. $\partial_3 \epsilon(\mathbf{x}, x_3^0) = 0$, and with $\sigma = 0$.

From (8.35), (8.125) and (8.130), one has

$$\begin{aligned}\gamma(\mathbf{x}, x_3^0) &= 2\mathbf{E}_T^-(\mathbf{x}, x_3^0, \tau) \\ &= \mathbf{E}_T(\mathbf{x}, x_3^0, \tau) - \frac{1}{\epsilon} N_{x_3^0} \mathbf{H}_T(\mathbf{x}, x_3^0, \tau) \\ &= \mathbf{E}_T(\mathbf{x}, x_3^0, \tau) - \overline{\mathbf{E}}_T(\mathbf{x}, x_3^0, \tau).\end{aligned} \tag{8.131}$$

The solution for the mixed problem in $x_3 \geq x_3^0$ has the following structure

$$\begin{aligned}\overline{\mathbf{E}}_T(x,t) = \overline{\mathbf{E}}_T^{(0)}(x)\delta(t-\overline{\tau}) + \overline{\mathbf{E}}_T^{(1)}(x)\Theta(t-\overline{\tau}) \\ +\overline{\mathbf{E}}_T^{(2)}(x)\cdot(t-\overline{\tau})\Theta(t-\overline{\tau}) + \ldots,\end{aligned} \tag{8.132}$$

$$\begin{aligned}\overline{\mathbf{H}}_T(x,t) = \overline{\mathbf{H}}_T^{(0)}(x)\delta(t-\overline{\tau}) + \overline{\mathbf{H}}_T^{(1)}(x)\Theta(t-\overline{\tau}) \\ +\overline{\mathbf{H}}_T^{(2)}(x)\cdot(t-\overline{\tau})\Theta(t-\overline{\tau}) + \ldots.\end{aligned} \tag{8.133}$$

Comparing (8.133) with (8.129) at $x_3 = x_3^0$, yields

$$\overline{\tau} = \tau, \quad x_3 = x_3^0, \tag{8.134}$$

$$\overline{\mathbf{H}}_T^{(i)} = \mathbf{H}_T^{(i)}, \quad i = 0,1,2,\ldots, \quad x_3 = x_3^0. \tag{8.135}$$

From the eikonal equation, one has

$$\partial_3\tau(x) = +(\mu_0\epsilon(x) - |\nabla^T\tau|^2)^{1/2},$$

$$\partial_3\overline{\tau}(x) = +(\mu_0\epsilon(\mathbf{x}, x_3^0) - |\nabla^T\overline{\tau}|^2)^{1/2}$$

($\nabla^T = (\partial_1, \partial_2)$ and $|\nabla^T\tau|^2 = \partial_1^2\tau + \partial_2^2\tau$), which indicate that (cf. (8.134))

$$\partial_3\overline{\tau} = \partial_3\tau, \quad x_3 = x_3^0. \tag{8.136}$$

Substituting (8.67) and (8.68) into (8.63) with η given by (8.58), and matching the coefficient of each singular term, yields

$$-(\partial_3\tau)\mathbf{E}_T^{(0)} + \mu_0 P\mathbf{H}_T^{(0)} - \tfrac{1}{\epsilon}J\mathbf{H}_T^{(0)} = 0, \tag{8.137}$$

$$\partial_3\mathbf{E}_T^{(0)} - (\partial_3\tau)\mathbf{E}_T^{(1)} + \mu_0 P\mathbf{H}_T^{(1)} - J_\epsilon^{(1)}\mathbf{H}_T^{(0)} - \tfrac{1}{\epsilon}J\mathbf{H}_T^{(1)} + \tfrac{\sigma}{\epsilon^2}J\mathbf{H}_T^{(0)} = 0, \tag{8.138}$$

where

$$J = \begin{bmatrix} -\partial_1\tau\partial_2\tau & (\partial_1\tau)^2 \\ -(\partial_2\tau)^2 & \partial_2\tau\partial_1\tau \end{bmatrix}, \tag{8.139}$$

$$J_\epsilon^{(1)}\mathbf{H}_T^{(0)} = \begin{bmatrix} \partial_1[\frac{1}{\epsilon}\partial_2\tau H_1^{(0)}] + \frac{1}{\epsilon}\partial_1\tau\partial_2 H_1^{(0)} & -\partial_1[\frac{1}{\epsilon}\partial_1\tau H_2^{(0)}] - \frac{1}{\epsilon}\partial_1\tau\partial_1 H_2^{(0)} \\ \partial_2[\frac{1}{\epsilon}\partial_2\tau H_1^{(0)}] + \frac{1}{\epsilon}\partial_2\tau\partial_2 H_1^{(0)} & -\partial_2[\frac{1}{\epsilon}\partial_1\tau H_2^{(0)}] - \frac{1}{\epsilon}\partial_2\tau\partial_1 H_2^{(0)} \end{bmatrix}. \tag{8.140}$$

Similarly, from (8.126), (8.132) and (8.133) one has

$$-(\partial_3\overline{\tau})\overline{\mathbf{E}}_T^{(0)} + \mu_0 P\overline{\mathbf{H}}_T^{(0)} - \tfrac{1}{\epsilon}\overline{J}\,\overline{\mathbf{H}}_T^{(0)} = 0, \tag{8.141}$$

$$\partial_3\overline{\mathbf{E}}_T^{(0)} - (\partial_3\overline{\tau})\overline{\mathbf{E}}_T^{(1)} + \mu_0 P\overline{\mathbf{H}}_T^{(1)} - \overline{J}_\epsilon^{(1)}\overline{\mathbf{H}}_T^{(0)} - \tfrac{1}{\epsilon}\overline{J}\,\overline{\mathbf{H}}_T^{(1)} = 0, \tag{8.142}$$

with similar definitions for $\overline{J}$ and $\overline{J}_{\frac{1}{\epsilon}}^{(1)}$ as for J and $J_{\frac{1}{\epsilon}}^{(1)}$ (replacing τ with $\overline{\tau}$). Comparing (8.141) with (8.137) at $x_3 = x_3^0$, one has (cf. (8.134)–(8.136)))

$$\overline{\mathbf{E}}_T^{(0)} = \mathbf{E}_T^{(0)}, \quad x_3 = x_3^0. \tag{8.143}$$

It thus follows from (8.131) and (8.143) that

$$\gamma(\mathbf{x}, x_3^0) = \mathbf{E}_T^{(1)}(\mathbf{x}, x_3^0) - \overline{\mathbf{E}}_T^{(1)}(\mathbf{x}, x_3^0). \tag{8.144}$$

From (8.135), (8.136), (8.138) and (8.142), one has

$$\mathbf{E}_T^{(1)} - \overline{\mathbf{E}}_T^{(1)} = (\partial_3\tau)^{-1}[(\partial_3\mathbf{E}_T^{(0)} - \partial_3\overline{\mathbf{E}}_T^{(0)}) + \frac{\sigma}{\epsilon^2}J\mathbf{H}_T^{(0)}], \quad x_3 = x_3^0. \tag{8.145}$$

Therefore, from the above two equations one obtains

$$\gamma(\mathbf{x}, x_3^0) = (\partial_3\tau)^{-1}[(\partial_3\mathbf{E}_T^{(0)} - \partial_3\overline{\mathbf{E}}_T^{(0)}) + \frac{\sigma}{\epsilon^2}J\mathbf{H}_T^{(0)}]. \tag{8.146}$$

From the $x_j, j = 1, 2$, component of the transport equation (8.79), one has

$$\sum_{i=1}^{3}(\partial_i\tau)\partial_i E_j^{(0)} = -\frac{1}{2}(\nabla^2\tau + \mu_0\sigma)E_j^{(0)} - \frac{\mathbf{E}^{(0)}\cdot\nabla\epsilon}{2\epsilon}\partial_j\tau, \quad j = 1, 2. \tag{8.147}$$

Similarly, one has (from the corresponding transport equation for $\overline{\mathbf{E}}_T^{(0)}$)

$$\sum_{i=1}^{3}(\partial_i\overline{\tau})\partial_i\overline{E}_j^{(0)} = -\frac{1}{2}(\nabla^2\overline{\tau})\overline{E}_j^{(0)} - \frac{\overline{\mathbf{E}}^{(0)}\cdot\nabla\epsilon}{2\epsilon}\partial_j\overline{\tau}, \quad j = 1, 2. \tag{8.148}$$

From the above two equations with $x_3 = x_3^0$, one obtains (cf. (8.144), (8.134) and (8.136))

$$\begin{aligned}&(\partial_3\tau)[\partial_3 E_j^{(0)} - \partial_3\overline{E}_j^{(0)}]\\ &= -\frac{1}{2}[(\partial_3^2\tau - \partial_3^2\overline{\tau}) + \mu_0\sigma]E_j^{(0)} - \frac{E_3^{(0)}\partial_3\epsilon}{2\epsilon}\partial_j\tau, \ j = 1, 2, \ x_3 = x_3^0.\end{aligned} \tag{8.149}$$

Since

$$\begin{aligned}\partial_3^2\tau &= +\frac{1}{2}(\mu_0\epsilon - |\nabla^T\tau|^2)^{-1/2}\{\mu_0\partial_3\epsilon - 2[(\partial_1\tau)\partial_3(\partial_1\tau) + (\partial_2\tau)\partial_3(\partial_2\tau)]\}\\ &= \frac{1}{2}(\partial_3\tau)^{-1}\{\mu_0\partial_3\epsilon - 2[(\partial_1\tau)\partial_1(\partial_3\tau) + (\partial_2\tau)\partial_2(\partial_3\tau)]\}\\ \partial_3^2\overline{\tau} &= \frac{1}{2}(\partial_3\overline{\tau})^{-1}\{-2[(\partial_1\overline{\tau})\partial_1(\partial_3\overline{\tau}) + (\partial_2\overline{\tau})\partial_2(\partial_3\overline{\tau})]\},\end{aligned}$$

one has

$$\partial_3^2\tau - \partial_3^2\overline{\tau} = \frac{1}{2}\mu_0(\partial_3\tau)^{-1}\partial_3\epsilon, \quad x_3 = x_3^0. \tag{8.150}$$

Therefore, from (8.147), (8.149) and (8.150) one obtains

$$\gamma = \begin{bmatrix}\omega_1\\ \omega_2\end{bmatrix} + \frac{\sigma}{\epsilon^2}(\partial_3\tau)^{-1}J\mathbf{H}_T^{(0)}, \tag{8.151}$$

where

$$\omega_j = -\frac{1}{2}(\partial_3\tau)^{-2}\left\{\left[\frac{\mu_0}{2}(\partial_3\tau)^{-1}E_j^{(0)} + \frac{1}{\epsilon}\partial_j\tau E_3^{(0)}\right]\partial_3\epsilon + \mu_0\sigma E_j^{(0)}\right\}, \quad j = 1, 2. \tag{8.152}$$

Exercises for Chapter 8

Exercise 8.1. Derive the local approximation to the absorbing boundary conditions as given by (8.123) and (8.124) by approximating the surface by a paraboloid of revolution, $y_3 = -(y_1^2 + y_2^2)/(2A)$.

Exercise 8.2. Derive the explicit expression (8.46) for the operator Γ_α for the case when ϵ, μ depend upon the parameter α only, i.e., $\epsilon = \epsilon(\alpha), \mu = \mu(\alpha)$.

(1) From (8.28) deduce that

$$\frac{\partial}{\partial \alpha}\left(\epsilon^{-1}(\alpha) N_\alpha\right) = \left[-\frac{\epsilon'}{\epsilon} + \mathbf{K}'_\alpha \mathbf{K}_\alpha^{-1}\right] \frac{1}{\epsilon} N_\alpha - \frac{2c'}{\epsilon c^3} \mathbf{K}_\alpha P \partial_t, \tag{8.153}$$

where the matrix P is given by (8.33), ϵ', c' are the derivatives of ϵ, c with respect to the parameter α, and $\mathbf{K}'_\alpha$ is the derivative of the operator $\mathbf{K}_\alpha$ with respect to the parameter α.

(2) From (8.29) and (8.43) deduce that

$$\Gamma_\alpha = \left[\frac{\partial}{\partial \alpha}\left(\epsilon^{-1}(\alpha) N_\alpha\right)\right] \mu^{-1}(\alpha) N_\alpha, \ \alpha = x_3^0. \tag{8.154}$$

Hence, employing the inverse relation (8.7) combined with (8.154) and the results of part (1), show that

$$\Gamma_\alpha = \left(\frac{\epsilon'}{\epsilon} - \mathbf{K}'_\alpha \mathbf{K}_\alpha^{-1}\right) I + \frac{2c'}{c} \mathbf{K}_\alpha^2 \begin{bmatrix} (\Box^T + \partial_1^2) & \partial_1 \partial_2 \\ \partial_1 \partial_2 & (\Box^T + \partial_2^2) \end{bmatrix}.$$

Hence (8.46) is obtained by using the relationship (7.32), and replacing α with x_3.

REFERENCES

Åberg, I. and Karlsson, A. (1997). The source problem – transient waves propagating from internal sources in non-stationary media. *Wave Motion*, **26**, (1), 43–68.

Åberg, I., Kristensson, G., and Wall, D. J. N. (1995). Propagation of transient electromagnetic waves in time-varying media – direct and inverse scattering problems. *Inverse Problems*, **11**, (1), 29–49.

Barnes, P. R. and Tesche, F. M. (1991). On the direct calculation of a transient plane wave reflected from a finitely conducting half-space. *IEEE Transactions on Electromagnetic Compatibility*, **33**, (2), 90–6.

Beezley, R. S. and Krueger, R. J. (1985). An electromagnetic inverse problem for dispersive media. *Journal of Mathematical Physics*, **26**, (2), 317–25.

Bellman, R. and Wing, G. M. (1975). An Introduction to Invariant Imbedding. Wiley, New York.

Bennia, A. and Riad, S. M. (1990). An optimization technique for iterative frequency-domain deconvolution. *IEEE Transactions on Instrumentation and Measurement*, **39**, (2), 368–72.

Bernekorn, P., Karlsson, A., and Kristensson, G. (1996). Propagation of transient electromagnetic waves in inhomogeneous and dispersive waveguides. *Journal of Electromagnetic Waves and Applications*, **10**, (9),1263–86.

Bleistein, N. (1984). Mathematical Methods for Wave Phenomena. Academic Press, New York.

Bloom, C.O. and Kazarinoff, N.D. (1976). Short Wave Radiation Problems in Homogeneous Media: Asymptotic Solutions. Springer, New York.

Bohren, C. F. and Huffman, D. R. (1983). Absorption and Scattering of Light by Small Particles. Wiley, New York.

Born, M. and Wolf, E. (1980). Principles of Optics: Electromagnetic Theory of Propagation, Interference and Diffraction of Light. Pergamon Press, Oxford.

Braga, A. M. B., Honein, B. Barbone, P. E., and Herrmann, G. (1992). Suppression of sound reflected from a piezoelectric plate. *Journal of Intelligent Material Systems and Structures*, **3**, 209–223.

Brillouin, L. (1960). Wave Propagation and Group Velocity. Academic Press, New York.

Buzdin, A. A. (1971). On an inverse problem in a medium with absorption. *Proceedings of the Steklov Institute of Mathematics*, No. 115, 77–85.

Cao, J. and He, S. (1994). Closed-form solution for the transient reflected pressure for a point source above an acoustic half-space with an exponentially stratified density. *Journal of the Acoustical Society of America*, **96**, (4), 2516–25.

Cao, J. and He, S. (1996). An exact absorbing boundary condition and its application to 3D scattering from thin dispersive structures. *Journal of the Acoustical Society of America*, **99**, (4), 1854–61.

Cao, J. and He, S. (1997). Reconstruction of the velocity and density in a stratified acoustic half-space using a short-pulse point source. *Journal of the Acoustical Society of America*, **102**, (2), 815–24.

Cao, J., He, S., and Weston, V. H. (1996). An ill-posed problem of the continuation of transient data for a hyperbolic equation in a three-dimensional inhomogeneous half-space. *Journal of Mathematical Physics*, **37**, (11), 5776–91.

Cao, J., He, S., and Weston, V. H. (1998). Parameter reconstruction for a hyperbolic equation in a three-dimensional inhomogeneous half-space. Technical Report TRITA-TET 98-3, Royal Institute of Technology, S-100 44 Stockholm, Sweden.

Chandrasekhar, S. (1960). Radiative Transfer. Dover, New York.

Chang, S. and Weston, V. H. On the inverse problem of the 3-D telegraph equation. *Inverse Problems*, **13**, (5), 1207–21.

Cheng, D. K. (1993). Fundamentals of Engineering Electromagnetics. Addison-Wesley, Massachusetts.

Chew, C. W. and Otto, G. P. (1992). Microwave imaging of multiple conducting cylinders using local shape functions. *IEEE Microwave and Guided Wave Letters*, **2**, (7), 284–6.

Colton, D. and Kress, R. (1992). Inverse Acoustic and Electromagnetic Scattering Theory. Springer-Verlag, Berlin.

Corones, J. and Krueger, R. (1983). Obtaining scattering kernels using invariant imbedding. *Journal of Mathematical Analysis and Applications*, **95**, (2), 393–415.

Corones, J. and Sun, Z. (1993). Simultaneous reconstruction of material and transient source parameters using the invariant imbedding method. *Journal of Mathematical Physics*, **34**, (5), 1824–45.

Corones, J., Davison, M. E., and Krueger, R. J. (1983a). Direct and inverse scattering in the time domain via invariant imbedding equations. *Journal of the Acoustical Society of America*, **74**, (5), 1535–41.

Corones, J., Davison, M. E., and Krueger, R. J. (1983b). Wave splittings, invariant imbedding and inverse scattering. In *Inverse optics* (proceedings of the SPIE, No. 413, edited by A. J. Devaney), 102–106, SPIE, Bellingham, WA.

Corones, J. P., Krueger, R. J., and Weston, V. H. (1985). Some recent results in inverse scattering theory. In *Inverse problems of acoustic and elastic waves*, (ed. F. Santosa, Y. Pao, W. Symes, and C. Holland), pp. 65–81, Proc. SPIE 413, SIAM Philadelphia, PA.

Corones, J. P., Kristensson, G., Nelson, P., and Seth, D. L. (eds.). (1992). Invariant imbedding and inverse problems. SIAM (Society for Industrial and Applied Mathematics) Proceedings, Philadelphia.

Cory, H. (1995). Chiral devices – an overview of canonical problems. *Journal of Electromagnetic Waves and Applications*, **9**, (5/6), 805–30.
Courant, R. and Hilbert, D. (1962a). Methods of Mathematical Physics. Part I. Interscience, New York,
Courant, R. and Hilbert, D. (1962b). Methods of Mathematical Physics. Part II. Interscience, New York.
de Hoop, M. V. (1996). Generalization of the Bremmer coupling series. *Journal of Mathematical Physics*, **37**, (7), 3246–82.
de Hoop, M. V. and de Hoop, A. T. (1994). Elastic wave up/down decomposition in inhomogeneous and anisotropic media: An operator approach and its approximations. *Wave Motion*, **20**, (1), 57–82.
Engquist, B. and Majda, A. (1977). Absorbing boundary conditions for the numerical simulation of wave. *Mathematics of Computation*, **31**, 629–51.
Fishman, L. and McCoy, J. J. (1984a). Derivation and application of extended parabolic wave theories. I. The factorized Helmholtz equation. *Journal of Mathematical Physics*, **25**, (2), 285–96.
Fishman, L. and McCoy, J. J. (1984b). Derivation and application of extended parabolic wave theories. II. Path integral representations. *Journal of Mathematical Physics*, **25**, (2), 297–308.
Fishman, L., Gautesen, A. K. and Sun, Z. (1997). Uniform high-frequency approximations of the square root Helmholtz operator symbol. *Wave Motion*, **26**, (2), 127–61.
Forslund, O. and He, S. (1998). Electromagnetic scattering from an inhomogeneous grating using a wave-splitting approach. *Progress in Electromagnetics Research*, **19**, 147–71.
Frank, M. S. and Balanis, C. A. (1987). A conjugate direction method for geophysical inversion problems. *IEEE Transactions on Geoscience and Remote Sensing*, **GE-25**, 691–701.
Friedlander, F. G. (1958). Sound Pulses. Cambridge University Press, Cambridge.
Fuks, P., Kristensson, G., and Larson, G. (1990). Permittivity profile reconstructions using transient electromagnetic reflection data. Technical Report TRITA-TET 90-6, Department of Electromagnetic Theory, Royal Institute of Technology, Stockholm, Sweden.
Fuks, P., Karlsson, A., and Larson, G. (1994). Direct and inverse scattering from dispersive media, *Inverse Problems*, (10), 555–71.
Garnero, L., Franchois, A., Hugonin, J-P., Pichot, C., and Joachimowicz, N. (1991). Microwave imaging – complex permittivity reconstruction by simulated annealing. *IEEE Transactions on Microwave Theory and Techniques*, **39**, (11), 1801–07.
Givoli, D. (1992). Numerical Methods for Problems in Infinite Domains. Elsevier, Amsterdam.
Günter, N.M. (1967). Potential Theory. Ungar, New York.
Gustafsson, M. and He, S. (1997). An optimization approach to multi-dimensional

time-domain acoustic inverse problems. Technical Report TRITA-TET 97-12, Royal Institute of Technology, Stockholm, Sweden.

Gustafsson, M. and He, S. (1998). An optimization approach to multi-dimensional time-domain inverse problems for Maxwell's equations. Technical Report TRITATET 98-4, Royal Institute of Technology, Stockholm, Sweden.

Harris, F. J. (1978). On the use of windows for harmonic analysis with the discrete Fourier transform. *Proceedings of the IEEE*, **66**, (1), 51–83.

He, S. (1992a). Factorization of a dissipative wave equation and the Green's functions technique for axially symmetric fields in a stratified slab. *Journal of Mathematical Physics*, **33**, (3), 953–66.

He, S. (1992b). A time-harmonic Green's function technique and wave propagation in a stratified nonreciprocal chiral slab with multiple discontinuities. *Journal of Mathematical Physics*, **33**, (12), 4103–10.

He, S. (1993a). A 'compact Green's function' approach to the time domain direct and inverse problems for a stratified dissipative slab. *Journal of Mathematical Physics*, **34**, (10), 4628–45.

He, S. (1993b). Frequency and time domain Green's function technique for nonuniform LCRG transmission lines with frequency-dependent parameters. *Journal of Electromagnetic Waves and Application*, **7**, (1), 31–48.

He, S. (1993c). Wave propagation through a dielectric-uniaxial bianisotropic interface and the computation of Brewster angles. *Journal of the Optical Society of America*, **10**, (11), 2402–9.

He, S. (1994a). An inverse approach for an anisotropic wave equation in a stratified half-space. *Journal of Inverse Ill-posed Problems*, **2**, (1), 49–60.

He, S. (1994b). The correct ETLC model for anisotropic lossy materials – a comment on 'Transient Propagation in Anisotropic Laminated Composites'. *IEEE Transactions on Electromagnetic Compatibility*, **36**, (4), 409–10.

He, S. (1994c). Time-domain eigenmodes of Maxwell's equations for stratified anisotropic media and reconstruction of the permittivity tensor using 3-D reflectivity. *Journal of Inverse Ill-posed Problems*, **2**, (2), 117–30.

He, S. (1995a). A useful fractional linear transformation for the Riccati equation for reflection coefficients. *Microwave and Optical Technology Letters*, **8**, (6), 312–4.

He, S. (1995b). Closed-form solution for lossy exponential transmission line problem in frequency and time domains. *Journal of Electromagnetic Waves and Applications*, **9**, (4), 521–540.

He, S. (1996a). Reconstruction of source distributions on a nonuniform LCRG line using an optimization approach. *Journal of Electromagnetic Waves and Applications*, **10**, (3), 405–22.

He, S. (1996b). A comment on 'Propagation of ultra-wide-band electromagnetic pulses through dispersive media'. *IEEE Transactions on Electromagnetic Compatibility*, **38**, (2), 202–5.

He, S. (1996c). Newton–Kantorovich approach to an inverse problem for a quasi-linear acoustic wave equation. *Journal of Inverse and Ill-posed Problems*, **4**,

(3), 223–32.

He, S. (1997a). Wave-splitting approach to a scattering problem for a laterally periodic inhomogeneous structure. *Journal of Electromagnetic Waves and Applications*, **11**, 633–44.

He, S. (1997b). A closed-form formula for scattering coefficients for a lossy multilayered structure, *Microwave and Optical Technology Letters*, **14**, (3), 159–63.

He, S. (1998). Electromagnetic surface waves for some artificial bianisotropic media. *Journal of Electromagnetic Waves and Applications*, **12**, 449–66.

He, S. An explicit asymptotic solution for scattering from an inhomogeneous conducting thin slab. (personal communication.)

He, S. and Hu, Y. (1993). Electromagnetic scattering for a stratified biisotropic (nonreciprocal chiral) slab: numerical calculations. *IEEE Transactions on Antennas and Propagation*, **41**, (8), 1057–62.

He, S. and Kabanikhin, S. I. (1995). An optimization approach to a three-dimensional acoustic inverse problem in the time domain. *Journal of Mathematical Physics*, **36**, (8), 4028–43.

He, S. and Karlsson, A. (1993). Time domain Green function technique for a point source over a dissipative stratified half-space. *Radio Science*, **28**, (4), 513–26.

He, S. and Ström, S. (1992). Time domain wave splitting approach to transmission along a nonuniform LCRG line. *Journal of Electromagnetic Waves and Applications*, **6**, (8), 995–1014.

He, S. and Ström, S. (1995a). Calculation of the transient reflection from a finitely conducting half-space using time domain propagating modes. *IEEE Transactions on Electromagnetic Compatibility*, **37**, (2), 277–82.

He, S. and Ström, S. (1995b). Reconstruction of depth-dependent flow in a moving half-space using transient acoustic plane waves. *Journal of the Acoustical Society of America*, **98**, (3), 1778–85.

He, S. and Ström, S. (1995c). An optimization approach to the time-domain electromagnetic inverse problem for a bi-isotropic slab. *Proceedings of the IEE: Microwaves, Antennas and Propagation*, **142**, (6), 429–34.

He, S. and Ström, S. (1996). Time-domain wave splitting and propagation in dispersive media. *Journal of the Optical Society of America*, **13**, (11), 2200–7.

He, S. and Weston, V. H. (1992). Inverse problem for the dissipative wave equation in a stratified half-space and linearization of the imbedding equation. *Inverse Problems*, **8**, (3), 435–55.

He, S. and Weston, V. H. (1995). Determination of the permittivity and conductivity in R^3 using wave splitting of Maxwell's equations. *Journal of Mathematical Physics*, **36**, (4), 1776–89.

He, S. and Weston, V.H. (1996). Wave-splitting and absorbing boundary conditions for Maxwell's equations on a curved surface. Technical Report TRITA-TET 96-13, Royal Institute of Technology, Stockholm, Sweden.

He, S. and Weston, V.H. (1997). Three-dimensional electromagnetic inverse scattering for bi-isotropic dispersive media, *Journal of Mathematical Physics*, **38**, (1), 182–95.

He, S., Hellberg, R., and Weston, V. H. (1993a). Simultaneous reconstruction for the telegraph equation in a stratified half-space using 3-D reflectivity. *IEICE Transactions on Communications*, **E76-B**, (12), 1538–45.

He, S., Hu, Y., and Ström, S. (1993b). Time domain Green's functions technique for a point source over a dissipative stratified half-space with a phase velocity mismatch at the surface. *Wave Motion*, **17**, 241–54.

He, S., Kabanikhin, I., Romanov, V. G., and Ström, S. (1993c). Analysis of the Green's function approach to one-dimensional inverse problems. Part I: One parameter reconstruction. *Journal of Mathematical Physics*, **34**, (12), 5724–46.

He, S. Hu, Y., and Ström, S. (1994a). Electromagnetic scattering from a stratified bianisotropic slab. *IEEE Transactions on Antennas and Propagation*, **42**, (6), 856–58.

He, S., Norgren, M., and Ström, S. (1994b). Wave propagation in a stratified chiral slab with multiple discontinuities: oblique incidence. PIER (Progress in Electromagnetics Research) 9: Bianisotropic and Bi-isotropic Media and Applications, Chapter 8, A. Priou (ed.), EMW Publishing, Massachusetts.

He, S., Romanov, V.G., and Ström, S. (1994c). Analysis of the Green's function approach to one-dimensional inverse problems. Part II: Simultaneous reconstruction of two parameters, *Journal of Mathematical Physics*, **35**, (5), 2315–35.

He, S., Fuks, P., and Larson, G. (1996). An optimization approach to time-domain electromagnetic inverse problem for a stratified dispersive and dissipative slab. *IEEE Transaction on Antennas and Propagation*, **44**, (9), 1277–82.

He, S., Norgren, M., and Takenaka, T. (1997) Trace formalism and explicit gradients for parameter reconstruction/design of a stratified bianisotropic slab/coating. Technical Report TRITA-TET 97-7, Royal Institute of Technology, Stockholm, Sweden.

Hellberg, R. (1995). Design of reflectless slabs for obliquely incident transient electromagnetic waves. Technical Report TRITA-TET 95-12, Royal Institute of Technology, Stockholm, Sweden.

Hellberg, R., Karlsson, A., and Thärning, P. (1992). Non-reflecting dispersive media. *Smart Materials and Structures*, **1**, 341–6.

Hille, E. (1969). Lectures on Ordinary Differential Equations. Addison-Wesley, Massachusetts.

Ishimaru, A. (1978). Wave Propagation and Scattering in Random Media. Academic Press, New York.

Jackson, J. D. (1975). Classical Electrodynamics. 2nd edition. Wiley, New York.

Jordan, A. K. and Lakshmanasamy, S. (1989). Inverse scattering theory applied to the design of single-mode planar optical waveguides. *Journal of the*

Optical Society of America, **6**, (8), 1206–12.

Joseph, R. M., Hagness, S. C., and Taflove, A. (1991). Direct time integration of Maxwell's equations in linear dispersive media with absorption for scattering and propagation of femtosecond electromagnetic pulses. *Optics Letters*, **16**, (18), 1412–4.

Kantorovich, L. V. and Akilov, G. P. (1964). Functional Analysis in Normed spaces. Trans. by D. E. Brown, ed. by A. P. Robertson, Pergamon Press, New York.

Karlsson, A. (1989). Direct and inverse electromagnetic scattering from a dispersive medium. Technical Report TRITA-TET 89-2, Royal Institute of Technology, Stockholm, Sweden.

Karlsson, A. (1996). Wave propagators for transient waves in one-dimensional media. *Wave Motion*, **24**, (1), 85–99.

Karlsson, A. and Kreider, K. (1996). Transient electromagnetic wave propagation in transverse periodic media. *Wave Motion*, **23**, (3), 259–77.

Karlsson, A. and Kristensson, G. (1992). Constitutive relations, dissipation and reciprocity for the Maxwell's equations in the time domain. *Journal of Electromagnetic Waves and Applications*, **6**, (5/6), 537–51.

Karlsson, A. and Rikte, S. (1997). The time-domain theory of forerunners. Technical Report LUTEDX/(TEAT-7054)/1-31/(1997), Department of Electromagnetic Theory, Lund Institute of Technology, S-21100 Lund, Sweden, 1997 (to appear in *Journal of the Optical Society of America, A*).

Karlsson, A. and Stewart, R. (1995). Wave propagators for transient waves in periodic media. *Journal of the Optical Society of America, A*, **12**, (9), 1513–21.

Karlsson, A., Otterheim, H., and Stewart, R. (1993a). Electromagnetic fields in an inhomogeneous plasma from obliquely incident transient waves. *Radio Science*, **28**, (3), 365–78.

Karlsson, A., Otterheim, H., and Stewart, R. (1993b). Transient wave propagation in composite media: A Green's functions approach. *Journal of the Optical Society of America, A*, 10, 886–95.

Kleinman, R. E. and van den Berg, P. M. (1990). Non-linearized approach to profile inversion. *International Journal of Imaging Systems and Technology*, **2**, 119–26.

Kline, M. and Kay, J.W. (1965). Electromagnetic Theory and Geometric Optics. Interscience Publishers.

Kong, J. A. (1986). Electromagnetic Wave Theory. Wiley, New York.

Krasnoselskii, M. A., Vainikko, G. M., Zabreiko, P. P., Rutitskii, Ya. B., and Stetsenko, V. Ya. (1972). Approximate Solution of Operator Equations. Wolters-Noordhoff Publishing Groningen.

Kreider, K. L. (1989a). Time-dependent inverse scattering from gradient-type interfaces using an exact solution. *Journal of Mathematical Physics*, **30**, (1), 53–8.

Kreider, K. (1989b). A wave splitting approach to the time dependent inverse problem for the stratified cylinder. *SIAM Journal of Applied Mathematics*, **49**, (3), 932–43.

Kristensson, G. (1995). Transient electromagnetic wave propagation in waveguides. *Journal of Electromagnetic Waves and Applications*, **9**, (516), 645–71.

Kristensson, G. and Krueger, R. J. (1986a). Direct and inverse scattering in the time domain for a dissipative wave equation. I. Scattering operators. *Journal of Mathematical Physics*, **27**, (6), 1667–82.

Kristensson, G. and Krueger, R. J. (1986b). Direct and inverse scattering in the time domain for a dissipative wave equation. II. Simultaneous reconstruction of dissipation and phase velocity profiles. *Journal of Mathematical Physics*, **27**, (6) 1683–93.

Kristensson, G. and Krueger, R. J. (1987). Direct and inverse scattering in the time domain for a dissipative wave equation. III. Scattering operators in the presence of a phase velocity mismatch. *Journal of Mathematical Physics*, **28**, (2), 360–70.

Kristensson, G. and Krueger, R. J. (1989). Direct and inverse scattering in the time domain for a dissipative wave equation. Part 4: use of phase velocity mismatches to simplify inversions. *Inverse Problems*, **5**, 375–89.

Kristensson, G. and Rikte, S. (1993). The inverse scattering problem for a homogeneous bi-isotropic slab using transient data. *Inverse Problems in Mathematical Physicas*, L. Päivärinta and E. Somersalo (eds.), 112–25, Springer, Berlin.

Krueger, R. J. and Ochs, R. L. (1989). A Green's function approach to the determination of internal fields. *Wave Motion*, **11**, (6), 525–43.

Ladyzhevskaya, O.A. (1985). The Boundary Value Problems of Mathematical Physics. Springer-Verlag, New York.

Levy, M.F. (1997). Parabolic wave propagation techniques for radiowave propagation. *The Radio Science Bulletin*, No. 282, September, 6–13.

Lindell, I. V. (1993). Application of the image concept in electromagnetics. URSI Review of Radio Science 1990–1992 (ed. W.R. Stone), Oxford University Press, 107–26.

Lindell, I. V., Sihvola, A. H., Tretyakov, S.A., and Viitanen, A. J. (1994). Electromagnetic Waves in Chiral and Bi-isotropic Media. Artech House, Boston.

Lu, S. Y., and Berryman, J. G. (1990). Inverse scattering, seismic traveltime tomography and neural networks. *International Journal of Imaging Systems and Technology*, 2, 112–8.

Luenberger, D. G. (1969). Optimization by Vector Space Methods. Wiley, New York.

Lundstedt, L. (1995a). Condition for distortionless transmission line with a nonuniform characteristic impedance. *IEEE Transactions on Microwave Theory and Techniques*, **43**, (6), 1386–9.

Lundstedt, L. (1995b). On the reconstruction of the electrical parameters of a nonuniform LCRG transmission line from its transient response. Technical Report TRITA-TET 95-11, Royal Institute of Technology, Stockholm,

Sweden.

Lundstedt, J. and He, S. (1994). Signal restoration after transmission through a nonuniform LCRG line. *IEEE Transactions on Microwave Theory and Techniques*, **42**, (11), 2087–92.

Lundstedt, J. and He, S. (1996). A time-domain optimization technique for the simultaneous reconstruction of the characteristic impedance, resistance and conductance of a transmission line. *Journal of Electromagnetic Waves and Application*, **10**, (4), 581–601.

Lundstedt, J. and He, S. (1997). Time domain direct and inverse problems for a nonuniform LCRG line with internal sources. *IEEE Transactions on Electromagnetic Compatibility*, **39**, (2), 79–88.

Lundstedt, J. and Ström, S. (1996). Simultaneous reconstruction of two parameters from the transient response of a nonuniform LCRG transmission line. *Journal of Electromagnetic Waves and Application*, **10**, (1),19–50.

Marx, E. and Maystre, D. (1982) Dyadic Green's Function for the time-dependent wave equation, *Journal of Mathematical Physics*, **23**, 1047–56.

Morse, P. M. and Feshbach, H. (1953). Methods of Theoretical Physics. Part I, McGraw-Hill, New York.

Nahman, S. N. (1986). Software correction of measured pulse data. in *Fast Electrical and Optical Measurements*, Vol. 1, Thompson, J. E. and Luessen, L. H. (eds.), Nijhoff, Dordrecht, 351–418.

Nahman, N. S. and Guillaume, M. E. (1981). Deconvolution of time domain waveforms in the presence of noise. NBS Technical note, No. 1047.

Nilsson, J. W. (1990). Electric Circuits. Addison-Wesley, Reading, Massachusetts.

Norgren, M. (1996). Optimal design using stratified bianisotropic media. Technical Report TRITA TET 96-8, Royal Institute of Technology, Stockholm, Sweden.

Norgren, M. (1997). General optimization approach to a frequency-domain inverse problem for a stratified bianisotropic slab. *Journal of Electromagnetic Waves and Applications*, **11**, 515–46.

Norgren, M. and He, S. (1994). Electromagnetic reflection and transmission for a dielectric-Ω interface and a Ω slab. *International Journal of Infrared and MM Waves*, **15**, (9), 1537–54.

Norgren, M. and He, S. (1995). A general scheme for the electromagnetic reflection and transmission for composite structures of complex media. *Proceedings of IEE: Microwaves, Antennas and Propagation*, **142**, (1), 52–6.

Norgren, M. and He, S. (1996a). Optimal designs for nonuniform LCRG transmission lines. *Journal of Electromagnetic Waves and Applications*, **10**, 1113–27.

Norgren, M. and He, S. (1996b). An optimization approach to the frequency-domain inverse problem for a nonuniform LCRG transmission line. *IEEE Transactions on Microwave Theory and Techniques*, **44**, (8), 1503–7.

Norgren, M. and He, S. (1997). On the possibility of a reflectionless coating of a homogeneous bianisotropic layer on a perfect conductor. *Electromagnetics*,

17, (4), 295–307.

Nucci, C. A., Rachidi, F., Ianoz, M. V., and Mazzetti, C. (1993). Lightning-induced voltages on overhead lines. *IEEE Transactions on Electromagnetic Compatibility*, **35**, (1), 75–86.

Oberhettinger, F. and Badii, L. (1973). Tables of Laplace Transforms. Springer-Verlag, New York.

Parruck, B. and Riad, S. M. (1984). Study and performance of two iterative frequency-domain deconvolution techniques. *IEEE Transactions on Instrumentation and Measurement*, **33**, (4), 281–7.

Petit, R. (ed.) (1980). Electromagnetic Theory of Gratings. Springer-Verlag, Berlin.

Pogorzelski, W. (1966). Integral Equations and Their Applications. Vol. I, Pergamon, Oxford.

Polack, E. (1971). Computational Methods in Optimization – a Unified Approach. Academic Press, New York.

Poovendran, R., Dorband, J. E., and Yagle, A. E. (1995). An algorithm for forward and inverse scattering in time domain. *Journal of the Acoustical Society of America*, **97**, (4), 2021–7.

Rabiner, L. R. and Schafer, R. W. (1969). The Chirp z-Transform Algorithm, *IEEE Transactions on Audio Electroacoust*, AU-17, 86–92.

Redheffer, R. M. (1962). On the relation of transmission-line theory to scattering and transfer. *Journal of Mathematics and Physics*, **41**, 1–41.

Ren, H. (1995). Numerical solution of the continuation problem for a hyperbolic differential equation. Ph. D. Thesis, Purdue University.

Ren, H. and Weston, V.H. (1995). A Cauchy–type problem for the wave equation. *Inverse Problems*, **11**, 439–62.

Richtmyer, R. D. and Morton, K. W. (1967). Finite-Difference Methods for Initial-Value Problems. Chapter 10, Interscience, New York.

Rikte, S. Reconstruction of bi-isotropic material parameters using transient electromagnetic fields. *Wave Motion*, (in press).

Robin, T., Souillard, B., Accolas, R., Theret, P., and Bolioli, S. (1994). Modelling of electromagnetic properties of helix-loaded chiral media. Chiral'94, 3rd International Workshop on chiral, bi-isotropic and bi-anisotropic media, pp. 103–108, F. Mariotte and J-P. Parneix (eds.), Perigueux, France, May 18–20.

Romanov, V. G. and He, S. (1995). An analysis of the time-domain electromagnetic inverse problem of determining the susceptibility kernel for a stratified dispersive slab. *Journal of Inverse and Ill-posed Problems*, **3**, (6), 469–94.

Romanov, V. G. and Kabanikhin, S. I. (1994). Inverse Problems of Geoelectrics. VNU Science Press, Utrecht.

Rothwell, E. J. and Sun, W. (1990). Time domain deconvolution of transient radar data. *IEEE Transactions on Antennas and Propagation*, **AP-38**, 470–5.

Saadoun, M. M. I. and Engheta, N. (1992). A reciprocal phase shifter using a

novel pseudochiral or Ω medium. *Microwave and Optical Technology Letters*, **5**, (4), 184–7.
Schulz-DuBois, E. O. (1970). Sommerfeld pre and postcursors in the context of waveguide transients. *IEEE Transactions on Microwave Theory and Techniques*, **18**, (8), 455–60.
Scott, M. R. (1973). Invariant imbedding and its application to ordinary differential equations. Addison Wesley, Massachusetts.
Silvester, P.P. and Ferrari, R.L. (1990). Finite Elements for Electrical Engineerings. 2nd edition, Cambridge University Press, Cambridge.
Stewart, R. (1989). Transient reflectivity for anisotropic inhomogeneous media. Ph. D. thesis, Iowa University.
Stratton, J. A. (1941). Electromagnetic Theory, McGraw-Hill, New York.
Sun, Z. (1992). Direct scattering and reconstruction of internal source. *Wave Motion*, **16**, 249–63.
Taflove, A. (1995). Computational Electrodynamics – The Finite Difference Time Domain Method, Artech House, Massachusetts.
Tai, C. (1971). Dyadic Green's Functions in Electromagnetic Theory. Intext, Scranton, PA.
Takenaka, T., Tanaka., T., Harada, H., and He, S. (1997). FDTD approach to time-domain inverse scattering problem for stratified lossy media. *Microwave and Optical Technology Letters*, **16**, (5), 292–6.
Tikhonov, A.N. and Arsenine, V.Y., (1977). Solutions of Ill-Posed Problems. Winston, Washington, DC.
Tricomi, F.G. (1957). Integral Equations. Interscience, New York.
Vladimirov, V.S. (1971). Equations of Mathematical Physics. Dekker, New York.
Vogel, C. R. (1986). A Banach space ODE framework for inverse scattering via invariant imbedding. *Journal of Mathematical Analysis and Applications*, **118**, (1), 107–14.
Weile, D. S. and Michielssen, E. (1997). Genetic algorithm optimization applied to electromagnetics: a review. *IEEE Transactions on Antennas and Propagation*, **45**, (3), 343–53.
Weston, V. H. (1987). Factorization of the wave equation in higher dimensions. *Journal of Mathematical Physics*, **28**, (5), 1061–8.
Weston, V. H. (1988a). Factorization of the wave equation in a non-planar stratified medium. *Journal of Mathematical Physics*, **29**, (1), 36–45.
Weston, V. H. (1988b). Factorization of the dissipative wave equation and inverse scattering. *Journal of Mathematical Physics*, **29**, (10), 2205–18.
Weston, V, H. (1989). Wave splitting and the reflection operator for the wave equation in R^3. *Journal of Mathematical Physics*, **30**, (11), 2545–62.
Weston, V. H. (1990). Invariant imbedding for the wave equation in three dimensions and applications to direct and inverse problems. *Inverse Problems*, **6**, (6), 1075–105.
Weston, V. H. (1992a). Invariant imbedding and wave splitting in R^3: II. The

Green function approach to inverse scattering. *Inverse Problems*, **8**, (6), 919–47.

Weston, V.H. (1992b). Root of a second order hyperbolic differential operator and wave-splitting. In *Invariant Imbedding and Problems* (SIAM Proceedings), edited by J. Corones, G. Kristensson, P. Nelson and D. Seth, 139–52.

Weston, V. H. (1993). Time domain wave splitting of Maxwell's equations. *Journal of Mathematical Physics*, **34**, (4), 1370–92.

Weston, V.H. Note on an ill-posed problem for the layer–stripping approach to the inverse problem. (Personal communication.)

Weston, V. H. and He, S. (1993). Wave splitting of the telegraph equation in R^3 and its application to inverse scattering. *Inverse Problems*, **9**, (6), 789–812.

Wholey, W. B. and Eldred, W. N. (1950). A new type of slotted line section. *Proceedinds of the IRE*, March, 244–8.

Wyns, P., Foty, D. P., and Oughstun, K. E. (1989). Numerical analysis of the precursor fields in linear dispersive pulse propagation. *Journal of the Optical Society of America*, **6**, (9), 1421–9.

BIBLIOGRAPHY

This bibliography contains references which, without being cited in text, describe methods and results concerning the wave-splitting approach which complements those that are presented in this book.

Åberg, I. (1996) Transient waves in nonstationary media with applications to lightning and nonlinear media. Ph.D thesis, Lund Institute of Technology, Lund, Sweden.

Åberg, I., Kristensson, G., and Wall, D. J. N. (1996). Transient waves in nonstationary media. *Journal of Mathematical Physics*, **37**, (5), 2229-52.

Ammicht, E., Corones, J. P., and Krueger, R. J. (1987). Direct and inverse scattering for viscoelastic media. *Journal of the Acoustical Society of America*, **81**, 827-34.

Ayoubi, I. S. (1989). An inverse scattering problem for hyperbolic systems of first-order in semi-infinite media. Ph.D thesis, Texas Technical University, USA.

Ayoubi, I. S. (1993). The Riemann–Green function and the invariant imbedding equations for hyperbolic systems of first order. *Applied Mathematics and Computation*, **55**, 101-29.

Ayoubi, I. S. and Nelson, P. (1984). A generalized invariant imbedding for wave propagation. *Arabian Journal for Science and Engineering*, **9**, 135-43.

Beezley, R. S. (1985). Electromagnetic direct and inverse problems for absorbing media. Ph.D thesis, University of Nebraska, Lincoln, Nebraska.

Borzdov, G. N. (1998). Frequency domain wave-splitting techniques for plane stratified bianisotropic media. *Journal of Mathematical Physics*, **38**, (12), 6328-66.

Brent, R. I. and Fishman, L. (1992). Derivation of extended parabolic theories for vector electromagnetic wave propagation. In *Invariant Imbedding and Problems* (SIAM Proceedings), edited by J. Corones, G. Kristensson, P. Nelson and D. Seth. pp. 90-102, SIAM, Philadelphia.

Bui, D.D. (1993). The inverse electromagnetic scattering problem for a spatially homogeneous, dispersive and dissipative medium. Ph.D thesis, Texas A&M University.

Cheney, M. and Kristensson, G. (1988). Three-dimensional inverse scattering: layer-stripping formulas and ill-posedness results. *Inverse Problems*, **4**, (3), 625-42.

Corones, J. P. and Karlsson, A. (1988). Transient direct and inverse scattering for inhomogeneous viscoelastic media: obliquely incident SH mode. *Inverse*

Problems, **4**, 643-60.

Corones, J. and Sun, Z. (1992). Transient reflection and transmission problems for fluid-saturated porous media. In *Mathematical and Numerical Aspects of Wave Propagation* (ed. R. Kleinman, T. Angell, D. Colton, F. Santosa, and I. Stakgold), pp. 153-68, SIAM, Philadelphia.

Corones, J. P., DeFacio, B., and Krueger, R. J. (1982). Parabolic approximations to the time-dependent elastic wave equation. *Journal of Mathematical Physics*, **23**, 577-86.

Corones, J. P., Davison, M. E., and Krueger, R. J. (1983). The effects of dissipation in one-dimensional inverse problems. In *Inverse optics* (Proc. SPIE 413, edited by A. J. Devaney), pp. 107-114, SPIE Bellingham, WA.

Corones, J. P., Davison, M. E., and Krueger, R. J. (1985). Dissipative inverse problems in the time domain. In *Inverse methods in electromagnetic imaging*, (ed. W. M. Boerner), NATO ASI series, Series C, No. 143, pp. 121-130, Reidel, Dordrecht.

Corones, J. P., Krueger, R. J., and Vogel, C. R. (1985). The effects of noise and bandlimiting on a one-dimensional time dependent inverse scattering technique. In *Review of Progress in Quantitative Nondestructive Evaluation*, (edited by D.O. Thompson and D.E. Chimenti), Plenum, New York.

Davison, M. E. and Winther, R. C. (1994). A general approach to splitting and invariant imbedding for linear wave equations. *Journal of Mathematical Analysis and Applications*, **188**, (1), 158-81.

de Hoop, M. V. (1992). Directional decomposition of transient acoustic wave fields. Ph.D thesis, Delft University Press, Delft.

de Hoop, M. V. and de Hoop, A. T. (1992). Scalar space-time waves in their spectral domain first- and second-order Thiele approximations. *Wave Motion*, **15**, 229-65.

de Hoop, M. V. and de Hoop, A. T. (1993). Interface reflection of spherical acoustic waves in the first- and second-order rational parabolic approximations and their artifacts. *Journal of the Acoustical Society of America*, **93**, 22-35.

Dougherty, R. P. (1986). Direct and inverse scattering of classical waves at oblique incidence to stratified media via invariant imbedding equations. Ph.D thesis, Iowa State University, Ames, Iowa.

Fishman, L. (1991). Exact and approximate solutions of the Helmholtz, Weyl composition equation in scalar wave propagation. *Wave Motion*, **14**, (3), 205-24.

Fishman, L. (1992). Exact and operator rational approximate solutions of the Helmholtz, Weyl composition equation in underwater acoustics – the quadratic profile. *Journal of Mathematical Physics*, **33**, (5), 1887-914.

Fishman, L. (1993). One-way wave propagation methods in direct and inverse scalar wave propagation modeling. *Radio Science*, **28**, (5), 865-76.

Fishman, L. McCoy, J. J. (1984). Factorization, path integral representations and the construction of direct and inverse wave propagation theories. *IEEE*

Transactions on Geoscience and Remote Sensing, **GE-22**, (6), 682-92.

Fishman, L., and Wales, S. C. (1987). Phase space methods and path integration: the analysis and computation of scalar wave equations. *Journal of Computational and Applied Mathematics*, **20**, 219-38.

Fishman, L., McCoy, J. J., and Wales, S. C. (1987). Factorization and path integration of the Helmholtz equation: Numerical algorithms. *Journal of the Acoustical Society of America*, **81**, (5), 1355-76.

Folkow, P. D., Kristensson, G., and Olsson, P. Time domain Green functions for the homogeneous Timoshenko beam. *Quarterly Journal of Mechanics and Applied Mathematics*, (in press).

Fridén, J. (1994). Inverse scattering for anisotropic mirror image symmetric media. *Inverse Problems*, **10**, (5), 1133-44.

Fridén, J. (1995). Bartholinian Pleasures – Time domain direct and inverse electromagnetic scattering for dispersive anisotropic media. Ph.D thesis, Chalmers University of Technology, Sweden.

Fridén, J. (1996). Inverse scattering for the homogeneous dispersive anisotropic slab using transient electromagnetic fields. *Wave Motion*, **23**, (4), 289-306.

Fridén, J. and Kristensson, G. (1997). Transient external 3D excitation of a dispersive and anisotropic slab. *Inverse Problems*, **13**, (3), 691-709.

Fridén, J., Kristensson, G., and Stewart, R. D. (1993). Transient electromagnetic wave propagation in anisotropic dispersive Media. *Journal of the Optical Society of America A*, **10**, (12), 2618-27.

Haines, A. J. and de Hoop, M. V. (1996). An invariant imbedding analysis of general wave scattering problems. *Journal of Mathematical Physics*, **37**, 3854-81.

He, S. (1991). Time domain direct and inverse scattering for higher dimensional fields propagating in dissipative stratified media. Ph.D thesis, Royal Institute of Technology, Stockholm, Sweden.

He, S. (1996). A uniformly approximate solution to the scattering and propagation problem for the time-harmonic electromagnetic wave in a stratified bi-anisotropic slab. *International Journal of Infrared and MM Waves*, **17**, (2), 415-31.

He, S. (1997a). Explicit expressions of the reflection and transmission for two coupled exponential lines. *IEEE Transactions on Microwave Theory and Techniques*, **45**, (5), 695-8.

He, S. (1997b). Calculation of the transient reflection from a two-port network with a nonlinear resistor load. *Microwave and Optical Technology Letters*, **15**, (2), 113-6.

He, S. and Lindell, I. (1993). Propagating eigenmodes for plane waves in a uniaxial bianisotropic medium and reflection from a planar interface. *IEEE Transactions on Antennas and Propagation*, **41**, (12), 1659-1664.

He, S. and Ström, S. (1991). The electromagnetic scattering problem in the time domain for a dissipative slab and a point source using invariant imbedding. *Journal of Mathematical Physics*, **32**, (12), 3529-3539.

He, S. and Ström, S. (1992). The electromagnetic inverse problem in the time domain for a dissipative slab and a point source using invariant imbedding: Reconstruction of the permittivity and conductivity. *Journal of Computational and Applied Mathematics*, **42**, 137-55.

He, S. and Ström, S. (1993). A simplified time-domain inverse approach to nonplanar stratified media. *Journal of Electromagnetic Waves and Applications*, **7**, (4), 495-512.

Hellberg, R. (1995) An optimization approach to the design of low-reflecting slabs for obliquely incident transient electromagnetic waves. Technical Report TRITA-TET 95-21, Royal Institute of Technology, Stockholm, Sweden.

Hellberg, R. (1997). A time-domain method for the design of non-reflecting media. Ph.D thesis, Royal Institute of Technology, Stockholm, Sweden.

Hellberg, R. and Karlsson, A. (1995). Design of reflectionless media for transient electromagnetic waves. *Inverse Problems*, **11**, 147-64.

Hellberg, R. and Karlsson, A. (1996). The direct and inverse problem for obliquely incident transient elastodynamic waves. Technical Report TRITA-TET 96-14, Royal Institute of Technology, Stockholm, Sweden.

Karlsson, A. (1987). Inverse scattering for viscoelastic media using transmission data. *Inverse Problems*, **3**, 691-709.

Karlsson, A. and Kristensson, G. (1990). Wave splitting in the time domain for a radially symmetric geometry. *Wave Motion*, **12**, (3), 197-211.

Karlsson, A., Kreider, K., and Kristensson, G. (1992). Wave splitting and imbedding equations for a spherically symmetric dispersive medium. In *Invariant Imbedding and Inverse Problems*, (ed. J. P. Corones, G. Kristensson, P. Nelson, and D. L. Seth), pp. 103-13, SIAM, Philadelphia.

Kreider, K. (1989). Time-dependent direct and inverse scattering for the dispersive cylinder. *Wave Motion*, **11**, 427-40.

Kreider, K. (1990). Time-dependent linear transport via Green's functions. *Transport Theory and Statistical Physics*, **18**, (1), 513-25.

Kristensson, G. (1991). Direct and inverse scattering problems in dispersive media Green's functions and invariant imbedding techniques. In *Direct and Inverse Boundary Value Problems*, Methoden und Verfahren der Mathematischen Physik, Band 37, (ed. R. Kleinman, R. Kress and E. Martensen), 105-19. Mathematisches Forschungsinstitut Oberwolfach, FRG.

Kristensson, G. and Krueger, R. J. (1992). Time domain inversion techniques for electromagnetic scattering problems. In *Invariant Imbedding and Problems* (SIAM Proceedings), edited by J. Corones, G. Kristensson, P. Nelson and D. Seth, 1-29.

Kristensson, G. and Rikte, S. (1992). Scattering of transient electromagnetic waves in reciprocal bi-isotropic media. *Journal of Electromagnetic Waves and Applications*, **6**, (11), 1517-35.

Kristensson, G. and Rikte, S. (1993). Transient wave propagation in reciprocal bi-isotropic media at oblique incidence. *Journal of Mathematical Physics*, **34**, (4), 1339-1359.

Kristensson, G., Krueger, J. K., and Winther, R. C. (1991). Existence and construction of solutions of dissipative inverse problems. *Journal of Mathematical Analysis and Applications*, **157**, (2), 542-54.

Kristensson, G., Karlsson, A., and Otterheim, H. (1992). Transient wave propagation in gyrotropic media. In *Invariant Imbedding and Problems* (SIAM Proceedings), edited by J. Corones, G. Kristensson, P. Nelson and D. Seth, 77-89.

Lu, Y. Y. and McLaughlin, J. R. (1996). The Riccati method for the Helmholtz equation. *Journal of the Acoustical Society of America*, **100**, (3), 1432-46.

McCoy, J. J. and Frazer, L. N. (1986). Propagation modelling based on wave field factorization and invariant imbedding. *Geophysical Journal of the Royal Astronomical Society*, **86**, 703-19.

McCoy, J. J., Fishman, L., and Frazer, L. N. (1986). Reflection and transmission at an interface separating transversely inhomogeneous acoustic half-spaces. *Geophysical Journal of the Royal Astronomical Society*, **85**, 543-62.

Norgren, M. and He, S. (1995). Reconstruction of the constitutive parameters for an Ω material in a rectangular waveguide. *IEEE Transactions on Microwave Theory Techniques*, **43**, (6), 1315-21.

Ochs, R. L. and Vogel, C. R. (1991). The optimization of electromagnetic energy within a dissipative slab. *Wave Motion*, **14**, 69-84.

Ochs, R. L. and Vogel, C. R. (1993a). The optimization of dissipation in dispersive dielectric slabs. *Wave Motion*, **17**, 349-63.

Ochs, R. L. and Vogel, C. R. (1993b). Analysis of signal optimization problems for dielectric media. *Journal of Optimization Theory and Applications*, **79**, 237-51.

Olsson, P. and Kristensson, G. (1994). Wave splitting of the Timoshenko beam equation in the time domain. *Zeitschrift für andgewandte Mathematik und Physik*, **45**, 866-81.

Otterheim, H. (1993). Time domain direct and inverse scattering for gyrotropic media. Ph.D thesis, Royal institute of technology, Stockholm.

Rikte, S. (1993). Transient inverse scattering in dispersive homogeneous biisotropic media. In *Mathematical and Numerical Aspects of Wave Propagation*, (ed. R. Kleinman, T. Angell, D. Colton, F. Santosa, and I. Stakgold), SIAM, 407-16.

Rikte, S. (1994). Propagation of transient electromagnetic waves in stratified bi-isotropic media and related inverse scattering problems. Ph.D thesis, Lund Institute of Technology, Lund, Sweden.

Rikte, S. Existence, uniqueness, and causality theorems for wave propagation in stratified, temporally dispersive, complex media. *SIAM Journal of Mathematical Analysis*, (in press).

Roberts, T. M. (1992). Causality theorems. In *Invariant Imbedding and Problems* (SIAM Proceedings), edited by J. Corones, G. Kristensson, P. Nelson and D. Seth, 114-128.

Roberts, T. M. (1993). An inverse scattering algorithm for layered electromagnetically dispersive media. In *Second International Conference on Mathematical and Numerical Aspects of Wave Propagation*, (ed. R. Kleinman, D. Colton, F. Santosa, and I. Stakgold), Chapter 44, pp. 417-25, SIAM, Philadelphia.

Roberts, T. M. and Hobart, M. (1992). Energy velocity, damping, and elementary inversion. *Journal of the Optical Society of America A*, **9**, 1091-101.

V. G. Romanov and He, S. (1996). An inverse problem for a 3D inhomogeneous plasma. *Journal of Inverse and Ill-posed Problems*, **4**, (2), 153-70.

Song, H. and Zhang, G.Q. (1994). Potential inversion of the two-dimensional plasma wave equation. *Inverse problems*, **10**, 464-84.

Stewart, R. D. (1989). Transient electromagnetic scattering on anisotropic media. Ph.D thesis, Iowa State University, Ames, Iowa.

Sun, Z. (1992). Time domain direct scattering and inverse source problems. In *Mathematical and Numerical Aspects of Wave Propagation*, (ed. R. Kleinman, T. Angell, D. Colton, F. Santosa, and I. Stakgold), pp. 64-73, SIAM, Philadelphia.

Sun, Z. and Corones, J. (1992). Invariant imbedding method and inverse source problems. *Inverse Problems in Scattering and Imaging*, SPIE Proc., **1767**, 13-20.

Sun, Z. and Wickham G. (1993). The inversion of transient reflection data to determine solid grain structure. *Wave Motion*, **18**, 143-162.

Wall, D. J. N. (1991). On some differential equations arising in time domain scattering problems for a dissipative wave equation. *Transport Theory and Statistical Physics*, **20**, 29-54.

Wall, D. J. N. (1992). On the numerical solution of a functional differential equation pertaining to a wave equation. In *Invariant Imbedding and Problems* (SIAM Proceedings), edited by J. Corones, G. Kristensson, P. Nelson and D. Seth, 169-86.

Wall, D. J. N. and Olsson, P. (1997). Invariant imbedding and hyperbolic heat waves. *Journal of Mathematical Physics*, **38**, (3), 1723-49.

Weston, V. (1974). On inverse scattering. *Journal of Mathematical Physics*, **15**, 209-13.

Weston, V. and Krueger, R. J. (1973). On the inverse problem for a hyperbolic dispersive partial differential equation, II. *Journal of Mathematical Physics*, **14**, 406-8.

Vogel, C.R. (1992). Wave splitting for some nonhyperbolic time-dependent PDEs. In *Invariant Imbedding and Problems* (SIAM Proceedings), edited by J. Corones, G. Kristensson, P. Nelson and D. Seth, pp. 129-39.

Zhang, G. Q., (1985). High order approximation of one-way wave equations. *Journal of Computational Mathematics*, **3**, 90-7.

Zhang, G. Q., (1988). A new algorithm for finite-difference migration of steep dips. *Geophysics*, **53**, 167-75.

INDEX

The manufacturer's authorised representative in the EU for product safety is Oxford University Press España S.A. of El Parque Empresarial San Fernando de Henares, Avenida de Castilla, 2 - 28830 Madrid (www.oup.es/en or product.safety@oup.com). OUP España S.A. also acts as importer into Spain of products made by the manufacturer.
Printed and bound by CPI Group (UK) Ltd, Croydon, CR0 4YY
14/09/2026
02216290-0007